FLORA OF TROPICAL EAST AFRICA

ACANTHACEAE (Part 1)

KAJ VOLLESEN[1]

Herbs, shrubs, twiners or (rarely) trees; branches often angled or ridged, with usually distinct transverse ridges at nodes. Stems and leaves often with cystoliths (linear intercellular calcium carbonate concretions) looking like small white streaks or rods. Leaves opposite, rarely whorled or in rosettes, without stipules. Inflorescences of single flowers or dichasia or cymules which are often aggregated into axillary and/or terminal paniculate to racemoid cymes; bracts from foliaceous to distinctly differentiated or minute; bracteoles usually present. Flowers nearly actinomorphic to zygomorphic, bisexual. Calyx 4–5-lobed (or in *Thunbergioideae* a subentire to undulate rim or of 10–20 irregular lobes), fused at least at base, with equal to unequal lobes or distinctly 2-lipped. Corolla sympetalous, with a narrow basal tube widening into a more or less distinct throat, limb from almost regularly 5-lobed to distinctly 2-lipped, more rarely 1-lipped, lobes imbricate or contorted in bud. Stamens 4, usually didynamous, or 2 and then with or without 2–3 staminodes, attached at base of throat or in narrow corolla tube, basal part of filaments often connate in two pairs; anthers 2- or 1-thecous, thecae usually parallel attached at same height or one above the other, at base rounded, mucronate to apiculate or with distinct appendages or spurs, dehiscing longitudinally (rarely with apical or basal pores). Ovary superior, 2-locular, at base surrounded by an annular disc, with 2 collateral (*Thunbergioideae*) or with 2–28 superposed ovules per locule; style simple, filiform, with 2 linear (one often reduced) to oblong or ovoid or capitate stigmatic lobes. Fruit a 2-locular loculicidal capsule (rarely a drupe), usually explosively dehiscent, the placentae sometimes separating from the capsule wall on opening. Seeds 1 to over 20 per locule, held on thickened curved laterally compressed hook-shaped outgrowths (retinaculae) from the funicle (retinaculae papilliform in *Nelsonioideae* and absent in *Thunbergioideae*), discoid to spheroid, reniform or cordiform to subglobose, surface often covered in hygroscopic hairs or glabrous and then smooth or variously sculptured.

About 250 genera and 2500–3000 species widely distributed in all tropical and subtropical regions, extending sparingly into warm temperate zones in Europe and Asia and as far north as Canada in temperate North America. The largest diversity at both genus and species level is in the dry woodland, bushland and grassland regions of tropical and southern Africa, Madagascar, India and Central and South America. In the Flora area 50 genera and 550–600 species.

The system used for arranging the genera here is an amalgam of those outlined by Bremekamp in Bull. Bot. Surv. India 7: 21–30 (1965), Scotland & Vollesen in Kew Bull. 55: 513–589 (2000) and Manktelow et al. in Syst. Bot. 26: 104–119 (2001).

---

[1] Royal Botanic Gardens Kew

Subfam. **Nelsonioideae** (*Nees*) *Pfeiff.*, Nomencl. Bot. 1(1): 10 (1871)
  Genera 1–4 (*Anisosepalum, Elytraria, Nelsonia, Saintpauliopsis*)
  Plants without cystoliths. Leaves opposite (rarely whorled) but bracts in inflorescence spirally arranged. Calyx 4–5-lobed. Stamens 4, with 2-thecous anthers. Ovules numerous, superposed. Seeds numerous, minute, borne on papilliform retinaculae.

Subfam. **Thunbergioideae** *Kostel.*, Allg. Med.-Pharm. Fl. 3: 923 (1834)
  Genera 5–7. (*Mendoncia, Pseudocalyx, Thunbergia*)
  Plants without cystoliths. Leaves and bracts opposite. Calyx a subentire to undulate rim or of 10–20 irregular lobes. Stamens 4, with 2-thecous anthers. Ovules 2 per locule, collateral. Retinaculae absent. Seeds 2–4, large, subglobose, hollow or with large scar on one side.

Subfam. **Acanthoideae** *Link*, Handbuch 1: 500 (1829)
  Genera 8–14 (*Acanthus, Blepharis, Crossandra, Crossandrella, Sclerochiton, Stenandrium, Streptosiphon*)
  Plants without cystoliths. Leaves opposite or whorled, bracts opposite. Calyx 4–5-lobed, chaffy or bony. Stamens 4, with 1-thecous anthers. Ovules 4 per locule, superposed. Retinaculae hook-shaped. Seeds 2–4, covered with large pectinate scales or tuberculate or with large flattened branched hygroscopic hairs or glabrous to puberulous with non-hygroscopic hairs.

Subfam. **Ruellioideae** *T.Anderson* in Thwaites, Enum. Pl. Zeyl.: 224 (1829)
  Plants with cystoliths. Leaves and bracts opposite. Calyx (4–)5-lobed, chartaceous. Stamens 2–4 with 1–2-thecous anthers, if 4 always with 2-thecous anthers. Ovules 2–28 per locule, superposed. Retinaculae hook-shaped. Seeds 2 to over 40, discoid with fine non-branched terete hygroscopic hairs or sphaeroid, reniform or cordiform to subglobose with smooth to variously sculptured surface.

  Tribe **Ruellieae** *Dumort.*, Anal. Fam. Pl.: 23 (1829)
    Genera 15–26 (*Acanthopale, Brillantaisia, Duosperma, Dyschoriste, Epiclastopelma, Eremomastax, Hygrophila, Mellera, Mimulopsis, Phaulopsis, Ruellia, Satanocrater*)
    Stamens 4, rarely (*Brillantaisia*) 2, with 2-thecous anthers. Stigma with linear lobes, dorsal usually smaller and often reduced to a tooth. Ovules 2–28 per locule. Seeds 2 to over 40, discoid, with hygroscopic hairs, rarely (*Satanocrater*) with thick clavate to almost globose hairs and with peltate scales.

  Tribe **Whitfieldieae** *Bremek.* in Bull. Bot. Surv. India 7: 27 (1965)
    Genera 27–29 (*Chlamydacanthus, Lankesteria, Whitfieldia*)
    Stamens 2–4, with 2-thecous anthers. Stigma with two equal capitate lobes. Ovules 1–2 per locule. Seeds 2–4, discoid, with hygroscopic hairs or with concentric ridges.

  Tribe **Barlerieae** *Nees* in Martius, Fl. Bras. 9: 7, 65 (1847)
    Genera 30–33 (*Barleria, Crabbea, Lepidagathis, Neuracanthus*)
    Stamens 2–4 with 2-thecous anthers or (*Neuracanthus* and *Lepidagathis*) one pair with 1-thecous and the other with 2-thecous anthers. Stigma with linear-oblong to rhomboid-rounded lower lobe and upper lobe missing or reduced to a small tooth or (*Barleria*) with 2 equal subcapitate to obconical lobes. Ovules 1–4 per locule. Seeds 2–8, with hygroscopic hairs or (rarely in *Barleria*) glabrate.

  Tribe **Justicieae** *Dumort.*, Anal. Fam. Pl.: 23 (1829)
    Genera 34–50 (*Anisotes, Asystasia, Brachystephanus, Cephalophis, Chlamydocardia, Dicliptera, Ecbolium, Hypoestes, Isoglossa, Justicia, Megalochlamys, Monothecium, Pseuderanthemum, Rhinacanthus, Ruspolia, Ruttya, Trichaulax*)
    Stamens 2–4, with 1–2-thecous anthers. Stigma with two equal oblong-ellipsoid lobes. Ovules 1–2 per locule. Seeds 2–4, discoid to sphaeroid, reniform or cordiform to subglobose, never with hygroscopic hairs.

  The tribes *Barlerieae* and *Justicieaea* will be treated in Part 2.

  Due to their often large and showy flowers a great number of Acanthaceae – both native and introduced – have been grown as ornamentals in East Africa. It is the intention to treat these in Part 2.

KEY TO NATIVE GENERA

1. Fruit a 1-seeded drupe .... **7. Mendoncia** (p. 76)
   Fruit a capsule; seeds more than 1 2
2. Seeds not borne on hook-shaped retinaculae, retinaculum lacking or papilliform; plant without cystoliths; anthers 2-thecous 3
   Seeds borne on prominent hook-shaped retinaculae; plant with or without cystoliths, if plant without cystoliths then anthers 1-thecous 8
3. Bracts alternate, each flower supported by a single small bract; calyx with four or five distinct lobes; capsule small (3–12 mm long), without or with sterile apical beak not longer and narrower than fertile part; seeds numerous, minute (up to 2 mm in diameter), usually angular 4
   Bracts opposite, each flower supported by two large bract-like bracteoles; calyx a subentire to undulate rim or of 10–20 irregular lobes; capsule large (1.5–4.5 cm long), with a sterile apical beak longer and narrower than fertile globose part or triangular; seeds two or four, large (5 mm or more in diameter), subglobose 7
4. Bracts fused to pedicels, not imbricate; bracteoles inserted on base of calyx; two lateral sepals minute; corolla distinctly 2-lipped; stamens four 5
   Bracts not fused to pedicels, imbricate; bracteoles absent or clearly inserted below calyx; all five sepals same length; corolla subactinomorphic; stamens two (rarely four) 6
5. Plant with trailing stems; corolla small (6–8 mm long), widened almost from base into a campanulate throat; flowers in an open cyme with small (1–2 mm long) bracts and long (2–13 mm long) pedicels **3. Saintpauliopsis** (p. 19)
   Plant with erect stems; corolla large (17–38 mm long), with a distinct linear basal tube and widening upwards into a campanulate throat; flowers in a dense racemoid cyme with large (8–20 mm long) bracts and short (0.5–2 mmlong) pedicels .... **4. Anisosepalum** (p. 21)
6. Leaves in a basal or apical rosette; peduncles with sterile bracts; fertile bracts scale-like; bracteoles present; calyx 5-lobed .... **1. Elytraria** (p. 11)
   Leaves dispersed along elongated stems; peduncles without sterile bracts; fertile bracts herbaceous; bracteoles absent; calyx 4-lobed .... **2. Nelsonia** (p. 15)

7. Anthers opening by apical pores; seed with a
   scar on the proximal face; indumentum
   stellate; anther bases not with stiff setose
   hairs . . . .   5. **Pseudocalyx** (p. 24)
   Anthers opening by longitudinal slits
   (sometimes only in apical half); seed
   hollow on the proximal face; indumentum
   usually not stellate, if so then anther bases
   with a boss with stiff setose hairs . . . .   6. **Thunbergia** (p. 27)
8. Fertile stamens four 9
   Fertile stamens two 38
9. Anthers 1-thecous; plant without cystoliths 10
   Anthers 2-thecous (rarely two 2-thecous and
   two 1-thecous anthers); plant with cystoliths 17
10. Stamens included in corolla tube, subsessile or
    with short thin filaments, usually not
    inserted on a thickened flange 11
    Stamens exserted, with long flattened bony
    filaments, inserted on a thickened flange 15
11. Calyx divided into 4 sepals, ventral 2-fid;
    bracteoles elliptic, foliaceous; mature
    capsule transversely cracking . . . .   14. **Crossandrella** (p. 151)
    Calyx divided into 5 sepals, ventral entire;
    bracteoles linear-lanceolate, glumaceous;
    mature capsule not transversely cracking 12
12. Corolla tube twisted through 180°, limb split
    dorsally to give a small entire or 3-lobed
    lip which due to the twisting of the tube is
    held dorsally; stamens inserted on a
    thickened flange . . . .   11. **Streptosiphon** (p. 131)
    Corolla tube not twisted, limb subequally 5-
    lobed or of a 5-lobed lower lip; stamens
    not inserted on a thickened flange 13
13. Dorsal sepal with two acute to cuspidate apical
    teeth and two strong veins from base . . . .   13. **Crossandra** (p. 137)
    Dorsal sepal with an entire, acute to
    cuspidate apex or indistinctly 2-toothed,
    with one strong vein from base or
    indistinctly veined or with several equally
    strong veins 14
14. Corolla tube split dorsally to give a 5-lobed
    lower lip with horizontal lobes; corolla
    hairy on the outside; leaf base cuneate to
    attenuate . . . .   13. **Crossandra** (p. 137)
    Corolla tube not split dorsally, limb equally
    or subequally 5-lobed with 2 upper lobes
    held vertically; corolla glabrous on the
    outside; leaf base subcordate to cordate .   12. **Stenandrium** (p. 133)
15. Calyx divided into 5 sepals; seeds with large
    pectinate scales or concentric ridges;
    leaves opposite . . . .   8. **Sclerochiton** (p. 78)
    Calyx divided into 4 sepals; seeds smooth to
    rugose, with hygroscopic hairs or glabrous
    (rarely puberulous); leaves opposite or
    whorled 16

16. Leaves opposite; all four filaments similar; seeds glabrous or sericeous, hairs not hygroscopic .. **9. Acanthus** (p. 89)

Leaves in whorls of four; two filaments flattened and with apical forwardly directed appendage; seeds with branched hygroscopic hairs ...... **10. Blepharis** (p. 95)

17. Calyx 4-lobed, the ventral and dorsal lobes usually much larger than the lateral lobes 18

Calyx 5-lobed 19

18. Inflorescence distinctly dorsiventral with flowers and bracts on one side and sterile bracts on the other side or of dense globose heads or 4-sided cymes; corolla strongly 2-lipped with hooded upper lip and 3-lobed lower lip .... **32. Lepidagathis** (part 2)

Inflorescence usually not dorsiventral or of dense heads or racemoid cymes; corolla subactinomorphic or slightly 2-lipped, but never with hooded upper lip or 3-lobed lower lip .... **33. Barleria** (part 2)

19. Stems, leaves and calyces with scattered to dense subsessile peltate scale-like glands; calyx large, inflated, longitudinally 5-ribbed or almost winged, shortly 5-lobed; seed densely covered with thick clavate to almost globose hairs and with peltate scales .... **26. Satanocrater** (p. 263)

Stems, leaves and calyces not with peltate scale-like glands; calyx deeply divided into 5 lobes or slightly to distinctly 2-lipped, never inflated or shortly 5-lobed; seed with long thin hygroscopic hairs all over or on rim only or glabrous with concentric ridges 20

20. Dorsal calyx lobe ovate to ellipsoid, several times wider than and covering the other four; inflorescences distinctly dorsiventral with flowers and bracts on one side and sterile bracts only on the other side .... **25. Phaulopsis** (p. 256)

Dorsal calyx lobe linear-lanceolate to narrowly triangular or narrowly oblong to spathulate, not much wider than and not covering the other four; inflorescences not distinctly dorsiventral 21

21. Calyx 2-lipped with an entire or finely 3-toothed dorsal lip and an entire or 2-toothed ventral lip **30. Neuracanthus** (part 2)

Calyx subequally 5-lobed (dorsal lobe often longer and/or wider than the other four) or with the two ventral lobes fused higher up than the three dorsal lobes 22

22. Two ventral calyx lobes fused higher up than three dorsal lobes; flowers not in dense axillary heads surrounded by hard spines; capsule with 2 seeds .... **24. Duosperma** (p. 239)

Calyx subequally 5-lobed (dorsal lobe often longer and/or wider than the other four), if two ventral lobes fused almost to apex then flowers in dense axillary heads surrounded by hard spines; capsule with 4 or more seeds (very rarely 2) 23

29. One or more anther thecae with 1–2 mm long curved basal spur 30

All anther thecae rounded or muticous at base or with spur up to 0.5(–1) mm long 32

30. Corolla resupinate, uniformly bright red; capsule 4-seeded .... **22. Epiclastopelma** (p. 228)

Corolla not resupinate, white to mauve or blue, lower lip often darker to dark blue or dark purple, throat and pleated area on lower lip yellow to orange or brownish; capsule 6–8-seeded 31

31. Lower corolla lip with numerous stiff retrorse hairs; corolla hairy on the outside, distinctly 2-lipped with 2-lobed upper lip and 3-lobed lower lip (rarely subactinomorphic) .... **23. Mellera** (p. 232)

Lower corolla lip without stiff retrorse hairs; corolla glabrous on the outside, subactinomorphic (two lobes in upper lip similar to two lateral lobes in lower lip, middle lobe in lower lip narrower and usually longer) .... **21. Mimulopsis** (p. 219)

32. Corolla crimson red or maroon; inflorescence an open much branched terminal panicle with long spreading branches and flowers in 3-flowered dichasial cymes or solitary towards apex of panicle .... **22. Epiclastopelma** (p. 228)

Corolla white to mauve or purple, never red; flowers solitary or in axillary clusters or in moderately dense to open axillary cymes 33

33. Calyx lobes linear to narrowly oblong or spathulate, not widened towards base; plant strongly aromatic when crushed; at least one anther theca with 0.5–1 mm long spur (rarely all apiculate) .... **23. Mellera** (p. 232)

Calyx lobes linear-lanceolate to narrowly triangular, widest near base; plant not aromatic when crushed; anther thecae all rounded or apiculate or with spurs less than 0.5 mm long 34

34. Corolla subactinomorphic, over 1.5 cm long; ventral stigma lobe flattened to spoon-shaped with wavy margins; seeds with distinctly raised rim .... **19. Ruellia** (p. 202)

Corolla distinctly 2-lipped or if actinomorphic usually less than 1 cm long; ventral stigma lobe linear with straight margins; seeds without distinctly raised rim 35

35. Flowers in dense heads, surrounded by hardened spiny modified branches; two ventral sepals fused almost to apex, other three only at base; capsule with 4 seeds ... **16. Hygrophila** (p. 163)

Flowers solitary or in axillary clusters or in moderately dense to open axillary cymes, not in heads surrounded by spines; capsule with 4 to over 20 seeds 36

36. Lower corolla lip with (rarely without) stiff
retrorse hairs; anther thecae rounded to
muticous, without spurs; capsule with 8 to
over 20 seeds . . . .   16. **Hygrophila** (p. 163)
Lower corolla lip without (very rarely with a
few) stiff retrorse hairs; anther thecae with
minute spurs (rarely without); capsule
with 4 seeds [or 2–3-seeded by abortion]   17. **Dyschoriste** (p. 179)
37. Flowers single or in 2–3-flowered cymules,
enveloped by two partly fused foliaceous
bracts; calyx small (under 1 cm long);
corolla small (under 1.5 cm long) . . . . . .   28. **Chlamydacanthus** (p. 271)
Flowers single, each supported by a single
bract; calyx large (1.3–4.7 cm long),
petaloid, often whitish; corolla large (over
3 cm long) . . . .   27. **Whitfieldia** (p. 266)
38. Mature seeds with long hygroscopic hairs 39
Mature seeds glabrous, smooth, rugose,
verrucose or glochidiate but not with
hygroscopic hairs 43
39. Calyx 4-lobed with decussate lobes, the
ventral and dorsal lobes usually much
larger than the lateral lobes, ventral lobe
sometimes 2-fid . . . .   33. **Barleria** (part 2)
Calyx 5-lobed 40
40. Flowers in dense racemoid cymes with
imbricate bracts; stigma of two capitate
lobes; capsule with 2 seeds . . . .   29. **Lankesteria** (p. 273)
Flowers solitary or in axillary clusters, cymes
or panicles, if inflorescence dense and
racemoid not with imbricate bracts;
stigma lobes linear, upper lobe usually
reduced to a small tooth; capsule with 4 to
over 20 seeds 41
41. Upper part of petiole distinctly winged;
corolla distinctly 2-lipped with large
strongly curved and laterally compressed
upper lip, lower lip with a distinct "hinge"
at base . . . .   15. **Brillantaisia** (p. 153)
Upper part of petiole not winged; corolla
subactinomorphic to distinctly 2-lipped
but not with laterally compressed upper
lip, nor with a distinct "hinge" at base of
lower lip 42
42. Capsule with 8 to over 20 seeds; lower corolla
lip with numerous stiff retrorse hairs;
anther thecae rounded to muticous,
without spurs . . . .   16. **Hygrophila** (p. 163)
Capsule with 4 seeds; lower corolla lip
without (very rarely with a few) stiff
retrorse hairs; anther thecae with minute
spurs (rarely without) . . . .   17. **Dyschoriste** (p. 179)

43. Calyx 4-lobed with decussate lobes, the
    ventral and dorsal lobes usually much
    larger than the lateral lobes, ventral lobe
    sometimes 2-fid . . . .   33. **Barleria** (part 2)
    Calyx 4- or 5-lobed, if 4-lobed with subequal
    linear-lanceolate lobes 44
44. Stamens with 1-thecous anthers 45
    Stamens with 2-thecous anthers 49
45. Flowers in cymules (rarely solitary) enclosed
    between two opposite bracts which are
    free or connate; corolla resupinate . . . . .   50. **Hypoestes** (part 2)
    Each flower supported by a single bract;
    corolla not resupinate 46
46. Corolla with the three lobes in lower lip
    strongly deflexed against the tube and with
    a large glossy black nectariferous patch on
    basal part of lower lip; seed smooth on
    both sides . . . .   38. **Ruttya** (part 2)
    Corolla not with strongly deflexed lobes in
    lower lip and not with large glossy black
    nectariferous patch; seed variously
    sculptured, never smooth on both sides 47
47. Corolla small (less than 1.5 cm long),
    strongly 2-lipped with narrow shortly bifid
    hooded upper lip and spreading shortly
    3-lobed oblong-obovate lower lip . . . . . .   39. **Monothecium** (part 2)
    Corolla with long (over 1.5 cm) linear basal
    tube and subactinomorphic limb with
    erect, spreading or deflexed lobes 48
48. Stamens long, much exserted from corolla
    tube; stigma lobes capitate . . . .   35. **Brachystephanus** (part 2)
    Stamens included in corolla tube or shortly
    exserted; stigma lobes oblong-ellipsoid . .   37. **Ruspolia** (part 2)
49. Flowers in cymules (rarely solitary) enclosed
    between two opposite bracts, rarely with
    reduced bracts; corolla resupinate . . . . . .   49. **Dicliptera** (part 2)
    Each flower supported by a single bract;
    corolla not resupinate (very rarely so in
    Justicia) 50
50. The two anther thecae attached at the same
    height, rounded at base 51
    The two anther thecae attached at different
    height, rounded, apiculate or with distinct
    basal appendage on lower theca 57
51. One anther theca about half the size of the
    other; corolla with reflexed lobes;
    bracteoles subulate to filiform, longer than
    the calyx . . . .   42. **Chlamydocardia** (part 2)
    Anther thecae roughly similar in size; corolla
    with erect or spreading lobes; bracteoles
    usually shorter than the calyx, if longer than
    not subulate to filiform 52
52. Lower corolla lip with well developed rugula
    differently coloured to rest of corolla . . .   41. **Isoglossa** (part 2)
    Lower corolla lip without rugula, corolla
    uniformly coloured 53

61. Corolla tube much longer than lobes; lower
    anther theca mucronate . . . .   **43. Rhinacanthus** (part 2)
    Corolla tube usually about the same length
    as limb, if distinctly longer then lower
    anther theca with distinct appendage 62
62. Lower (rarely also upper) anther theca
    distinct appendage (rarely apiculate),
    thecae never completely offset, upper
    theca always smaller than lower . . . .   **40. Justicia** (part 2)
    Lower anther theca rounded to muticous,
    thecae often completely offset, upper theca
    larger than lower or the two subequal . . . .   **41. Isoglossa** (part 2)

1. ELYTRARIA

Michaux, Fl. Bor. Am. 1: 8 (1803), *nom. conserv.*; Nees in DC., Prodr. 11: 62 (1847);
C.B. Clarke in F.T.A. 5: 27 (1899); Bremekamp in Reinwardtia 3: 249 (1955);
E. Hossain in Not. Roy. Bot. Gard. Edinb. 31: 378 (1972)

Tubiflora Gmelin, Syst. Nat. (ed. 13) 2: 27 (1791), *nom. rej.*; Benoist in Fl.
Madagascar, Acanthacées 1: 24 (1967)

Acaulous or short-stemmed annual or perennial herbs (rarely taller shrubs). Leaves
in basal or apical rosettes. Flowers single, in dense cylindrical spikes from the rosettes,
spikes sessile, unbranched or with branches from base of the terminal spike; peduncles
with scattered to imbricate scale-like sterile bracts; fertile bracts imbricate, spirally
arranged, scale-like, persistent, without or with indistinct midrib; bracteoles 2, narrowly
ovate-elliptic. Calyx divided almost to the base into 5 sepals; dorsal much wider than the
rest, the two lateral slightly wider than the two ventral which are united higher up than
the rest. Corolla weakly 2-lipped, sometimes cleistogamous; tube linear, cylindric,
widened into a short throat; lower lip 3-lobed, lobes often bifid. Stamens 2 (rarely 4 or
2 plus 2 staminodes), included in the throat; filaments short; anthers 2-thecous, thecae
at same height, oblong or broadly elliptic, without basal appendages, connective
produced apically. Ovary 2-locular, with 6–10 ovules per locule; stigma oblong-
spathulate or rounded, entire or 2-lobed. Capsule 10–20-seeded, ellipsoid, sessile,
rostrate (sterile from apex) with recurved apical part when ripe, retinaculae absent.
Seed minute, ovoid-ellipsoid, often with ± truncate ends, rugose and pitted.

A pantropical genus of approximately 25 species. Mostly in America, but with 5–10 species in
Africa, Madagascar and SE Asia.

1. Annual or short-lived perennial herb, basal part of stems
   creeping and rooting at nodes, apical part erect, 2–18 cm
   long, with leaves in an apical rosette; peduncle with
   clearly spaced bracts . . . .   *3. E. marginata*
   Acaulous perennial herb with short subterranean root-
   stock and leaves in a basal rosette, often pressed to the
   ground; peduncle with imbricate bracts 2
2. Apical part of rootstock (where leaves attached) with long
   (up to 3 mm) strigose hairs; leaves elliptic to obovate,
   sometimes sinuate-incised towards base (rarely lyrate);
   upper sepal 4–5 mm long, acute or subacute corolla tube
   5–7 mm long; capsule 5.5–6.5 mm long . . . .   *1. E. acaulis*
   Apical part of rootstock with short (under 1 mm)
   puberulous hairs; leaves lyrate; upper sepal 3–4 mm
   long, broadly rounded to truncate (rarely sub-acute);
   corolla tube 2–3.5 mm long; capsule 4–5 mm long . . . . .   *2. E. minor*

1. **Elytraria acaulis** (*L.f.*) *Lindau* in E.P. Nachtr. II–IV: 304 (1897); F.W.T.A. 2: 261 (1931) pro parte; Bremekamp in Reinwardtia 3: 251 (1955); Vollesen in Opera Bot. 59: 80 (1980); Thulin in Fl. Somalia 3: 375 (2006). Type: India, Tranquebar, *Koenig* 77 (LINN, holo.; C!, iso.)

Acaulescent perennial herb with a short creeping sometimes branched rootstock; apical part of rootstock (where leaves attached) with long (up to 3 mm) strigose hairs. Leaves in a basal rosette, subsessile (rarely with petiole up to 4(–6) cm long); lamina elliptic to obovate, largest 5–18 × 2.5–6.5 cm, apex subacute to broadly rounded, base attenuate, decurrent, margin subentire to crenate-dentate on apical part, subentire to sinuate-incised on basal part; below crisped-puberulous to pilose or sparsely so along midrib (sparser along lateral veins), glabrous on lamina, above with similar indumentum and ± scattered hairs on lamina. Spikes 2–10 cm long, branched or unbranched; peduncle 6–20 cm long; sterile bracts imbricate, ovate, 4.5–7 mm long, acuminate, glabrous but for the finely ciliate margin; fertile bracts pale green with white scarious margin, broadly ovate 5–8 mm long, acuminate to cuspidate, with similar indum.; bracteoles elliptic or narrowly so, 3.5–5 mm long, acuminate, crisped-puberulous in a central band and ciliate towards apex. Sepals ciliate near apex, otherwise glabrous; dorsal ovate-elliptic, 4–5 mm long, subacute to acute; lateral and ventral lanceolate to narrowly elliptic, 4–5.5 mm long, acute to acuminate, sometimes with a small dorsifixed mucro. Corolla white, when fully developed with tube 5–7 mm long and lobes 2–3 mm long; cleistogamous flowers not seen in East African material, but common in southern Africa. Stamens 2; filaments ± 1 mm long; anthers ± 1 mm long, with sagittate base, slightly smaller in cleistogamous flowers. Stigma oblong-spathulate and erect in open flowers, rounded or reniform and bent in cleistogamous flowers. Capsule 5.5–6.5 mm long, glabrous. Seed ± 1 mm long.

UGANDA. Bunyoro District: Waki River, 14 Jan. 1907, *Bagshawe* 1433! & Rabongo Forest, 31 Aug. 1964, *H.E. Brown* 2125!
KENYA. Northern Frontier District: Ijara, 13 Jan. 1943, *Bally* 2081!; Tana River District: N of Mlango ya Simba, 6 Nov. 1957, *Greenway & Rawlins* 9460! & Galole-Garsen road, 8 km towards Garsen after turn-off to Wenje, 12 July 1974, *Faden* 74/1059!
TANZANIA. Ulanga District: Malendula, 12 June 1932, *Schlieben* 2352!; Lindi District: 10 km S of Mbemkuru River crossing, 5 Dec. 1955, *Milne-Redhead & Taylor* 7561!
DISTR. **U** 2; **K** 1, 7; **T** 6, 8; Ghana, Nigeria, Congo-Kinshasa, Somalia, Angola, Zambia, Malawi, Mozambique, Zimbabwe, Botswana, South Africa; India
HAB. Alluvial grassland with *Acacia-Combretum-Dobera* thicket clumps on seasonally flooded grey-brown clay, termite mounds, dry riverine thickets; 25–800 m

SYN. *Justicia acaulis* L.f., Suppl.: 84 (1781); Vahl, Symb. Bot. 2: 3 (1791)
    J. acaulis L.f. var. *lyrata* Vahl, Symb. Bot. 2: 3 (1791). Type: India, *Herb. Vahl* s.n. (C!, holo.)
    Elytraria lyrata (Vahl) Vahl, Enum. 1: 106 (1804); Vollesen in Opera Bot. 59: 80 (1980); Lebrun & Stork, Enum. Pl. Afr. Trop. 4: 481 (1997)
    E. crenata Vahl, Enum. 1: 106 (1804); Nees in DC., Prodr. 11: 63 (1847); T. Anderson in J.L.S. 7: 20 (1863) pro parte; C.B. Clarke in F.T.A. 5: 27 (1899) pro parte. Type as for *E. acaulis*.
    E. indica Pers., Syn. 1: 23 (1807). Type as for *E. acaulis*.
    E. crenata Vahl var. *lyrata* (Vahl) Nees in DC., Prodr. 11: 63 (1847)
    Tubiflora acaulis (L.f.) Kuntze, Rev. Gen. 2: 500 (1891); Lindau in P.O.A. C: 365 (1895)
    Elytraria acaulis (L.f.) Lindau var. *lyrata* (Vahl) Bremek. in Reinwardtia 3: 251 (1955); Iversen in Symb. Bot. Ups. 29(3): 161 (1991)
    [*E. marginata* sensu Heine in F.W.T.A. (ed. 2) ?: 418 (1963) pro parte, *non* Vahl (1804)]

NOTE. The Kenyan material has fertile bracts 5–6 mm long against 7–8 mm in Tanzanian material. Material from southern Africa shows a continuous variation from 6–8 mm. The Kenyan material also has shorter bracteoles (3.5–4 mm) than the Tanzanian (4–5 mm), and shows a slight tendency towards shorter sepals. In these characters the Kenyan material seems to approach *E. minor*. A case could possibly be made for treating the two groups as subspecies, but the material is otherwise so homogeneous that this approach is rather unappealing.

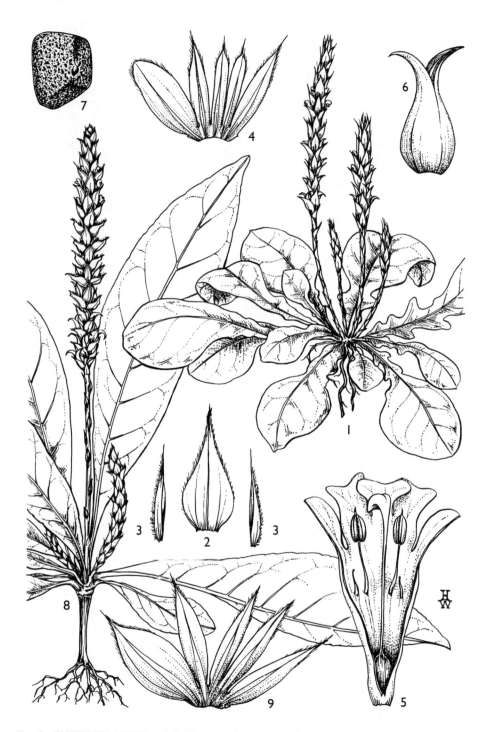

FIG. 1. *ELYTRARIA MINOR* — **1**, habit, × 1; **2**, bract, × 5; **3**, bracteoles, × 5; **4**, calyx opened up, × 7; **5**, corolla opened up, × 10; **6**, capsule, × 6; **7**, seed, × 20. *E. MARGINATA* — **8**, habit, × 1; **9**, calyx opened up, × 7. 1–7 from *Drummond & Hemsley* 3508, 8–9 from *Liebenberg* 937. Drawn by Heather Wood.

2. **Elytraria minor** *Dokosi* in Adansonia 18: 433 (1979). Type: Kenya, Lamu District: Mambasasa, Utwani Forest Reserve, *Greenway & Rawlins* 9366 (EA!, holo.; K!, iso.)

Acaulescent perennial herb with a short creeping or suberect sometimes branched rootstock, apical part (where leaves attached) with short (under 1 mm long) puberulous hairs. Leaves in a basal rosette, appressed to the ground, often variegated with pale green areas along midrib and lateral veins, subsessile or with petiole up to 1.5(–2) cm long; lamina lyrate, obovate-spathulate in outline, terminal segment elliptic to orbicular, largest 2.5–10 × 1–4.2 cm, apex subacute to truncate, base attenuate, decurrent, margin on terminal segment entire to crenate; below sparsely puberulous along midrib and lateral veins and often with very fine scattered hairs on lamina, above with sparse curly hairs on midrib and usually scattered hairs on lamina. Spikes 1.5–6.5(–9) cm long, unbranched or branched; peduncle 1–7.5(–9.5) cm long; sterile bracts imbricate, ovate, 3–5.5(–6) mm long, acuminate to cuspidate, glabrous but for a finely ciliate margin, very finely rugose on back, indistinctly keeled; fertile bracts green or pale green, broadly ovate, 5–6.5 mm long, acuminate to cuspidate, ± a narrow scarious margin, glabrous but for a finely ciliate margin, sometimes indistinctly keeled; bracteoles lanceolate, 3–4 mm long (including a dorsally attached mucro up to 1 mm long), acuminate, crisped-puberulous in a central band and finely ciliate near apex. Sepals ciliate near apex, otherwise glabrous; dorsal elliptic, 3–4 mm long, broadly rounded to truncate (rarely subacute); lateral and ventral lanceolate to narrowly elliptic, 3.5–4.5 mm long (including a dorsally attached mucro 0.5–1 mm long). Corolla white or cream, often cleistogamous or apparently only opening partially; tube 2–3.5 mm long; lobes 1.5–3 mm long, sometimes bifid. Stamens 2, subsessile; anthers ± 0.5 mm long. Capsule 4–5 mm long, glabrous. Seed ± 0.75 mm long, pitted. Fig. 1, 1–7, p. 13.

KENYA. Lamu District: Utwani Forest, 31 Aug. 1956, *Rawlins* 34! & Mambasasa, 30 Jan. 1958, *Verdcourt* 2131!; Kilifi District: Sokoke Forest, near Jilore Forest Station, 28 Aug. 1971, *Faden* 71/798!

TANZANIA. Tanga District: 8 km SE of Ngomeni, 29 July 1953, *Drummond & Hemsley* 3508!; Lushoto District: E Usambara Mts, Sigi, 30 May 1917, *Peter* 60525!; Morogoro District: Nguru Mts, Turiani, Diwali River, 23 Nov. 1955, *Milne-Redhead & Taylor* 7360!

DISTR. **K** 7; **T** 3, 6; not known elsewhere

HAB. Dry evergreen or semi-deciduous forest on sandy, loamy or rocky soils; 15–450(–650) m

NOTE. This is obviously closely related to the widespread *E. acaulis*, but deserves to be treated as a distinct species. Superficially most similar to forms of *E. acaulis* with lyrate leaves, but distinguished by the smaller flowers, differently shaped and smaller sepals, smaller capsule and rootstock with short puberulous hairs. In Kenya and Tanzania where both species occur there are also differences in habitat with *E. acaulis* usually growing on clayey floodplains and *E. minor* on lighter soils in coastal forest.

3. **Elytraria marginata** *Vahl*, Enum. 1: 108 (1804); Nees in DC., Prodr. 11: 63 (1847); Heine in F.W.T.A. (ed. 2) 2: 418 (1963) pro parte & in Fl. Gabon. 13: 155, pl. 31 (1966); Hepper, Herb. Isert & Thonning: 17 (1976); Lebrun & Stork, Enum. Pl. Afr. Trop. 4: 481 (1997); Friis & Vollesen in Biol. Skr. 51(2): 443 (2005). Type: "ad Senegal in Guinea", *Thonning* s.n. (C!, holo.)

Annual or short-lived perennial herb, basal part of stem creeping and rooting, apical part erect, 2–18 cm long, subglabrous to crisped-puberulous. Leaves sometimes mottled, in a terminal rosette (rarely an extra whorl of leaves on middle part of stem), subsessile or petiole up to 5(–10) mm long; lamina elliptic-obovate, largest 7–17 × 2.5–5.5 cm, apex acute to subacute, base attenuate to cuneate, margin entire to faintly crenate; below sparsely crisped-puberulous along veins, above with sparse long hairs along midrib and uniformly scattered hairs on lamina. Spikes 2–13 cm long, unbranched (rarely branched); peduncle 1–8 cm long; sterile bracts not imbricate or ± imbricate towards apex of peduncle, broadly ovate-elliptic, 3–6 mm long, acute to acuminate, glabrous but for finely ciliate margin, indistinctly keeled; fertile bracts green with scarious white margin, broadly ovate-elliptic, 4.5–7 mm long, acute to acuminate, glabrous but for finely ciliate margin, indistinctly keeled; bracteoles lanceolate to narrowly elliptic, 3–4 mm long, subacute to acute, crisped-puberulous in a central band and finely ciliate near apex. Sepals 4–5 mm long, ciliate in apical half, otherwise glabrous; dorsal elliptic or broadly so, acute; lateral elliptic or narrowly so, with a dorsally attached seta ± 1 mm long; ventral lanceolate to narrowly elliptic, acute to acuminate. Corolla white, often cleistogamous or only opening partially; tube 3–4 mm long; lobes 2–3 mm long. Stamens 2, subsessile; anthers ± 0.5 mm long. Stigma erect and spathulate or rounded and ± bent in cleistogamous flowers. Capsule 4–5 mm long, glabrous. Seed ± 0.75 mm long. Fig. 1, 8–9, p. 13.

UGANDA. Toro District: Bwamba, Semliki Forest, 23 Sep. 1969, *Lye* 4267!; Bunyoro District: Bujenje, Budongo Forest, 27 Oct. 1971, *Synnott* 717!; Mengo District: Gaba, July–Aug. 1915, *Dümmer* 2641!
DISTR. **U** 2, 4; West Africa to S Sudan and south to Angola
HAB. Wet evergreen forest ground floor; 700–1300 m

SYN. *Tubiflora paucisquamosa* De Wild. & T.Durand in Bull. Soc. Roy. Bot. Belg. 38, 2: 42 (1899). Types: Congo-Kinshasa, Bania-Lecoula, *Dewèvre* s.n. (BR, syn.) & Gandayanga, *Cabra* s.n. (BR, syn.)
[*Elytraria crenata* sensu T. Anderson in J.L.S. 7: 20 (1863) pro parte; C.B. Clarke in F.T.A. 5: 27 (1899) pro parte, *non* Vahl (1804)]
[*E. squamosa* sensu Lindau in Z.A.E.: 291 (1911), *non Tubiflora squamosa* (Jacq.) Kuntze (1891)]
[*E. acaulis* sensu Robyns in F.P.N.A. 2: 264 (1947), *non* (L.f.) Lindau (1897)]

2. NELSONIA

R.Brown, Prodr. Fl. Nov. Holl.: 480 (1810); Nees in DC., Prodr. 11: 65 (1847); C.B. Clarke in F.T.A. 5: 28 (1899); Bremekamp in Reinwardtia 3: 247 (1955); Morton in K.B. 33: 401 (1979); E. Hossain in Willdenowia 14: 397 (1984); Barker in Journ. Adelaide Bot. Gard. 9: 52 (1986); Vollesen in Proc. XIIIth Plen. Meet. AETFAT 1: 315 (1994)

Annual or perennial herbs without cystoliths. Leaves opposite. Flowers single, in dense ovate to cylindric terminal or axillary many-flowered spikes, pedicellate; bracts imbricate, spirally arranged, herbaceous, persistent, with a number of indistinct veins from base; bracteoles absent. Calyx of 4 thin and herbaceous unequal sepals, dorsal wider, 5-veined from base; ventral bifid, 2-veined; lateral narrower, 1-veined. Corolla subactinomorphic or weakly 2-lipped with 2-lobed upper lip and 3-lobed lower lip, glabrous outside; tube short, cylindric, straight. Stamens 2, inserted in upper part of tube, ± included; filaments short; anthers 2-thecous, thecae ellipsoid, at same height, with or without small basal flanges; staminodes absent. Ovary 2-locular, with 4–15 ovules per locule; style minutely bifid. Capsule (2–)8–30-seeded, ovoid-ellipsoid, sessile, rostrate (sterile from apex) with recurved apical part when ripe, retinaculae absent. Seed ellipsoid-globose, flattened, covered with minute anchor-shaped hooks.

5 species. Three in Africa one of which extends to India and SE Asia and is introduced in America, one in Australia and one in South America.

NOTE. I find it impossible to agree with E. Hossain (l.c.) that this genus only contains one pantropical species. *N. campestris* R. Br. from Australia is clearly distinct as pointed out by Barker (l.c.). As pointed out by Morton (l.c.) there are two distinct species over most of Africa (as separated below), and there is an additional species in Southern Africa. There is also one endemic species in South America, but the picture here has been muddled by the introduction of *N. canescens* from the Old World. At least part of the type material of *N. pohlii*, the only name available for the native South American species, belongs to *N. canescens*.

Stems creeping and rooting at base, ascending apically, no central
    rootstock; young stems puberulous to pubescent or sparsely so
    with thick (flattened when dry and collapsed) curly hairs;
    sepals 3.5–5.5 mm long; corolla tube 2.5–4 mm long, lobes
    2–3 mm long; capsule 3.5–5 mm long, glabrous . . . .   1. *N. smithii*
Stems radiating from a central taproot or rootstock, not rooting at
    lower nodes; young stems pilose or densely so with long (up to
    4 mm) thin straight not collapsing hairs; sepals 4.5–7 mm
    long; corolla tube (3–)4–6 mm long, lobes 2.5–5 mm long;
    capsule 4–7 mm long, sparsely glandular near apex . . . .   2. *N. canescens*

1. **Nelsonia smithii** *Oerst.* in Vidensk. Medd. Naturhist. Foren. Kjöb. 1854: 119
(1854); Morton in K.B. 33: 401 (1979); Vollesen in Proc. XIIIth Plen. Meet. AETFAT
1: 322 (1994); Friis & Vollesen in Biol. Skr. 51(2): 451 (2005); Ensermu in F.E.E. 5:
347 (2006). Type: "Congo", *Smith* s.n. (C!, holo.; BM!, iso.)

Perennial herb without central rootstock; stems usually trailing and ascending
apically, but sometimes semi-erect when growing in dense vegetation, rooting at lower
nodes, up to 50 cm long, puberulous to pubescent or sparsely so with thick (flat and
collapsed when dry) many-celled curly hairs, sometimes with long thin pilose hairs
intermixed on uppermost node. Leaves with petiole 0.5–3 cm long; lamina ovate to
elliptic, below middle abruptly narrowed into an attenuate to cuneate or rounded base,
largest 3.5–8.7 × 1.8–4 cm, apex subacute to rounded; beneath subglabrous or with
sparse to dense ± appressed thick hairs on veins, glabrous or with sparser hairs on
lamina, above subglabrous or with scattered thick many-celled hairs. Spikes
1–4.5(–7) cm long, axis pilose; peduncle 2–5 mm long, densely pilose with long thin
hairs and with long-stalked capitate glands; bracts elliptic or broadly so, 5–9 × 3–5.5 mm,
narrowing gradually or concave (with distinct "shoulders") below the acute tip,
puberulous and with few to many long pilose hairs and with usually dense long-stalked
capitate glands; pedicel ± 0.5 mm long, with dense long pilose hairs. Sepals ± equal in
length, 3.5–5.5 mm long, acute; dorsal elliptic; ventral narrowly elliptic and divided
$^{1}/_{4}$–$^{1}/_{3}$ down; lateral lanceolate; all with long pilose hairs (up to 3 mm long) all over or at
base only and with stalked capitate glands towards apex. Corolla pink to mauve (rarely
white or white with a pink tinge), subactinomorphic, ± hairy in throat; tube 2.5–4 mm
long, narrowing slightly upwards and then widened into a short ventral pouch; lobes ±
spreading, 2–3 mm long, slightly emarginate to rounded. Stamens held in the pouch,
subsessile; anthers ± 0.5 mm long, thecae parallel, basal flanges absent or minute.
Capsule 3.5–5 mm long, glabrous. Seeds numerous (usually over 20), ± 0.5 mm long.

UGANDA. Kigezi District: Ruhinda, May 1946, *Purseglove* 2048!; Busoga District: Butembe Bunya,
    15 km E of Jinja, Linya Valley Forest, 4 Dec. 1952, *G.H.S. Wood* 549!; Mengo District: near Sisa,
    Sep. 1937, *Chandler* 1927!
TANZANIA. Lushoto District: E Usambara Mts, Kwamkoro to Amani, 15 Dec. 1916, *Peter* 60482!;
    Mpanda District: Mt Kungwe, Kasoje, 16 July 1959, *Newbould & Harley* 4384!; Rungwe District:
    Kyimbila, 12 Oct. 1910, *Stolz* 334!
DISTR. U 2–4; T 1, 3, 4, 7; widespread in the forest regions of Africa from Guinea to SW Ethiopia
    and south to N Malawi, N Zambia and Angola
HAB. Wet evergreen forest ground floor, usually in swampy areas or along streams, riverbanks,
    riverine forest, lake shores; 700–1300(–1500) m

SYN. *Nelsonia canescens* (Lam.) Spreng. var. *smithii* (Oerst.) E. Hossain in Willdenowia 14: 403
    (1984); Lebrun & Stork, Enum. Pl. Afr. Trop. 4: 496 (1997)

NOTE. Until Morton (l.c.) resurrected *N. smithii* it had long been general practice to treat all
    African material as one very variable species. Even though some of the separating characters
    used by Morton break down others can be found, and I agree with him that there are two
    distinct species over most of Africa. One (*N. smithii*) is a wet forest species and the other (*N.
    canescens*) is a dry bushland species. The latter has obviously benefited from man's presence
    and has in recent times become more widely dispersed as a result of human activity.

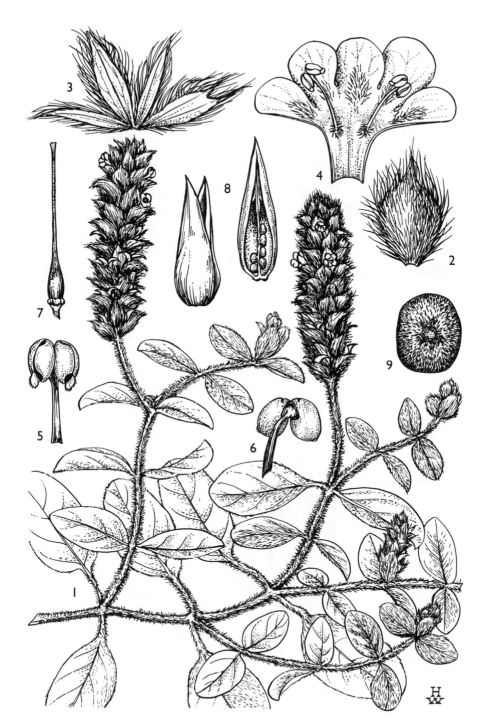

FIG. 2. *NELSONIA CANESCENS* — **1**, habit, × 1; **2**, bract, × 4; **3**, calyx opened up, × 4; **4**, corolla opened up with stamens, × 4; **5**, filament and anther, front view, × 10; **6**, idem, back view, × 10; **7**, ovary, style and stigma, × 5; **8**, capsule, whole and opened up, × 5; **9**, seed, × 20. 1–7 from *Burnett* s.n., 8–9 from *Bally* 7646. Drawn by Heather Wood.

2. **Nelsonia canescens** (*Lam.*) *Spreng.*, Syst. Veg. (ed. 16) 1: 42 (1824); Nees in DC.,
Prodr. 11: 67 (1847); Richard, Tent. Fl. Abyss. 2: 140 (1850); Milne-Redhead in Mem.
N. Y. Bot. Gard. 9: 19 (1954); F.P.S. 3: 184 (1956); Heine in F.W.T.A. (ed. 2) 2: 418
(1963) & in Fl. Gabon 13: 158, pl. 32 (1966); Bremekamp in Reinwardtia 3: 248
(1955); Morton in K.B. 33: 399 (1979); Vollesen in Opera Bot. 59: 81 (1980); E.
Hossain & Emumwen in K.B. 36: 565 (1981); E. Hossain in Willdenowia 14: 402
(1984); Iversen in Symb. Bot. Ups. 29(3): 162 (1991); Vollesen in Proc. XIIIth Plen.
Meet. AETFAT 1: 322 (1994); Lebrun & Stork, Enum. Pl. Afr. Trop. 4: 496 (1997);
Friis & Vollesen in Biol. Skr. 51(2): 451 (2005); Ensermu in F.E.E. 5: 347 (2006).
Type: Senegal, *Roussillon* 53 (P!, iso.)

Perennial herb with several trailing stems from a central taproot or rootstock;
stems not (or rarely) rooting at lower nodes, up to 60 cm long, pilose or densely so
with thin hairs up to 4 mm long, and with subsessile capitate glands. Leaves with
petiole 0.2–2(–2.5) cm long; lamina ovate to elliptic, not (or very rarely) abruptly
narrowed below middle, largest 1.5–7(–10) × 1–4(–5) cm, apex acute to rounded,
base attenuate to truncate; beneath sparsely to densely pubescent to pilose with thin
hairs (rarely with thick hairs), densest on veins but lamina always uniformly hairy,
above with similar or sparser indumentum. Spikes 1–12 cm long, axis pilose;
peduncle 0.2–4.5 cm long, indumentum as stems; bracts ovate-elliptic or broadly so
to orbicular, 5–9 × 3.5–6 mm (lowermost pair sometimes slightly leafy and larger),
narrowing gradually or concave (with distinct "shoulders") below the acute tip, pilose
and with stalked capitate glands; pedicel ± 0.5 mm long, with dense pilose hairs.
Sepals ± equal in length, 4.5–7 mm long, acute; dorsal elliptic; ventral elliptic,
divided ± ¹/₃ down; lateral lanceolate; usually with long pilose hairs up to 3 mm long
all over (very rarely towards base only) and with stalked capitate glands. Corolla pale
clear blue to bright deep blue or bluish purple, subactinomorphic, with hairs in
throat; tube (3–)4–6 mm long, widened into a campanulate throat; lobes 2.5–5 mm
long, ± spreading, slightly emarginate to rounded. Stamens included to slightly
exserted; filaments 0.5–2 mm long; anthers ± 1 mm long, thecae diverging, basal
flanges conspicuous. Capsule 4–7 mm long, sparsely glandular near apex. Seeds
numerous (but usually less than 20), ± 0.5 mm long. Fig. 2, p. 17.

UGANDA. Karamoja District: Napenyenya, Nakyranyet River, 6 Jan. 1937, *A.S. Thomas* 2192! &
    Kidepo River, Apr. 1960, *Wilson* 993!; Bunyoro District: Rabongo Forest, 18 Feb. 1964, *H.E.*
    Brown 2009!
KENYA. Meru District: Tana River, Adamsons Falls, 21 Nov. 2000, *Luke et al.* 7091!; North
    Kavirondo District: Kakamega, Buyangu, 19 Feb. 2005, *Malombe* 951!
TANZANIA. Mpanda District: Mahali Mts, Sobogo, 1 Aug. 1958, *Newbould & Jefford* 1258!; Tabora
    District: Urambo, 9 Oct. 1949, *Bally* 7646!; Iringa District: Jongomeru, 2 Nov. 1970, *Greenway*
    & Kanuri 14634!
DISTR. **U** 1, 2; **K** 4/7, 5; **T** 1, 3–8; widespread in tropical Africa (but rare in the northeast and
    absent from South Africa); Madagascar, India and SE Asia, introduced into central and south
    America
HAB. Woodland, bushland and grassland, often disturbed or secondary or along paths and in
    clearings, riverbanks, old fields, lawns, ruderal; 50–1750 m

SYN. *Justicia canescens* Lam., Tabl. Encycl. Meth. Bot. 1: 41 (1791); Vahl, Enum. Pl. 1: 122 (1804);
    Roem. & Schult., Syst. Veg. (ed. 15) 1: 145 (1817)
    [*Nelsonia campestris* sensu C.B. Clarke in F.T.A. 5: 28, non R.Br. (1810)]
    Nelsonia canescens (Lam.) Spreng. var. *vestita* (Roem. & Schult.) E.Hossain in Willdenowia
        14: 403 (1984); Lebrun & Stork, Enum. Pl. Afr. Trop. 4: 496 (1997)

NOTE. This now basically weedy species' original habitat is probably grassland and riverbanks
    in drier and more open places than *N. smithii*. It is without any doubt also native in India and
    SE Asia (where *N. smithii* does not occur) where it also shows great morphological variation.
    It is most likely that it is introduced in America where it shows a much more restricted range
    of morphological variation.
        The fresh leaves have a sour lemon-like taste and are often reported to be eaten.

3. SAINTPAULIOPSIS

Staner in B.J.B.B. 13: 8 (1934); B.L. Burtt in Not. Roy. Bot. Gard. Edinb. 22: 313 (1958); Champluvier in B.J.B.B. 61: 154 (1991)

Perennial herb without cystoliths. Leaves opposite. Flowers one per bract, alternate, in open terminal racemes, pedicellate; bracts glumaceous, fused to the pedicel, persistent; bracteoles glumaceous, similar to the larger sepals, inserted on and fused to base of calyx, covering the two lateral sepals. Calyx deeply divided into 5 unequal sepals, two lateral small and hyaline, dorsal and two ventral larger, glumaceous. Corolla 2-lipped, campanulate; tube widened almost from base; upper lip 2-lobed, hooded; lower lip 3-lobed, with flat lateral lobes and spoon-shaped median lobe without rugula. Stamens 4, didynamous, included in corolla-tube; anthers 2-thecous, thecae elliptic, with a filiform appendage at base, divergent, inserted at same height on a broad connective; staminode absent. Ovary partly 2-locular, with numerous ovules; style linear; stigma 2-lobed, upper lobe small, lower larger, truncate. Capsule sessile oblong-triangular, obtuse, apical part sterile and recurved at maturity, ± included in calyx, glabrous; retinaculae papilliform, unindurated. Seed papillose.

One species in Central Africa.

Saintpauliopsis lebrunii *Staner* in B.J.B.B. 13: 8 (1934); Champluvier in Fl. Rwanda 3: 484 (1985) & in B.J.B.B. 61: 154 (1991) & in Distrib. Pl. Afr. 40: map 1321 (1994); Lebrun & Stork, Enum. Pl. Afr. Trop. 4: 502 (1997). Type: Congo-Kinshasa, Kivu, Walikale to Kalehe, *Lebrun* 5270 (BR!, holo.)

Perennial herb, sometimes mat-forming; stems creeping and rooting at nodes, up to 10 cm long, puberulous to pubescent with many-celled purplish tinged hairs. Leaves pale beneath, crowded at points of rooting or towards end of stems, petiole up to 4 cm long, lamina ovate to elliptic or broadly so to orbicular, largest 2–3.3 × 1–2 cm, apex broadly rounded to subacute, base truncate to cordate; puberulous to pubescent or sparsely so (dense along veins) with many-celled purplish tinged hairs. Raceme 2–10 cm long (rarely reduced to a single flower) of which the peduncle (0–)1–6 cm; peduncle and axis puberulous; pedicels 2–13 mm long; bracts lanceolate, 1–2 mm long, ciliate; bracteoles lanceolate, 2–3 mm long, with many-celled crisped hairs. Calyx divided to within 0.5 mm of base; dorsal and two ventral sepals lanceolate, 3–5 mm long, with many-celled crisped hairs, two lateral sepals narrowly triangular, 1–2 mm long. Corolla white to pale blue, 6–8 mm long, glabrous outside; tube 3–4 mm long, glabrous inside; upper lip 2–3 mm long; lateral lobes of lower lip held vertically, oblong, 3–4 mm long, median lobe deflexed, broadly elliptic to circular, 3–4 mm long. Filaments glabrous, 2–3 mm long; anthers glabrous, 0.5–1 mm long; tails ± 0.5 mm long, minutely puberulous. Ovary with a few hairs apically. Capsule ± 3 mm long. Fig. 3, p. 20.

Tanzania. Morogoro District: Mikumi National Park, Uvinda Mts, above Uvinda, 26 Nov. 1997, *Frimodt-Møller* 161!; Iringa District: Udzungwa Mts, Mwanihana Forest Reserve, above Sanje, Apr. 1981, *Rodgers & Homewood* 996! & 10 Oct. 1984, *D.W. Thomas* 3805!
Distr. **T** 6, 7; Gabon, Congo-Kinshasa, Rwanda, Burundi
Hab. Forest floor, tree roots or mossy banks in wet montane forest; 1400–1800 m

Syn. *Saintpauliopsis lebrunii* Staner var. *obtusa* Staner in B.J.B.B. 13: 9 (1934). Type: Congo-Kinshasa, W of Lake Kivu, Mt Biega, *Humbert* 7625 (BR!, holo.; BR!, K!, iso.)
　　Staurogyne lebrunii (Staner) B.L.Burtt in Not. Roy. Bot. Gard. Edinb. 22: 313 (1958)

FIG. 3. *SAINTPAULIOPSIS LEBRUNII* — **1**, habit, small plant, × ²/₃; **2**, habit, large plant, × ²/₃; **3**, bract, × 10; **4**, bracteole, × 10; **5**, large sepal, × 10; **6**, small sepal, × 10; **7**, corolla with bracteoles and calyx, × 5; **8**, corolla opened up, × 6; **9**, whole capsule, × 8; **10**, capsule valve, dorsal view, × 8; **11**, seed, × 20. 1 from *Lovett* 310, 2 from *Reekmans* 10955, 3–8 from *D. Thomas* 3805, 9–11 from *Reekmans* 8119. Drawn by Eleanor Catherine.

4. ANISOSEPALUM

E. Hossain in Not. Roy. Bot. Gard. Edinb. 31: 377 (1972); Champluvier in B.J.B.B.
61: 127 (1991)

Staurogyne sensu Heine in Fl. Gabon 13: 120 (1966) pro parte, quoad *S. alboviolacea*,
non Nees (1831)

Perennial or shrubby herbs without cystoliths. Leaves opposite or in whorls, with
scattered peltate subsessile orange glands beneath. Flowers one per bract, alternate,
in terminal lax or dense racemoid cymes, pedicellate; bracts fused to the pedicel,
persistent, 3-veined from base; bracteoles similar to the larger sepals, inserted on
base of calyx and covering the lateral sepals. Calyx deeply divided into 5 unequal
sepals, two lateral small, hyaline, 1-veined, dorsal and two ventral herbaceous, 3-
veined. Corolla 2-lipped; tube linear, broadening gradually upwards or abruptly
campanulate; upper lip flat or ± hooded, entire or emarginate, lower deeply 3-lobed
with two narrow lateral lobes and broad rounded central lobe with 2–4 bosses at
base. Stamens 4, didynamous, inserted deep in the tube, included or shortly
exserted; filaments linear; anthers 2-thecous, thecae elliptic, with small apiculus at
base, divergent, inserted at same height on a broad connective; staminode absent.
Ovary 2-locular, with 6–8 ovules per locule; style linear; stigma deeply 2-lobed with
2 linear lobes, or lower lobe reduced, the upper lobe bifid. Capsule sessile, oblong-
triangular, obtuse, glabrous, apical part sterile and ± recurved when mature;
retinaculae papilliform, unindurated. Seed black, ovoid, flattened, rugose, with a
hilar excavation.

3 species in Central Tropical Africa.

1. Leaves in whorls of 3, crenate-serrate; longer sepals
   1.5–1.8 cm long; corolla 3.5–3.8 cm long; capsule ±
   12 mm long ....   3. *A. lewallei*
   Leaves opposite, entire or crenate; longer sepals 0.4–1.3 cm
   long; corolla 1.7–3.3 cm long; capsule 5–10 mm long 2
2. Bracts, bracteoles and calyx glabrous; corolla hairy,
   17–20 mm long; largest leaf 12–15 cm long; peduncle
   1.5–5.5 cm long; bracteoles and larger sepals 4–6 mm
   long ....   1. *A. alboviolaceum*
   Bracts, bracteoles and calyx hairy; corolla glabrous,
   20–33 mm long; largest leaf 5–11 cm long; peduncle
   0–1(–1.5) cm long; bracteoles and larger sepals 9–13 mm
   long ....   2. *A. humbertii*

1. **Anisosepalum alboviolaceum** (*Benoist*) *E.Hossain* in Not. Roy. Bot. Gard. Edinb.
31: 378 (1972); Champluvier in Fl. Rwanda 3: 433 (1985) & in B.J.B.B. 61: 128, fig. 9
(1991) & in Distrib. Pl. Afr. 40: map 1314 (1994); Lebrun & Stork, Enum. Pl. Afr.
Trop. 4: 467 (1997). Type: Gabon, Lastoursville, Nnyounga na Pounga, *Le Testu* 7442
(P!, holo.; P!, iso.)

Erect or scandent shrubby herb; stems up to 1 m long, glabrous. Leaves opposite;
petiole up to 6 cm long; lamina ovate to elliptic, largest 12–15 × 5–6 cm, apex
acuminate with obtuse to acute tip, base cuneate to attenuate, margin slightly
crenate, glabrous or with a few scattered hairs above. Raceme lax, 4–10 cm long,
glabrous; peduncle 1.5–5.5 cm long; bracts with orange glands, narrowly
oblanceolate to elliptic, 8–13 mm long, glabrous; pedicels 1–2 mm long; bracteoles
linear-lanceolate, 4–6 mm long, glabrous. Dorsal and two ventral sepals linear-
lanceolate, 4–6 mm long, glabrous; two lateral triangular, ± 1 mm long. Corolla white
with 2 purple bosses on lower lip, 17–20 mm long; tube 14–16 mm long, widening
gradually near apex, glandular-pubescent outside, puberulous inside and with ring

FIG. 4. *ANISOSEPALUM ALBOVIOLACEUM* — **1**, habit, × ²/₃; **2**, bract, × 4; **3**, bracteole, × 4;
4, calyx opened up, × 4; **5**, corolla opened up with stamens, × 2; **6**, anther, × 15; **7**, stigma,
× 15; **8**, capsule, × 4; **9**, seed, × 15. *A. HUMBERTII* — **10**, flowering branch, × ²/₃; **11**, bract
× 4; **12**, bracteole, × 4; **13**, calyx opened up, × 4; **14**, corolla, × 2; **15**, corolla, × 2. 1–7 from
Purseglove 2661, 8–9 from *Yamada* 406, 10 from *Lovett* 299, 11–14 from *Mwasumbi* 11872, 15
from *Eggeling* 4178. Drawn by Eleanor Catherine.

of hairs 3–4 mm above base; upper lip hooded, emarginate, lower lip spreading, lateral lobes lanceolate, 2–3 mm long, median lobe 3–4.5 mm long. Stamens included; filaments 3–4 mm long, sparsely puberulous near base; anthers ± 1 mm long, glabrous; stigma lobes similar. Capsule 5–8 mm long. Seed not seen. Fig. 4, 1–9, p. 22.

UGANDA. Kigezi District: Impenetrable Forest, Apr. 1948, *Purseglove* 2661! & Ishasha Forest, June 1947, *Purseglove* 2467! & 8 June 1952, *Lind* 63A!
DISTR. U 2; Gabon, Congo, Central African Republic, Congo-Kinshasa, Rwanda, Zambia
HAB. Wet evergreen montane forest; 1200–1500 m

SYN. *Staurogyne alboviolacea* Benoist in Notul. Syst. 11: 151 (1944); Heine in Fl. Gabon 13: 121, pl. 25, fig. 6–8 (1966)
   S. alboviolacea Benoist subsp. *grandiflora* Napper in K.B. 24: 342, fig. 5 (1970). Type: Uganda, Kigezi District: Impenetrable Forest, *Purseglove* 2661 (K!, holo.; EA!, iso.)
   Anisosepalum alboviolaceum (Benoist) E.Hossain subsp. *grandiflorum* (Napper) E.Hossain in Not. Roy. Bot. Gard. Edinb. 31: 378 (1972)

NOTE. The Ugandan material (subsp. *grandiflora*) at first sight looks strikingly different from the Central African material, but really only differs in the larger corolla. Plants from Gabon to Congo and Rwanda have a corolla 8–11 mm long while in Kivu Province the corolla is 13–17 mm long. Champluvier in B.J.B.B. 61: 131 (1991) divides the species into a number of informal groups also partly based on corolla size. The Uganda material falls into her "groupe 5".

2. **Anisosepalum humbertii** (*Mildbr.*) *F.Hossain* in Not. Roy. Bot. Gard. Edinb. 31: 377 (1972); Champluvier in Fl. Rwanda 3: 433 (1985) & in B.J.B.B. 61: 142, figs. 13 & 14 (1991) & in Distrib. Pl. Afr. 40: map 1317 (1994); Lebrun & Stork, Enum. Pl. Afr. Trop. 4: 467 (1997). Type: Congo-Kinshasa, Kivu, Tshibinda, *Humbert* 7489bis (B†, holo.; BR!, K!, P!, iso.)

Shrubby herb or shrub, forming large bushes or ± scandent; stems up to 1.5(–3) m long, with transverse bands of hairs at upper nodes, otherwise glabrous. Leaves opposite; petiole up to 3 cm long, pubescent adaxially; lamina ovate to elliptic, abruptly narrowed into a decurrent base or attenuate, largest 5–11 × 2.5–4 cm, apex acuminate to cuspidate with acute or obtuse tip, margin entire to crenate, subglabrous to crisped-puberulous, densest along veins. Racemes dense and subcapitate in flower, elongating in fruit, 1–6 cm long; axes glabrous to puberulous; peduncle 0–1(–1.5) cm long, glabrous; bracts oblanceolate or narrowly so, 8–15 mm long, puberulous on midrib and ciliate, ± scattered peltate glands; pedicel 0.5–1(–2) mm long, glabrous or puberulous; bracteoles linear-lanceolate, 9–13 mm long, crisped-ciliate. Dorsal and two ventral sepals linear-lanceolate, 9–12 mm long, ciliate, lateral sepals similar or narrowly triangular, 1.5–4 mm long. Corolla dark blue, bluish violet, purple or dark purple, with 4 white bosses on lower lip, 2–3.3 cm long, glabrous outside; tube 1.6–2 cm long, campanulate above, glabrous inside but for band of hairs 3–4 mm above base; upper lip hooded or ± flat, entire or emarginate, 6–15 mm long; lateral lobes of lower lip spreading, obovate-spathulate, 5–10 mm long, median lobe deflexed, broadly oblong, 8–13 mm long. Stamens held under upper lip; filaments 8–20 mm long, glabrous; anthers ± 1.5 mm long, crisped-puberulous; stigma lobes similar. Capsule 8–10 mm long. Seed pyriform, reticulate, 1.5–2 mm long. Fig. 4, 10–15, p. 22.

subsp. **humbertii**

Corolla tube 1.6–2 cm long; corolla glabrous inside but for band of hairs 3–4 mm above base.

UGANDA. Kigezi District: Kasotoro, Sep. 1947, *Dale* 485! & Impenetrable Forest, Sep. 1936, *Eggeling* 3304! & idem, Oct. 1940, *Eggeling* 4178!
TANZANIA. Iringa District: Luluzi Hill, July 1953, *Carmichael* 197! & Udzungwa Escarpment, Boma la Mzinga, 19 June 1979, *Mwasumbi* 11872! & Udzungwa Mts, Mwanihana Forest Reserve, above Sanje, 14 June 1984, *Lovett* 299!

DISTR. **U** 2; **T** 7; Congo-Kinshasa, Rwanda, Burundi
HAB. Wet evergreen montane forest; 1250–2100 m

SYN. *Staurogyne humbertii* Mildbr. in B.J.B.B. 14: 353 (1937); F.P.N.A. 2: 262, pl. 25 (1947)

NOTE. The material from Tanzania has larger and darker corollas than the material from Uganda. It also has longer filaments. But I do not find the populations sufficiently different to recognise two subspecies and looking further west, where other variation patterns occur, the two areas are more or less connected by the populations in Rwanda, Burundi and eastern Congo. There is therefore a real possibility of the species occurring in W Tanzania.
    Subsp. *zambiense* Champl. from NW Zambia and Congo has a longer corolla tube with the ring of hairs higher up. This could also occur in SW Tanzania.

3. **Anisosepalum lewallei** *P.Bamps* in B.J.B.B. 42: 301, fig. 1 (1972); Champluvier in B.J.B.B. 61: 152 (1991) & in Distrib. Pl. Afr. 40: map 1320 (1994); Lebrun & Stork, Enum. Pl. Afr. Trop. 4: 467 (1997). Type: Burundi, Kumuyange, *Lewalle* 6746 (BR!, holo.; K!, iso.)

Perennial herb with numerous stems from creeping and branched rhizomes, forming large clumps; stems ascending to erect, to 1 m tall, rounded (but triangular at nodes), glabrous but for thin lines of hairs at nodes. Leaves in whorls of 3, with pale yellow sessile glands; petiole pubescent-ciliate, up to 7 mm long; lamina ovate, abruptly narrowed to truncate above a decurrent wavy base, largest 8–9.5 × 3–4 cm, apex acute to subacuminate with obtuse tip, margin crenate-dentate, wavy, above glabrous, beneath pubescent along midrib and larger veins. Racemes dense, pyramidal in flower, elongating and cylindrical in fruit, 6–18(–23) cm long; peduncle 0–12 mm long, glabrous; axis finely puberulous; bracts lanceolate to narrowly elliptic, 1–2 cm long (lowermost pair sometimes foliaceous), glabrous, with sessile glands; pedicels 1–2 mm long, glabrous; bracteoles linear-lanceolate, 1.5–2 cm long, ciliate apically. Dorsal and two ventral sepals linear-lanceolate, 1.5–1.8 cm long, ciliate apically, two lateral similar, ± 5 mm long. Corolla pale mauve on the outside, dark velvety vine-red on the inside, 3.5–3.8 cm long, glabrous on the outside; tube ± 2.5 cm long, with inner band of hairs ± 5 mm above base; upper lip slightly hooded, 1–1.3 cm long, lateral lobes in lower lip spreading, elliptic, 0.8–1 cm long, median lobe deflexed, broadly ovate, 1.2–1.5 cm long. Stamens held under upper lip; filaments 1.5–2 cm long, glabrous; anthers ± 1.5 mm long, crisped-puberulous; ventral stigma lobe reduced. Capsule ± 1.2 cm long. Seed pyriform, 1.5–2 mm long.

TANZANIA. Mpanda District: Mpanda–Uvinza road, Uzondo Plateau, 30 May 2000, *Bidgood et al.* 4570! & 23 June 2000, *Bidgood et al.* 4728!
DISTR. **T** 4; Burundi
HAB. Border zone between grassland and *Syzygium owariense* groundwater forest; ± 1600 m

NOTE. The peculiar leaf arrangement with the leaves in whorls of three is – to my knowledge – unique in the Acanthaceae. As mentioned in the description the stems are more or less rounded but become triangular at the nodes.

5. PSEUDOCALYX

Radlk. in Abh. Naturw. Verein. Bremen 8: 416 (1883); Baillon, Hist. Pl. 10: 423 (1890); Benoist in Notul. Syst. 11: 138, 149 (1944); Breteler in K.B. 49: 809 (1994) & in Adansonia, ser. 3, 20: 271 (1998); Schönenberger & Endress in Int. J. Plant Sci. 159: 446 (1998); Champluvier in Syst. Geogr. Pl. 69: 199 (1999)

Scandent shrubs or woody twiners with stellate indumentum; cystoliths absent. Leaves opposite, entire or crenate. Flowers solitary or in fascicles merging into terminal pseudoracemes; bracteoles 2, large, bract-like, coriaceous, connate for $^2/_3$

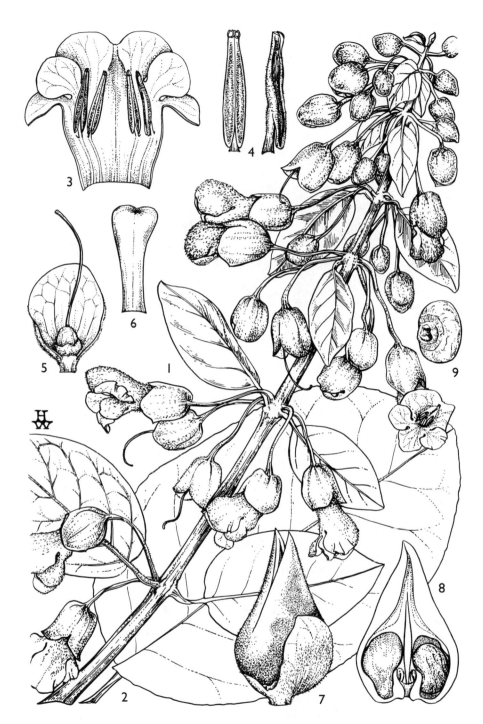

FIG. 5. *PSEUDOCALYX SACCATUS* — **1**, flowering stem, × 1; **2**, leaf, × 1; **3**, corolla opened up with stamens, × 1.5; **4**, anther, side and front views, × 3; **5**, bracteole, calyx, ovary and style, × 1.5; **6**, stigma, × 20; **7**, capsule, × 1.5; **8**, section of young capsule, × 1.5; **9**, seed, × 3. 1–6 from *Richards* 9008, 7–9 from *Fanshawe* 4814. Drawn by Heather Wood.

of their length. Calyx an undulate rim or irregularly lobed or toothed. Corolla contorted in bud, tube cylindric; limb indistinctly 2-lipped, upper lip 2-fid, lower 3-fid. Stamens 4, didynamous, included in tube and inserted near middle, subsessile, sometimes also a staminode present; anthers 2-thecous, hairy or glabrous, thecae oblong-linear, sagittate at base, opening by apical pores. Disc conspicuous, annular, fleshy. Ovary 2-locular, 2 ovules per locule; stigma 2-lobed. Fruit a woody thick-walled tardily dehiscent 4-seeded capsule without or with a short poorly defined beak; retinacula absent. Seed subglobose with a large scar on one side.

6 species. The others in Liberia, Gabon (3 species) and Congo.

Pseudocalyx saccatus *Radlk.* in Abh. Naturw. Verein Bremen 8: 417 (1883); Baillon, Hist. Pl. 10: 423 (1890); Lindau in E. & P. Pf. IV, 3b: 293 (1895); Benoist in Bull. Mus. Hist. Nat. Paris 32: 150 (1926) & in Notul. Syst. 11: 149 (1944) & in Fl. Madagascar 182, 1: 20 (1967); Breteler in K.B. 49: 812 (1994); Lebrun & Stork, Enum. Pl. Afr. Trop. 4: 500 (1997); Breteler in Adansonia, ser. 3, 20: 277 (1998). Type: Madagascar, Nossi-be, *Rutenberg* s.n. (BREM, holo.)

Strong scandent shrub or woody twiner to 15 m or more; young stems terete or quadrangular, densely yellowish to reddish stellate-pubescent, glabrescent. Petiole 0.5–2 cm long, stellate-pubescent; lamina elliptic to obovate, largest 7.5–13 × 4–9 cm, apex acute to acuminate or obtuse with a mucronate tip, base cuneate to subcordate, margin entire to crenate-dentate, above subglabrous to stellate-pubescent, densest along midrib, glabrescent, below similar or denser. Flowers solitary or in 2–3-flowered axillary fascicles, often merging towards apex into pseudo-racemes up to 15(–20) cm long with leafy bracts gradually decreasing in size upwards; flowers erect in bud, pendulous when open and becoming erect in fruit; pedicels 0.5–2.7 cm long, densely yellowish to reddish stellate-pubescent, some hairs with a stalk up to 1 mm long and a "crown" of hairs; bracteoles broadly ovate, 1.3–1.8 × 1.1–1.7 cm, broadly rounded, densely orange to reddish stellate-pubescent outside, glabrous or with scattered stellate hairs inside. Calyx an undulate rim, ± 2 mm high, densely yellowish stellate-pubescent. Corolla pink to reddish purple, paler on the inside, covered with a dense floccose-stellate yellowish indumentum which rubs off easily; tube 1.4–2 cm long, widening slightly upwards; lobes 5–7 × 4–8 mm, broadly ovate-reniform. Anthers 7–8 mm long, with bulbous-based hairs and sessile glands along their whole length. Disc slightly lobed, ± 1 mm high. Ovary ± 5 mm long, densely yellowish stellate-pubescent; style ± 1.5 cm long, glabrous; stigma with 2 equal lobes ± 1 mm long. Capsule broadly triangular in outline, acute, beak not defined, 2.5–3 × 1.5–2 cm, when young densely floccose-stellate, glabrescent. Seed dark grey to blackish, finely hairy with simple hairs when young, 6–8 mm in diameter. Fig. 5, p. 25.

TANZANIA. Lindi District: Rondo Plateau, Mchinjiri, March 1952, *Semsei* 722! & Rondo Forest Reserve, 22 Nov. 1966, *Gillett* 17989! & 4 Nov. 1984, *Mwasumbi* 12692!
DISTR. **T** 8; Congo-Kinshasa, Mozambique, Angola, Zambia, Zimbabwe; Madagascar
HAB. Semideciduous lowland forest; 500–800 m

SYN. *Pseudocalyx africanus* S.Moore in J.L.S. 40: 156 (1911); White in F.F.N.R.: 383 (1962); Breteler in K.B. 49: 812 (1994). Type: Zimbabwe, Chirinda Forest, *Swynnerton* 97 (BM!, holo.; K!, iso.)

NOTE. Gillett (l.c.) records that the fibrous bark is used for rope making.

6. THUNBERGIA

Retz., Phys. Saelsk. Handl. 1, 3: 163 (1780); Nees in DC., Prodr. 11: 54 (1847); Roulet in Bull. Herb. Boiss. 2: 259 (1894); S. Moore in J.B. 32: 219 (1894); Burkill in F.T.A. 5: 8 (1899); Benoist in Notul. Syst. 2: 287 (1912) & in Notul. Syst. 11: 144 (1944); Bremekamp in Verh. K. Nederl. Akad. Wetensk., Afd. Natuurk., Sect. 2, Part 50, No. 4: 1 (1955); Schönenberger & Endress in Int. J. Plant Sci. 159: 446 (1998), *nom. conserv., non Thunbergia* Montin (1773)

Valentiana Rafin., Specchio 1: 87 (1814)

Endomelas Rafin., Fl. Tellur. 4: 67 (1836)

Erect, trailing or twining annual or perennial herbs, or erect or scandent shrubs or woody twiners; cystoliths absent. Leaves opposite, entire to crenate or dentate; venation pinnate or palmate. Flowers axillary, solitary or paired or in racemes; bracteoles 2, large, hyaline to leathery, connate for $^2/_3$–$^3/_4$ dorsally and about $^1/_4$ ventrally (rarely free to base). Calyx a subentire to undulate rim or with 10–20 triangular to linear teeth. Corolla contorted in bud; tube straight or curved, cylindric or campanulate, widening at point of insertion of stamens, with subsessile capitate glands on the inside; limb spreading, subequally 5-lobed. Stamens 4, didynamous or subequal, inserted at transition from throat to tube, included (rarely exserted); anthers 2-thecous, one or both thecae with a setosely bristled boss or a spur at base (rarely muticous); thecae linear to ovoid, opening by apical slits. Disc conspicuous, annular, fleshy. Ovary 2-locular with 2 collateral ovules per locule; stigma funnel-shaped or with two large unequal lobes (one erect the other horizontal) or subequally 2-lobed with two erect flattened lobes. Fruit a 2–4-seeded thick-walled woody capsule, triangular with weakly defined beak or subglobose with sharply defined beak. Seed hemispherical, hollow on the proximal face, smooth to reticulate, tuberculate, lamellate or otherwise sculptured; retinaculae absent.

100–150 species, mostly in Eastern and Southern Tropical Africa, also from India to SE Asia and Australia, introduced in America. A number of species are widely grown as ornamentals.

Several Indian and SE Asian species have been cultivated as ornamentals but, apart from *T. laurifolia*, never seem to have become naturalized.

A number of the indigenous species have at one time or other also been cultivated. They include *T. alata, battiscombei, erecta, gibsonii, gregorii, holstii, kirkii, petersiana* and *vogeliana*.

1. Anthers with distinctly unequal thecae, all
   thecae at the base with a boss with many glossy
   setose hairs or with 1–2 bristles 2
   Anthers with equal or subequal thecae, one or
   more thecae with a curved or bent spur at
   base or all thecae muticous 9
2. Herbaceous twiner; leaves palmately veined;
   bracteoles free to base; corolla-with long
   cylindric tube and short throat; stigma with
   subequal erect lobes (III. Subgen. *Macrosiphon*)   4. *T. guerkeana* (p. 37)
   Erect, scandent or twining shrubs or shrubby
   herbs; leaves pinnately (rarely palmately)
   veined; bracteoles partly fused; corolla with
   short cylindric tube and long campanulate
   throat 3

3. Leaves with palmate venation; flowers in 1–6-flowered axillary racemes; thecae at the base with 1–2 bristles (II. Subgen. *Thamnidium*) ....   3. *T. kirkii* (p. 37)

      Leaves with pinnate venation; flowers solitary; thecae at the base with a boss with many glossy setose hairs 4

4. Young branches, bracteoles and/or corolla with stellate indumentum; corolla limb white or yellow; stigma with subequal erect curved lobes forming a funnel (I. Subgen. *Stellatae*) 5

      Young branches, bracteoles and corolla glabrous or with simple or glandular hairs; corolla limb mauve, blue or deep purple; stigma with unequal flat lobes, upper lobe erect, lower horizontal (IV. Subgen *Coniostephanus*) 6

5. Young branches uniformly stellate-pubescent; bracteoles stellate-pubescent on the outside; corolla limb yellow ....   1. *T. heterochondros* (p. 36)

      Young branches glabrous but for thin line of stellate hairs at nodes; bracteoles glabrous on the outside; corolla limb white ....   2. *T. stelligera* (p. 36)

6. Erect shrub; longitudinal ridges on stems excurrent as small pseudo-stipular spines ....   8. *T. holstii* (p. 40)

      Scandent shrubs or woody twiners; longitudinal ridges at most forming rounded "ears" at nodes 7

7. Young branches with sessile glands; leaf-margin dentate with numerous teeth; capsule puberulous ....   5. *T. vogeliana* (p. 38)

      Young branches glabrous but for thin line of non-glandular hairs at nodes; leaf-margin entire (but usually strongly undulate) or with a single large tooth; capsule glabrous 8

8. Young branches with pale yellow glossy bark; leaf-margin with a single large triangular tooth near or above middle, at base not folded and clasping pedicel ....   6. *T. erecta* (p. 39)

      Young branches with brown bark; leaf-margin not with single large tooth, strongly undulate, basal part folded upwards and clasping pedicel ....   7. *T. crispa* (p. 39)

9. All thecae muticous; evening- or night-flowering; cylindric part of corolla-tube long, throat short, indistinct (VIII. Subgen. *Thunbergia*) 10

      One or more thecae with a sharply bent or gently curving spur; day-flowering; cylindric part of corolla-tube short, throat long and campanulate 11

10. Cylindric part of corolla tube 5–7 cm long; anthers ± 4 mm long; petiole 0.5–4(–6) mm long; capsule beak 11–15 mm long ....   48. *T. cycnium* (p. 74)

      Cylindric part of corolla tube 1.5–2.5 cm long; anthers ± 2.5 mm long; petiole (2–)4–25 mm long; capsule beak 8–12 mm long ....   49. *T. schimbensis* (p. 74)

11. At least one theca on all four anthers with long gently curved spur; stigma with subequal flat lobes 12

One theca on anterior pair of anthers with a sharply bent spur, other thecae muticous; stigma funnel-shaped, shortly 2-lobed with slighty unequal erect lobes, with two tufts of downwardly directed setose hairs below sinuses (VII. Subgen. *Hypenophora*) 32

12. Anther spurs flattened, 4–5 mm long; anthers with long hairs with small lateral spinules; style bent apically, with two erect stigmatic lobes; strong woody climber (V. Subgen. *Hexacentris*)   9. *T. laurifolia* (p. 42)

Anther spurs terete, 1–2 mm long; anthers with shorter club-shaped hairs without lateral spinules; style straight, upper stigmatic lobe erect, lower horizontal; herbs or herbaceous twiners (VI. Subgen. *Parahexacentris*) 13

13. Flowers in 2–6-flowered pedunculate fascicles concealed between two large foliaceous bracts; corolla purple or deep bluish purple . .   10. *T. fasciculata* (p. 43)

Flowers solitary (rarely paired) in axils of vegetative leaves; corolla white, yellow or orange (rarely brick red), sometimes with purple throat 14

14. Corolla with dark red to deep purple throat 15

Corolla without dark red to deep purple throat, but throat often a darker shade of yellow than limb 20

15. Petiole not winged 16

Petiole winged 18

16. Corolla throat sharply bent near middle; leaves usually about as wide as long; cylindric corolla tube 2–3 mm long . . . .   26. *T. reniformis* (p. 55)

Corolla throat straight or very slightly bent; leaves distinctly longer than wide; cylindric corolla tube 3–12 mm long 17

17. Capsule glabrous; cylindric corolla tube (5–) 8–12 mm long; anther spurs 2–2.5 mm long; seed ± 7 mm in diameter . . . .   23. *T. mildbraediana* (p. 53)

Capsule puberulous; cylindric corolla tube 3–6 mm long; anther spurs 1–1.5 mm long; seed 4–6 mm in diameter . . . .   28. *T. alata* (p. 56)

18. Leaf lamina lanceolate to narrowly ovate or oblong, passing gradually into the broad undulate petiole-wing, lamina more than twice as long as wide . . . .   24. *T. napperae* (p. 53)

Leaf lamina triangular or triangular-ovate, petiole and wing clearly distinct, lamina less than twice as long as wide 19

27. Twining herbs; petiole 0.8–6 cm long; largest leaf
6.5–7.5 cm long; calyx glabrous or ciliate,
without capitate glands, lobes broadly triangular,
0.5–1 mm long 28
Erect, decumbent or trailing herbs; petiole 0–5
mm long; largest leaf 1.7–6.2 cm long; calyx
puberulous and with capitate glands, lobes
linear to narrow triangular, 2–5 mm long 29
28. Pedicels 3.5–4 cm long; bracteoles 1.7–2.3 cm
long; calyx with ciliate lobes; corolla throat
2.5–3 cm long, lobes 1–1.7 × 1.4–2 cm, anthers
± 4 mm long . . . . 14. *T. schliebenii* (p. 47)
Pedicels 2–3 cm long; bracteoles 1.4–1.7 cm
long; calyx glabrous; corolla throat 1.5–1.8 cm
long, lobes 0.7–1 × 0.8–1.2 cm, anthers ± 2.5 mm
long . . . . 15. *T. minziroensis* (p. 47)
29. Annual herb; pedicels 3–6 mm long in flower;
corolla throat 7–10 mm long, lobes 3–5 ×
3–5 mm, anthers ± 1.5 mm long; seed smooth 16. *T. annua* (p. 48)
Perennial herb; pedicels 0.5–3.5 cm long in
flower; corolla throat 1–2 cm long, lobes
0.6–1.5 × 0.7–1.7 cm, anthers 2–3.5 mm long;
seed with lamellate or echinate sculpturing 30
30. Stems tomentose; leaves sericeous-pubescent to
tomentose; bracteoles not purple veined, not
4-angled at base; anthers ± 2 mm long; seed
with lamellate sculpturing . . . . 17. *T. fischeri* (p. 48)
Stems sericeous-pubescent or sparsely so; leaves
subglabrous to sparsely pubescent; bracteoles
with conspicuous purple venation, distinctly
4-angled at base; anthers 3–3.5 mm long; seed
with echinate sculpturing . . . . 18. *T. laborans* (p. 49)
31. Petiole 2–8 mm long; largest leaf 3 or more times
longer than wide; pedicels 0.8–2(–3.2) cm long;
bracteoles 14–17(–19 in fruit) mm long . . . . . . 12. *T. huillensis* (p. 45)
Petiole (10–)12–27 mm long; largest leaf less than
3 times longer than wide, if more then pedicels
over 5 cm long; pedicels (3–)4–11 cm long;
bracteoles (16–)18–27(–32 in fruit) mm long 11. *T. kirkiana* (p. 43)
32. Flowers in axillary racemes (or solitary in upper
axils and racemose in lower axils) 33
All flowers solitary or paired (very rarely in 3's) 37
33. Bracteoles puberulous, without long glossy
capitate glands . . . . 29. *T. battiscombei* (p. 58)
Bracteoles with long (longest 2–3 mm) glossy
capitate glands on veins and margins 34
34. All flowers in racemes 35
Flowers solitary or paired in upper axils, in
racemes in lower axils . . . . 35. *T. petersiana* (p. 63)
35. Erect or trailing herbs from woodland 36
Herbaceous twiner from forest . . . . 32. *T. microchlamys* (p. 61)

36. Leaves 4–6.5 cm wide, with petiole 5–25 mm
long; bracts 13–20 mm long; bracteoles dark
green to purple with obscure reticulation ... 　30. *T. verdcourtii* (p. 59)
Leaves 1–2.5(–4) cm wide, with petiole 0–3 mm
long; bracts 0.5–6 mm long; bracteoles pale
green with conspicuous raised dark green
reticulation .... 　31. *T. racemosa* (p. 61)
37. Vegetative leaves below flowers with petiole
1–13 cm long 38
Vegetative leaves with petiole up to 0.7(–1.2) cm
long or subsessile (pedicels with long (to 2 mm)
glossy glandular hairs if petiole over 0.7 cm) 41
38. Pedicels glabrous or hairy and/or glandular near
apex; bracteoles pale green with conspicuous
raised darker green reticulation, glabrous or
with sparse to dense long glossy hairs and/or
stalked capitate glands 39
Pedicels uniformly puberulous or densely so;
bracteoles dark green, without raised dark green
reticulation, densely uniformly puberulous .... 　34. *T. bogoroensis* (p. 62)
39. Strong herbaceous twiner to 5 m; bracteoles
2.7–3.3 cm long; corolla lobes 1.7–2.8 cm
long; filaments ± 18 mm long; capsule with
beak ± 28 mm long .... 　33. *T. masisiensis* (p. 62)
Erect or straggling perennial herbs or
subshrubs to 1.5 m tall; bracteoles 0.9–2.2 cm
long; corolla lobes 1–1.8(–2) cm long;
filaments 7–13 mm long; capsule with beak
12–18 mm long 40
40. Corolla limb white or pale lilac; tube 2–3.2 cm
long; filaments 7–9 mm long; usually from
montane forest, (1300–)1650–2450 m ...... 　36. *T. usambarica* (p. 64)
Corolla limb dark bluish purple or royal blue;
tube (3.2–)3.5–5 cm long; filaments 9–13 mm
long; usually from lowland forest, 600–1500
(–2200) m .... 　35. *T. petersiana* (p. 63)
41. Leaf-base truncate to cordate, with large rounded
to hastate lobes 42
Leaf-base attenuate to cuneate (rarely truncate),
without basal lobes or with small sub-
auriculate lobes 49
42. Whole plant glabrous; leaves glaucous, with
sagittate (rarely rounded) basal lobes 43
Plant more or less hairy; leaves green, with
rounded or hastate lobes 44
43. Leaves held erect, parallel to the stem, largest
less than 10 × 3.5 cm; corolla limb deep
purple or bluish purple .... 　42. *T. lathyroides* (p. 68)
Leaves spreading from the stem, largest more
than 15 × 10 cm; corolla limb white .... 　37. *T. natalensis* (p. 65)
44. Bracteoles green, with obscure reticulation 45
Bracteoles pale green, with conspicuous raised
dark green reticulation 46

45. Bracteoles with sparse to dense (to 2 mm) glossy
capitate glands on main veins and margins;
leaves subglabrous to puberulous. . . . .   37. *T. natalensis* (p. 65)
Bracteoles uniformly pubescent to lanate with
long (to 4 mm) glossy non-glandular hairs;
leaves pubescent to lanate (rarely sparsely
pubescent) . . . .   38. *T. ciliata* (p. 66)

46. Corolla uniformly lemon yellow; tube ± 2.2 cm
long; filaments ± 4 mm long . . . .   41. *T. austromontana* (p. 68)
Corolla white, blue, mauve or deep purple; tube
2.5–5 cm long; filaments 8–13 mm long 47

47. Pedicels with long (to 2 mm) glossy capitate
glands; usually from medium altitude dry
woodland; (800–)1000–1650 m . . . .   39. *T. richardsiae* (p. 66)
Pedicels glabrous or hairy, but not with long
glossy capitate glands; usually from higher
altitude wetter grassland or woodland;
1400–2450(–2625) m 48

48. Largest leaf (2.7–)3–5.8(–10.5) cm wide, usually
less than twice as long as wide; petiole 2–5 mm
long; corolla limb white to blue or mauve . . .   37. *T. natalensis* (p. 65)
Largest leaf 1–3 cm wide, usually more than
twice as long as wide; petiole 1–3 mm long;
corolla limb deep blue to deep purple . . . . . .   40. *T. mufindiensis* (p. 67)

49. Corolla white (rarely very pale blue); tube
1.8–2.5 cm long; filaments 4–6 mm long;
bracteoles with purple main veins and
margins; stems greyish, slender, 1–2 mm
diameter at base . . . .   43. *T. stellarioides* (p. 69)
Corolla pale blue to mauve, purple to maroon
or almost black (rarely white but then tube
over 3 cm long); tube 2–4.5(–5.5) cm long;
filaments 8–13 mm long; bracteoles not with
purple main veins and margins; stems
greenish, stout, 2–9 mm diameter at base 50

50. Bracteoles with long (1–2 mm) glossy setose
non-glandular hairs; plant flowering in the
rainy season . . . .   47. *T. barbata* (p. 73)
Bracteoles glabrous to puberulous or densely
glandular-pubescent, but hairs or glands
always under 1 mm long; plant flowering in
the dry season, usually shortly after burning 51

51. Corolla limb and upper part of tube dark
purple to deep maroon (also described as
burgundy or deep wine red) . . . .   45. *T. lancifolia* (p. 70)
Corolla limb and upper part of tube white or
pale blue to blue or mauve 52

52. Leaves distinctly ciliate (at least when young)
with long curly glossy hairs, penninerved or
with two weak lateral veins; stems glabrous to
puberulous or pubescent, if glabrous with
bands of long glossy setose hairs at nodes . . .   46. *T. oblongifolia* (p. 73)
Leaves totally glabrous, even when young, with
2(–4) strong rib-like lateral veins running
almost to tip of leaf (or only $\frac{1}{2}$–$\frac{3}{4}$ up); stems
glabrous or sparsely puberulous in thin lines
at nodes . . . .   44. *T. graminifolia* (p. 69)

The species in the Flora area fall into the following infrageneric groups:

I. Subgen. **Stellatae** *Vollesen* **subgen. nov.** Caulis atque bracteolae corollaque capsulaque pilos stellatos gerentes; thecae antherarum ad basin umbone setis brevibus rigidis nitentibus obtecto instructae; stigma lobis subaequalibus erectis complanatis infundibulum formentibus instructum; capsula sensim in rostrum angustata; semen leave et stellato-puberulum. Type: *Thunbergia heterochondros*.

Scandent shrubs or woody twiners. Stems with stellate hairs, at least at nodes. Leaves with pinnate venation. Bracteoles connate, with stellate hairs, at least on the inside. Calyx cupular, irregularly lobed. Corolla with stellate hairs and sessile glands on the outside; basal tube short, cylindric; throat long, campanulate. Stamens didynamous, included; anthers linear-oblong, one theca distinctly longer than the other, apiculate at apex, all thecae at base with a boss covered with short stiff glossy setose hairs. Stigma included, with subequal erect flattened curved lobes forming a funnel. Capsule triangular, narrowing gradually into the beak, when young with a dense yellow stellate floccose indumentum, glabrescent. Seed smooth, stellate puberulous. Species 1–2.

II. Subgen. **Thamnidium** *Bremekamp* in Verh. Kon. Nederl. Akad. Wetensch., Afd. Natuurk., 2nd. Ser., 50(4): 15 (1955)

Erect shrubby herb. Stems glabrous or with simple hairs. Leaves with palmate venation. Bracteoles connate, glabrous or with simple hairs. Calyx a shallowly lobed rim with broadly triangular lobes. Corolla with scattered capitate glands; tube short, cylindric; throat long, campanulate. Stamens didynamous; anthers oblong, one theca distinctly longer than the other, apiculate at apex, all thecae at base with 1–2 straight sometimes bifurcate bristles otherwise glabrous. Stigma included, with erect curved lobes forming a funnel. Capsule triangular, narrowing gradually into the beak, minutely puberulous. Seed tuberculate. Species 3.

III. Subgen. **Macrosiphon** *Bremekamp* in Verh. Kon. Nederl. Akad. Wetensch., Afd. Natuurk., 2nd. Ser., 50(4): 15 (1955)

Strong herbaceous twiner. Stems glabrous or with simple hairs. Leaves with palmate venation. Bracteoles free, glabrous or with simple hairs. Calyx with long linear to spathulate segments. Corolla with scattered capitate glands; tube long, cylindric; throat short. Stamens subsequal, partly exserted; anthers linear-oblong, theca subequal, apiculate at apex, all thecae at base with a boss covered with short stiff glossy setose hairs and with setose hairs along line of dehiscence. Stigma exserted, with subequal erect flattened curved lobes forming a funnel. Capsule triangular, narrowing gradually into the beak, glabrous. Seed honey-combed. Species 4.

IV. Subgen. **Coniostephanus** Bremekamp in Verh. Kon. Nederl. Akad. Wetensch., Afd. Natuurk., 2nd. Ser., 50(4): 15 (1955)

Erect or scandent shrubs or subshrubs. Stems glabrous or with simple hairs or capitate glands. Leaves with pinnate venation. Bracteoles connate, glabrous or with simple hairs or capitate glands. Calyx with linear or narrowly triangular segments. Corolla with scattered capitate glands; tube short, cylindric; throat long, campanulate. Stamens didynamous; anthers linear-oblong, one theca distinctly longer than the other, apiculate at apex, all thecae at base with a boss covered with short stiff glossy setose hairs and with setose hairs along line of dehiscence. Stigma included, with unequal flat lobes, upper lobe erect, lower horizontal. Capsule triangular, narrowing gradually into the beak, glabrous or with minute glandular hairs. Seed rugose. Species 5–8.

The East African species of this subgenus are all closely related, and future monographic work may well show them to be only subspecifically distinct. But such work would also need to include a number of taxa not occurring in the Flora area, and is considered outside the scope of the present work.

V. Subgen. **Hexacentris** *Benth. & Hook.* emend. Bremek. in Verh. Kon. Nederl. Akad. Wetensch., Afd. Natuurk., 2nd. Ser., 50(4): 16 (1955)

Woody climbers. Leaves with palmate venation. Flowers in many-flowered racemoid cymes. Bracteoles free or connate. Calyx a truncate or slightly undulate rim. Corolla large, without capitate glands; tube short; throat long, broadly campanulate. Stamens didynamous; anthers oblong, with long gently curved spurs which are flattened in basal part. Style bent apically; stigma with two subequal flattened erect lobes. Capsule with globose basal part narrowing abruptly into a long parallel-sided beak. Seed verrucose-reticulate. Species 9.

VI. Subgen. **Parahexacentris** *Bremek.* in Verh. Kon. Nederl. Akad. Wetensch., Afd. Natuurk., 2nd. Ser., 50(4): 15 (1955)

Erect, trailing or twining herbs. Stems quadrangular. Leaves with palmate venation. Bracteoles connate. Calyx a distinct rim with broadly triangular to linear segments. Corolla with scattered capitate glands; tube short, cylindric; throat long, campanulate, straight, more rarely curved. Stamens didynamous; filaments glabrous; anthers oblong or oblong-elliptic, thecae subequal, usually indistinctly apiculate, all thecae (or only one on shorter pair) with long gently curving spurs. Stigma with two flattened lobes, upper lobe erect, lower horizontal. Capsule with a subglobose basal part, narrowing abruptly into a parallel-sided beak. Seed, minutely puberulous, from almost smooth to reticulate, lamellate or echinate. Species 10–28.

VII. Subgen. **Hypenophora** *Bremek.* in Verh. Kon. Nederl. Akad. Wetensch., Afd. Natuurk., 2nd. Ser., 50(4): 17 (1955)

Erect, scrambling or twining herbs. Stems glabrous or with simple hairs, with two deep longitudinal furrows. Leaves with pinnate or palmate venation. Bracteoles connate, glabrous or with simple hairs or capitate glands. Calyx a shallowly lobed rim with broadly or irregularly triangular lobes. Corolla with scattered capitate glands; tube short, cylindric; throat long, campanulate, distinctly curved. Stamens equal; anthers ovoid, thecae subequal, with a long apiculus at apex, one theca on anterior pair with a sharply bent spur, other thecae muticous. Stigma funnel-shaped, shortly 2-lobed with slighty unequal erect lobes, with two tufts of downwardly directed setose hairs below sinuses. Capsule with a globose basal part, narrowing abruptly into a narrowly triangular beak, densely finely puberulous. Seed reticulate to almost smooth on back, sometimes almost spinulose towards apex, with a distinct entire to crenate lateral rim. Species 29–47.

VIII. Subgen. **Thunbergia**

SYN. Subgen. *Eu-thunbergia* Bremek. in Verh. Kon. Nederl. Akad. Wetensch., Afd. Natuurk., 2nd. Ser., 50(4): 17 (1955)

Erect or scrambling evening or night flowering herbs with fragrant flowers. Stems quadrangular, with simple hairs. Leaves with palmate venation. Bracteoles free, with simple hairs. Calyx with long linear segments. Corolla with scattered capitate glands; tube long, cylindric, widening only slightly into a short indistinct throat; limb horizontal. Stamens didynamous, inserted in upper half of tube; anthers linear, glabrous, thecae subequal, all muticous at base, connective with a long apical apiculus. Stigma with two erect subequal dorsiventrally flattened broadly elliptic lobes, the upper often partially folded around the lower. Capsule with globose basal part narrowing abruptly into the narrowly triangular beak, glabrous. Seed with large lamellate-pectinate scales. Species 48–49.

1. **Thunbergia heterochondros** (*Mildbr.*) *Vollesen* **comb. nov.** Type: Tanzania, Ulanga District: Ruhudje, Massagati, Mnjera, *Schlieben* 1113 (B†, holo.; BM!, BR!, P!, iso.)

Woody twiner to 10 m; young stems terete, floccose stellate-pubescent or sparsely so, glabrescent. Leaves with petiole 1–2.5 cm long, stellate-pubescent; lamina elliptic to broadly so or obovate, largest (7–)8.5–16 × (3–)4.5–7.5 cm, apex acute to acuminate, base cuneate to truncate, margin entire to crenate-dentate, often wavy, floccose stellate-pubescent or sparsely so along midrib and larger veins, glabrescent. Flowers solitary; pedicels 1–2.5(–3.5) cm long, apical part stellate-pubescent; bracteoles pale green, ovate to elliptic or broadly so, 2–3.5 × 1.5–2.5 cm (to 4 × 3 cm in fruit), obtuse, stellate-pubescent outside. Calyx 1.5–3 mm high. Corolla yellow; tube 3.5–5 cm long; lobes 1.5–3 cm long. Filaments 10–12 and 12–16 mm long, with capitate glands along the whole length; anthers 4–5 mm long, truncate with connective apiculate at apex, curved at base, with a line of stiff hairs ventrally between thecae and capitate glands dorsally. Capsule 2.5–3 × 1.8–2.2 cm. Seed dark brown, ± 1 cm in diameter.

TANZANIA. Morogoro District: Nguru Mts, Turiani, Diwali River, March 1956, *Mgaza* 93! & Mikumi National Park, Vuma Hills, 26 June 1977, *Wingfield* 4015!; Lindi District: Makonde Plateau, Newala, 1 May 1935, *Schlieben* 6470!
DISTR. **T** 6–8; not known elsewhere
HAB. Evergreen or semi-evergreen lowland forest, riverine forest; 150–700 m

SYN. *Pseudocalyx heterochondros* Mildbr. in N.B.G.B. 11: 409 (1932); T.T.C.L. 2: 14 (1949); Lebrun & Stork, Enum. Pl. Afr. Trop. 4: 500 (1997)

NOTE. This species with its yellow stellate indumentum shares some characters with *Pseudocalyx*, but is, in my opinion, best kept in *Thunbergia*. It has solitary axillary flowers; the anthers open by apical slits and have a setose boss at base and the seeds are hollowed on the proximal face.

2. **Thunbergia stelligera** *Lindau* in E.J. 33: 185 (1902); T.T.C.L. 2: 18 (1949); Luke & Robertson, Kenya Coast. For. 2. Checklist Vasc. Pl.: 84 (1993); Ruffo *et al.*, Cat. Lushoto Herb. Tanzania: 11 (1996); Lebrun & Stork, Enum. Pl. Afr. Trop. 4: 507 (1997). Type: Tanzania, Lindi District: Mtua, *Busse* 1119 (B†, holo.; EA!, iso.)

Vigorous semi-woody twiner or scandent shrub to 8 m tall; young branches quadrangular or narrowly 4-winged, with narrow band of yellowish stellate hairs at nodes and sometimes a few on stems below nodes, glabrescent. Leaves dark green, glossy; petiole 0.5–3 cm long, with stellate hairs at base; lamina elliptic or broadly so, largest 11–15 × 4–7.5 cm, apex acuminate to cuspidate, base cuneate, margin entire, slightly wavy, glabrous. Flowers solitary, sometimes congested into a pseudo-raceme by reduction of leaves towards apex of stem; pedicels 1.5–2.5 cm long, glabrous; bracteoles whitish green, broadly oblong, 2–3 × 1–1.7 cm, acute or subacute, glabrous outside, with large stellate hairs inside. Calyx 1.5–3.5 mm high. Corolla white with yellow to apricot throat; tube (2.5–)3–4 cm long. Filaments 12–14 and 14–18 mm long, with a knee ± 5 mm above base, with capitate glands below anther; anthers 7–8 mm long, truncate with apiculate connective at apex, curved at base, with capitate glands all along connective, occasionally with a small staminode. Capsule ± 2.5 × 1.8 cm. Seed not seen.

KENYA. Kilifi District: Kaya Jibana, 14 Dec. 1990, *Luke & Robertson* 2645!; Kwale District: Shimba Hills, Risley's Ridge, 1 May 1992, *Luke* 3115!
TANZANIA. Uzaramo District: Pugu Hills, 19 May 1970, *B.J. Harris* 4672!; Lindi District: north slopes of Rondo [Muera] Plateau, 24 Feb. 1935, *Schlieben* 6065! & Rondo Plateau, St. Cyprians College, 14 Feb. 1991, *Bidgood et al.* 1574!
DISTR. **K** 7; **T** 6, 8; Mozambique
HAB. Semi-evergreen lowland forest, thickets regenerating after forest clearance; 100–650 m

3. **Thunbergia kirkii** *Hook.f.* in Bot. Mag. 109: t. 6677 (1883); U.O.P.Z.: 472 (1949); Bremekamp in Verh. Kon. Nederl. Akad. Wetensch., Afd. Natuurk., 2nd. Ser., 50(4): 39 (1955); Iversen in Symb. Bot. Ups. 29, 3: 162 (1991); K.T.S.L.: 610 (1994). Types: Kenya, Mombasa, *Wakefield* s.n. (K!, syn.); cult. at Kew from material sent from Zanzibar, *Kirk* s.n. (K!, syn.)

Erect shrubby herb to 1.5 m tall (rarely scandent to 3 m); branches quadrangular, narrowly winged, the wings forming rounded "ears" at nodes, glabrous or with thin line of hairs at nodes. Leaves dark green glossy; petiole 1–6 mm long, glabrous or puberulous; lamina lanceolate to broadly elliptic, largest 3–11(–15) × 0.5–7.5 cm, apex acute to acuminate, base cuneate to rounded (rarely attenuate), margin on some or all usually with one large triangular tooth near middle, otherwise entire to crenate-dentate or undulate, glabrous. Flowers in 1–6-flowered axillary racemes (rarely 2 racemes per axil or some flowers solitary); peduncles and pedicels angular, glabrous; peduncle and axis 0.2–3(–7.5) cm long; bracts minute, caducous; pedicels 0.7–1.8(–2.3) cm long; bracteoles green, ovate to oblong-elliptic, 8–14(–21) × 3–6(–8) mm, acute to obtuse (rarely subacuminate), apiculate, glabrous to finely puberulous and finely ciliate. Calyx to 1(–5 in fruit) mm high of which the broadly triangular lobes about half, glabrous to finely puberulous. Corolla limb mauve, purple or blue (rarely white), tube whitish, throat yellow; tube 1.6–2.2(–3.5) cm long; lobes 0.5–1(–1.5) cm long; filaments 3–6 and 6–9 mm long, glabrous; anthers 3–5 mm long. Capsule 1.5–2(–2.3) × 0.8–1 cm, minutely puberulous. Seed dark reddish brown, 5–6 mm in diameter.

KENYA. Kwale District: Gongoni Forest, Gazi, 1927, *Gardner* 1407! & Shimba Hills, Sheldrick's Falls, 2 Apr. 1968, *Magogo & Glover* 619! & Mwachi Forest Reserve, 17 May 1990 (flowering in Hort. Robertson, Malindi 15 May 1991), *Robertson & Luke* 6454!
TANZANIA. Tanga District: Kange Gorge, 27 Aug. 1955, *Faulkner* 1689! & 8 June 1956, *Faulkner* 1873!; Morogoro District: Kimboza Forest Reserve, 1 Apr. 1983, *Mwasumbi et al.* 12426!; Zanzibar, Muyuni, 16 July 1933, *Vaughan* 2145!
DISTR. **K** 7; **T** 3, 6, 7; **Z**; Mozambique
HAB. Evergreen or semi-evergreen lowland and lower montane forest; near sea level to 1250 m

SYN. *Thunbergia hookeriana* Lindau in E.J. 17, Beibl. 41: 38 (1893); Burkill in F.T.A. 5: 13 (1899); Lebrun & Stork, Enum. Pl. Afr. Trop. 4: 505 (1997). Type: as for *T. kirkii*
  T. amanensis Lindau in E.J. 38: 67 (1905); T.T.C.L.: 18 (1949); Lebrun & Stork, Enum. Pl. Afr. Trop. 4: 505 (1997). Type: Tanzania, Lushoto District: Amani, *Warnecke* 408 (B†, holo.; EA!, iso.)

NOTE. *Drummond & Hemsley* 1883 from Morogoro District: Nguru Mts, Turiani, Mkobwe, is a scandent shrub with much larger flowers than any other specimens seen (maximum dimensions in description). It is also described as having a white corolla. The corolla of the type of *T. amanensis* is also said to be white, but all other material from the East Usambaras have mauve to purple corolla. The sparse material from Mozambique has equally large white flowers.
  For the time being I am keeping this material as an aberrant form of *T. kirkii*, but further collections from the Nguru Mts and further south are needed to clarify the status of this form.
  There is also a very large-leaved purple-flowered form from Morogoro District: Kimboza Forest. This is e.g. represented by *Luke et al.* 7648 and 8796. Vegetatively and in corolla size it is similar to the above mentioned. For the time being I prefer to keep it under *T. kirkii*.

4. **Thunbergia guerkeana** *Lindau* in P.O.A. C: 366 (1895); Burkill in F.T.A. 5: 18 (1899); S. Moore in J.B. 40: 342 (1902); Engler, Pflanzenw. Afr. 1: 180, fig. 148 (1910); E.P.A.: 929 (1964); Blundell, Wild Fl. E. Afr.: 397, fig. 129 (1987); K.T.S.L.: 609 (1994); Lebrun & Stork, Enum. Pl. Afr. Trop. 4: 506 (1997); Ensermu in F.E.E. 5: 350 (2006); Thulin in Fl. Somalia 3: 377 (2006). Types: Kenya, Machakos District: Ulu, *Fischer* 465 (B†, syn.); Tanzania, Tanga District: Nyika Steppe, *Holst* 2410 (B†, syn.)

Vigorous herbaceous twiner to 4 m tall, with large tuberous rootstock; stems terete to quadrangular, glabrous, with inconspicuous tufts of brownish hairs in axils. Leaves glossy; petiole 1.5–10 cm long, glabrous; lamina ovate to cordiform, largest 4–9.5 × 3–8.5 cm, apex acute to acuminate, base subcordate to deeply cordate (rarely truncate or cuneate), margin entire, slightly undulate, glabrous. Flowers solitary (rarely paired); pedicels 3–14 cm long, glabrous; bracteoles whitish or greenish white, oblong to elliptic or slightly obovate, 1.5–3.5(–4) × 0.7–1.7 cm, obtuse to subacuminate, glabrous. Calyx sparsely to densely glandular, rim 2–5(–8 in fruit) mm high, segments linear to slightly spathulate, 1.2–3.5(–4) cm long. Corolla fragrant, pure white; tube 9–12 cm long; lobes 1–2.5 cm long, densely glandular and with ciliate margins. Filaments 1–2 mm long, glabrous; anthers 7–9 mm long. Style glandular and finely puberulous; stigma lobes ellipsoid in transection, flattened at top. Capsule 3–4 × 1.5–2 cm. Seed brown, 10–12 mm in diameter.

KENYA. Northern Frontier District: Moyale, 20 Oct. 1952, *Gillett* 14073!; Machakos District: Mtito Andei, 1962, *J. G. Williams* EAH12540!; Teita District: Voi, 21 Dec. 1953, *Verdcourt* 1112!
TANZANIA. Pare District: Kisuani to Gonja, 4 Feb. 1930, *Greenway* 2144!; Lushoto District: Umba Steppe, below Mlalo, Dec. 1966, *Procter* 3409!
DISTR. **K** 1, 4, 7; **T** 3; Ethiopia, Somalia
HAB. *Acacia-Commiphora* bushland; 400–1050 m

SYN. *Thunbergia longisepala* Rendle in J.B. 34: 129 (1896); Burkill in F.T.A. 5: 11 (1899); Lebrun & Stork, Enum. Pl. Afr. Trop. 4: 506 (1997). Type: Kenya, Teita District: Teita Plains, *Scott Elliot* 6166 (BM!, holo.; K!, iso.)
　　T. glandulifera Lindau in E.J. 33: 184 (1902); Engler, Pflanzenw. Afr. 1: 180, fig. 148 (1910); E.P.A.: 928 (1964); Lebrun & Stork, Enum. Pl. Afr. Trop. 4: 506 (1997). Type: Ethiopia, Boran, Fiuno, *Ellenbeck* 2181 (B†, holo.; FT, iso.)
　　T. chiovendae Fiori in Bull. Soc. Bot. Ital. 1915: 51 (1915); E.P.A.: 928 (1964); Lebrun & Stork, Enum. Pl. Afr. Trop. 4: 505 (1997). Types: Somalia, Baidoa, *Stefanini & Paoli* 1113 (FT, syn.), Bur Eibe, *Stefanini & Paoli* 1151 (FT, syn.), Bur Acaba, *Stefanini & Paoli* 1224 (FT, syn.)

5. **Thunbergia vogeliana** *Benth.* in Hook., Niger Fl.: 476 (1849); T. Anderson in J.L.S. 7: 18 (1863); Burkill in F.T.A. 5: 10 (1899); De Wildeman & T. Durand, Contr. Fl. Congo 2: 47 (1900); De Wildeman, Étud. Fl. Congo 1: 314 (1906); Durand, Syll. Fl. Congo.: 415 (1909); De Wildeman, Étud. Fl. Congo 3: 264 (1910) & Pl. Beq. 1: 451 (1921) & Contrib. Fl. Katanga: 194 (1921); F.P.N.A. 2: 265 (1947); Heine in F.W.T.A. (ed. 2) 2: 402 (1963); Lebrun & Stork, Enum. Pl. Afr. Trop. 4: 507 (1997); Friis & Vollesen in Biol. Skr. 51(2): 456 (2005). Type: Bioko [Fernando Po], *Vogel* 147 (K!, holo.; K!, iso.)

Erect or scandent shrub to 4 m tall; young branches subquadrangular, with scattered to dense sessile glands, with dense tufts of brownish hairs in axils. Leaves with petiole 3–8 mm long, glabrous; lamina elliptic to slightly obovate, largest 8–19(–22) × 3.5–7.5 cm, apex acuminate to cuspidate, base cuneate to rounded (rarely attenuate), margin dentate (often irregularly) along whole length, glabrous. Flowers solitary (Uganda) or in 2–6-flowered racemes on short dwarf-shoots (Tanzania); pedicels 2.3–5 cm long, glabrous (Uganda) or glandular-puberulous (Tanzania); bracteoles pale green or greenish white with pink tinge, ovate to elliptic, 2.7–4.5 × 1.2–1.5 cm, acuminate to cuspidate (rarely obtuse), glabrous with finely ciliate margin (Uganda) or glandular-puberulous (Tanzania). Calyx rim to 3 mm high, segments to 1.7 cm long. Corolla limb pale mauve to pale violet or blue, tube whitish, throat yellow; tube 4–6 cm long; lobes 2–2.5 cm long. Filaments 8–10 and 12–14 mm long, with scattered glands along whole length; anthers 4–6 mm long. Capsule ± 3 × 1.5 cm, finely glandular-puberulous. Seed dark brown.

UGANDA. West Nile District: Zoka Forest, 15 Nov. 1941, *Thomas* 3769!; Toro District: Kibale Forest, Aug. 1936, *Eggeling* 3116! & 5 Sep. 1941, *Thomas* 3932!

TANZANIA. Buha District: Gombe National Park, Upper Kakombe River, 5 May 1992, *Mbago* 1104!; Mpanda District: Kungwe-Mahali Peninsula, Lubugwe River, 11 July 1958, *Juniper & Jefford* 104! & S of Pasagulu, Kabwe River, 7 Aug. 1959, *Harley* 9194!

DISTR. U 1, 2; **T** 4; Ghana, Nigeria, Cameroon, Bioko, Sudan, Congo-Kinshasa, Rwanda

HAB. Evergreen forest and riverine forest; 1000–1850 m

SYN. *Meyenia vogeliana* (Benth.) Hook. in Bot. Mag. 89: t. 5389 (1863)
    Thunbergia thonneri De Wild. & T.Durand in B.S.B.B. 38: 41 (1900) & Pl. Thonn. Congol.: 38, t. 8 (1900); Burkill in F.T.A. 5: 506 (1900); Lebrun & Stork, Enum. Pl. Afr. Trop. 4: 507 (1997). Type: Congo-Kinshasa, near Gali, Bobi, *Thonner* 34 (BR!, holo.; BR! iso.)
    [*T. affinis* sensu F.P.S. 3: 188 (1956), *non* S.Moore (1880)]

6. **Thunbergia erecta** (*Benth.*) *T.Anderson* in J.L.S. 7: 18 (1863); Lindau in E.J. 17, Beibl. 41: 37, 39 (1893) & in P.O.A. C: 366 (1895); Burkill in F.T.A. 5: 12 (1899); Durand, Syll. Fl. Congo: 414 (1909); Benoist in Notul. Syst. 2: 286 (1912); De Wildeman, Pl. Beq. 1: 449 (1921); Bremekamp in Verh. Kon. Nederl. Akad. Wetensch., Afd. Natuurk., 2nd. Ser., 50(4): 37 (1955); Heine in F.W.T.A. (ed. 2) 2: 402 (1963) & in Fl. Gabon. 13: 60, pl. 12 (1966); Lebrun & Stork, Enum. Pl. Afr. Trop. 4: 506 (1997). Type: Ghana, *Vogel* 14 (K!, lecto.)

Scandent shrub to 3 m tall; young branches quadrangular, with pale yellow glossy bark, narrowly winged with the wings forming small "ears" at nodes, glabrous but for thin lines of hairs at youngest nodes. Leaves with petiole 2–6 mm long, glabrous or with a few hairs; lamina elliptic, largest 4–12.5 × 2–4.5 cm, apex acuminate to cuspidate, base cuneate to rounded (rarely attenuate), margin on some or all with a single large triangular tooth near or above middle, usually undulate towards base, glabrous or with a few hairs on midrib. Flowers solitary; pedicels (3.5–)4–8 cm long, glabrous, basal part not clasped by leaf; bracteoles pale green, ovate to elliptic, 2–3.5 × 0.8–1.6 cm, acute to acuminate (rarely obtuse), glabrous. Calyx rim to 3 mm high, segments to 3 mm long (to 1 cm in fruit), with sessile glands. Corolla limb blue to deep bluish purple, tube whitish, throat yellow; tube 3–5 cm long; lobes 2–2.5 cm long. Filaments 10–12 and 12–14 mm long, with scattered glands along whole length; anthers 4–5 mm long. Capsule 2–3 × 1–1.3 cm, glabrous. Seed dark brown, ± 8 mm in diameter.

UGANDA. West Nile District: Zoka Forest, June 1933, *Eggeling* 1340!; Toro District: Bwamba Forest, 14 Aug. 1938, *Thomas* 2368!; Masaka District: 6 km SSW of Katera, Malabigambo Forest, 2 Oct. 1953, *Drummond & Hemsley* 4572!

TANZANIA. Bukoba District: Minziro Forest Reserve, Bulembe Hill, 17. Nov. 1999, *Mwiga et al.* 204! & Minziro Forest Reserve, 4 July 2000, *Bidgood et al.* 4843!

DISTR. U 1, 2, 4; **T** 1; Senegal, Guinea, Sierra Leone, Liberia, Ghana, Nigeria, Cameroon, Gabon, Central African Republic, Congo-Kinshasa, Sudan, Rwanda

HAB. Wet evergreen forest, swamp forest; 800–1150 m

SYN. *Meyenia erecta* Benth. in Hooker, Niger Fl.: 475 (1849); Hooker in Bot. Mag. 83: t. 5013 (1857)
    Thunbergia ikbaliana De Wild. in Pl. Nov. Herb. Hort. Then. 2: 95, t. 90 (1910); F.P.S. 3: 189 (1956); Lebrun & Stork, Enum. Pl. Afr. Trop. 4: 506 (1997). Type: Congo-Kinshasa, Sankuru, *Luja* s.n. (BR!, holo.)
    T. mestdaghii De Wild. in B.J.B.B. 5: 9 (1915); Lebrun & Stork, Enum. Pl. Afr. Trop. 4: 506 (1997). Type: Congo-Kinshasa, Libenge, *Mestdagh* 35 (BR!, holo.)

NOTE. As pointed out by Heine in Fl. Gabon this species is very closely related to the Angolan *T. affinis* which has entire leaves and shorter pedicels.

7. **Thunbergia crispa** *Burkill* in F.T.A. 5: 12 (1899); White, F.F.N.R.: 383 (1962); Binns, Checklist Herb. Fl. Malawi: 16 (1968); Bolnick, Common Wild Fl. Zambia: 40, pl. 18 (1995); Lebrun & Stork, Enum. Pl. Afr. Trop. 4: 505 (1997); White *et al.*, For. Fl. Malawi: 119 (2001). Types: Malawi, "Zambesiland", *Kirk* s.n. (K!, syn.), Manganja Hills, *Waller* s.n. (K!, syn.), Shire Highlands, *Buchanan* 27 (K!, syn.), without locality, *Buchanan* 866, 1156 and 1173 (all K!, syn.)

Woody twiner or scandent shrub to 4 m tall; young branches quadrangular, with brownish bark, narrowly winged with the wings forming small "ears" at nodes, glabrous but for thin lines of hairs at youngest nodes, with dense tufts of brownish hairs in axils; older woody stems to 1 cm diam. Leaves pale green beneath; petiole 3–8 mm long, glabrous, with narrow undulate wings adaxially; lamina elliptic, largest 8.5–12.5 × 3.5–4.5 cm, apex acuminate to cuspidate, base attenuate, margin on some or all undulate, never with a single large triangular tooth near or above middle (rarely all entire), glabrous. Flowers solitary; pedicels 4–9 cm long, glabrous, near base held in a fold created by the basal convolutions of the leaf-margin; bracteoles pale green, ovate to elliptic, 2–3.6 × 1.1–1.6 cm, acute to obtuse, glabrous. Calyx rim 1–3 mm high, segments to 6 mm long, always some over 3 mm, with a few sessile glands and finely ciliate segments. Corolla limb and upper part of tube deep violet to rich purple, lower part of tube tube whitish, throat yellow; tube 4–6 cm long; lobes 2–3 cm long. Filaments 10–13 and 12–15 mm long, with scattered glands at base and apically; anthers 4–5 mm long. Capsule 2.5–3 × 1.5–1.8 cm, glabrous. Seed dark brown, 7–8 mm in diameter.

TANZANIA. Mpanda District: 48 km on Sitalike–Sumbawanga road, 12 Feb. 1962, *Richards* 16099!; Chunya District: near Mbangala Village, 14 Feb. 1994, *Bidgood et al.* 2277!; Songea District: Gumbiro, 24 Jan. 1956, *Milne-Redhead & Taylor* 8511!
DISTR. T 4, 7, 8; Congo-Kinshasa, Angola, Zambia, Malawi, Mozambique, Zimbabwe
HAB. Termite mounds in *Brachystegia* woodland, rocky outcrops, riverine scrub; (200–)800–1200 m

SYN. *Thunbergia homblei* De Wild. in F.R. 13: 105 (1914) & Not. Fl. Katanga 4: 74 (1914) & Contrib. Fl. Katanga: 193 (1921); Lebrun & Stork, Enum. Pl. Afr. Trop. 4: 505 (1997). Type: Congo-Kinshasa, Katanga, Lubumbashi [Elisabethville], *Homblé* 165 (BR!, holo.; BR!, iso.)
    T. variabilis De Wild. in F.R. 13: 106 (1914) & Not. Fl. Katanga 4: 76 (1914) & Contrib. Fl. Katanga: 194 (1921); Lebrun & Stork, Enum. Pl. Afr. Trop. 4: 505 (1997). Type: Congo-Kinshasa, Katanga, Lubumbashi [Elisabethville], *Hock* s.n. (BR!, holo.; BR!, iso.)

8. **Thunbergia holstii** *Lindau* in E.J. 17: 95 (1893) & in P.O.A. C: 366 (1895); Blundell, Wild Fl. E. Afr.: 397, fig. 614 (1987), U.K.W.F., ed. 2: 266, pl. 114 (1994); K.T.S.L.: 609 (1994); Ruffo *et al.*, Cat. Lushoto Herb. Tanz.: 11 (1996); Thulin in Fl. Somalia 3: 376 (2006). Type: Tanzania, Lushoto District: Usambara, Kumusha Valley, *Holst* 543 (B†, holo.)

Erect shrub or subshrub to 1.5(–2) m tall; young branches quadrangular, with greyish to brownish bark, narrowly winged with the wings excurrent at the nodes as pseudo-stipular spines up to 2 mm long, glabrous but for thin lines of hairs at youngest nodes, with dense tufts of brownish hairs in axils. Leaves with petiole 1–4 mm long, glabrous or with a few hairs; lamina ovate to elliptic or broadly so, largest 1.5–8 × 0.7–3(–3.8) cm, apex acuminate to rounded, base cuneate to rounded (rarely attenuate), margin entire but usually strongly undulate, glabrous or hairy along midrib beneath. Flowers solitary; pedicels 0.5–3 cm long, glabrous, near base sometimes held in a fold created by the basal convolutions of the leaf-margin; bracteoles pale green, ovate to elliptic or broadly so, (1.2–)1.7–3.5 × (0.5–)0.8–2(–2.3) cm, acute to obtuse, glabrous. Calyx rim 1–2(–4) mm high, segments to 1 cm long, always some over 3 mm, glandular. Corolla limb and upper part of tube purple to dark purple or dark blue, lower part of tube tube whitish, throat yellow; tube (2.5–)3–4.5 cm long; lobes 1–2.5 cm long. Filaments 10–12 and 12–14 mm long, with scattered glands at base and apically or along whole length; anthers 4–5 mm long. Capsule 2.5–3.2 × 1–1.2 cm, glabrous. Seed dark brown, ± 8 mm in diameter. Fig. 6, p. 41.

KENYA. Machakos District: Sultan Hamud, 3 Aug. 1963, *Verdcourt* 3694!; Masai District: 155 km on Nairobi–Namanga road, 15 Dec. 1959, *Verdcourt* 2590!; Teita District: Worssera, 15 Dec. 1966, *Greenway & Kanuri* 12773!

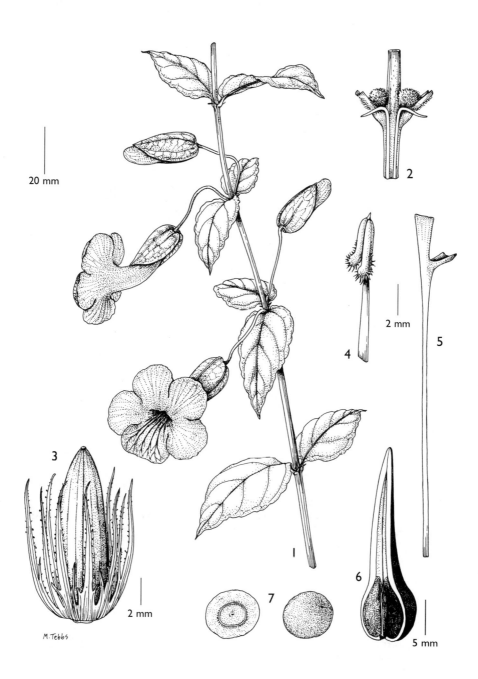

FIG. 6. *THUNBERGIA HOLSTII* — **1**, habit, × ²/₃; **2**, node with spinules, × 3; **3**, calyx and ovary; **4**, anther; **5**, style and stigma; **6**, capsule valve; **7**, seed, outer and inner view, × 3. 1 from *Richards* 23383, 2 from *Kokwaro* 2936, 3 from *Greenway* 2169, 4–5 from *Gillett* 17220, 6–7 from *Bally* 16805. Drawn by Margaret Tebbs.

TANZANIA. Arusha District: Kerekenyi Gorge, 21 Dec. 1969, *Richards* 24984!; Pare District: Mkomazi Game Reserve, Ibaya Hill, 26 Apr. 1995, *Abdallah & Vollesen* 95/14!; Lushoto District: Mashewa, 6 March 1952, *Faulkner* 919!; Zanzibar, without locality, *Kirk* s.n.! [see note]
DISTR. **K** 4, 6, 7; **T** 1–6; Somalia
HAB. *Acacia* and *Acacia-Commiphora* bushland, often on rocky hills, coastal bushland; near sea level to 1700 m

SYN. *Thunbergia affinis* S.Moore var. *pulvinata* S.Moore in J.B. 18: 6 (1880); Lindau in P.O.A. C: 366 (1895); Burkill in F.T.A. 5: 12 (1899); S. Moore in J.B. 40: 342 (1902); Chiovenda, Racc. Bot. Miss. Consol. Kenya: 94 (1935); T.T.C.L.: 18 (1949); K.T.S.: 19 (1961); E.P.A.: 927 (1964); Lebrun & Stork, Enum. Pl. Afr. Trop. 4: 505 (1997). Type: Kenya, Kitui, *Hildebrandt* 2749 (BM!, holo.; K!, iso.)
    [*T. affinis* sensu S. Moore in J.B. 18: 5 (1880) quoad *Hildebrandt* 2004b; Hook. in Bot. Mag. 114: t. 6975 (1888); Lindau in P.O.A. C: 366 (1895) pro parte; Burkill in F.T.A. 5: 11 (1899) pro parte; T.T.C.L.: 18 (1949); Bremekamp in Verh. Kon. Nederl. Akad. Wetensch., Afd. Natuurk., 2nd. Ser., 50(4): 37 (1955); K.T.S.: 19 (1961); E.P.A.: 927 (1964); Iversen in Symb. Bot. Ups. 29, 3: 162 (1991), *non* S.Moore (1880), sensu stricto]
    T. spinulosa Chiov., Fl. Somala 2: 346, fig. 198 (1932); E.P.A.: 929 (1964); Lebrun & Stork, Enum. Pl. Afr. Trop. 4: 505 (1997). Type: Somalia, Oltregiuba, *Senni* 543 (FT, holo.)
    [*T. erecta* sensu T.T.C.L.: 18 (1949); U.O.P.Z.: 471 (1949), *non* (Benth.) T. Anderson (1863)]

NOTE. The two *Kirk* collections labelled "Zanzibar" and "Mainland" are almost certainly from cultivated plants. I have not seen any wild specimens from either Zanzibar or the Tanzania coast opposite Zanzibar, but it is certainly widely grown as an ornamental. It is also widely grown in the rest of tropical Africa as well as in India and SE Asia.

9. **Thunbergia laurifolia** *Lindl.* in Gard. Chron. 1856: 260 (1856); Hook. in Bot. Mag. 83: t. 4985 (1857); Bremek. in Verh. Kon. Nederl. Akad. Wetensch., Afd. Natuurk., 2nd. Ser., 50(4): 47 (1955); Lebrun & Stork, Enum. Pl. Afr. Trop. 4: 507 (1997). Type: Cult. in Britain, no specimen preserved. Could be lectotypified with t. 4985 in Bot. Mag.

Vigorous woody twiner to 25 m or more, forming large tangles and often completely covering large trees; young branches glabrous or puberulous at nodes. Leaves glossy; petiole 1.5–5 cm long, glabrous; lamina lanceolate to triangular ovate, largest 13–20.5 × 4–10.5 cm, apex acuminate, base truncate to cordate, without or with rounded to hastate lobes, margin subentire or with a few large teeth, glabrous, with whitish pustules along major veins above. Flowers in pendulous racemoid cymes to 30 cm long; peduncle to 11 cm long, glabrous, with a pair of leafy sessile bracts to 8 × 4.5 cm at base of cyme; pedicels 2–4.5 cm long, glabrous; bracteoles oblong to obovate, 2.5–4 × 1–2 cm, subacute to rounded, apiculate, truncate at base, glabrous. Calyx an entire or slightly undulate puberulous rim. Corolla pale mauve to mauve or purple; cylindric tube ± 1 cm long; throat broadly campanulate, 3–4 cm long, 2–3 cm in diameter apically; lobes 3–4 × 3–4 cm. Filaments 9–15 and 11–17 mm long, glabrous; anthers narrowly oblong, 7–9 mm long, indistinctly apiculate, bearded at base and almost to apex along one side with long hairs with small lateral spinules; all thecae spurred, spurs 4–5 mm long, flattened. Capsule subglobose, 13–15 mm in diameter, glabrous, beak 25–30 mm long, parallel-sided. Seed 8–12 mm in diameter.

UGANDA. Mengo District: Kisugu–Muyenga, 12 Aug. 1987, *Rwaburindore* 2512! & Busiro, Bbanda, near Kawanda, 14 March 1995, *Rwaburindore* 3876!
TANZANIA. Lushoto District: East Usambara Mts, Amani, near Forest House, no date, *Ruffo & Mmari* 2079! & 25 Oct. 1986, *Borhidi et al.* 86133!
DISTR. **U** 4; **T** 3; native of India and SE Asia, widely cultivated in East Africa and occasionally naturalized
HAB. Margins and clearings of wet lowland rainforest, sometimes disturbed; 850–1200 m

SYN. *Thunbergia grandiflora* (Rottl.) Roxb. var. *laurifolia* (Lindl.) Benoist in Fl. Gén. Indo-Chine 4: 618 (1935)

T. harrisii Hook. in Bot. Mag. 83: t. 4998 (1857). Type: Cult. in Hort. Veitch (not seen)
[*T. grandiflora* sensu Iversen in Symb. Bot. Ups. 29, 3: 162 (1991); Ruffo *et al.*, Cat. Lushoto
Herb. Tanzania: 10 (1996), as *grabdiflora, non* (Rottl.) Roxb. (1819)]

10. **Thunbergia fasciculata** *Lindau* in E.J. 17: 97 (1893) & in E. & P. Pf. IV, 3b: 292
(1895); Burkill in F.T.A. 5: 15 (1899); S. Moore in J.B. 38: 269 (1908); Lindau in
Z.A.E.: 292 (1911); F.P.N.A. 2: 265 (1947); F.P.S. 3: 189 (1956); Heine in F.W.T.A. (ed.
2) 2: 400 (1963); Lebrun & Stork, Enum. Pl. Afr. Trop. 4: 506 (1997); Friis & Vollesen
in Biol. Skr. 51(2): 455 (2005); Ensermu in F.E.E. 5: 455 (2006). Type: Cameroon,
Buea, *Preuss* 987 (B†, holo.; K!, iso.)

Herbaceous twiner to 3(–5) m; stems sparsely sericeous. Leaves with petiole 3–10.5
cm long, indumentum as stems; lamina cordiform or broadly so, largest 6.5–9 ×
4.5–6.5 cm, apex acuminate, mucronate, base deeply cordate, margin irregularly and
coarsely dentate, sparsely sericeous-puberulous, densest along veins. Flowers in 2–6-
flowered pedunculate axillary fascicles, concealed between two large foliaceous
subsessile bracts; peduncle 4.5–10(–13) cm long, sericeous-puberulous or sparsely so;
bracts broadly ovate to cordiform, 3.5–5 × 2.5–4 cm, acuminate to cuspidate,
truncate to subcordate at base, dentate to serrate, indumentum as leaves; pedicels
6–13 mm long, puberulous and with scattered to dense stalked capitate glands;
bracteoles ovate-elliptic, 1.6–1.9 × 0.9–1.1 cm, acute or subacute, apiculate, truncate
at base, puberulous or sparsely so and with scattered to dense stalked capitate glands.
Calyx with long pilose hairs, no glands, rim ± 0.5 mm high, segments broadly
triangular, ± 1 mm long. Corolla purple to deep bluish purple with yellow throat;
cylindric tube 1–2 cm long; throat broadly campanulate, 2.5–3 cm long, 1–1.5 cm
diameter apically; lobes 1.5–2 × 2–2.5 cm. Filaments ± 8 and 10 mm long; anthers 2–4
mm long, indistinctly apiculate, densely bearded at base; all thecae spurred, spurs
1.5–2 mm long. Immature capsule ± 8 mm in diameter, densely pilose, beak ± 2 cm
long, parallel-sided. Seed not seen.

UGANDA. Bunyoro District: Budongo Forest, Sep. 1933, *Eggeling* 1432! & 28 Nov. 1938, *Loveridge*
   120!; Mengo District: Mabira Forest, Kyagwe, Sep. 1933, *Brasnett* 1378!
KENYA. North Kavirondo District: Bukura, 26 Dec. 1943, *Graham* 49! & Kakamega Forest,
   Ikuywa River, 27 Nov. 1969, *Faden* 69/2056! & Kakamega Forest, 23 Nov. 1974, *J.G. Williams*
   74/36!
DISTR. **U** 2, 4; **K** 5; Togo, Nigeria, Cameroon, Congo-Kinshasa, Sudan, Ethiopia
HAB. Wet lowland and lower montane forest; 800–1600 m

SYN. *Thunbergia fasciculata* Lindau var. *orientalis* De Wild., Pl. Beq. 1: 449 (1921). Type: Congo-
   Kinshasa, Masisi to Walikale, *Bequaert* 6454 (BR!, holo.)

11. **Thunbergia kirkiana** *T.Anderson* in J.L.S. 7: 19 (1863); S. Moore in Trans. Linn.
Soc., ser. 2, 4: 29 (1894); Lindau in P.O.A. C: 366 (1895); Burkill in F.T.A. 5: 19
(1899); Lindau in Schwed. Rhod.-Congo Exp. 1911–12, 1, Bot.: 303 (1916); Binns,
Checklist Herb. Fl. Malawi: 16 (1968); Richards & Morony, Checklist Fl. Mbala Distr.:
235 (1969); Moriarty, Wild Fl. Malawi: 85, pl. 43, 5 (1975); Lebrun & Stork, Enum.
Pl. Afr. Trop. 4: 506 (1997). Type: Malawi, Mt Sochi, *Kirk* s.n. (K!, holo.)

Perennial herb with several erect to decumbent stems from creeping woody
rootstock; stems to 50 cm long, subglabrous to sericeous or pubescent. Leaves with
petiole (1–)1.2–2.7 cm long, not winged, indumentum as stems; lamina narrowly
ovate to ovate or triangular (rarely ovate-oblong or reniform), largest 3.5–8.5(–11) ×
1.5–5.7 cm, less than three times longer than wide (very rarely more but then
pedicels over 5 cm long), apex acute to broadly rounded, base truncate to cordate
with rounded hastate lobes, margin entire to crenate; puberulous to pubescent or
sparsely so or sericeous (rarely subglabrous). Flowers axillary, solitary; pedicels
(3–)4–11 cm long, sparsely sericeous to pubescent (rarely subglabrous); bracteoles

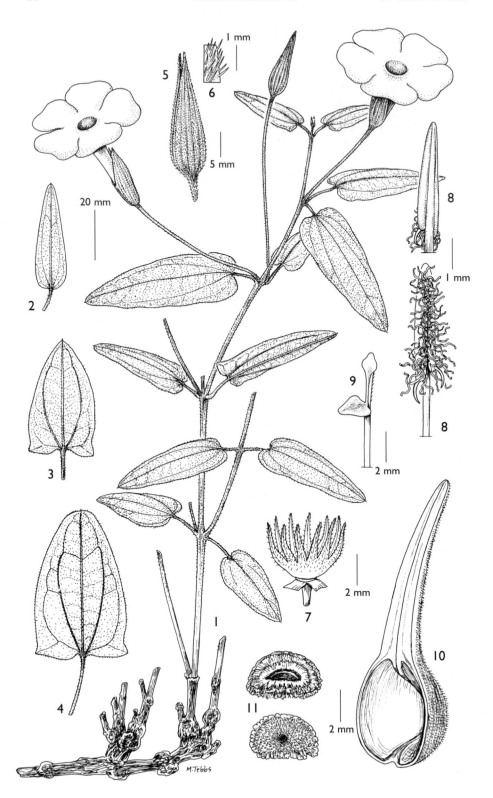

leathery, green with brownish often rib-like veins, narrowly ovate or oblong, (1.6–)1.8–2.7(–3.2 in fruit) × 0.4–0.7(–1 in fruit) cm, acuminate, cuneate to rounded at base, not angular, sericeous to pubescent. Calyx glabrous to puberulous, with subsessile glands on edges of lobes, rim 1–2 mm high, segments narrowly triangular to linear, 2–5 mm long. Corolla white with pale yellow centre to lemon yellow with darker yellow centre, without purple throat; cylindric tube 5–7 mm long; throat straight, narrowly campanulate, (1.8–)2.2–2.7 cm long and 0.5–0.7 cm in diameter apically; lobes (1.3–)1.5–2.5 × 1.5–2.5 cm, truncate to broadly emarginate. Filaments 3–5 and 4–6 mm long; anthers 3.5–5 mm long, apiculate, bearded at base and along whole side of one theca; spurs ± 1 mm long. Capsule densely puberulous, subglobose, 9–11 mm in diameter, beak narrowly triangular, not constricted at base, 13–15 mm long. Seed reddish brown, 5–6 mm in diameter, lamellate-pectinate near edges to reticulate near centre. Fig. 7, p. 44.

TANZANIA. Sumbawanga District: Tatanda, Mbaa Hill, 29 Oct. 1992, *Harder et al.* 1315!; Mbeya District: Tunduma, 14 Oct. 1956, *Richards* 6438!; Songea District: Songea, 27 Dec. 1956, *Milne-Redhead & Taylor* 7912!

DISTR. **T** 4, 6–8; Congo-Kinshasa, Angola, Zambia, Malawi, Mozambique

HAB. *Brachystegia-Julbernardia* woodland and wooded grassland, montane grassland, often on red loamy lateritic soil; 450–2100 m

SYN. *Thunbergia micheliana* De Wild. in Ann. Mus. Congo Bot., ser. 4, 1: 136, pl. 35 (1903) & Contrib. Fl. Katanga: 193 (1921); Lebrun & Stork, Enum. Pl. Afr. Trop. 4: 506 (1997). Type: Congo-Kinshasa, Lukafu, *Verdick* 318 pro parte (BR!, holo.)
     T. proxima De Wild. in Ann. Mus. Congo Bot., ser. 4, 1: 137, pl. 34 (1903) & Ann. Mus. Congo Bot., ser. 4, 2: 144 (1913) & Notes Fl. Katanga 2: 70 (1913) & Notes Fl. Katanga 3: 24 (1914) & Notes Fl. Katanga 4: 75 (1914) & Contrib. Fl. Katanga: 194 (1921) & Contrib. Fl. Katanga, suppl. 3: 139 (1930); Lebrun & Stork, Enum. Pl. Afr. Trop. 4: 506 (1997). Type: Congo-Kinshasa, Lukafu, *Verdick* 231 (BR!, holo.)
     T. verdickii De Wild. in Ann. Mus. Congo Bot., ser. 4, 1: 138, pl. 35 (1903) & Contrib. Fl. Katanga: 194 (1921); Lebrun & Stork, Enum. Pl. Afr. Trop. 4: 507 (1997). Type: Congo-Kinshasa, Lukafu, *Verdick* 318 pro parte (BR!, holo.)
     T. subfulva S.Moore in J.B. 51: 186 (1913); De Wildeman in Ann. Mus. Congo Bot., ser. 4, 2: 144 (1913) & Contrib. Fl. Katanga: 194 (1921); Lebrun & Stork, Enum. Pl. Afr. Trop. 4: 507 (1997). Type: Congo-Kinshasa, Katanga, *Rogers* 10400 (BM!, holo.)
     T. acutibracteata De Wild. in F.R. 13: 104 (1914) & Notes Fl. Katanga 4: 70 & 75 (1914) & Contrib. Fl. Katanga: 192 (1921) & Contrib. Fl. Katanga, suppl. 3: 134 (1930); Lebrun & Stork, Enum. Pl. Afr. Trop. 4: 505 (1997). Type: Congo-Kinshasa, Katanga, *Homblé* 132 (B!, holo.; BR! iso.)
     T. proximoides De Wild. in F.R. 13: 105 (1914) & Notes Fl. Katanga 4: 75 (1914) & Contrib. Fl. Katanga: 194 (1921); Lebrun & Stork, Enum. Pl. Afr. Trop. 4: 506 (1997). Type: Congo-Kinshasa, Katanga, *Bequaert* 214 (BR!, holo.)

12. **Thunbergia huillensis** *S.Moore* in J.B. 18: 194 (1880); Engl., Hochgebirgsfl. Trop. Afr.: 387 (1892); Burkill in F.T.A. 5: 15 (1899); Hiern, Cat. Afr. Pl. Welw. 4: 802 (1900); De Wildeman, Etud. Fl. Katanga 2: 143 (1913) & Notes Fl. Katanga 2: 70 (1913) & Contrib. Fl. Katanga: 193 (1921); Lebrun & Stork, Enum. Pl. Afr. Trop. 4: 506 (1997). Type: Angola, Huilla, *Welwitsch* 5025 (BM!, holo.; K!, iso.)

Perennial herb with several trailing (rarely erect) stems from woody rootstock; stems to 1.2 m long, sericeous to pubescent or sparsely so. Leaves glossy; petiole 2–8 mm long, not winged, indumentum as stems; lamina narrowly oblong-elliptic, largest 6.5–9 × 1.5–2.5 cm, three or more times longer than wide, apex subacute to retuse,

FIG. 7. *THUNBERGIA KIRKIANA* — **1**, flowering plant; **2–4** leaf variation; **5**, bracteoles; **6**, detail of bract indumentum; **7**, calyx; **8**, stamens; **9**, stigma; **10**, capsule valve; **11**, seed, two views. 1 & 5–9 from *Milne-Redhead & Taylor* 7912, 2 & 10–11 from *Milne-Redhead* 2572, 3 from *Richards* 20583, 4 from *Richards* 6890. Drawn by Margaret Tebbs.

base rounded to truncate with subhastate lobes, margin entire or shallowly crenate, sparsely sericeous to pubescent. Flowers axillary, solitary; pedicels 0.8–2.5(–4) cm long, sparsely sericeous to pubescent; bracteoles leathery, green with brownish often rib-like veins, narrowly to broadly (in fruit) ovate, 1.4–1.7(–1.9 in fruit) × 0.4–0.6(–0.9 in fruit) cm, acute to acuminate, cuneate to rounded at base, not angular, sericeous to pubescent. Calyx puberulous, no capitate glands, rim 1–2 mm high, segments narrowly triangular to linear, 2–4 mm long. Corolla lemon yellow, sometimes with darker yellow centre, without purple throat; cylindric tube 5–6 mm long; throat straight, narrowly campanulate, 1.8–2.3 cm long, 0.4–0.7 cm in diameter apically; lobes 1.2–1.5 × 1.5–1.7 cm, truncate to broadly emarginate. Filaments ± 5 and 6 mm long; anthers ± 2.5 mm long, apiculate, bearded at base and along whole side of one theca; spurs ± 1 mm long. Capsule densely puberulous, subglobose, 7–9 mm in diameter, beak narrowly triangular, not constricted at base, 10–12 mm long. Seed reddish brown, ± 5 mm in diameter, lamellate near edges to reticulate near centre.

TANZANIA. Kigoma District: N of Kigoma, Mukaraganga Village, 28 Apr. 1994, *Bidgood et al.* 3220! & 43 km on Uvinza–Mpanda road, 11 March 1994, *Bidgood et al.* 2763!; Mpanda District: Mahali Mts, Selambula, 14 Sep. 1958, *Newbould & Jefford* 2410!
DISTR. T 4; Congo-Kinshasa, Angola, Zambia, Zimbabwe
HAB. *Brachystegia-Julbernardia* woodland on sandy to gravelly or loamy soil, persisting in cultivated areas; 800–1500 m

SYN. *Thunbergia hanningtonii* Burkill in F.T.A. 5: 19 (1899); Lebrun & Stork, Enum. Pl. Afr. Trop. 4: 506 (1997). Type: Tanzania, Tabora District: Urambo, *Hannington* s.n. (K!, holo.)
   T. kassneri S.Moore in J.B. 51: 210 (1913); De Wildeman, Etud. Fl. Katanga 2: 143 (1913) & Contrib. Fl. Katanga: 193 (1921); Lebrun & Stork, Enum. Pl. Afr. Trop. 4: 506 (1997). Type: Zambia, Bwana Mkubwa, *Kässner* 2280 (BM!, holo.)
   T. angustata De Wild. in F.R. 13: 106 (1914) & Notes Fl. Katanga 4: 71 (1914) & Contrib. Fl. Katanga: 192 (1921); Lebrun & Stork, Enum. Pl. Afr. Trop. 4: 505 (1997). Type: Congo-Kinshasa, Katanga, *Homblé* 1267 (BR!, holo.; BR!, iso.)

13. **Thunbergia hamata** *Lindau* in E.J. 20: 2 (1894) & in P.O.A. C: 366 (1895); Burkill in F.T.A. 5: 18 (1899); T.T.C.L. 2: 17 (1949); Iversen in Symb. Bot. Ups. 29, 3: 162 (1991); Lebrun & Stork, Enum. Pl. Afr. Trop. 4: 506 (1997). Type: Tanzania, Lushoto District: W Usambara Mts, Kwa Mshusa, *Holst* 9092 (B†, holo.; COI!, K!, iso.)

Herbaceous twiner to 6 m; stems aculeate on edges with upwardly or downwardly directed prickles, otherwise glabrous. Leaves with petiole 1–4.5 cm long, not winged, aculeate on edges, otherwise glabrous; lamina triangular, largest 5.5–10 × 3.5–5.7 cm, apex subacuminate with a long apiculate tip, base cordate with subhastate lobes, margin subentire or with fine apiculate teeth, sometimes also with a few large triangular teeth, beneath finely aculeate along main veins, otherwise glabrous, above with short scabrid hairs. Flowers axillary, solitary; pedicels 4–12 cm long, aculeate, otherwise glabrous; bracteoles green, narrowly ovate-oblong or narrowly obovate with a long linear tip, 2.6–3.3 × 0.6–0.8 cm, rounded at base, densely glossy yellowish setose and with a dense "undercoat" of puberulous hairs. Calyx puberulous and with scattered longer hairs and scattered large subsessile glands, rim ± 1 mm high, segments with triangular base and linear tips, 3–4 mm long. Corolla bright primrose yellow, without purple throat; cylindric tube 4–6 mm long; throat straight, narrowly campanulate, 3.7–4.5 cm long, 0.8–1.3 cm in diameter apically; lobes 2–3 × 2.5–3.5 cm, broadly emarginate. Filaments ± 5 and 7 mm long; anthers 4–5 mm long, not apiculate, bearded at base and finely hairy along lower half, spurs ± 1 mm long. Ovary with subsessile glands. Capsule and seed not seen.

TANZANIA. Lushoto District: W Usambara Mts, Kwa Mshusa, Aug. 1893, *Holst* 9092! & Kwai Valley, 25 Apr. 1953, *Drummond & Hemsley* 2234! & Magamba Forest Reserve, 28 Feb. 1964, *Semsei* 3647!

DISTR. **T** 3; not known elsewhere

HAB. Evergreen montane forest; 1500–1800 m

NOTE. This very distinctive species is only known from the above-mentioned collections from the W Usambara Mts. It is easily recognised by the aculeate stems, leaves and pedicels, the narrow bracteoles with long linear tip and glossy setose indumentum and the large bright yellow corolla. The very few collections of a species with such large showy flowers would seem to indicate a genuine rarity.

14. **Thunbergia schliebenii** *Vollesen* **sp. nov.** a *T. hamata* caulibus atque petiolis pedicellisque non aculeatis, bracteolis ad marginem tantum piloso-ciliatis alibi glabris nec extus omnino nitente flavido-setosis et ad apicem acuminatis nec linearibus, corolla alba nec flava, tubo corollae cylindrico atque longiore (7–10 mm nec 4–6 mm) et faux breviore (2.5–3 cm nec 3.7–4.5 cm) differt. Type: Tanzania, Morogoro District: Uluguru Mts, NW side, *Schlieben* 3785 (K!, holo.; MO!, iso.)

Vigorous herbaceous twiner to 2 m; stems sparsely pilose near nodes. Leaves glossy; petiole 0.8–6 cm long, not winged, sparsely pubescent to pilose, densest apically; lamina ovate-triangular, largest 6.5–9 × 3–6.5 cm, apex acuminate, apiculate, base cordate with subhastate lobes, margin dentate, sparsely pilose on both sides, densest along veins. Flowers axillary, solitary; pedicels 3.5–4 cm long, sparsely pubescent near base, otherwise glabrous; bracteoles green, ovate, 1.7–2.3 × 0.6–0.7 cm, acuminate, truncate and slightly angular at base, pilose-ciliate along edges and midrib, otherwise glabrous. Calyx with ciliate lobes, otherwise glabrous, no glands, rim ± 0.5 mm high, segments broadly triangular, acute, ± 1 mm long. Corolla white, sometimes pale yellow towards centre; cylindric tube 7–10 mm long; throat straight, narrowly campanulate, 2.5–3 cm long and 0.8–1.5 cm in diameter apically; lobes 1–1.7 × 1.4–2 cm, rounded. Filaments ± 5 and 7 mm long; anthers ± 4 mm long, apiculate, bearded at base, spurs ± 1.5 mm long. Ovary glabrous. Capsule glabrous, globose, ± 8 × 8 mm, beak ± 12 mm long, parallel-sided, not constricted at base. Seed not seen.

TANZANIA. Morogoro District: NW side of Uluguru Mts, 17 Apr. 1933, *Schlieben* 3785!; Iringa District: Mufindi, Lulanda Forest, 28 March 1988, *Lovett* 3181! & Udzungwa Mts National Park [7°43'S 36°53'E], 7 June 2002, *Luke et al.* 8760!

DISTR. **T** 6, 7; not known elsewhere

HAB. Intermediate and lower montane evergreen forest; (700–)950–1450 m

15. **Thunbergia minziroensis** *Vollesen* **sp. nov.** a *T. schliebenii* pedicellis brevioribus (2–3 cm nec 3.5–4 cm), bracteolis minoribus (1.4–1.7 cm nec 1.7–2.3 cm), calyce glabro nec ciliato, faux corollae breviore (1.5–1.8 cm nec 2.5–3 cm), lobis corollae minoribus (0.7–1 × 0.8–1.2 cm nec 1–1.7 × 1.4–2 cm) et antheris minoribus (± 2.5 mm nec 4 mm) differt. Type: Tanzania, Bukoba District: Minziro Forest Reserve, Kele Hill, *Congdon* 347 (K!, holo.)

Slender herbaceous twiner, but also flowering as a short erect herb; stems to 2 m long, glabrous to sparsely pilose near nodes. Leaves not glossy; petiole 2.5–6.5 cm long, not winged, glabrous; lamina ovate-triangular, largest 6–8 × 4–7 cm, apex acuminate to cuspidate, base subcordate to cordate with subhastate lobes, margin subentire to dentate, below sparsely pilose on on veins, above uniformly so. Flowers axillary, solitary; pedicels 2–3.5 cm long, glabrous; bracteoles green, ovate, 1.4–1.7 × 0.4–0.7 cm, acuminate, truncate and rounded at base, not angular, pilose-ciliate along edges and and with a few hairs on midrib, otherwise glabrous. Calyx glabrous, rim ± 1 mm high, segments broadly triangular, rounded, ± 0.5 mm long. Corolla white, pale yellow towards centre; cylindric tube 7–10 mm long; throat straight, narrowly campanulate, 1.5–1.8 cm long and 0.6–0.9 cm in diameter apically; lobes 0.7–1 × 0.8–1.2 cm, truncate or slightly emarginate. Filaments ± 4 and 6 mm long; anthers ± 2.5 mm long, with apiculus ± 0.5 mm long, bearded at base, spurs ± 1 mm long. Ovary glabrous. Capsule and seed not seen.

TANZANIA. Bukoba District: Minziro Forest Reserve, Kele Hill, 7 Apr. 1994, *Congdon* 347! & Minziro Forest Reserve, Kinwa Kyaishemweru, 12 July 2001, *Festo et al.* 1608! & Kalagara-Kayunga Village, 20 July 2001, *Festo et al.* 1668!
DISTR. **T** 1; not known elsewhere
HAB. In deep shade in undergrowth of swamp forest; 1100–1200 m

16. **Thunbergia annua** *Nees* in DC., Prodr. 11: 55 (1847); T. Anderson in J.L.S. 7: 20 (1863); Solms in Schweinfurth, Beitr. Fl. Aeth.: 113 (1867); Lindau in E. & P. Pf. IV, 3b: 292 (1895); Burkill in F.T.A. 5: 13 (1899); F.P.S. 3: 189 (1956); E.P.A.: 927 (1964); Lebrun & Stork, Enum. Pl. Afr. Trop. 4: 505 (1997); Ensermu in F.E.E. 5: 355 (2006). Types: Sudan, Kordofan, *Kotschy* 97 (K!, syn.; BM!, iso.) & *Kotschy* 109 (K!, syn.; BR!, iso.)

Annual herb with single erect to decumbent unbranched to branched stems; stems to 40 cm long, sericeous-pilose or sparsely so. Leaves with petiole 0–2(–4) mm long, not winged, indumentum as leaves; lamina oblong-obovate or narrowly so, largest 3–6.2 × 1–2.5 cm, apex acute to broadly rounded, base cuneate, decurrent, margin with a pair of large rounded teeth in lower third and sometimes coarsely and irregularly dentate towards apex; sericeous-pilose or sparsely so, densest on veins beneath. Flowers axillary, solitary; pedicels 3–6(–13 in fruit) mm long, puberulous and sometimes with longer pilose hairs; bracteoles green, ovate, 1–1.7 × 0.4–0.8 cm, acute, truncate at base, pilose or sparsely so, mostly on midrib and margins, and with puberulous hairs. Calyx puberulous and with scattered capitate glands, rim ± 1 mm high, segments narrowly triangular, 2–3(–5 in fruit) mm long. Corolla white, with or without pale yellow centre; cylindric tube ± 2 mm long; throat straight, narrowly campanulate, 7–10 mm long and 3–5 mm in diameter apically; lobes 3–5 × 3–5 mm. Filaments ± 2 and 3 mm long; anthers ± 1.5 mm long, distinctly apiculate, bearded at base; spurs ± 1 mm long. Capsule densely puberulous, depressed globose, 5–8 × 6–10 mm, beak 8–12 mm long, parallel-sided and constricted at base. Seed 3.5–5 mm in diameter, dark reddish brown, smooth.

UGANDA. Karamoja District: Matheniko, Rupa, Oct. 1958, *Wilson* 574! & Sep. 1958, *Wilson* 610!
KENYA. South Kavirondo District: Homa Bay to Lambwe, July 1934, *Napier* 6615! & Mbita Point Field Station, 30 Nov. 1981, *Gachati & Opon* 109/81!; Masai District: Chyulu Plains, 8 June 2003, *Luke* 9477!
TANZANIA. Shinyanga District: Huruhuru Plain, 26 Jan. 1936, *Burtt* 5592!; Arusha District: Lower Nduruma River, June 1928, *Haarer* 1427!; Dodoma District: 24 km on Dodoma–Kondoa road, 19 Apr. 1988, *Bidgood et al.* 1213!
DISTR. **U** 1; **K** 3, 5, 6; **T** 1, 2, 4, 5; Niger, Sudan, Ethiopia, Botswana
HAB. Grassland and *Acacia* bushland on black cotton soil, old cultivations; 600–1350 m

17. **Thunbergia fischeri** *Engl.*, Hochgebirgsfl. Trop. Afr.: 387 (1892); Lindau in E.J. 17, Beibl. 41: 41 (1893) & in P.O.A. C: 366 (1895); Burkill in F.T.A. 5: 14 (1899); U.K.W.F.: 578 & 590 (1974); Blundell, Wild Fl. E. Afr.: 397, fig. 405 (1987); U.K.W.F., ed. 2: 266, pl. 114 (1994); Lebrun & Stork, Enum. Pl. Afr. Trop. 4: 506 (1997). Type: Kenya, S Kavirondo District: Gucha [Igutscha] River to Kamiana, *Fischer* 492 (B†, holo.)

Perennial herb with several erect to procumbent stems from thick sometimes branched woody rootstock; stems stout (erect plants) to slender (creeping plants), to 55 cm long, whitish tomentose. Leaves with petiole 1–5 mm long, not winged, indumentum as stems; lamina ovate to elliptic or broadly so, largest 1.7–5.5 × 0.8–4 cm, apex acute to broadly rounded, base truncate to cordate, margin entire to indistinctly sinuate-dentate; sericeous-pubescent to tomentose, densest beneath and sometimes sparsely puberulous or pubescent above. Flowers axillary, solitary; pedicels 0.5–2(–2.5) cm long, densely whitish sericeous to tomentose; bracteoles ovate to elliptic or narrowly so, 1.4–2.2 × 0.6–1 cm, acute, rounded to truncate at base, not angled, whitish sericeous to tomentose. Calyx sparsely glandular-puberulous and with scattered eglandular hairs, rim 0.5–1.5 mm high, segments narrowly triangular, 2–5 mm long. Corolla lemon yellow with darker yellow centre, without purple throat; cylindric tube 2–5 mm long; throat straight, narrowly campanulate, 1–2 cm long and 0.3–0.7 cm in diameter apically; lobes 0.6–1(–1.5) × 0.7–1.2(–1.5) cm, broadly emarginate. Filaments ± 2 and 4 mm long; anthers ± 2 mm long, indistinctly apiculate, densely bearded at base; spurs ± 1 mm long. Capsule depressed globose, densely puberulous, 6–9 × 7–12 mm, beak 10–15 mm long, parallel-sided and slightly narrowed at base. Seed 4–5.5 mm in diameter, dark reddish brown, lamellate.

KENYA. Northern Frontier District: Marsabit, 14 Feb. 1953, *Gillett* 15095!; Naivasha District: Kinangop, 1 March 1932, *Napier* 1798!; Nairobi District: main entrance to Nairobi National Park, 26 May 1961, *Verdcourt* 3169!
TANZANIA. Musoma District: Serengeti National Park, 32 km on Seronera–Ngorongoro road, 15 Oct. 1960, *Greenway* 9730!; Masai District: Ngorongoro, Endulen, 13 Nov. 1956, *Tanner* 3229!; Buha District: 135 km on Kibondo–Kasulu road, 15 July 1960, *Verdcourt* 2847!
DISTR. **K** 1, 3–6; **T** 1, 2, 4; Burundi
HAB. Grassland, often on black cotton soil but also on rocky hillsides, in W Tanzania in *Brachystegia* woodland, roadsides; 1200–2500 m

SYN. *Thunbergia brewerioides* Schweinf. in von Höhnel, Zu Rudolph-See und Stephanie-See: 858 (1892); Engl., Hochgebirgsfl. Trop. Afr.: 387 (1892); Lindau in E.J. 17, Beibl. 41: 41 (1893) & in P.O.A. C: 366 (1895); Burkill in F.T.A. 5: 14 (1899); Lebrun & Stork, Enum. Pl. Afr. Trop. 4: 505 (1997). Type: Kenya, N Nyeri District: Ndoro, *von Höhnel* 89 (B†, holo.)
T. sericea Burkill in F.T.A. 5: 14 (1899); Lebrun & Stork, Enum. Pl. Afr. Trop. 4: 507 (1997). Types: Kenya, 'Nandi', *Scott Elliot* 6903 (K!, syn.; BM!, iso.); Machakos, *Scott Elliot* 6696 (K!, syn.); Laikipia, *Gregory* 78 (BM!, syn.)
T. primulina Hemsl. in Bot. Mag. 130: t. 7969 (1904); Lebrun & Stork, Enum. Pl. Afr. Trop. 4: 506 (1997). Type: cultivated at Kew from seeds sent by Kirk from Kenya (K!, holo.)

NOTE. The closely related *T. ruspolii* Lindau, which is widespread in the Ethiopian highlands, differs in its yellowish strigose indumentum and dark yellow to orange corolla. The two should possibly be treated as subspecies.

18. **Thunbergia laborans** *Burkill* in F.T.A. 5: 507 (1900); Champluvier in Fl. Rwanda 3: 489, fig. 149, 1 (1985); Lebrun & Stork, Enum. Pl. Afr. Trop. 4: 506 (1997). Type: Burundi, *Scott Elliot* 8371 (K!, holo.; BM!, iso.)

Perennial herb with several trailing (or erect when young) stems from woody rootstock; stems to 60 cm long, sericeous pubescent or sparsely so. Leaves fleshy; petiole 1–2 mm long, not winged, indumentum as stems; lamina ovate-oblong or narrowly so, largest 2.5–4.2 × 1–1.6 cm, apex subacute to obtuse, base cuneate, truncate or subcordate, margin purple, entire or with a pair of large teeth in lower third, venation impressed above, subglabrous (puberulous along veins) to sparsely pubescent. Flowers axillary, solitary; pedicels 2–4(–6) cm long, pubescent; bracteoles green with conspicuous purple venation, narrowly ovate, 1.5–2.5 × 0.5–1.1 cm, subacute to obtuse, cordate at base and distinctly 4-angled, puberulous, ciliate along edges and midrib. Calyx puberulous with glandular and eglandular hairs, rim 0.5–1.5 mm high, segments narrowly triangular, 2–3(–5 in fruit) mm long. Corolla white to bright lemon yellow with darker yellow centre, without purple throat; cylindric tube ± 5 mm long; throat straight, narrowly campanulate, 1.5–1.8 cm long and 0.6–0.8 cm in diameter apically; lobes 1–1.5 × 1.2–1.7 cm, truncate to broadly emarginate. Filaments ± 4 and 5 mm long; anthers 3–3.5 mm long, distinctly apiculate, bearded at base and along ⅔ of one theca of upper anthers; spurs ± 1 mm long. Capsule depressed globose, densely puberulous, 7–9 × 9–10 mm, beak 10–15 mm long, parallel-sided and slightly narrowed at base. Seed 4–6 mm in diameter, brown, echinate.

TANZANIA. Musoma District: 45 km on Musoma–Mwanza road, Butiama, 18 July 1960, *Verdcourt* 2908!; Mwanza District: Geita, Uzunza, Nyamililo, 10 July 1953, *Tanner* 1578!; Biharamulo District: 47 km on Biharamulo–Muleba road, 11 July 2000, *Bidgood et al.* 4899!
DISTR. **T** 1; Rwanda, Burundi
HAB. *Brachystegia* woodland and bushland on sandy soil or on lateritic loam, roadsides; 1150–1550 m

19. **Thunbergia citrina** *Vollesen* **sp. nov.** lamina ovato-oblonga plusquam duplo longior quam latior atque sensim in petiolum alatum angustata; corolla citrina fauce purpurea carenti. *T. napperae* aliquantum similis sed foliis latioribus et fauce corollae non purpureo differt. Type: Tanzania, Tabora District: 6 km on Ipole–Inyonga road, *Bidgood et al.* 6048 (K!, holo.; CAS!, DSM!, EA!, MO!, NHT!, iso.)

Perennial herb with several erect to decumbent or (in late season forms) trailing stems from woody rootstock; stems to 0.3(–1 in trailing forms) m long, pubescent or sericeous-pubescent. Leaves not fleshy; petiole 4–15 mm long, apart from basal 1–2 mm with a flat wing widening gradually into the lamina; lamina ovate-oblong or narrowly so, largest 2–6 × 0.8–2.2 cm, apex subacute to broadly rounded, margin entire or with a large tooth near base, sparsely puberulous and pubescent-ciliate. Flowers axillary, solitary; pedicels 2.5–11 cm long, pubescent or sparsely so; bracteoles green with indistinct to conspicuous pale brown venation, ovate to ovate-oblong, 1.3–2.3 × 0.7–1.2 cm, obtuse to broadly rounded, cordate and distinctly 4-angular at base, pubescent and ciliate on midrib and margins. Calyx puberulous and with scattered capitate glands, rim 0.5–1.5 mm high, segments narrowly triangular, 2–4 mm long. Corolla lemon yellow, without purple throat; cylindric tube 5–7 mm long; throat straight, narrowly campanulate, 1.5–2.2 cm long and 0.5–0.9 cm in diameter apically; lobes 1.2–1.8 × 1.3–2 cm, broadly emarginate. Filaments 4–5 and 5–6 mm long; anthers 4–5 mm long, apiculate, densely bearded at base and almost to apex along one side; spurs ± 1.5 mm long. Capsule depressed globose, densely puberulous, ± 7 × 8 mm, beak ± 12 mm long, parallel-sided and slightly narrowed at base. Seed not seen.

TANZANIA. Mpanda District: Kapapa Camp, 28 Oct. 1959, *Richards* 11616! & Rungwe village–Inyanga road, 25 Oct. 1960, *Richards* 13386!; Tabora District: 6 km on Ipole–Inyonga road, 17 May 2006, *Bidgood et al.* 6048!
DISTR. **T** 4; not known elsewhere
HAB. *Brachystegia* woodland on sandy to gravelly soil, appearing after burning; 1050–1150 m

20. **Thunbergia paulitschkeana** *Beck* in Paulitschke, Ber. Bot. Ergebn. Exp. Harar: 459, t. 12 (1888); Lindau in Ann. Ist. Bot. Roma 6: 67 (1896); Burkill in F.T.A. 5: 13 (1899); Lugard in K.B. 1933: 94 (1934); E.P.A.: 929 (1964); Lebrun & Stork, Enum. Pl. Afr. Trop. 4: 506 (1997); Ensermu in F.E.E. 5: 351 (2006). Type: Ethiopia, Harar, *Hardegger & Paulitschke* s.n. (W, holo.)

Perennial herb with several trailing (rarely twining) stems from large woody rootstock; stems to 0.6(–1) m long, sericeous-pubescent to pilose or densely so (rarely tomentose). Leaves with petiole 3–15(–25) mm long, apart from basal 1–2 mm with a flat broad wing widening upwards but clearly distinct from lamina; lamina ovate to triangular or broadly so, largest 1–4.8(–5.5) × 0.7–4(–5.5) cm, apex subacute to obtuse, margin entire to coarsely dentate; sparsely pubescent to densely pilose (rarely tomentose). Flowers axillary, solitary (rarely paired); pedicels 0.7–9.5(–11) cm long, pubescent or sparsely so or pilose; bracteoles reddish brown with green veins, ovate to ovate-oblong (rarely broadly so), 1.5–3(–3.5) × 0.5–1.3(–1.6) cm, subacute to subacuminate, truncate to cordate and slightly (rarely distinctly) 4-angular at base, sparsely pilose to tomentose, densest on veins and margins. Calyx puberulous and with scattered capitate glands, rim 1.5–3 mm high, segments narrowly triangular, 2–4(–6) mm long. Corolla golden yellow to bright orange (rarely lemon yellow), without purple throat; cylindric tube 3–6 mm long; throat straight or slightly curved, narrowly campanulate, 1–2.2(–2.5) cm long and 0.5–1 cm in diameter apically; lobes 0.7–1.5 × 0.7–1.8 cm, broadly emarginate. Filaments 3–5(–6) and 4–6(–7) mm long; anthers 2.5–4.5 mm long, apiculate or not, densely bearded at base and almost to apex along one side; spurs ± 1 mm long. Capsule pubescent or densely so, depressed globose, 7–9(–11) × 10–12 mm, beak 11–15(–18) mm long, constricted at base. Seed 4–5.5 mm in diameter, dark reddish brown, reticulate.

UGANDA. Karamoja District: Napak, June 1950, *Eggeling* 5963!; Mbale District: Mt Elgon, Mutusyet, 4 July 1971, *Lye & Katende* 6407!
KENYA. Uasin Gishu District: Kaptaget, 1 May 1958, *Verdcourt* 2147!; Trans-Nzoia District: SE Elgon, Seboti, Aug. 1969, *Tweedie* 3674!; Masai District: 19 km on Narok–Olokurto road, Orengitok, 17 May 1961, *Glover et al.* 1216!
TANZANIA. Mbulu District: Dongobesh, Oct. 1925, *Haarer* 102B!; Masai District: Mt Monduli, 5 Dec. 1944, *Bally* 4118!; Arusha District: Sakila Swamp, 14 Sep. 1971, *Richards & Arasululu* 27224!
DISTR. U 1, 3; K 3–6; T 2, 3; Ethiopia
HAB. Upland grassland and bushland, streambanks, glades in montane forest, roadsides, abandoned cultivations; (1350–)1450–2800 m

SYN. *Thunbergia elliotii* S.Moore in J.B. 39: 300 (1901); U.K.W.F.: 575 & 578 (1974); Blundell, Wild Fl. E. Afr.: 396 (1987); U.K.W.F., ed. 2: 266, pl. 114 (1994); Lebrun & Stork, Enum. Pl. Afr. Trop. 4: 506 (1997). Type: Kenya, Nandi, *Scott Elliot* 6969 (BM!, holo.; K!, iso.)

NOTE. There is a local form around Arusha (e.g. *Richards & Arasululu* 27224) which is much coarser and hairier than the rest of the material and has large leaves, bracteoles and corollas. But typical forms also occur in Tanzania and the differences are not clear cut.

21. **Thunbergia gibsonii** S.*Moore* in J.B. 32: 131 (1894); Lindau in P.O.A. C: 366 (1895); Burkill in F.T.A. 5: 15 (1899); Jex-Blake, Gard. E. Afr. (ed. 4): 141 (1957); U.K.W.F.: 577 (1974) & U.K.W.F., ed. 2: 267, pl. 115 (1994); Lebrun & Stork, Enum. Pl. Afr. Trop. 4: 506 (1997). Type: Kenya, Mau, *Gibson* s.n. (BM!, holo.)

Perennial herb with several trailing or twining stems from large woody rootstock; stems to 2 m long, sparsely to densely whitish to yellowish pubescent to pilose. Leaves with petiole 1–3 cm long, not winged, indumentum as stems; lamina triangular-cordiform, largest 2.3–7.8 × 2–5.5 cm, apex acute or subacute, base cordate with subhastate lobes, margin coarsely dentate; sparsely pubescent to tomentose or velutinous, densest on veins. Flowers axillary, solitary; pedicels 5–25 cm long, indumentum as stems; bracteoles ovate to oblong or broadly so, (2–)2.5–4 × (0.8–)1–2.5 cm, acute to acuminate, truncate to subcordate and slightly 4-angular at base, yellowish pubescent to tomentose. Calyx densely puberulous and with capitate glands, rim 2–5 mm high, segments triangular or narrowly so, 1–3 mm long. Corolla bright orange, without purple throat; cylindric tube ± 5 mm long; throat sharply curved, campanulate, 2.5–4 cm long and 1–1.7 cm in diameter apically; lobes 1.3–2.5 × 1.5–2.5 cm, broadly emarginate. Filaments 6–8 and 8–10 mm long; anthers 4–6 mm long, not apiculate, densely bearded at base and half way along one side; spurs ± 1.5 mm long. Capsule densely pubescent, subglobose, 8–10 mm in diameter, beak (12–)15–20 mm long, constricted at base. Seed ± 5 mm in diameter, dark brown, reticulate.

KENYA. Uasin Gishu District: 57 km on Nakuru–Eldoret road, 18 May 1969, *Napper* 2100!; Kisumu-Londiani District: Londiani–Fort Ternan road, Limutit Hill, 26 Sep. 1953, *Drummond & Hemsley* 4443!; Kericho District: 10 km N of Kericho, 7 Aug. 1969, *Kokwaro* 2071!
DISTR. K 3, 5; not known elsewhere
HAB. Montane grassland and bushland with *Euclea* and *Protea*, forest glades, roadsides; (?1500–)2000–2900 m

SYN. *Thunbergia prostrata* Turrill in K.B. 1920: 25 (1920); Lebrun & Stork, Enum. Pl. Afr. Trop. 4: 506 (1997). Type: Cultivated at Kew from seeds sent by Battiscombe from Kenya (K!, holo.)
    T. robertii Mildbr. in N.B.G.B. 9: 491 (1926); Lebrun & Stork, Enum. Pl. Afr. Trop. 4: 506 (1997). Type: Kenya, Kisumu-Londiani District: Lumbwa, *R.E. & Th.C.E. Fries* 2815 (B†, holo.; BM!, K!, iso.)

22. **Thunbergia gregorii** *S.Moore* in J.B. 32: 130 (1894); Lindau in P.O.A. C: 366 (1895); Burkill in F.T.A. 5: 17 (1899); T.T.C.L. 2: 17 (1949); Jex-Blake, Gard. E. Afr. (ed. 4): 142 (1957); E.P.A.: 928 (1964); U.K.W.F.: 579 (1974); Cribb & Leedal, Mountain Fl. S. Tanz.: 127, pl. 32 (1982); Champluvier in Fl. Rwanda 3: 489 (1985); U.K.W.F., ed. 2: 267 (1994); Lebrun & Stork, Enum. Pl. Afr. Trop. 4: 506 (1997); Ensermu in F.E.E. 5: 350 (2006). Type: Kenya, Machakos District: Ukambani, Kilungu, *Gregory* s.n. (BM!, holo.)

Perennial herb with several trailing or twining stems from large woody rootstock; stems to 1 m long in trailing plants and to 3(–5) m in twining plants, sparsely to densely orange setose (hairs to 5 mm long) when young (rarely whitish pubescent or pilose). Leaves with petiole 1.5–6.5 cm long, with a flat broad wing widening upwards but clearly distinct from lamina, indumentum as stems; lamina triangular-cordiform or broadly so, largest 2.5–11 × 2–6.7(–8) cm, apex subacute to broadly rounded, base cordate with hastate lobes, margin entire to coarsely and irregularly dentate; pilose or sparsely so or setose on veins (rarely pubescent). Flowers axillary, solitary; pedicels 4–20(–26) cm long, indumentum as stems; bracteoles ovate-oblong, 2.2–4.2 × 1.2–2.7 cm, acute to subacuminate, subcordate to deeply cordate and slightly 4-angular at base, orange setose or densely so (rarely whitish pubescent). Calyx puberulous and with capitate glands, rim 1–2 mm high, segments triangular or narrowly so, 0.5–2 mm long. Corolla bright orange, without purple throat; cylindric tube 3–7 mm long; throat sharply curved, campanulate, 1.5–3.5 cm long and 0.8–1.5 cm in diameter apically; lobes 1–2.5 × 1.2–2.8 cm, broadly emarginate. Filaments 5–6 and 6–7 mm long; anthers 4–7 mm long, apiculate or not, densely bearded at base and $^2/_3$ along one side; spurs ± 2 mm long. Capsule densely pubescent or pilose, subglobose, 8–14 mm in diameter, beak 10–20 mm long, constricted at base. Seed 5–7 mm in diameter, dark reddish brown, reticulate and with large pectinate scales along edge.

UGANDA. Mbale District: Mt Elgon, 25 Feb. 1993, *Katende & Sheil* 1921!
KENYA. Machakos District: Ol Doinyo Sapuk, 12 Aug. 1974, *Faden* 74/1312!; Masai District: Ngong Hills, Oct. 1930, *Napier* 766! & 28 May 1995, *Vollesen* 95/219!
TANZANIA. Masai District: Ngorongoro, Masikio, 14 Dec. 1961, *Msuya* 9!; Ufipa District: 29 km on Namanyere–Karonga road, 5 March 1994, *Bidgood et al.* 2640!; Mbeya District: 8 km E of Mbeya, 19 Nov. 1986, *Brummitt et al.* 18056!
DISTR. **U** 3; **K** 4–6; **T** 2, 4, 7; Ethiopia, Congo-Kinshasa, Burundi
HAB. In Uganda, Kenya and N Tanzania in montane grassland and bushland with *Euclea* and *Protea* and on forest margins, in S and W Tanzania also in *Brachystegia-Julbernardia* and *Isoberlinia* woodland; (900–)1350–2300(–2450) m

SYN. *Thunbergia exasperata* Lindau in E.J. 30: 408 (1901); T.T.C.L. 2: 17 (1949); Lebrun & Stork, Enum. Pl. Afr. Trop. 4: 506 (1997). Type: Tanzania, Mbeya District: Unyika, Nkanka [Unkana], *Goetze* 1382 (B†, holo.; BR!, iso.)
    T. longepedunculata De Wild. in Ann. Mus. Congo Bot., ser. 4, 1: 136 (1903) & Contrib. Fl. Katanga: 193 (1921); Lebrun & Stork, Enum. Pl. Afr. Trop. 4: 506 (1997). Type: Congo-Kinshasa, Katanga, *Verdick* s.n. (BR!, holo.)
    [*T. gibsonii* sensu Turrill in Bot. Mag. 141: t. 8604 (1915), *non* S.Moore (1894)]
    T. aureosetosa Mildbr. in N.B.G.B. 11: 408 (1932); T.T.C.L. 2: 17 (1949); Lebrun & Stork, Enum. Pl. Afr. Trop. 4: 505 (1997). Type: Tanzania, Njombe District: Lupembe, Ruhudje River, *Schlieben* 212 (B†, holo.; BM!, BR!, iso.)

NOTE. Specimens from S and W Tanzania generally have larger leaves which are usually entire, and larger corollas. But there are too many intermediates with the material from Kenya and N Tanzania to contemplate separating out two subspecies.

Some of the S Tanzanian collections also lack the long orange setose hairs normally so characteristic for *T. gregorii*. But there are transitional collections with orange hairs on the peduncles but not on the stems and leaves and vice versa. I am treating this form (*T. exasperata*, s. str.) as no more than a local variant.

23. **Thunbergia mildbraediana** *Lebrun & Toussaint* in B.J.B.B. 17: 85 (1943); F.P.N.A. 2: 267 (1947); Champluvier in Fl. Rwanda 3: 489 (1985); Lebrun & Stork, Enum. Pl. Afr. Trop. 4: 506 (1997). Type: Congo-Kinshasa, Mikeno, *Lebrun 7293* (BR!, holo.)

Herbaceous twiner; stems to 2 m long, pilose or sparsely so, densest in two bands. Leaves with petiole not winged, 2–6 cm long, indumentum as stems; lamina triangular-ovate in outline, largest 4.5–10.5 × 2–5.5 cm, apex acuminate, base deeply cordate, with sagittate lobes, margin subentire to coarsely dentate; subglabrous to sparsely pilose, below densest on veins, above uniformly. Flowers axillary, solitary; pedicels (2–)4–9 cm long, subglabrous to sparsely pilose; bracteoles green with conspicuous main veins and reticulation, ovate or narrowly so, 2.7–3.5 × 1.1–1.8 cm, acute to acuminate, cordate and distinctly 4-angular at base, subglabrous to sparsely pilose, dense along edges and veins. Calyx puberulous with stalked capitate glands only or with intermixed hairs (rarely with hairs only), rim 1–2 mm high, segments linear, 3–6 mm long. Corolla white with reddish-purple throat (? rarely pale yellow); cylindric tube (5–)8–12 mm long; throat straight or slightly curved, campanulate, 1.3–2.5 cm long and 1–1.5 cm in diameter apically; lobes 1–2.2 × 1–2.4 cm, deeply emarginate. Filaments 5–8 mm long; anthers ± 3 mm long, rounded or indistinctly apiculate, densely bearded at base and about half way up one side; spurs 2–2.5 mm long. Capsule glabrous, depressed globose, 9–12 × 11–14 mm, beak 13–25 mm long, parallel-sided or slightly constricted at base. Seed dark brown, ± 7 mm in diameter, lamellate at edges, reticulate towards centre.

UGANDA. Kigezi District: Behungi Hill, 1 Dec. 1930, *Burtt* 2948! & Kinaba, Luhiza, March 1947, *Purseglove* 2356!; Toro District: Ruwenzori Mts, Minimba Camp, 21 Jan. 1962, *Loveridge* 361!
TANZANIA. Mpanda District: Mahali Mts, Sisaga Mt, 29 Aug. 1958, *Newbould & Jefford* 1917! & Mahali Mts, Myako, 12 Apr. 1978, *Uehara* 591!
DISTR. U 2; T 4; Congo-Kinshasa, Rwanda, Burundi
HAB. Evergreen montane forest, often in the bamboo zone, secondary forest at lake level; (750–)2250–2600 m

24. **Thunbergia napperae** *Mwachala, Malombe & Vollesen* **sp. nov.** Lamina lanceolata vel anguste ovata vel anguste oblonga plus quam duplo longior quam latior atque sensim in petiolum alatum angustata; corolla citrina vel sulphurea fauce purpurea instructa. *T. richardsiae* aliquante similis sed foliis angustioribus et fauce corollae purpurea differt. Type: Kenya, Machakos District: Makueni, Ngutwa Village, Malivani Hill, *Malombe et al.* 970 (EA!, holo.; K!, iso.)

Perennial herb with several erect (when young) to decumbent stems from woody rootstock; stems to 50 cm long, pubescent or sparsely so or sericeous-pubescent. Leaves slightly fleshy; petiole 8–25(–30) mm long, apart from basal few mm with a wide undulate wing passing gradually into the lamina; lamina lanceolate to narrowly ovate or oblong, largest 2–7 × 0.8–1.5 cm, apex obtuse, margin entire to undulate (rarely with a single large triangular tooth on each side near base), sericeous-pubescent or sparsely so, usually densest along veins. Flowers axillary, solitary; pedicels 3.5–7.5 cm long, indumentum as stems; bracteoles dark green, narrowly ovate to ovate or oblong, 1.7–2.4 × 0.5–1 cm, obtuse to subacute, cordate and distinctly 4-angular at base, sericeous-pubescent or sparsely so, densest on veins. Calyx puberulous and with scattered subsessile capitate glands, rim ± 1 mm high, segments linear, 1–3(–8 in fruit) mm long. Corolla lemon or sulphur yellow with purple throat; cylindric tube 5–8 mm long; throat straight, narrowly campanulate, 1.5–2.5 cm long and 0.7–1 cm in diameter apically; lobes 1.2–1.8 × 1.4–1.9 cm, broadly emarginate. Filaments 4–5 and 5–6 mm long; anthers 3–4 mm long, indistinctly apiculate, densely bearded at base and about half way up one side; spurs ± 1 mm long. Capsule densely puberulous, depressed globose, 8–12 × 9–12 mm, beak 13–14 mm long, slightly constricted at base. Seed (immature) ± 5 mm in diameter, lamellate near edge to echinate towards centre.

KENYA. Machakos District: Donyo Sabuk, 15 Dec. 1931, *Napier* 2378! & Nzaui Hill, Feb. 1957, *Bally* 11417!; Masai District: Emali Hill, 24 March 1940, *van Someren* 247!
DISTR. **K** 4, 6; not known elsewhere
HAB. *Acacia-Combretum* wooded grassland, *Acacia* bushland, grassland, roadsides; 1400–1800 m

SYN. *Thunbergia sp. C* of U.K.W.F.: 578 (1974) & U.K.W.F., ed. 2: 267 (1994)

25. **Thunbergia tsavoensis** *Vollesen* **sp. nov.** *T. alatae* valde affinis sed ala petioli undulata atque deorsum (nec sursum) dilatata, corolla plerumque alba (nec flava), tubo corollae cylindrico longiore (7–10 mm nec 3–6 mm longo) et seminibus echinato-sculptis (nec lamellatis neque reticulatis) differt. Type: Kenya, Kwale District: Mackinnon Road, *Drummond & Hemsley* 4104 (K!, holo.; BR!, EA!, K!, iso.)

Annual or perennial herb, when annual single-stemmed and erect, when perennial with several erect to decumbent stems from large woody rootstock; stems to 50 cm long, pubescent or sericeous-pubescent. Leaves with petiole 5–17(–26 in annual plants) mm long, with an undulate wing which on some or all leaves widens distinctly towards base and is often conspicuously constricted just below lamina, indumentum as stems; lamina triangular-ovate, largest 1.8–4(–6) × 1.1–2.2(–5) cm, apex acute to rounded, base truncate to subcordate (rarely cuneate), often with subhastate lobes, margin subentire to coarsely, dentate, pubescent or sericeous-pubescent or sparsely so, dense along veins. Flowers axillary, solitary; pedicels 2–9 cm long, indumentum as stems; bracteoles green, sometimes with purple venation, narrowly ovate to ovate or oblong, (1.3–)1.5–2.7(–3) × (0.4–)0.6–1.4(–1.8) cm, acute to obtuse, truncate to cordate and distinctly 4-angular at base, pubescent or sericeous-pubescent or sparsely so, densest on veins. Calyx puberulous with mixture of hairs and stalked capitate glands, rim 1–2 mm high, segments linear, 2–5 mm long. Corolla white to cream (rarely yellow) with purple throat; cylindric tube 7–10 mm long; throat straight, campanulate or narrowly so, (1.5–)1.8–3 cm long and (0.5–)0.7–1.4 cm in diameter apically; lobes (1–)1.5–2.2 × (1–)1.4–2.3 cm, broadly emarginate. Filaments 4–5 and 6–7 mm long; anthers 3–4 mm long, indistinctly apiculate, densely bearded at base and along $^2/_3$ of one side; spurs ± 1 mm long. Capsule densely puberulous, depressed globose, 6–7 × 7–9 mm, beak 9–17 mm long, slightly constricted at base. Seed 3–5 mm in diameter, dark brown, echinate.

UGANDA. Karamoja District: Lalachat, no date, *Eggeling* 2780! & Kukumongole, May 1956, *Wilson* 234!
KENYA. Northern Frontier District: Samburu, Wamba, 28 Nov. 1958, *Newbould* 2923!; Kitui District: Garissa road, 24 km E of Ukazzi, 12 Jan. 1972, *Gillett* 19483!; Masai District: 17 km on Sultan Hamud–Loitokitok road, 26 Feb. 1969, *Napper & Abdallah* 1904!
TANZANIA. Same District: 5 km SW of Same Railway Station, 6 Apr. 1971, *Wingfield* 1492! & Mkomazi Game Reserve, SE of Ndea Hill, 27 Apr. 1995, *Abdallah & Vollesen* 95/21! & Mkomazi Game Reserve, Kamakota to Kifukua, 13 June 1996, *Abdallah, Mboya & Vollesen* 96/212!
DISTR. **U** 1; **K** 1–4, 6, 7; **T** 2, 3; not known elsewhere
HAB. *Acacia* and *Acacia-Commiphora* bushland, grassland, on a variety of soils from sandy to stony or loamy or on rocky hills; (150–)300–1350(? –1800) m

SYN. *Thunbergia sp. D* of U.K.W.F.: 578 (1974) & U.K.W.F., ed. 2: 267 (1994)

26. **Thunbergia reniformis** *Vollesen* **sp. nov.** *T. gibsonii* fauce corollae abrupte flexuosa similis sed tubo corollae cylindrico breviore (2–3 mm nec 5 mm longo), foliis latioribus et indumento caulis foliorumque crispo (nec recto) differt. Type: Kenya, Kiambu District: Limuru, *van Someren* 2658 (K!, holo.; BM!, BR!, EA!, K!, iso.)

Perennial herb with several trailing stems from thick woody rootstock; stems becoming sub-woody at base, to 50 cm long, whitish pubescent to tomentose with curly hairs. Leaves with petiole not winged, 0.5–2.5 cm long, indumentum as stems; lamina triangular-cordiform to reniform, largest 1.5–4 × 1.8–3.5 cm, apex acute to broadly rounded, base cordate, with or without subhastate lobes, margin coarsely dentate, tomentose below, pubescent to tomentose (rarely pilose) above. Flowers axillary, solitary; pedicels 3.5–7 cm long, indumentum as stems; bracteoles ovate to oblong, 1.6–3 × 0.8–1.3 cm, acute or subacute, cordate and distinctly 4-angular at base, pubescent to tomentose or pilose. Calyx puberulous and with stalked capitate glands, rim 1–2 mm high, segments linear, 4–6 mm long. Corolla bright orange with deep purple throat; cylindric tube 2–3 mm long; throat sharply curved, broadly campanulate, 1.7–2.8 cm long and 0.7–1.3 cm in diameter apically; lobes 1–2 × 1–2.2 cm. Filaments ± 6 and 7 mm long; anthers 5–6 mm long, indistinctly apiculate, densely bearded at base and about half way up one side; spurs ± 2 mm long. Capsule densely puberulous, 8–10 mm in diameter, beak 12–18 mm long, parallel-sided, not narrowed at base. Seed not seen.

KENYA. Kiambu District: Limuru, 20 June 1933, *van Someren* 2658! & 1.5 km N of Uplands turning on Njalini road, 25 Jan. 1969, *Napper & J. Stewart* 1814!; Masai District: 38 km on Old Kajiado road, Feb. 1940, *Bally* 750!
DISTR. **K** 4, 6; not known elsewhere
HAB. Grassland, often on stony soil, roadsides, gravel pits; 1800–2400 m

SYN. *Thunbergia sp.* E of Agnew, U.K.W.F.: 578 (1974); Blundell, Wild Fl. E. Afr. 396, pl. 404 (1987); U.K.W.F., ed. 2: 267 (1994)

NOTE. It has been suggested that this species might be a hybrid between *T. gibsonii* and *T. alata*. However, it occurs in an area of Kenya where *T. gibsonii* has never been recorded, and in the part of Kenya where both *T. gibsonii* and *T. alata* occur hybrids have never been seen. The broad leaves, indumentum of thin curly hairs and short cylindric corolla tube distinguish it from both these species.

27. **Thunbergia reticulata** *Nees* in DC., Prodr. 11: 58 (1847); Richard, Tent. Fl. Abyss. 2: 139 (1850); Martelli, Fl. Bogos: 64 (1886); Ensermu in F.E.E. 5: 353 (2006). Type: Ethiopia, Semien Mts, Gapdia, *Schimper* II.758 (K!, holo.; K!, iso.)

Erect unbranched single-stemmed annual herb; stems to 20 cm long, sparsely sericeous-pubescent. Leaves with petiole 1.5–3 cm long, with a flat wing which widens upwards, indumentum as stems; lamina triangular-ovate, largest 3.5–4.5 × 1.2–2.2 cm, apex subacute, base truncate to subcordate, with subhastate lobes, margin entire or with a few teeth in lower part, sparsely sericeous-pubescent, below densest along veins, above uniformly. Flowers axillary, solitary; pedicels 1.5–3 cm long, indumentum as stems; bracteoles green, with conspicuous purple venation, ovate, 1.5–2.2 × 0.6–1.3 cm, subacute, cordate and distinctly 4-angular at base, sparsely sericeous-pubescent, densest on veins and margins. Calyx puberulous with mixture of hairs and stalked capitate glands, rim ± 1 mm high, segments linear, 2–4 mm long. Corolla yellow, without purple throat; cylindric tube 3–4 mm long; throat straight, campanulate, 1–1.5 cm long and 0.7–0.8 cm in diameter apically; lobes ± 5 × 5 mm, truncate to broadly emarginate. Filaments ± 5 and 6 mm long; anthers ± 2 mm long, not apiculate, bearded at base only; spurs ± 1 mm long. Capsule puberulous, depressed globose, 6–7 × 8–9 mm, beak 10–11 mm long, constricted at base. Seed ± 4 mm in diameter, blackish, echinate.

KENYA. Machakos District: 7.5 km. from Hunters Lodge on Nairobi road, 6 Jan. 1972, *Faden 72/20!*; Masai District: Chyulu Hills, Noginyaa Kopje, 23 Dec. 2000, *Luke* 7201!
DISTR. **K** 4, 6; Sudan, Eritrea, Ethiopia, Zambia, Zimbabwe, Botswana, Namibia, South Africa
HAB. In grass in seepage area at base of large rocky outcrops; 950–1150 m

SYN. *Thunbergia angulata* R.Br. in Salt, Abyss., App.: 65 (1814), *nom. nud., non* Hilsenb. & Bojer (1827)
  T. saltiana Steud., Nom., ed. 2, 2: 683 (1841), *nom. nud.*
  T. alata Sims var. *reticulata* (Nees) Burkill in F.T.A. 5: 17 (1899)
  T. aurea N.E.Br. in K.B. 1909: 127 (1909); Meyer in Prodr. Fl. Südw. Afr. 129: 1 (1968); Lebrun & Stork, Enum. Pl. Afr. Trop. 4: 505 (1997). Types: Botswana, Kwebe Hills, *Lugard* 107 (K!, syn.) & *Lugard* 114 (K!, syn.)

NOTE. The large disjunction from NE Africa to southern Africa is unusual but is – to a lesser extent – also shown by *Thunbergia annua*, and is known from other Acanthaceae, e.g. *Crossandra mucronata*.

28. **Thunbergia alata** *Sims* in Bot. Mag. 52: t. 2591 (1825); Hook., Exotic Fl.: t. 166 (1827); Nees in DC., Prodr. 11: 58 (1847); Klotzsch in Peters, Reise Mossamb., Bot.: 196 (1861); Burkill in F.T.A. 5: 16 (1899); De Wild., Pl. Beq. 1: 443 (1921); Lugard in K.B. 1933: 93 (1934); F.P.N.A. 2: 266 (1947); U.O.P.Z.: 470 & 472 (1949); F.P.S. 3: 189, fig. 51 (1956); Heine in F.W.T.A. (ed. 2) 2: 400 (1963); E.P.A.: 927 (1964); Benoist in Fl. Madag. 182, Acanthacées 1: 10 (1967); Binns, Checklist Herb. Fl. Malawi: 16 (1968); U.K.W.F.: 575 & 578 (1974); Clutton-Brock & Gillett in Afr. J. Ecol. 17: 154 (1979); Champluvier in Fl. Rwanda 3: 489 (1985); Blundell, Wild Fl. E. Afr.: 396, pl. 128 & 403 (1987); Iversen in Symb. Bot. Ups. 29(3): 162 (1991); U.K.W.F., ed. 2: 267, pl. 115 (1994); Lebrun & Stork, Enum. Pl. Afr. Trop. 4: 505 (1997); White *et al.*, For. Fl. Malawi: 119 (2001); Friis & Vollesen in Biol. Skr. 51(2): 455 (2005); Ensermu in F.E.E. 5: 353 (2006); Thulin in Fl. Somalia 3: 376 (2006). Type: Zanzibar, *Bojer* s.n. (K!, holo.; K!, iso.)

Annual or perennial trailing or twining herb, when perennial with single or several stems from woody rootstock; stems to 5 m long, yellowish to whitish sparsely sericeous or sericeous-pubescent to pilose or tomentose. Leaves with petiole 0.8–7(–9.5) cm long, with a flat (rarely undulate) wing which on some or all leaves widens distinctly upwards or is parallel-sided, sometimes very narrow or absent on some (more rarely all) leaves, indumentum as stems; lamina triangular or triangular-ovate, largest 2–10(–12) × 1.2–7(–8.5) cm, apex acute or subacute, apiculate, base subcordate to deeply cordate (rarely truncate) with hastate or rounded basal lobes, margin entire to coarsely dentate, sericeous-pubescent or sparsely so to tomentose, below densest along veins, above uniformly. Flowers axillary, solitary (rarely paired or in 3's); pedicels 2–9.5(–11) cm long, indumentum as stems; bracteoles green, sometimes with purple venation or mottled with purple, ovate to oblong, (1–)1.4–3.5 × 0.5–1.6(–2.1) cm, acute to rounded, truncate to cordate and distinctly 4-angular at base, whitish to yellowish pubescent or sericeous-pubescent or sparsely so, densest on veins. Calyx puberulous and with scattered to dense stalked capitate glands, rim 0.5–2 mm high, segments linear, 1–5 mm long. Corolla pale yellow to brillant orange or brick red (rarely white) with deep purple to almost black throat; cylindric tube 3–6 mm long; throat straight or slightly curved, campanulate or narrowly so, (1.2–)1.5–2.8(–3.2) cm long and 0.7–1.8 cm in diameter apically; lobes 0.8–2 × 1–2.5 cm, broadly emarginate. Filaments 2–7 and 4–9 mm long; anthers 3–5 mm long, indistinctly apiculate, densely bearded at base and along ²/₃ of one side; spurs 1–1.5 mm long. Capsule densely puberulous, depressed globose, 6–9 × 7–12 mm, beak 8–17(–19) mm long, constricted at base. Seed 4–6 mm in diameter, dark brown, echinate-lamellate or reticulate. Fig. 8, p. 57.

FIG. 8. *THUNBERGIA ALATA* — **1**, flowering branch; **2**, leaf variation; **3**, bracteoles; **4**, detail of bracteole indumentum; **5**, calyx; **6**, stamens; **7**, stigma; **8**, capsule valve; **9**, seed, two views. 1, 3–4 & 6–7 from *Polhill* 1690, 2 from *Cribb* 10334 (small leaf) & *Mlangwa* 1043 (large leaf), 5 & 8–9 from *Lovett* 4128. Drawn by Margaret Tebbs.

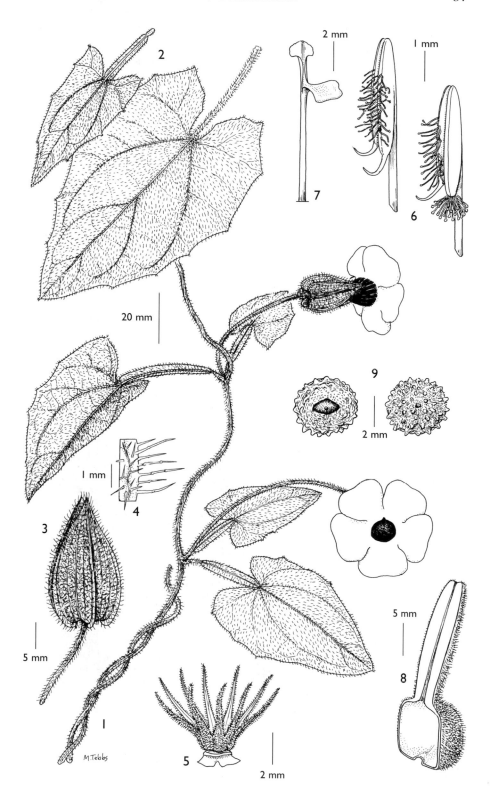

UGANDA. Karamoja District: Mt Napak, 24 Jan. 1957, *Dyson-Hudson* 137!; Kigezi District: Kachwekano Farm, May 1949, *Purseglove* 2822!; Mengo District: Kampala, Makerere University Campus, 18 Sep. 1987, *Katende* 3226!

KENYA. Northern Frontier District: Marsabit, 14 Jan. 1972, *Bally & Radcliffe-Smith* 14784!; Masai District: Ngerendei, 13 Apr. 1961, *Glover et al.* 490!; Teita District: Taita Hills, Mgange to Bura, 1 Sep. 1969, *Bally* 13506!

TANZANIA. Arusha District: Mt Meru, 12 Dec. 1966, *Richards* 21694!; Ufipa District: 5 km W of Mtai on Mbuza road, 20 Nov. 1994, *Goyder et al.* 3755!; Njombe District: 15 km S of Njombe, 5 July 1956, *Milne-Redhead & Taylor* 10964!; Zanzibar, Kisimboni, 10 July 1961, *Faulkner* 2866!

DISTR. U 1–4; K 1–7; T 1–8; Z; P; throughout tropical Africa, but especially common in the east, also in Madagascar, India and SE Asia and introduced in tropical America

HAB. A wide range of habitats from wet forest (usually on margins) to dry *Acacia-Commiphora* bushland, often in disturbed places and secondary vegetation; near sea level to 2700 m

SYN. *Valentia volubilis* Rafin., Specchio 1: 87 (1814), *non Thunbergia volubilis* Pers. (1806). Type: Ethiopia, *Salt* s.n. (BM!, neo; selected by Friis in Taxon 35: 360 (1986)
  Endomelas alata (Sims) Rafin., Fl. Tellur. 4: 67 (1836)
  Thunbergia alata Sims var. *albiflora* Hook. in Bot. Mag. 63: t. 3512 (1836); Gordon in Gard. Chron. 1845: 169 (1845). Type: Cult. in Glasgow Bot. Gard. (not seen)
  T. alata Sims var. *alba* Paxton, Mag. Bot. 3: 28 (1837); Nees in DC., Prodr. 11: 58 (1847). Type: Cultivated by Mrs. Lawrence of Ealing (not seen)
  T. manganjensis Lindau in E.J. 17: 92 (1893) & in P.O.A. C: 366 (1895); Fries in Schwed. Rhod. Congo Exp. Bot.: 302 (1916). Type: Malawi, Manganja Hills, *Kirk* s.n. (B†, holo.; K!, iso.)
  T. fuscata Lindau in E.J. 17, Beibl. 41: 40 (1893); Oliv. in Trans. Linn. Soc., ser. 2, 2: 345 (1886), *nom. nud.*; Engl., Hochgebirgsfl. Trop. Afr.: 386 (1892), *nom. nud.*; Lindau in P.O.A. C: 366 (1895); Chiov., Racc. Bot. Miss. Consol. Kenya: 93 (1935). Type: Malawi, Manganja Hills, *Meller* s.n. (B†, holo.; K!, iso.)
  T. alata Sims var. *vixalata* Burkill in F.T.A. 5: 16 (1899). Type: as for *T. fuscata.*
  T. alata Sims var. *retinervia* Burkill in F.T.A. 5: 17 (1899). Types: Malawi, Nyika Plateau, *Whyte* s.n. (K!, syn.); Mpata, *Whyte* s.n. (K!, syn.); Zambesi Land, *Kirk* s.n. (K!, syn.)
  T. delamerei S.Moore in J.B. 38: 205 (1900); C.B. Clarke in F.T.A. 5: 507 (1900); Lebrun & Stork, Enum. Pl. Afr. Trop. 4: 506 (1997). Type: Kenya, Northern Frontier District: Lake Marsabit, *Lord Delamere* s.n. (BM!, holo.)
  T. alata Sims var. *minor* S.Moore in J.B. 39: 300 (1901). Type: Kenya, Mombasa, *W.E.Taylor* s.n. (BM!, holo.)
  T. nymphaeifolia Lindau in E.J. 33: 184 (1902); T.T.C.L.: 17 (1949); Lebrun & Stork, Enum. Pl. Afr. Trop. 4: 506 (1997). Types: Tanzania, Lushoto District: Usambara, *Eick* 412 (B†, syn.); Iringa/Mbeya District: Uhehe, Utschungwe, *Frau Prince* s.n. (B†, syn.)
  T. oculata S.Moore in J.L.S. 38: 269 (1908); Lebrun & Stork, Enum. Pl. Afr. Trop. 4: 506 (1997). Type: Uganda/Congo-Kinshasa, Ruwenzori Mts, *Wollaston* s.n. (BM!, holo.)
  T. bikamaensis De Wild. in Pl. Beq. 1: 446 (1922); Lebrun & Stork, Enum. Pl. Afr. Trop. 4: 505 (1997). Type: Congo-Kinshasa, Bikama, *Bequaert* 2998 (BR!, holo.; BR!, iso.)
  T. kamatembica Mildbr. in B.J.B.B. 17: 84 (1943); Lebrun & Stork, Enum. Pl. Afr. Trop. 4: 506 (1997). Type: Congo-Kinshasa, Kamatembe, *de Witte* 1589 (BR!, holo.)
  T. sp. F of U.K.W.F.: 579 (1974); Iversen in Symb. Bot. Ups. 29, 3: 162 (1991); U.K.W.F., ed. 2: 267 (1994)

NOTE. This widespread species contains a number of forms which I have found impossible to separate into any meaningful taxa. Certain forms can be distinct locally, but there are always intermediate specimens to other forms, and a lot of the variation seems correlated to the wide range of habitats occupied by this species. The typical form from Zanzibar, much of Tanzania, western Kenya and Uganda is quite uniform. In northern Tanzania and central, eastern and northern Kenya dry country plants (e.g. *Richards* 27208) become very hairy with narrow or absent petiole wings and shallowly cordate leaves. An upland forest form (e.g. *Greenway* 12157) from northern Tanzania and central Kenya has large leaves, unwinged petioles and large flowers, but large flowered specimens also occur in typical material.

29. **Thunbergia battiscombei** *Turrill* in Hook., Ic. Pl. 31: t. 3041 (1915); F.P.S. 3: 191 (1956); Turrill in Bot. Mag. 173: t. 383 (1962); Blundell, Wild Fl. E. Afr.: 396, fig. 613 (1987); U.K.W.F., ed. 2: 266, pl. 115 (1994); Lebrun & Stork, Enum. Pl. Afr. Trop. 4: 505 (1997); Friis & Vollesen in Biol. Skr. 51(2): 455 (2005); Ensermu in F.E.E. 5: 349 (2006). Type: Kenya, 'Nyanza Basin', *Battiscombe* 667 (K!, holo.; EA!, iso.)

Perennial herb from woody rootstock; stems erect, to 75 cm long (rarely twining to 3 m), terete or quadrangular, glabrous or sparsely puberulous, usually with lines of hairs at nodes. Leaves slightly fleshy, palmately veined; petiole 1–4 cm long, glabrous or sparsely puberulous; lamina ovate to elliptic or slightly obovate, largest 8–17.5 × 5.5–9.5 cm, apex acuminate to broadly rounded, apiculate, base attenuate to truncate, decurrent, margin entire or with a few large irregular teeth, glabrous (rarely sparsely puberulous). Flowers in many-flowered axillary racemes (rarely two racemes per axil or one raceme and a solitary flower); peduncle and axis 2.5–30 cm long, glabrous to finely puberulous; bracts caducous, lower foliaceous, upper bracteole-like; pedicels 0.8–3.5 cm long, puberulous; bracteoles pale green with conspicuous raised dark green reticulation, ovate to elliptic, 2–2.7 × 0.8–1.2 cm, acute to obtuse, apiculate, puberulous. Calyx puberulous, 2–4 mm high of which the broadly triangular lobes about half. Corolla limb and upper part of tube purple to royal blue, lower part of tube tube whitish, throat yellow; tube (2.5–)3–4.5 cm long; lobes 1–1.5 × 1.5–2 cm. Filaments 10–15 mm long, glabrous; anthers 2–3 mm long. Capsule 8–10 mm in diameter, beak 16–20 mm long. Seed pale brown, ± 7 mm in diameter with reticulate surface and strong lateral ridge.

UGANDA. West Nile District: West Madi, Metu, Anua River, Oct. 1959, *E. M. Scott* EA11784!; Acholi District: Imatong Mts, Mt Lomwaga, 4 Apr. 1945, *Greenway & Hummel* 7269!; Mbale District: near Mbale, Jan. 1918, *Dümmer* 3749!
KENYA. Trans Nzoia District: foothills of Mt Elgon, no date, *Tweedie* 281!; Uasin Gishu District: Soy, June 1933, *Mainwaring* 2661!
DISTR. **U** 1–3; **K** 3, 5; Congo-Kinshasa, Sudan, Ethiopia
HAB. *Combretum* and other broad-leaved woodland and grassland subject to annual burning, riverine thicket; 750–2300 m

SYN. *Thunbergia adjumaensis* De Wild., Pl. Beq. 4: 419 (1928); Lebrun & Stork, Enum. Pl. Afr. Trop. 4: 505 (1997). Type: Congo-Kinshasa, Adjuma to Aba, *Claessens* 1641 (BR!, syn.) & 1656 (BR!, syn.)

30. **Thunbergia verdcourtii** *Vollesen* **sp. nov.** flores in racemis axillaribus pedunculo 18–45 cm longo instructis aggregati; folia 4–6.5 cm lata petiolo 5–25 mm longo instructa; bracteae 13–20 mm longae; bracteolae atrovirides vel purpureae obscure reticulatae. Type: Tanzania, Buha District: Kasulu, *Verdcourt* 3328 (K!, holo.; BR!, EA!, iso.)

Perennial herb with 1–5 erect or trailing stems from woody rootstock; stems to 60 cm long, quadrangular, glabrous or with sparse long curly hairs. Leaves palmately veined; petiole 0.5–2.5 cm long, sparsely to densely pubescent; lamina ovate to elliptic, largest 7.5–20.5 × 4–6.5 cm, apex acute to obtuse, apiculate, base truncate to cordate with rounded or hastate lobes, margin entire or slightly crenate, glabrous or sparsely pubescent along midrib. Flowers in erect 6–16-flowered axillary racemes; peduncle and axis 18–45 cm long, glabrous or with scattered long hairs at base, minutely puberulous towards apex; bracts caducous, mauve to purple, 13–20 mm long, with long (to 2 mm) glossy glandular hairs; pedicels 5–23 mm long, glabrous to finely puberulous (rarely with a few glossy glandular hairs); bracteoles dark green to purple, with obscure reticulation, ovate to ovate-elliptic, 2–2.7 × 0.8–1.2 cm, acute, apiculate, with long (to 2 mm) glossy glandular hairs on veins (rarely also minutely puberulous). Calyx glabrous, 1–2 mm high of which the broadly triangular lobes about half. Corolla limb and upper part of tube deep bluish purple, lower part of tube tube whitish, throat deep yellow; tube 2.5–4.5 cm long; lobes ± 1.5 × 1.5–2.5 cm. Filaments 10–12 mm long, glabrous; anthers 2.5–3 mm long. Immature capsule ± 10 mm in diameter, beak ± 14 mm long. Immature seed pale brown, ± 6 mm in diameter, with reticulate surface and raised lateral ridge. Fig. 9, p. 60.

TANZANIA. Buha District: 37 km on Kibondo–Kasulu road, 15 July 1960, *Verdcourt* 2846! & near Kasulu, 16 Nov. 1962, *Verdcourt* 3328!

Fig. 9. *THUNBERGIA VERDCOURTII* — **1**, flowering branch, × ²/₃; **2**, leaf, × ²/₃; **3**, bracteole, × 1.5; **4**, corolla, × 1; **5**, section of corolla with stamens, × 1; **6**, stamens, × 5; **7**, ovary, style and stigma, × 2; **8**, stigma, × 3; **9**, section of young capsule, × 1.5. 1 & 3–8 from *Verdcourt* 2846, 2 & 9 from *Verdcourt* 3328. Drawn by Heather Wood.

Distr. **T** 4; Burundi
Hab. *Brachystegia-Julbernardia* woodland, termite mounds in woodland, sandy to loamy or stony soil; 1200–1650 m

31. **Thunbergia racemosa** *Vollesen* **sp. nov.** *T. verdcourtii* floribus in racemis axillaribus aggregatis similis, sed pedunculo breviore ((2–)6–15(–27) cm nec 18–45 cm longo), foliis angustioribus (1–2.5(–4) cm nec 4–6.5 cm latis), petiolo breviore (0–3 mm nec 5–25 mm longo) instructis; bracteis brevioribus (0.5–6 mm nec 13–20 mm longis); bracteolis pallide viridibus (nec atroviridibus neque purpureis) et conspicue prominenterque atroviride reticulati (nec obscure reticulati) differt. Type: Tanzania, Rungwe District: S of Tukuyu, Chivanje Tea Estate, *Cribb, Grey-Wilson & Mwasumbi* 11287 (K!, holo.; K!, iso.)

Perennial herb with 1–3 erect stems from woody rootstock; stems to 40(–75) cm long, quadrangular, glabrous to densely pubescent. Leaves pinnately veined or with indistinct lateral veins from base; petiole 0–3 mm long, glabrous to densely pubescent; lamina lanceolate or narrowly so (rarely elliptic to obovate), largest (8–)13–27 × 1–2.5(–4) cm, apex acute to acuminate (rarely obtuse), apiculate, base cuneate to attenuate or sessile with subamplexicaul base, margin entire or slightly crenate, glabrous to pubescent, densest along midrib. Flowers in erect (1–)2–10-flowered axillary racemes; peduncle and axis (2–)6–15(–27) cm long, glabrous to densely pubescent; bracts caducous, linear-lanceolate, 0.5–6 mm long; pedicels 1–3 cm long, glabrous to densely pubescent; bracteoles pale green, with conspicuous raised dark green reticulation, ovate-oblong to elliptic, 1.8–2.9 × 0.7–1.3 cm, acute to obtuse, apiculate, with long (longest 2–3 mm) glossy glandular hairs on veins and edges. Calyx glabrous, ± 1 mm high of which the broadly triangular lobes about half. Corolla limb and upper part of tube deep maroon or deep purple, lower part of tube tube whitish, throat deep yellow; tube 3–4.5 cm long; lobes 1–2 × 2–3 cm. Filaments 11–13 mm long, glabrous; anthers 2–3 mm long. Style with scattered glands. Capsule 8–10 mm in diameter, beak 12–15 mm long. Seed chestnut, 5–6 mm in diameter, with spinose-reticulate surface, no lateral ridge.

Tanzania. Rungwe District: S of Tukuyu, Chivanje Tea Estate, 5 Feb. 1979, *Cribb et al.* 11287! & June 1986, *Congdon* 95! & 21 Jan. 2002, *Congdon* 621!
Distr. **T** 7; Zambia, Malawi
Hab. *Brachystegia-Uapaca* woodland on heavy loamy soil; 750–1100 m

Note. An exceedingly characteristic species which is probably closest to the *T. lancifolia* complex. It differs from the species in this group by having flowers in racemes, but also in the long glossy glandular hairs on the bracteoles.

32. **Thunbergia microchlamys** *S.Moore* in J.B. 45: 88 (1907); Lebrun & Stork, Enum. Pl. Afr. Trop. 4: 506 (1997). Type: Uganda, Toro District: Durro Forest, *Bagshawe* 1048 (BM!, holo.)

Herbaceous twiner to 3 m long; stems terete, glabrous or with thin lines of setose hairs at nodes. Leaves palmately veined; petiole 2.5–8 cm long, glabrous or puberulous apically; lamina ovate-cordiform, largest 6.5–17 × 3.5–10.5 cm, apex acute to acuminate, apiculate, base deeply cordate with hastate or rounded lobes, margin entire or slightly crenate, glabrous or finely puberulous along major veins. Flowers in (1–)4–8-flowered axillary racemes; peduncle and axis (1–)4.5–25 cm long, glabrous but for lines of glossy setose hairs at nodes or sparsely puberulous upwards; bracts persistent, linear to ovate-cordiform, up to 2 mm long, more rarely leafy and up to 2.5 cm long; pedicels 1–2.5 cm long, glabrous or with a few glandular hairs apically or puberulous; bracteoles pale green, with conspicuous raised dark green reticulation, ovate to elliptic, 1.1–2.2 × 0.4–0.7 cm, acute to acuminate, apiculate, glabrous to puberulous and with long (longest 2–3 mm) glossy glandular hairs on veins and edges. Calyx glabrous, ± 1 mm high of which the broadly triangular lobes about half. Corolla limb and upper part of tube purple or bluish purple, lower part of tube tube whitish, throat deep yellow; tube 2.5–4.5 cm long; lobes 1–2 × 2–3 cm. Filaments ± 12 mm long, glabrous; anthers ± 3 mm long. Style with scattered glands apically. Capsule and seed not seen (see note).

UGANDA. Toro District: Durro Forest, 8 June 1906, *Bagshawe* 1048!; Ankole District: Buhweju, Ngarwambu to Rugongo, 18 Feb. 1993, *Rwaburindore* 3550!
TANZANIA. Mpanda District: Mahali Mts, Lumpululu, 7 June 1992, *Uehara* 92–13!; & Kasoge area, 15 March 1995, *Turner* 40! & Kasye Forest, 19 March 1994, *Bidgood et al.* 2815!
DISTR. **U** 2; **T** 4; Congo-Kinshasa
HAB. Evergreen forest and riverine forest, often overhanging streams; 800–1550 m

SYN. *Thunbergia beninensioides* De Wild., Pl. Beq. 1: 444 (1922); Lebrun & Stork, Enum. Pl. Afr. Trop. 4: 505 (1997). Type, Congo-Kinshasa, Beni to Lesse, *Bequaert* 4110 (BR!, holo.).
　　T. beniensis De Wild., Pl. Beq. 1: 445 (1922); Lebrun & Stork, Enum. Pl. Afr. Trop. 4: 505 (1997). Type, Congo-Kinshasa, Beni to Kasindi, *Bequaert* 5161 (BR!, holo.; BR!, iso.)

NOTE. The type specimen has attached to the sheet an envelope containing a corolla and two loose seeds. The specimen itself is a flowering branch with no traces of thickened fruiting pedicels. It is not certain that the two seeds belong to the rest, and they have not been included in the description.

33. **Thunbergia masisiensis** *De Wild.*, Pl. Beq. 1: 450 (1922); Lebrun & Stork, Enum. Pl. Afr. Trop. 4: 506 (1997). Type: Congo-Kinshasa, Masisi, *Bequaert* 6415 (BR!, holo.; BR!, iso.)

Strong herbaceous twiner to 5 m long with a strong unpleasant smell; stems terete, with band of setose hairs at nodes, otherwise glabrous or with thin bands of hairs downwards from nodes. Leaves palmately veined; petiole 1.5–3.5 cm long on vegetative leaves, pilose; lamina ovate-cordiform, largest 6–9 × 4–5 cm, apex acuminate, apiculate, base cordate with rounded to hastate lobes, margin entire to irregularly dentate, pilose along major veins and on edges, sometimes sparsely so on lamina. Flowers solitary in upper leaf-axils; pedicels 1.5–3.5 cm long, glabrous or hairy at the very apex; bracteoles green with conspicuous darker green reticulation, ovate, 2.7–3.3 × 0.8–1.2 cm, acuminate, with sparse long broad glossy hairs and long-stalked capitate glands on margings and veins, densest towards base. Calyx glabrous, ± 2.5 mm high of which the broadly triangular lobes about half. Corolla limb and upper part of tube deep purple, lower part of tube whitish, throat deep yellow; tube 4–5 cm long; lobes 1.7–2.8 × 1.2–2.2 cm. Filaments ± 18 mm long; anthers and stigma not seen. Capsule ± 18 mm in diameter, beak ± 2.8 cm long. Seed not seen.

UGANDA. Ankole District: Lake Lutoto, Oct. 1940, *Eggeling* 4116!
DISTR. **U** 2; Congo-Kinshasa, Burundi
HAB. Wet evergreen forest, riverine forest; ± 1250 m

34. **Thunbergia bogoroensis** *De Wild.*, Pl. Beq. 1: 447 (1922); Lebrun & Stork, Enum. Pl. Afr. Trop. 4: 505 (1997). Type: Congo-Kinshasa, Bogoro, *Bequaert* 4964 (BR!, holo.; BR!, iso.)

Erect perennial herb with several stems from woody rootstock; stems to 45 cm long, quadrangular, uniformly puberulous or densely so. Leaves palmately veined; petiole 1–4 cm long on vegetative leaves, uniformly puberulous; lamina ovate, largest 8–11 × 4–6.5 cm, apex acute to acuminate, apiculate, base subcordate to cordate with rounded to hastate lobes, margin entire to crenate, puberulous or sparsely so, densest along major veins. Flowers solitary; pedicels 1.5–2.8 cm long, uniformly puberulous or densely so; bracteoles green, with obscure reticulation, ovate, 1.2–1.5 × 0.4–0.6 cm, acute, densely uniformly puberulous, with or without a few long glossy glandular hairs. Calyx minutely puberulous, ± 1.5 mm high of which the broadly triangular lobes about half. Corolla limb and upper part of tube royal blue to dark purple, lower part of tube whitish, throat deep yellow; tube 3–4 cm long; lobes ± 1 × 1.5–2 cm. Filaments ± 12 mm long; anthers ± 2.5 mm long. Style with scattered glands apically. Capsule 8–10 mm in diameter, beak 12–13 mm long. Seed dark brown, 4–5 mm in diameter, reticulate.

UGANDA. Toro District: Ruwenzori Mts, Kirabawah, Dec. 1925, *Maitland* 1029!
TANZANIA. Bukoba District: Karagwe, Nyashozi, Dec. 1931, *Haarer* 2398!; Mwanza District: Geita, Kafunzo River, Oct. 1954, *Carmichael* 447!
DISTR. **U** 2; **T** 1; Congo-Kinshasa, Rwanda, Burundi
HAB. Grassland along streams; 900–1550 m

SYN. *Thunbergia claessensii* De Wild., Pl. Beq. 4: 420 (1928); Lebrun & Stork, Enum. Pl. Afr. Trop. 4: 505 (1997). Type: Congo-Kinshasa, Bogoro, *Claessens* 1272 (BR!, holo.)
Thunbergia sp. A of Champluvier in Fl. Rwanda 3: 490 (1985)

NOTE. Superficially similar to *T. petersiana* but differs in the uniformly puberulous indumentum which is never seen in that species.

35. **Thunbergia petersiana** *Lindau* in E.J. 17: 89 (1893) & in E.J. 17, Beibl. 41: 35 (1893) & in P.O.A. C: 366 (1895); Burkill in F.T.A. 5: 23 (1899); Clutton-Brock & Gillett in Afr. J. Ecol. 17: 154 (1979); Iversen in Symb. Bot. Ups. 29, 3: 162 (1991); Ruffo *et al.*, Cat. Lushoto Herb. Tanz.: 11 (1996) pro parte; Lebrun & Stork, Enum. Pl. Afr. Trop. 4: 506 (1997). Type: Mozambique, Boror, *Peters* s.n. (B†, holo.)

Erect or straggling perennial or shrubby herb from creeping rootstock with fleshy roots; stems to 1.5 m long, quadrangular, glabrous to puberulous, with or without conspicuous bands of glossy setose hairs at nodes. Leaves thin, glossy, palmately veined; petiole (0.5–)1–13 cm long on vegetative leaves, glabrous to puberulous, distinctly ciliate or not; lamina ovate to cordiform or broadly so, largest (5–)6.5–17 × (3–)4.5–10 cm, apex acute to acuminate (rarely obtuse), apiculate, base subcordate to deeply cordate with rounded, hastate or sagittate lobes (rarely truncate), margin entire to grossly and irregularly dentate, glabrous to puberulous, often only on veins, rarely with longer glossy setose hairs. Flowers solitary (rarely paired) or from lower axils in racemes with small ovate sessile bracts; pedicels 1.5–5(–10 in fruit) cm long in solitary flowers and 0.5–2(–5 in fruit) cm in racemes, glabrous to puberulous (rarely with a few long glossy glandular hairs); bracteoles green, with conspicuous raised dark green reticulation, ovate-oblong, 0.9–2.2 × 0.4–1.2 cm, acute, with sparse to dense long (to 1.5 mm) glossy glandular hairs on veins and margins. Calyx glabrous or minutely puberulous, 1–2(–5 in fruit) mm high of which the broadly triangular lobes about half. Corolla limb and upper part of tube dark bluish purple or royal blue, lower part of tube white, throat yellow; tube (3.2–)3.5–5 cm long; lobes 1–1.8 × 1.5–2.5 cm. Filaments 9–13 mm long, glabrous; anthers 2–2.5 mm long. Style with scattered glands. Capsule 8–13 mm in diameter, beak 13–18 mm, long. Seed dark reddish brown, ± 5 mm in diameter, reticulate to spinulose near apex.

UGANDA. Bunyoro District: Budongo Forest, Kanyo, June 1932, *C.M. Harris* in *Brasnett* 830! & Rabongo Forest, 24 July 1964, *H.E. Brown* 2103!; Mengo District: Entebbe, 1909, *Fyffe* 51!
KENYA. Northern Frontier District: Marsabit, 14 May 1970, *Magogo* 1310!; Meru District: Meru, 16 June 1932, *Graham* 1757! & Nyambeni [Jombeni] River, Dec. 1939, *Copley* in *Bally* 515!
TANZANIA. Lushoto District: E Usambara Mts, Derema–Singale, 30 Jan. 1932, *Greenway* 2909!; Buha District: Gombe National Park, Rutanga Valley, 21 Jan. 1964, *Pirozynski* 264!; Songea District: Matengo Hills, 5 km E of Ndengo, 6 March 1956, *Milne-Redhead & Taylor* 9043!
DISTR. **U** 2, 4; **K** 1, 4, 7; **T** 3, 4, 6–8; Sudan, Congo-Kinshasa, Burundi, Zambia, Malawi, Mozambique, Zimbabwe
HAB. Lowland evergreen forest, often in clearings or on edges, riverine forest, in S and W Tanzania in montane forest; 600–1550(–2200) m

SYN. [*Thunbergia chrysops* sensu Klotzsch in Peters, Reise Mossamb.: 196 (1861), *non* Hook. (1844)]

T. *stuhlmanniana* Lindau in E.J. 17: 91 (1893); Lebrun & Stork, Enum. Pl. Afr. Trop. 4: 507 (1997). Type: Uganda, Kigezi District: Butumbi, Kawanda, *Stuhlmann* 2181 (B†, holo.)

T. *mollis* Lindau in E.J. 17, Beibl. 41: 35 (1893), *nom. nud.* & in E.J. 20: 2 (1894) & in P.O.A. C: 366 (1895); Burkill in F.T.A. 5: 22 (1899); Lindau in Z.A.E.: 292 (1911); Binns, Checklist Herb. Fl. Malawi: 16 (1968); Lebrun & Stork, Enum. Pl. Afr. Trop. 4: 506 (1997). Types: Malawi, *Buchanan* 263 (B†, syn.; BM!, K!, iso.) & *Buchanan* 1092 (B†, syn.; K!, iso.)

T. *katangensis* De Wild. in Ann. Mus. Congo Bot., Ser. 4, 1: 135 (1903) & Contrib. Fl. Katanga: 193 (1921); Lebrun & Stork, Enum. Pl. Afr. Trop. 4: 506 (1997). Type: Congo-Kinshasa, Katanga, Lukafu, *Verdick* 292 (BR!, holo.)

T. *swynnertonii* S.Moore var. *cordata* S.Moore in J.L.S. 40: 158 (1911); R.E. Fries in Wiss. Ergebn. Schwed. Rhod.-Congo-Exp. 1911–12. Bot. 1: 303 (1916). Types: Mozambique, Kurumadzi River, Jihu, *Swynnerton* 1924 (BM!, syn.; K!, iso.) & *Swynnerton* 1926 (BM!, syn.; K!, iso.)

T. *zernyi* Mildbr. in N.B.G.B. 15: 635 (1941); Lebrun & Stork, Enum. Pl. Afr. Trop. 4: 507 (1997). Type: Tanzania, WSW of Songea, Ugano, *Zerny* 205 (W!, holo.)

T. *torrei* Benoist in Notul. Syst. 11: 147 (1944); Lebrun & Stork, Enum. Pl. Afr. Trop. 4: 507 (1997). Type: Mozambique, Nampula, *Torre* 1300 (COI!, holo.; COI!, LISC!, iso.)

T. *sp.* near *cordata* sensu Jex-Blake, Gard. E. Afr. (ed. 3): 209 & pl. 19, fig 4 (1950) & Gard. E. Afr. (ed. 4): 228 (1957)

[T. *natalensis* sensu U.K.W.F.: 578 (1974) & U.K.W.F., ed. 2: 266 (1994), *non* Hook. f. (1858)]

NOTE. Some collections from K 4 have petioles only 0.5–1 cm long. They differ from *T. natalensis* in being completely glabrous. The only totally glabrous form of *T. natalensis* has a white corolla.

36. **Thunbergia usambarica** *Lindau* in E.J. 17: 89 (1893) & in P.O.A. C: 366 (1895); Burkill in F.T.A. 5: 23 (1899); Iversen in Symb. Bot. Ups. 29, 3: 162 (1991); U.K.W.F., ed. 2: 266 (1994); Ruffo *et al.,* Cat. Lushoto Herb. Tanzania: 11 (1996); Lebrun & Stork, Enum. Pl. Afr. Trop. 4: 507 (1997). Type: Tanzania, Lushoto District: Usambara Mts, *Holst* 215 (B†, holo.; K!, iso.)

Erect or straggling perennial herb from creeping woody rootstock; stems to 1 m long, quadrangular, glabrous to puberulous (rarely pubescent), with conspicuous bands of glossy setose hairs at nodes. Leaves thin, glossy, palmately veined; petiole 1–10 cm long, puberulous and distinctly ciliate; lamina ovate or broadly so (rarely oblong-lanceolate), largest (4–)6–14(–16.5) × (2.3–)3–7.5(–10.5) cm, apex acute to acuminate (rarely obtuse), apiculate, base subcordate to deeply cordate with rounded to hastate lobes, margin entire to crenate (rarely sharply dentate), glabrous to puberulous, mainly on veins or with longer glossy setose hairs on veins. Flowers solitary (rarely paired or very rarely in 3's); pedicels 2–6(–7) cm long, glabrous (rarely sparsely puberulous near apex); bracteoles green, with conspicuous raised dark green reticulation, lanceolate- or ovate-oblong, 1–1.9 × 0.4–0.8(–1) cm, acute, with sparse to dense long (to 2 mm) glossy glandular hairs on veins and margins, more rarely glabrous. Calyx minutely puberulous, 1–3(–5 in fruit) mm high of which the broadly triangular lobes about half. Corolla white with yellow throat, more rarely pale lilac; tube 2–3.2 cm long; lobes 1–1.5(–2) × 1.5–2(–2.5) cm. Filaments 7–9 mm long, basal part glandular; anthers 2–2.5 mm long. Style with scattered glands. Capsule 8–9 mm in diameter, beak 14–16 mm, long. Seed reddish brown, ± 6 mm in diameter, reticulate to spinulose near apex.

UGANDA. Toro District: Fort Portal, 20 Oct. 1906, *Bagshawe* 1267!; Kigezi District: Luhiza, June 1951, *Purseglove* 3670! & 20 km S of Nyaiguru, 15 Dec. 1982, *Katende* 1479!

KENYA. Nandi District: Kapsabet, Yala River, no date, *G.R. Williams & Piers* 594!; North Kavirondo District: Kakamega Forest, near Yala River, 25 Nov. 1969, *Bally* 13680! & Kakamega Forest, Kisieni–Ikuywa, 7 May 1971, *Mabberley & Tweedie* 1107!

TANZANIA. Lushoto District: W Usambara Mts, Magamba Forest, 1 March 1953, *Drummond & Hemsley* 1360!; Iringa District: Dabaga Highlands, Kibengu, 13 Feb. 1962, *Polhill & Paulo* 1462!; Njombe District: Livingstone Mts, Madunda Mission, 2 Feb. 1951, *Richards* 14097!

DISTR. **U** 2; **K** 3, 5; **T** 2, 3, 5, 7, 8; Congo-Kinshasa, Rwanda, Burundi, Malawi, Mozambique, Zimbabwe, South Africa

HAB. Wet evergreen montane forest, often on margins, in clearings and in disturbed areas; (1350–)1650–2450 m

SYN. *Thunbergia alba* S.Moore in J.B. 48: 250 (1910); Lebrun & Stork, Enum. Pl. Afr. Trop. 4: 505 (1997). Type: Uganda, Toro District: Fort Portal, *Bagshawe* 1267 (BM!, holo.)

T. swynnertonii S.Moore in J.L.S. 40: 157 (1911); Binns, Checklist Herb. Fl. Malawi: 16 (1968); Lebrun & Stork, Enum. Pl. Afr. Trop. 4: 507 (1997). Type: Zimbabwe, Chirinda Forest, *Swynnerton* 339 (BM!, holo.; K!, iso.)

T. cordata Lindau in P.O.A. C: 365 (1895) & in E.J. 24: 310 (1897); Burkill in F.T.A. 5: 22 (1899); Lebrun & Stork, Enum. Pl. Afr. Trop. 4: 506 (1997), *nom. illeg.*, *non* Colla (1824). Type: Tanzania, Kilimanjaro, Useri, *Volkens* 1968 (B†, holo.)

T. humbertii Mildbr. in B.J.B.B. 14: 354 (1937), *nom. illeg.*, *non* Benoist (1926). Type: Congo-Kinshasa, W of Lake Kivu, Tshibinda, *Humbert* 7480 (BR!, holo.; BR!, iso.)

[*T. petersiana* sensu Ruffo *et al.*, Cat. Lushoto Herb. Tanzania: 11 (1996), quoad *Mmari* 12 and *Shabani* 1291, *non* Lindau (1893)]

NOTE. Very similar to *T. petersiana* from which it differs in a smaller usually white corolla. Ecologically the two are more or less separate with *T. usambarica* growing at higher altitudes.

The only two collections (*Hornby* 498 & 2062) seen from **T** 5 (Kiboriani Mt) differ from all other material in the very wide shortly petiolate leaves but are otherwise quite typical. This locality is very isolated from the rest of the distribution area in Tanzania, and further collections are needed to prove whether this form might be worthy of taxonomic recognition.

37. **Thunbergia natalensis** *Hook.* in Bot. Mag. 84: t. 5082 (1858); T. Anderson in J.L.S. 7: 18 (1863); C.B. Clarke in Dyer, Fl. Cap. 5: 4 (1901); Cribb & Leedal, Mountain Fl. S. Tanz.: 129, pl. 32 (1982); Compton, Fl. Swaziland: 547 (1976); Walker, Wild Fl. Kwazulu-Natal: 150 (1996); Germishuizen, Wild Fl. N. S. Afr.: 392 (1997); Ensermu in F.E.E. 5: 355 (2006). Type: Cultivated in Hort. Veitch from South African material, *Herb. Hooker* s.n. (K!, holo.)

Perennial herb with 1–3 erect stems from large creeping woody rootstock with fleshy roots; stems to 75 cm long, quadrangular, sparsely to densely pubescent with curly non-glandular hairs, with distinct bands of longer hairs at nodes (rarely glabrous). Leaves dark green, glossy, palmately veined; petiole 2–5 mm long on vegetative leaves, indumentum as on stems; lamina ovate or broadly so (rarely elliptic), largest (3.5–)4.5–9(–19) × (2.7–)3–6(–10.5) cm, usually less than twice as long as wide, apex acute to obtuse, apiculate, base subcordate to cordate with rounded (rarely hastate) lobes, margin entire to slightly crenate, subglabrous to puberulous, densest on veins (rarely glabrous). Flowers solitary; pedicels 2–7.5 cm long, glabrous to puberulous (rarely pubescent); bracteoles pale green, with conspicuous raised dark green reticulation, more rarely with obscure reticulation, ovate-oblong, 1.5–3(–3.5) × 0.6–1.1 cm, acute to obtuse, with sparse to dense long (to 2 mm) glossy glandular hairs on veins and margins (rarely pubescent without glands or glabrous). Calyx glabrous to minutely puberulous, 1–2(–4 in fruit) mm high of which the broadly triangular lobes about half. Corolla limb and upper part of tube white to blue or mauve, lower part of tube white, throat yellow; tube 2.5–3.5 cm long; lobes 1–1.7 × 1.5–2 cm. Filaments 8–10 mm long, glabrous or glandular at base; anthers 2–2.5 mm long. Style with scattered glands. Capsule 7–9 mm in diameter, beak 13–16 mm long. Seed reddish brown, ± 5 mm in diameter, reticulate to spinulose near apex.

TANZANIA. Mbulu District: Mt Hanang, Nangwa, 6 Feb. 1946, *Greenway* 7616!; Rungwe District: Usafwa, 19 July 1913, *Stolz* 2395!; Mbeya District: Poroto Mts, Igali, 25 March 1988, *Bidgood et al.* 707!

DISTR. **T** 2, 4, 5, 7; Ethiopia, Malawi, Zimbabwe, Swaziland, South Africa

HAB. Montane grassland, usually on rich loamy volcanic soils, rarely in *Brachystegia* woodland or in *Terminalia-Pterocarpus* wooded grassland on steep rocky slopes; (1100–)1700–2450(–2650) m

SYN. *Thunbergia mellinocaulis* Burkill in F.T.A. 5: 23 (1899); Binns, Checklist Herb. Fl. Malawi: 16 (1968); Lebrun & Stork, Enum. Pl. Afr. Trop. 4: 506 (1997). Type: Malawi, Mt Zomba, *Whyte* s.n. (K!, holo.; K! iso.)

T. squamuligera Lindau in E.J. 30: 406 (1901); Lebrun & Stork, Enum. Pl. Afr. Trop. 4: 507 (1997). Type: Tanzania, Rungwe District: Mt Mbogo, *Goetze* 1453 (B†, holo.; BM!, BR!, iso.)

NOTE. The two collections from **T** 2 and 5 are both somewhat abnormal but seem to fit better into this than any other species. *Burtt* 1162 from Kondoa District differs in the total absence of stalked glands and in the pubescent indumentum. It also comes from a low altitude and from woodland. *Greenway* 7616 (cited above) is subglabrous and has very large bracteoles combined with a small corolla.

The only collection from **T** 4 (*Bidgood et al.* 4272) is from a very low altitude and a very dry habitat. It also has exceptionally large leaves, is totally glabrous and has a white corolla. Further collections may show this to be a distinct taxon.

38. **Thunbergia ciliata** *De Wild.* in F.R. 13: 105 (1913) & Not. Fl. Katanga 4: 73 (1914) & Contrib. Fl. Katanga: 192 (1921); Lebrun & Stork, Enum. Pl. Afr. Trop. 4: 505 (1997). Type: Congo-Kinshasa, Katanga, Katentania, *Homblé* 810 (BR!, holo.; BR!, iso.)

Perennial herb with several erect stems from large woody rootstock; stems to 40(–50) cm long, quadrangular, sparsely hirsute to lanate with long (to 4 mm) curly glossy non-glandular hairs. Leaves palmately veined; petiole 2–6 mm long on vegetative leaves, indumentum as on stems; lamina ovate to oblong or elliptic, largest 3–8 × 1.5–4.2 cm, usually less than twice as long as wide, apex acute to obtuse, apiculate, base truncate to cordate with rounded to hastate lobes, margin entire to coarsely and irregularly dentate, pubescent to lanate (rarely sparsely pubescent). Flowers solitary; pedicels 1–4(–5) cm long, hirsute to lanate with long (to 4 mm) glossy non-glandular hairs; bracteoles dark green, with obscure reticulation, ovate or ovate-oblong, (1.2–)1.5–3.2 × 0.5–1.4 cm, acute or subacute, pubescent to densely lanate with long (to 4 mm) glossy non-glandular hairs. Calyx glabrous to minutely puberulous, 1–2(–4 in fruit) mm high of which the broadly triangular lobes about half. Corolla limb and upper part of tube deep blue to deep purple, lower part of tube white, throat yellow; tube 3–5(–5.5) cm long; lobes (1–)1.5–2.3 × 1.5–2.7 cm. Filaments 8–12 mm long, glabrous; anthers 2–3 mm long. Style glabrous. Capsule 8–10 mm in diameter, beak 12–20 mm long. Seed reddish brown, ± 6 mm in diameter, reticulate to spinulose near apex.

TANZANIA. Ufipa District: Mbizi Forest, 27 Nov. 1954, *Richards* 2356! & 17 km on Sumbawanga–Mbala road, 18 Nov. 1986, *Brummitt et al.* 18039! & Tatanda Mission, 22 Feb. 1994, *Bidgood et al.* 2376!
DISTR. **T** 4; Congo-Kinshasa, Zambia
HAB. Montane grassland, *Brachystegia* woodland, often on rocky hillsides and subject to regular burning; 1500–2500 m

NOTE. Closely related to *T. natalensis* from which it differs in the dense pubescent to lanate indumentum without stalked capitate glands and in the different texture of the bracteoles.

39. **Thunbergia richardsiae** *Vollesen* **sp. nov.** a *T. ciliata* et *T. natalensi* pedicellis pilos longos (usque 2 mm longos) nitentes capitatos glandulosos gerentibus differt. A *T. ciliata* etiam indumento multo magis sparso et bracteolis reticulo venularum conspicuo viridique ornatis differt. Type: Tanzania, Mpanda District: Mlala Hills, *Richards* 11587 (K!, holo.; BR!, iso.)

Perennial herb with several erect (or trailing in fruit) stems from woody rootstock; stems to 55 cm long, quadrangular, hirsute or sparsely so with long (to 2 mm) curly glossy non-glandular hairs (rarely glabrous). Leaves palmately veined; petiole 2–7(–12) mm long on vegetative leaves, indumentum as on stems; lamina ovate or narrowly so (rarely lanceolate or ovate-elliptic), largest 5.5–13 × 1.5–3 cm, 3–4(–7) times as long as wide (rarely down to 2 times), apex acute to acuminate, apiculate, base subcordate to cordate with rounded to hastate lobes, margin entire, sparsely pubescent or sparsely hirsute (rarely glabrous). Flowers solitary; pedicels 2–4(–9 in fruit) cm long, hirsute or sparsely so with long (to 2 mm) glossy glandular hairs (rarely without); bracteoles pale green, with conspicuous raised dark green reticulation, ovate-oblong or narrowly so, (1.7–)2–3 × 0.5–1(–1.3) cm, acute to acuminate, with dense long (to 2 mm) glossy glandular hairs (rarely few and short or with sparse puberulous indumentum intermixed). Calyx glabrous to minutely puberulous, 1–3(–5 in fruit) mm high of which the broadly triangular lobes about half. Corolla limb and upper part of tube deep bluish purple, lower part of tube white, throat yellow; tube 3–5 cm long; lobes 1.8–2.7 × 2.2–4 cm. Filaments 9–13 mm long, glabrous; anthers 2.5–3 mm long. Style with scattered glands. Capsule and seed not seen.

TANZANIA. Mpanda District: Katuma to Mwese, 2 Dec. 1956, *Richards* 7112! & Silkcub Highlands, 4 Dec. 1956, *Richards* 7134! & Mlala Hills, 27 Oct. 1959, *Richards* 11587!
DISTR. T 4; not known elsewhere
HAB. *Brachystegia-Julbernardia* and *Brachystegia-Uapaca* woodland on sandy to loamy soil; (800–)1000–1650 m

NOTE. Differs from the preceeding species and from *T. natalensis* in the long capitate glands on the pedicels and grows in much drier vegetation at lower altitude.
    Richards 16060 from Lake Katavi is considered to be an abnormal late season form of this species. The bracteoles are small (17 mm long) and lack capitate glands.

40. **Thunbergia mufindiensis** *Vollesen* **sp. nov.** a *T. natalensi* foliis angustioribus (1–3 cm nec (2.7–)3–6(–10.5) cm latis), lobis basalibus hastatis (nec rotundatis), petiolis brevioribus (1–3 mm nec 2–5 mm longis), corolla magis intense colorata (atrocoerulea vel atropurpurea nec alba neque coerulea malvinave), lobis corollae longioribus (1.5–2 cm nec 1–1.7 cm longis) et capsula rostro breviore (10–13 mm nec 13–16 mm longo) ornata differt. Type: Tanzania, Iringa District: Mufindi, Lake Ngwazi, *J.Lovett* 1247 (K!, holo.; EA!, MO!, iso.)

Perennial herb with 1–5 erect stems from large creeping rootstock; stems to 75 cm long, quadrangular, glabrous to puberulous or pubescent with curly non-glandular hairs, with distinct bands of longer hairs at nodes. Leaves dark green, glossy, palmately veined; petiole 1–3 mm long on vegetative leaves, indumentum as on stems; lamina ovate-lanceolate to ovate (rarely elliptic), largest 3–6.5 × 1–3 cm, usually more than twice as long as wide, apex acute to obtuse, apiculate, base truncate to cordate with hastate (rarely rounded) lobes, margin entire to irregularly dentate in basal half, glabrous to puberulous, densest on veins (rarely glabrous). Flowers solitary; pedicels 1.5–6 cm long, puberulous or sparsely so, sometimes only apically; bracteoles pale green, with conspicuous raised dark green reticulation, more rarely with obscure reticulation, ovate-oblong, 1.3–2.3 × 0.5–0.9 cm, acute to obtuse, with a mixture of glandular and non-glandular hairs along veins and margins, more rarely uniformly puberulous without glands. Calyx glabrous to minutely puberulous, 1–2(–4 in fruit) mm high of which the broadly triangular lobes about half. Corolla limb and upper part of tube deep blue to deep purple, lower part of tube white, throat yellow; tube 2.5–4 cm long; lobes 1.5–2 × 1.7–2.5 cm. Filaments 8–10 mm long, glabrous or glandular at base; anthers 2–2.5 mm long. Style with scattered glands. Immature capsule 6–8 mm in diameter, beak 10–13 mm long. Seed not seen.

TANZANIA. Iringa District: Sao Hill, Ipogoro to M'kawa, 12 Dec. 1961, *Richards* 15568! & 20 km on Mafinga–Madibira road, 25 Nov. 1986, *Brummitt et al.* 18163! & 40 km on Mafinga–Madibira road, 27 Jan. 1991, *Bidgood et al.* 1288!
DISTR. T 7; not known elsewhere
HAB. *Brachystegia* woodland, often on rocky hills, montane grassland, persisting along tracks in pine plantations; 1450–1950 m

Note. This species, which is endemic in Iringa District, differs from *T. natalensis* in its narrower leaves with hastate basal lobes and shorter petiole, its darker corolla with larger lobes and capsule with shorter beak. It also generally grows at lower altitudes.

41. **Thunbergia austromontana** *Vollesen* **sp. nov.** ab omnibus ceteris speciebus subgeneris *Hypenophorae* corolla parva citrina differt. Type: Tanzania, Njombe, *Richards* 7872 (K!, holo.)

Perennial herb with (?) solitary erect stem from creeping rootstock; stems to 35 cm long, quadrangular, puberulous in furrows and with distinct bands of long glossy hairs at nodes. Leaves dark green, glossy, palmately veined; petiole 1–2 mm long on vegetative leaves, puberulous in dorsal groove; lamina narrowly ovate-elliptic, largest ± 6 × 1.5 cm, apex subacute, apiculate, base truncate to subcordate with hastate lobes to 7 mm long, margin entire, minutely puberulous on edges, otherwise glabrous. Flowers solitary; pedicels 4–5 cm long, sparsely puberulous towards apex; bracteoles pale green, with conspicuous raised dark green reticulation, ovate-oblong, 1.5–2 × 0.4–0.6 cm, obtuse, with sparse non-glandular glossy hairs along veins and margins, densest towards base. Calyx glabrous to minutely puberulous, ± 1 mm high of which the broadly triangular lobes about half. Corolla uniformly lemon yellow; tube ± 2.2 cm long; lobes ± 1 × 1 cm. Filaments ± 4 mm long, glabrous or glandular at base; anthers ± 2 mm long. Style with scattered glands. Capsule and seed not seen.

Tanzania. Njombe District: near Njombe, 17 Jan. 1957, *Richards* 7872! & 24 Dec. 1965, *B.J. Harris* 10238!
Distr. **T** 7; not known elsewhere
Hab. Montane grassland; ± 2100 m

Note. Known only from these two collections. Superficially this is quite similar to *T. lathyroides*, and at one point the author considered that it might possibly be a hybrid between that species and either *T. natalensis* or *T. mufindiensis*. But it is difficult to imagine how a hybrid between two species with large purple corollas could produce offspring with a small lemon yellow corolla. The corolla size and colour immediately distinguishes this species from all others in the Subgenus.

42. **Thunbergia lathyroides** *Burkill* in F.T.A. 5: 24 (1899); De Wildeman, Etud. Fl. Katanga 1: 136 (1903) & Etud. Fl. Katanga 2: 144 (1913) & Not. Fl. Katanga 2: 70 (1913) & Not. Fl. Katanga 4: 75 (1914) & Contrib. Fl. Katanga: 193 (1921) & Suppl. 3: 140 (1930); Richards & Morony, Checklist Fl. Mbala & Distr.: 235 (1969); Lebrun & Stork, Enum. Pl. Afr. Trop. 4: 506 (1997). Types: Zambia, Fwambo, *Carson* 4 (K!, syn.) & *Nutt* s.n. (K!, syn.) & *Scott Elliot* 8269 (K!, syn.)

Perennial herb with 1–2 erect stems from woody rootstock with fleshy roots; stems to 75 cm long, quadrangular, glabrous. Leaves palmately veined, glaucous, held erect parallel to the stem; petiole 0–1 mm long, glabrous; lamina narrowly ovate-elliptic, largest 4–10 × 1.2–2.8(–3.5) cm, apex acute or subacute, apiculate, base cordate with sagittate lobes (or some rounded), margin entire, glabrous. Flowers solitary; pedicels 1–2.2(–3.2 in fruit) cm long, glabrous; bracteoles dark green, sometimes purplish tinged, with obscure reticulation, ovate-oblong or narrowly so, 1.8–3.3(–3.7) × 0.7–1.1(–1.6) cm, subacute or obtuse, glabrous. Calyx glabrous to minutely puberulous, 1–2(–4 in fruit) mm high of which the broadly triangular lobes about half. Corolla limb and upper part of tube deep purple or deep bluish purple, lower part of tube white, throat yellow; tube 3–4 cm long; lobes 1.5–2 × 1.8–2.2 cm. Filaments 10–12 mm long, glabrous; anthers 2–2.5 mm long. Style glabrous. Capsule 8–10 mm in diameter, beak 12–15 mm long. Seed reddish brown, ± 7 mm in diameter, slightly reticulate on back to spinulose along edges.

TANZANIA. Ufipa District: Lake Kwela, 4 Nov. 1956, *Richards* 6866!; Mbeya District: Mbosi Circle, Mchembo Estate, 11 Jan. 1961, *Richards* 13846a! & 5 km on Tunduma–Mbeya road, 15 Dec. 1962, *Richards* 17069!

DISTR. **T** 4, 7; Congo-Kinshasa, Zambia, Malawi

HAB. *Brachystegia-Julbernardia* woodland, grassland, roadsides, often on heavy loamy soil; (1000–)1350–1800 m

SYN. *Thunbergia rumicifolia* Lindau in E.J. 43: 349 (1909); Lebrun & Stork, Enum. Pl. Afr. Trop. 4: 507 (1997). Type: Tanzania, Ufipa District: Kasanga [Bismarcksburg], Mtamba River, *von Wangenheim* 15 (B†, holo.)

43. **Thunbergia stellarioides** *Burkill* in F.T.A. 5: 26 (1899); Richards & Morony, Checklist Fl. Mbala & Distr.: 236 (1969); Lebrun & Stork, Enum. Pl. Afr. Trop. 4: 507 (1997). Types: Zambia, Fwambo, *Carson* 85, 92 & 101 (all K!, syn.), Stevenson Road, *Scott Elliot* 8298bis (K!, syn.)

Perennial herb with several slender erect greyish stems from woody rootstock with thick fleshy roots; stems to 50 cm long, 1–2 mm in diameter at base, quadrangular, glabrous to sparsely puberulous and with thin transverse lines at nodes. Leaves greyish, with central vein and two strongly developed rib-like lateral veins running almost whole length of lamina; petiole 0–1 mm long, glabrous to sparsely puberulous; lamina linear, largest 4–10 × 0.1–0.5 cm, apex acute, base attenuate, margin entire, glabrous to sparsely puberulous along veins. Flowers solitary; pedicels 1.5–4.5 cm long, glabrous to puberulous, densest apically; bracteoles green with purple tinge and with purple main veins and margins, with obscure reticulation, narrowly ovate-elliptic, 1.2–2.2 × 0.3–0.5 cm (to 2.4 × 0.9 cm in fruit), acute to obtuse, puberulous or sparsely so, no capitate glands. Calyx glabrous to minutely puberulous, 1–2 mm high of which the broadly triangular lobes about half. Corolla white (rarely very pale blue), sometimes with pale yellow throat; tube 1.8–2.5 cm long and 0.6–0.9 cm in diameter at mouth; lobes 1–1.5 × 1.3–1.8 cm. Filaments 4–6 mm long, glabrous; anthers 1.5–2 mm long. Style with long (to 1 mm) stiff non-glandular hairs near apex. Immature capsule ± 8 mm in diameter, beak ± 12 mm long. Seed not seen.

TANZANIA. Mpanda District: Uzondo Plateau, 15 Apr. 2006, *Bidgood et al.* 5492!; Chunya District: Mbogo, 17 Oct. 1932, *Geilinger* 3086!; Mbeya District: Piseki Village, 1899, *Goetze* 1425!

DISTR. **T** 4, 7; Congo-Kinshasa, Zambia

HAB. Seasonally wet short seepage grassland on sandy to peaty soils, seepage areas at edges of dambos; 1400–1550 m

SYN. *Thunbergia argentea* Lindau in E.J. 30: 407 (1901) & in Wiss. Ergebn. Schwed. Rhod.-Congo-Exp. 1. Bot.: 304 (1916); T.T.C.L. 2: 18 (1949); Lebrun & Stork, Enum. Pl. Afr. Trop. 4: 505 (1997). Type: Tanzania, Mbeya District: Piseki Village, *Goetze* 1425 (B†, holo.; BR!, iso.)

44. **Thunbergia graminifolia** *De Wild.* in Ann. Mus. Congo, Bot., Ser. 4, 1: 134 (1903) & Contrib. Fl. Katanga: 193 (1921) & Suppl. 3: 135 & 137 (1930) & Suppl. 4: 91 (1932); Lebrun & Stork, Enum. Pl. Afr. Trop. 4: 506 (1997). Type: Congo-Kinshasa, Katanga, *Verdick* s.n. (BR!, holo.)

Perennial herb with several stout erect green stems from large woody rootstock; stems to 60 cm long, 2–4 mm in diameter at base, subquadrangular, glabrous, sometimes sparsely puberulous in thin transverse lines at nodes. Leaves green, with central vein and two (rarely four) strongly developed rib-like lateral veins running almost whole length of lamina, more rarely only $^1/_2$–$^3/_4$ up; petiole 0–1 mm long, glabrous; lamina linear to linear-lanceolate (rarely narrowly obovate), largest (3–)6–16.5 × 0.2–0.7(–1.3) cm, usually more than 10 times as long as wide, apex acuminate (rarely acute), base attenuate, margin entire, glabrous. Flowers solitary; pedicels 1.5–4(–6 in fruit) cm long, glabrous; bracteoles green or dark green or tinged dull purple, with obscure reticulation, ovate-elliptic or narrowly so, 2.3–3.8 × 0.6–1.2 cm, acute, glabrous. Calyx minutely puberulous, 2–3(–5 in fruit) mm high of which the broadly triangular lobes about half. Corolla limb and upper part of tube pale blue to blue or pale mauve, lower part of tube white to pale yellow, throat yellow; tube 3–4.5 cm long and 1.2–1.7 cm in diameter at mouth; lobes 1.5–2.5 × 1.5–2.5 cm. Filaments 10–12 mm long, glabrous; anthers 2–2.5 mm long. Style glabrous. Capsule 12–15 mm in diameter, beak 18–25 mm long. Seed chestnut brown, 10–12 mm in diameter, smooth on back, with a few spinules near apex and with an entire or laciniate lateral wing.

TANZANIA. Mpanda District: Mahali Mts, Itemba, 1 Oct. 1958, *Newbould & Jefford* 2827!; Mbeya District: 3 km on Tunduma-Mbeya road, 8 Nov. 1950, *Richards* 13538! & 30 km N of Tunduma, Mbozi Plateau, 17 Nov. 1958, *Napper* 982!
DISTR. T 4, 7; Congo-Kinshasa, Zambia, Malawi
HAB. *Brachystegia-Julbernardia*, *Brachystegia-Isoberlinia*, *Brachystegia-Uapaca* woodland, grassland (? secondary), *Protea* bushland, on yellowish to reddish sandy to gravelly or stony soil in areas subject to regular burning; 900–1700 m

SYN. *Thunbergia stellarioides* Burkill var. *graminea* Burkill in F.T.A. 5: 27 (1899). Type: Zambia, Stevenson Road, *Scott Elliot* 8392 (K!, holo.; BM!, iso.)
   T. oblongifolia Oliv. var. *glaberrima* Burkill in F.T.A. 5: 25 (1899), quoad *Thomson* s.n. (K!, syn.)
   [*T. lamellata* sensu Lindau in E.J. 30: 408 (1901), *non* Hiern (1900)]
   T. collina S.Moore in J.B. 51: 186 (1913); Lebrun & Stork, Enum. Pl. Afr. Trop. 4: 505 (1997). Type: Zambia, Broken Hill, *Rogers* 8642 (BM!, holo.)
   T. glaucina S.Moore in J.B. 51: 187 (1913); De Wildeman, Contrib. Fl. Katanga: 192 (1921); Lebrun & Stork, Enum. Pl. Afr. Trop. 4: 506 (1997). Type: Congo-Kinshasa, Katanga, Lubumbashi [Elisabethville], *Rogers* 10320 pro parte (BM!, holo.; K!, iso.)
   T. glaucina S.Moore var. *latifolia* S.Moore in J.B. 51: 187 (1913). Type: Congo-Kinshasa, Katanga, Lubumbashi [Elisabethville], *Rogers* 10320 pro parte (BM!, holo.)
   T. fasciculata De Wild. in F.R. 13: 105 (1914) & Not. Fl. Katanga 4: 73 (1914) & Contrib. Fl. Katanga: 192 (1921), *nom. illeg.*, *non* Lindau (1893). Type: Congo-Kinshasa, Katanga, Katentania, *Homblé* 826 (BR!, holo.; BR!, iso.)
   T. stenophylla Lindau in Wiss. Ergebn. Schwed. Rhod.Congo-Exp. 1911–12, Bot. 1: 304 (1916); Lebrun & Stork, Enum. Pl. Afr. Trop. 4: 507 (1997), *nom. illeg.*, *non* C.B.Clarke (1901). Type: Zambia, Luera River, *R.E. Fries* 588 (UPS, holo.)
   T. katentaniensis De Wild. in Pl. Beq. 1: 450 (1922); Lebrun & Stork, Enum. Pl. Afr. Trop. 4: 507 (1997). Type: as for *T. fasciculata*.
   T. trinervis S.Moore in J.B. 67: 227 (1929); Lebrun & Stork, Enum. Pl. Afr. Trop. 4: 507 (1997). Type: Congo-Kinshasa, Katanga, Lubumbashi [Elisabethville], *Rogers* 10364 (BM!, holo.; K!, iso.)
   T. trinervis S.Moore var. *angustifolia* S.Moore in J.B. 67: 228 (1929). Type: Congo-Kinshasa, Katanga, Kambwe Mine, *Burtt Davy* 18029 (BM!, holo.)

45. **Thunbergia lancifolia** *T.Anderson* in J.L.S. 7: 19 (1863); S. Moore in J.B. 18: 195 (1880); Engler, Hochgebirgsfl. Trop. Afr.: 387 (1892); Lindau in E.J. 17, Beibl. 41: 35 (1893) & in P.O.A. C: 366 (18195); Burkill in F.T.A. 5: 25 (1899); Hiern, Cat. Welw. Afr. Pl. 4: 804 (1900); Lindau in E.J. 30: 407 (1901) & in Wiss. Ergebn. Schwed. Rhod.-Congo-Exp. Bot.: 303 (1916); De Wildeman, Contrib. Fl. Katanga: 193 (1921) & Suppl. 3: 139 (1930); Binns, Checklist Herb. Fl. Malawi: 16 (1968); Moriarty, Wild Fl. Malawi: 85 (1975); Vollesen in Opera Bot. 59: 81 (1980); Cribb & Leedal, Mountain Fl. S. Tanz.: 129, pl. 32 (1982); Ruffo *et al.*, Cat. Lushoto Herb. Tanz.: 11 (1996) pro parte; Lebrun & Stork, Enum. Pl. Afr. Trop. 4: 506 (1997). Type: Malawi, Tshinsunze, *Kirk* s.n. (K!, holo.)

Perennial herb with 1-several stout erect stems from large woody rootstock; stems to 0.7(–1) m long, 2–9 mm in diameter at base, subquadrangular, glabrous or finely puberulous at and below nodes and often also with band of longer setose hairs. Leaves penninerved or with two weak lateral veins running less than half way up; petiole 0–3 mm long, glabrous; lamina linear-lanceolate to elliptic or obovate, largest

5–13.5(–18.5) × (0.3–)0.6–5.5(–6.7) cm, apex acuminate to obtuse, apiculate, base attenuate to cuneate or subauriculate, margin entire, glabrous or with minute bulbous-based upwardly directed hairs, especially along margin (rough to the touch). Flowers solitary; pedicels 1–3(–6 in fruit) cm long, with sparse to dense capitate glands, more rarely glabrous; bracteoles pale green to dark green or tinged purple, with obscure reticulation, ovate to ovate-elliptic or ovate-oblong, 1.8–3.2 × 0.6–1.4 cm, acute to obtuse, with sparse to dense short (to 0.5(–1) mm long) capitate glands (rarely glabrous or with glands confined to main veins). Calyx glabrous to finely puberulous, 1–3(–6 in fruit) mm high of which the broadly triangular lobes about half. Corolla limb and upper part of tube dark purple to deep maroon (also described as burgundy or deep wine red), lower part of tube white to pale yellow, throat yellow; tube 3–4.5 cm long and 1–2 cm in diameter at mouth; lobes 1–2 × 1.5–2.5 cm. Filaments 9–13 mm long, glabrous; anthers 2–2.5 mm long. Style with scattered capitate glands. Capsule 10–12 mm in diameter, beak 17–22 mm long. Seed brown, 6–9 mm in diameter, smooth to slightly pitted on back and with an entire or laciniate lateral wing. Fig. 10, p. 72.

TANZANIA. Ufipa District: Mbisi Forest Reserve, 25 Nov. 1994, *Goyder et al.* 3820!; Mbeya District: Mbosi Circle, Mchembo Estate, 11 Jan. 1961, *Richards* 13839!; Masasi District: Ndanda Mission, 1 Feb. 1991, *Bidgood et al.* 1334!

DISTR. **T** 1, 4, 6–8; Congo-Kinshasa, Burundi, Angola, Zambia, Malawi, Mozambique

HAB. Montane grassland, *Brachystegia* and *Acacia* woodland, often in areas subject to regular burning, persisting in degraded woodland and in plantations, in sandy to stony or loamy soil; 200–2250 m

SYN. *Thunbergia lancifolia* T.Anderson var. *laevis* S.Moore in J.B. 18: 195 (1880); Vollesen in Opera Bot. 59: 82 (1980). Types: Angola, Lobati Quilambo, *Welwitsch* 5110 (BM!, syn.; BM!, K!, iso.); Izanga, *Welwitsch* 5161 (BM!, syn.; BM!, K!, iso.)

[*T. oblongifolia* sensu Burkill in F.T.A. 5: 25 (1899) quoad *Cameron* s.n., *non* Oliv. (1873)]

T. glaberrima Lindau in E.J. 33: 185 (1902); Lebrun & Stork, Enum. Pl. Afr. Trop. 4: 506 (1997). Types: Tanzania, Songea District: Ungoni, Kwa Kihingi, *Busse* 763 (B†, syn.; EA!, iso.) & Songea, *Busse* 1324 (B†, holo.; EA!, iso.)

T. lancifolia T.Anderson var. *rhodesica* Turrill in K.B. 1912: 362 (1912). Types: Zambia, Broken Hill, *Rogers* 8540 (K!, syn.); between Broken Hill and Bwana Mkubwa, *Allen* 298 & 320 (both K!, syn.)

T. puberula Lindau in E.J. 49: 399 (1913); Lebrun & Stork, Enum. Pl. Afr. Trop. 4: 506 (1997). Types: Rwanda, Russiga Mts, *H. Meyer* 1003 (B†, syn.); Tanzania, Mwanza District: Usambiro, *H. Meyer* 1113 (B†, syn.)

T. valida S.Moore in J.B. 51: 210 (1913); Lebrun & Stork, Enum. Pl. Afr. Trop. 4: 507 (1997). Type: Zambia, Malangushi River, *Kässner* 2038a (BM!, holo.)

T. hockii De Wild. in F.R. 11: 545 (1913) & in Ann. Mus. Congo, Bot., Ser. 4, 2: 142 (1913) & Contrib. Fl. Katanga: 193 (1921) & Suppl. 3: 137 (1930); Lebrun & Stork, Enum. Pl. Afr. Trop. 4: 506 (1997). Type: Congo-Kinshasa, Katanga, Lubumbashi [Elisabethville], *Hock* s.n. (BR!, holo.)

T. bequaertii De Wild. in F.R. 13: 104 (1914) & Not. Fl. Katanga 4: 72 (1914) & Contrib. Fl. Katanga: 192 (1921); Lebrun & Stork, Enum. Pl. Afr. Trop. 4: 505 (1997). Type: Congo-Kinshasa, Katanga, Sankisia, *Bequaert* 189 (BR!, holo.)

T. friesii Lindau in Wiss. Ergebn. Schwed. Rhod.-Congo-Exp. 1911–12, Bot. 1: 303 (1916); Lebrun & Stork, Enum. Pl. Afr. Trop. 4: 506 (1997). Type: Zambia, Broken Hill, *R.E. Fries* 319 (UPS, holo.)

T. elskensii De Wild., Pl. Beq. 4: 422 (1922); Lebrun & Stork, Enum. Pl. Afr. Trop. 4: 506 (1997). Type: Burundi, Lake Kanzigi, Mt Sure, *Elskens* 34 (BR!, holo.)

T. sp. of Bolnick, Common Wild Fl. Zambia: 40, pl. 18 (1995)

NOTE. Variable especially in the shape and size of the leaves, but not unusually variable for such a widespread *Brachystegia* woodland species. It is usually easily distinguished by the combination of the very dark corolla and the dense short glands on the bracteoles.

Two collections (*Bidgood et al.* 1334 and *Vollesen* MRC4856) both from **T** 8 have very narrow leaves and long (to 1 mm) glands on the bracteoles. Narrow leaves also occur in SW Tanzania (e.g. *Brummitt* 17968), but the long glands are unique. The picture in **T** 8 is muddled further by the occurrence of forms with glabrous bracteoles, but glabrous forms also occur in SW Tanzania and are common in Zambia. When more material becomes available the form with long glands may be found worthy of taxonomic recognition.

FIG. 10. *THUNBERGIA LANCIFOLIA* — **1**, flowering branch, × 1; **2**, rootstock, × 1; **3**, variation in leaf shape, × 1; **4**, basal part of bracteole, × 6; **5**, calyx, × 6; **6**, stamens, × 6; **7**, stigma, × 6; **8**, capsule valve, × 3; **9**, seed, inner and outer view, × 3. 1 from *Richards* 12859, 2 from *Goyder* 3820, 3 (broad leaf) from *Richards* 13839, 3 (narrow leaf) from *Vollesen* MRC4856, 4–7 from *Richards* 13231, 8–9 from *Drummond* 6188. Drawn by Margaret Tebbs.

46. **Thunbergia oblongifolia** *Oliv.* in Trans. Linn. Soc. 29: 125, t.123 (1873); Lindau in P.O.A. C: 366 (1895); Burkill in F.T.A. 5: 24 (1899); De Wildeman, Not. Fl. Katanga 4: 75 (1914) & Contrib. Fl. Katanga: 193 (1930); Ruffo *et al.*, Cat. Lushoto Herb. Tanz.: 11 (1996); Lebrun & Stork, Enum. Pl. Afr. Trop. 4: 506 (1997). Type: Tanzania, Mpwapwa District: Usagara Mts, Robeho, *Speke & Grant* s.n. (K!, holo.)

Perennial herb with 1-several stout erect stems from large woody rootstock, sometimes forming large clumps; stems to 0.7(–1) m long, 2–8 mm in diameter at base, subquadrangular, glabrous to puberulous or pubescent with curly glossy hairs, if glabrous with band of setose glossy hairs at nodes. Leaves penninerved or with two weak lateral veins running less than half way up; petiole 0–1 mm long, glabrous to puberulous or pubescent; lamina narrowly elliptic to elliptic or narrowly obovate to obovate, largest 3–10 × 0.7–4 cm, apex acuminate to obtuse, apiculate, base attenuate to subauriculate, margin entire, glabrous to puberulous, distinctly ciliate with long curly glossy hairs at least when young. Flowers solitary; pedicels 0.7–2.5(–4.5 in fruit) cm long, glabrous to puberulous or pubescent, rarely with intermixed capitate glands or of glands only; bracteoles pale green to dark green or tinged purple, with obscure (rarely conspicuous) reticulation, ovate to elliptic or oblong, 1.5–3.5(–4) × 0.5–1.7 cm, acute to obtuse, with sparse to dense (to 1 mm long) capitate glands, usually with intermixed non-glandular hairs or with non-glandular hairs only or glabrous but for distinctly ciliate margin. Calyx glabrous to finely puberulous, 1–3(–5 in fruit) mm high of which the broadly triangular lobes about half. Corolla limb and upper part of tube white, pale blue to blue or mauve, lower part of tube white to pale yellow, throat yellow; tube 2–4.5(–5.5) cm long and 0.7–2 cm in diameter at mouth; lobes 1–2.2(–2.8) × 1.2–2.8 cm. Filaments 8–13 mm long, glabrous; anthers 2–2.5 mm long. Style with scattered capitate glands. Capsule 10–13(–15) mm in diameter, beak 17–24(–28) mm, long. Seed brown, 6–9 mm, in diameter, slightly reticulate on back and with a broad entire or laciniate lateral wing.

TANZANIA. Mpanda District: Inyonga road, Rungwe Village, 25 Oct. 1960, *Richards* 13384!; Ufipa District: Mbisi Mts, 23 Nov. 1949, *Bullock* 1927!; Iringa District: Madibira road, 15 Nov. 1966, *Richards* 21595!; Songea District: 19 km W of Songea, Likuyu River, 30 Dec. 1955, *Milne-Redhead & Taylor* 7952!
DISTR. **T** 1, 2, 4–8; Congo-Kinshasa, Burundi, Zambia, Zimbabwe
HAB. *Brachystegia-Julbernardia* woodland, montane grassland, appearing shortly after burning, on sandy to stony or loamy soil or on rocky slopes; 750–2250 m

SYN. *Thunbergia oblongifolia* Oliv. var. *glaberrima* Burkill in F.T.A. 5: 25 (1899). Type: Zambia, *Carson* 87 (K!, syn.)
  T. manikensis De Wild. in F.R. 11: 546 (1913) & Contrib. Fl. Katanga: 193 (1921); Lebrun & Stork, Enum. Pl. Afr. Trop. 4: 506 (1997). Type: Congo-Kinshasa, Katanga, Manika Plateau, *Hock* s.n. (BR!, holo.)

NOTE. Included here is an array of specimens varying from almost glabrous to densely hairy. What keeps them together and separate from *T. lancifolia* is a pale corolla combined with distinctly ciliate leaves. Glabrous specimens are also quite similar to *T. graminifolia* but always have a distinct line of glossy setose hairs at the nodes and lack the very conspicuous rib-like veins of that species.

47. **Thunbergia barbata** *Vollesen* **sp. nov.** a *T. lancifolia* et *T. oblongifolia* nodis atque pedicellis bracteolisque pilos longos nitentes multicellulares gerentibus differt. A *T. lancifolia* etiam corolla malvina differt. Inter species affines insignis propter flores in tempo pluviali praesentes. Type: Tanzania, Morogoro District: Mlali, *E.A. Bruce* 882 (K!, holo.; K!, iso.)

Perennial herb with erect stems from woody rootstock; stems to 0.5 m long, ± 4 mm in diameter at base, subquadrangular, glabrous to pilose and with thick bands of long (to 2 mm) glossy hairs at nodes. Leaves penninerved; petiole 0–1 mm long, glabrous to pilose; lamina narrowly oblong-obovate, largest 11–15 × 1.8–4 cm, apex acuminate to subacute, apiculate, base cuneate, margin entire, with long glossy hairs along midrib and scattered on lamina, distinctly ciliate with similar hairs. Flowers solitary; pedicels 1–4 cm long, with long (to 2 mm) glossy setose hairs, densest near apex; bracteoles dark green, with obscure reticulation, ovate-oblong, 2.5–3.3 × 0.6–0.9 cm, subacute, with long (to 2 mm) glossy setose non-glandular hairs, densest at base (almost looking like a beard) and on edges and midrib. Calyx finely puberulous, ± 1(–4 in fruit) mm high of which the broadly triangular lobes about half. Corolla limb and upper part of tube mauve, lower part of tube white, throat yellow; tube ± 4 cm long and 1.5–2 cm in diameter at mouth; lobes ± 1.5 × 2 cm. Flower not dissected. Immature capsule ± 10 mm in diameter, beak ± 20 mm long. Seed not seen.

TANZANIA. Morogoro District: Uluguru Mts, Mlali, 11 March 1935, *Bruce* 882!; Ulanga District: Selous Game Reserve, Msolwa Camp, 8 Feb. 1977, *Vollesen* MRC4445!
DISTR. **T** 6; not known elsewhere
HAB. *Brachystegia* woodland on sandy to loamy soil, persisting in cultivated areas; 250–700 m

SYN. [*Thunbergia lancifolia* T. Anderson var. *ciliata* (De Wild.) Napper sensu Vollesen in Opera Bot. 59: 81 (1980), *non. T. ciliata* De Wild. (1913)]

NOTE. Close to *T. lancifolia* from which it differs most conspicuously in the long glossy hairs on the bracteoles and the mauve corolla. It shares the corolla colour with *T. oblongifolia*, but again the bracteole-indumentum is wrong. It differs from both these species by flowering in the rainy season rather than after burning in the dry season.

48. **Thunbergia cycnium** *S.Moore* in J.B. 18: 194 (1880); Lindau in E. & P. Pf. IV, 3b: 292 (1895); Burkill in F.T.A. 5: 17 (1899); Hiern, Cat. Afr. Pl. Welw. 4: 803 (1900); Lebrun & Stork, Enum. Pl. Afr. Trop. 4: 506 (1997). Type: Angola, Huila, *Welwitsch* 5009 (BM!, holo.; C!, K!, iso.)

Perennial herb with several erect or ascending stems from woody rootstock with fleshy roots; stems to 50 cm long, pale yellowish to whitish pilose or densely so. Leaves rough; petiole 0.5–4(–6) mm long, indumentum as stems; lamina narrowly to broadly ovate or elliptic or round, largest 4–7.5 × 2.2–4.5 cm, apex subacute to broadly rounded, base cuneate to subcordate, margin entire to irregularly dentate, with sparse to dense long pilose hairs, below densest on veins, above uniformly so, and with shorter appressed (rarely erect) hairs. Flowers axillary, solitary or 2 per exil; pedicels 2–6 mm long, indumentum as stems; bracteoles green, ovate to oblong or narrowly so, (1.5–)1.8–3 × 0.3–0.8 cm, acute to obtuse, sparsely to densely pilose and with short appressed hairs. Calyx with sparse subsessile capitate glands, rim 1.5–2.5 mm high, segments 4–6 mm long. Corolla pure white; tube cylindric, straight, 5–7 cm long and ± 5 mm in diameter at throat; lobes 1.2–2.5 × 1.5–2 cm. Filaments ± 7 and 10 mm long, glabrous; anthers ± 4 mm long, apiculus ± 1 mm long. Capsule 8–10 mm in diameter, beak 11–15 mm long. Seed dark brown, ± 6 mm in diameter, with large lamellate-pectinate scales along edges becoming smaller towards centre.

UGANDA. Teso District: Serere, Feb. 1933, *Chandler* 1072! & Soroti, 15 Sep. 1954, *Lind* 335!
KENYA. Masai District: Masai Mara Game Reserve, 27 Feb. 1972, *Taiti* s.n.!
TANZANIA. Musoma District: Serengeti National Park, Kleins Camp to Wogakuria Hill, 30 Dec. 1964, *Greenway & Turner* 11797!; Kilosa District: Mikumi National Park, 25 km on Mikumi–Morogoro road, 8 Jan. 1975, *Brummitt & Polhill* 13604!; Chunya District: 150 km N of Mbeya, Lupa North Forest Reserve, 28 Dec. 1962, *Boaler* 787!
DISTR. **U** 3; **K** 6; **T** 1, 4–7; Central African Republic, Sudan, Angola, Zambia
HAB. *Brachystegia* woodland, *Acacia-Combretum* wooded grassland, grassland, in areas subject to regular burning; (550–)1100–1700 m

49. **Thunbergia schimbensis** *S.Moore* in J.B. 40: 342 (1902); Vollesen in Opera Bot. 59: 82 (1980); Champluvier in Fl. Rwanda 3: 490, fig. 149, 2 (1985); Lebrun & Stork, Enum. Pl. Afr. Trop. 4: 507 (1997). Type: Kenya, Kwale District: Shimba Hills, *Kässner* 174 (BM!, holo.)

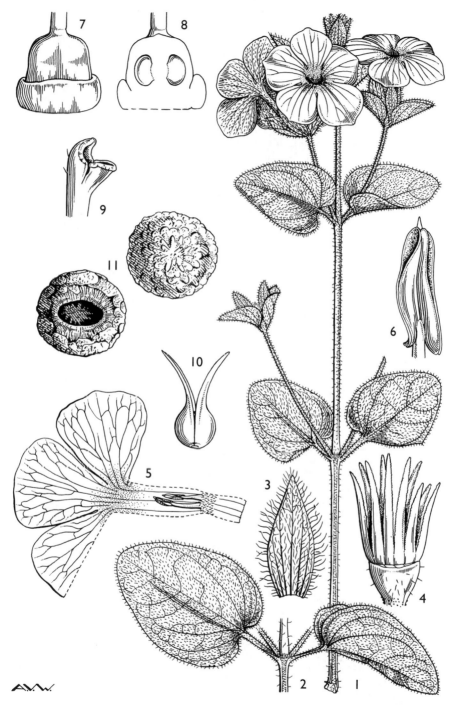

FIG. 11. *THUNBERGIA SCHIMBENSIS* — **1**, flowering branch, × 1; **2**, leaves, × 1; **3**, bracteole, × 2; **4**, calyx, × 6; **5**, section of corolla with stamens, × 2; **6**, anther, × 12; **7**, ovary and disc × 12; **8**, idem, section, × 12; **9**, stigma, × 12; **10**, capsule, × 2; **11**, seed, inner and outer view, × 6. 1–2 from *Faulkner* 1112, 3–9 from *Faulkner* 1222, 10–11 from *Drummond & Hemsley* 1136. Drawn by Ann Webster.

Perennial herb with several erect, ascending, trailing or climbing stems from creeping rootstock with fleshy roots; stems to 50 cm long, sparsely to densely pilose (rarely pubescent). Leaves with petiole (2–)4–25 mm long, indumentum as stems; lamina ovate to cordiform or broadly so, largest 3–8(–11) × 1.3–6 cm, apex subacute to retuse, base subcordate to cordate, margin entire to irregularly crenate-dentate (rarely with two small basal lobes), with sparse to dense long pilose hairs, below densest on veins, above uniformly so, and with (or without below) shorter appressed to erect hairs. Flowers axillary, solitary (rarely paired or in 3's or in 2-flowered cymes); pedicels 2–9(–11) mm long, indumentum as stems; bracteoles green, ovate to oblong or narrowly so, 1.1–2(–2.3) × 0.3–0.8 cm, acute or subacute, sparsely to densely pilose and with short appressed to erect hairs, distinctly pilose-ciliate. Calyx with subsessile capitate glands, rim 1–2.5 mm high, segments 4–6 mm long. Corolla pure white to pale mauve; tube cylindric, straight, 1.5–2.5 cm long, 2–3 mm in diameter at throat; lobes 1–2 × 1–2 cm. Filaments ± 6 and 9 mm long, glabrous; anthers ± 2.5 mm long, apiculus ± 0.5 mm long. Capsule 6–8(–9) mm in diameter, beak 8–12 mm long. Seed chestnut brown, 4–5(–6) mm in diameter, with large lamellate-pectinate scales along edges becoming smaller towards centre. Fig. 11, p. 75.

KENYA. Kwale District: Samburu to Mackinnon Road, 1 Sep. 1953, *Drummond & Hemsley* 4085! & S of Mrima Hill, 8 Sep. 1957, *Verdcourt* 1934!; Teita District: Taita Hills, Mbololo Hill, no date, *Faden et al.* 71/988!
TANZANIA. Lushoto District: Magunga Estate, 15 Sep. 1953, *Faulkner* 1222!; Uzaramo District: Kibamba, Oct. 1977, *Mwasumbi* 11453!; Chunya District: near Mbangala Village, 13 Feb. 1994, *Bidgood et al.* 2226!
DISTR. **K** 7; **T** 3, 6–8; Rwanda, Congo-Kinshasa, Zambia, Malawi, Mozambique, Zimbabwe
HAB. *Acacia-Commiphora*, *Combretum-Terminalia* and *Brachystegia* woodland, wooded grassland and bushland, grassland, glades in coastal forest; 25–900(–1050) m

SYN. [*Thunbergia sessilis* sensu Burkill in F.T.A. 5: 21 (1899), quoad *Buchanan* 745 and *Scott* s.n., *non* Lindau (1893)]

7. MENDONCIA

Vandelli, Fl. Lusit. & Bras.: 43, t. 3, fig. 22 (1788); Nees in DC., Prodr. 11: 50 (1847); G.P. 2: 1072 (1876); Benoist in Bull. Mus. Hist. Nat. 31: 386 (1925) & in Notul. Syst. 11: 139 (1944); Schönenberger & Endress in Int. J. Plant Sci. 159: 446 (1998)

Monachochlamys Baker in J.L.S. 20: 217, t. 26 (1883); Baill., Hist. Pl. 10: 424 (1890); Lindau in Nat. Pflanzenfam, IV, 3b: 291 (1895); S. Moore in J.B. 67: 225 (1929)

Afromendoncia Gilg in E.J. 17: 111 (1893) & in Ber. Deutsch. Bot. Ges. 11: 351 (1893); Lindau in E. & P. Pf. IV, 3b: 291 (1895); Burkill in F.T.A. 5: 6 (1899); Benoist in Notul. Syst. 2: 285 (1911); Turrill in K.B. 1919: 407 (1919)

Erect or scandent shrubs or woody twiners; cystoliths absent. Leaves opposite, entire. Flowers solitary or in axillary fascicles or on old branches; bracteoles 2, large, bract-like. Calyx a narrow undulate rim or shallowly 5-lobed. Corolla contorted in bud, tubular, subcylindric below, widening above with 5 small rounded lobes, all spreading or two upper broader and reflexed. Stamens 4, didynamous, all included or 2 longer exserted; anthers 2-thecous, thecae slightly unequal, slightly diverging below, apiculate at apex, opening by short apical slits. Disc conspicuous, annular, fleshy. Ovary becoming 1-locular (of 2 fused locules), 2-ovulate; ovules ascending from the base but affixed by ventral suture nearly to apex; stigma funnel-shaped, slightly 2-lobed. Fruit an indehiscent 1–2-seeded drupe; retinacula absent. Seed with thick stony exocarp.

60–70 species in tropical America, 4 species in tropical Africa and 3 species in Madagascar.

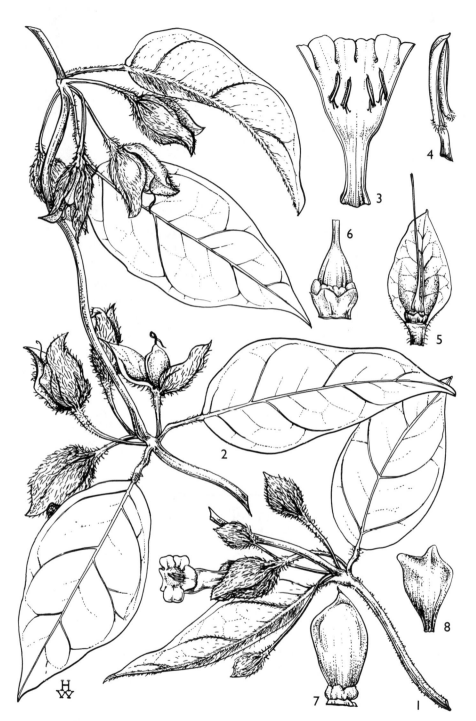

FIG. 12. *MENDONCIA GILGIANA* — **1**, flowering stem, × 1; **2**, fruiting stem, × 1; **3**, corolla opened up with stamens, × 1.5; **4**, anther, × 5; **5**, bracteole, ovary and style, × 1.5; **6**, calyx, disc and ovary, × 5; **7**, fruit, × 2; **8**, seed, × 2. 1 & 3–6 from *E. Brown* 314, 2 & 7–8 from *Loveridge* 106. Drawn by Heather Wood.

Mendoncia gilgiana (*Lindau*) *Benoist* in Bull. Soc. Bot. France 85: 679 (1939) & in Notul. Syst. 11: 143 (1944); Heine in F.W.T.A. (ed. 2) 2: 403 (1963) & in Fl. Gabon 13: 70, pl. 14 (1966); Lebrun & Stork, Enum. Pl. Afr. Trop. 4: 494 (1997); Friis & Vollesen in Biol. Skr. 51(2): 450 (2005). Types: Cameroon, Barombi, *Preuss* 481 (B†, syn.; BM!, K!, P!, iso.); Congo-Kinshasa, Ituri, *Stuhlmann* 2690 (B†, syn.).

Slender woody twiner to 10 m; young stems terete or quadrangular, tawny pubescent to sericeous or sparsely so, densest on two sides and here also with shorter curved hairs and sessile glands, glabrescent. Petiole 1–4.5 cm long, tawny pubescent to pilose or sericeous, densest apically; lamina elliptic to slightly obovate, largest 5–10 × 3–6 cm, apex acute to acuminate, base truncate to cordate, above glabrous or with scattered long tawny hairs, pubescent along midrib, below denser and with long dense tawny hairs along midrib and major lateral veins. Flowers solitary or in 2–4-flowered axillary fascicles on young stems, often with additional non-developing buds; pedicels 1–2.5 cm long, quadrangular, widening apically, pilose with tawny hairs, densest apically; bracteoles white or flushed mauve, ovate, 1.2–2.1 × 0.8–1.2 cm, acute to acuminate, pilose with tawny hairs, with sessile glands on the inside. Calyx a glabrous undulate or slightly lobed rim 0.5–1 mm high. Disc slightly lobed, up to 1.5 mm high. Corolla white to mauve; tube ± 2 cm long, ± 2 mm in diameter at base and ± 6 mm at apex, glabrous, on the inside with a band of glandular hairs; limb slightly zygomorphic, lower lobes ± 5 × 5 mm, upper ± 5 × 7 mm. Stamens included in tube, anterior filaments ± 4 mm long, posterior ± 2 mm long; anthers ± 4 mm long, with tufts of hairs at base. Ovary ± 2 mm long, glabrous; style ± 2 cm long; stigma slightly 2-lobed. Drupe bright green, 7–11 × 6–10 mm, obovoid, zygomorphic (produced to one side and angular apically), glabrous, hardly fleshy. Seed 6–10 × 5–9 mm. Fig. 12, p. 77.

UGANDA. Bunyoro District: Budongo Forest, 24 July 1971, *Synnott* 633!; Mengo District: Entebbe, Aug. 1905, *E. Brown* 314! & Kajanzi Forest Reserve, Sep. 1937, *Chandler* 1898!
KENYA. North Kavirondo District: Yala River Forest, Quarry Hill, 26 Jan. 1982, *M.G. Gilbert* 6886!
TANZANIA. Bukoba District: Minziro Forest Reserve, Nyakabanga, Kagera River, 13 Aug. 1999, *Festo et al.* 266! & Minziro Forest Reserve, 5 July 2000, *Bidgood et al.* 4858! & Minziro Forest Reserve, Mtukula, 10 Apr. 2001, *Festo et al.* 1291!
DISTR. U 2–4; **K** 5; **T** 1; Liberia, Ivory Coast, Ghana, Cameroon, Bioko, Gabon, Central African Republic, Congo-Kinshasa, Sudan
HAB. Wet evergreen forest, often secondary and in clearings, swamp forest; 1050–1600 m

SYN. *Afromendoncia gilgiana* Lindau in E.J. 20: 1 (1894) & in P.O.A. C: 365 (1895) & in E. & P. Pf. IV, 3b: 290, fig. 115, D-M (1895); Burkill in F.T.A. 5: 7 (1899); Benoist in Notul. Syst. 2: 285 (1911); Lindau in Z.A.E.: 291 (1911); De Wildeman in Pl. Beq. 4: 18 (1926); F.W.T.A. 2: 250 (1931); Robyns in Fl. Parc Nat. Albert 2: 264 (1947)
Monachochlamys gilgiana (Lindau) S.Moore in J.B. 67: 227 (1929)
Mendoncia gilgiana (Lindau) Benoist var. *tisserantii* Benoist in Bull. Soc. Bot. France 85: 679 (1939); Heine in F.W.T.A., ed. 2, 2: 403 (1963). Types: Central African Republic, 15 km NE of Bambari, Gbatemoze River, *Tisserant* 599 (P, syn.) & 20 km N of Bambari, Yamwa River, *Tisserant* 2055 (P, syn.); Congo-Kinshasa, Libange-Ubangi, *Lebrun* 1613 (P, syn.)

8. SCLEROCHITON

Harvey in London Journ. Bot. 1: 27 (1842); Nees in DC., Prodr. 11: 279 (1847); G.P. 2: 1090 (1873); Lindau in E. & P. Pf. IV, 3b: 316 (1895); C.B. Clarke in F.T.A. 5: 109 (1899); Vollesen in K.B. 46: 7 (1991)

Isacanthus Nees in DC., Prodr. 11: 278 (1847)

Pseudoblepharis Baill. in Bull. Soc. Linn. Paris 2: 837 (1890); P.O.A. C: 370 (1895); Lindau in E. & P. Pf. IV, 3b: 319 (1895)

Butayea De Wild. in Ann. Mus. Congo, Bot., IV, Etud. Fl. Katanga 1: 149, pl. 42 (1903)

Shrubs or small trees. Leaves opposite, anisophyllous, apex usually drawn out into an obtuse tip, base usually attenuate and decurrent on petiole. Flowers in usually terminal racemiform cymes or solitary, usually supported by bracts; rachis flattened; bracts glumaceous or the lower foliaceous, usually caducous, usually not imbricate; bracteoles 2, large, glumaceous, midrib often thickened and horny at base. Calyx large, divided to the base in 5 many-veined glumaceous sepals which are thickened and horny at base; ventral and lateral sepals subequal, 1-toothed, dorsal broader and usually longer, 1–5-toothed apically. Corolla tube wide, usually shorter than limb, glabrous, constricted at insertion of stamens, below insertion with band of downwardly directed hairs; limb split dorsally to give a single broad 5-lobed lower lip, below glabrous or hairy, above usually with central glabrous band flanked by two bands of stiff bulbous-based retrorsely directed hairs, these becoming shorter outwards; lobes oblong, broadly rounded. Stamens 4, inserted just inside tube on top of a broad thickened flange; filaments bony, flattened, subequal; anthers elliptic to oblong, exserted, 1-thecous, usually bearded, usually with a small thorn near base. Ovary glabrous, with broad annular disk at base; style usually glabrous, bent near tip; stigma with two subequal acute lobes. Capsule 4-seeded, woody, sessile, ellipsoid, usually glabrous; valves with distinct dorsal furrow. Seed discoid, irregularly ovate-triangular in outline, with concentric ridges or rings of large pectinate scales.

19 species in tropical and South Africa.

NOTE. The leaves of all species of *Sclerochiton* are anisophyllous to a smaller or larger degree. In some species (e.g. *S. bequaertii*) one of the two leaves of a pair is constantly reduced to a small orbicular prophyll-looking leaf. In other species (e.g. *S. vogelii* subsp. *holstii*) this condition is seen in some specimens while in others there is only a slight difference. In most species this latter condition is the norm.

1. Corolla limb yellow to orange, shorter than tube; bracts,
    bracteoles and calyx pink to purplish (rarely green) . . .   3. *S. boivinii*
   Corolla limb white, pale mauve to mauve, pale blue to
    bright blue or salmon pink, longer than tube (except
    S. uluguruensis); bracts, bracteoles and calyx greenish
    to straw-coloured or dark brown 2
2. Bracts and bracteoles 1–2 cm wide, broadly rounded to
    truncate, glossy; rachis glabrous; corolla limb 25–50 mm
    long; anthers 6–10 mm long 3
   Bracts and bracteoles 2–7 mm wide, acute to acuminate,
    not glossy; rachis usually hairy; corolla limb 15–26 mm
    long; anthers 3–6 mm long 4
3. Leaf base decurrent on petiole; bracts and bracteoles
    about same length; upper sepal 18–27 mm long;
    corolla limb 25–38 mm long; filaments 9–12 mm long;
    anthers 6–7 mm long . . . .   1. *S. kirkii*
   Base on some or all leaves subamplexicaul; bracteoles
    distinctly longer than bracts; upper sepal 37–43 mm
    long; corolla limb ± 50 mm long; filaments ± 20 mm
    long; anthers ± 10 mm long . . . .   2. *S. insignis*
4. Upper sepal oblong to spathulate, parallel-sided for
    most of its length or widest apically, 2–5-toothed at
    apex (rarely 1-toothed), usually only with midrib
    thickened and horny at base 5
   Upper sepal lanceolate, gradually narrowed from base,
    1-toothed, acute to acuminate, basal $^1/_5$–$^2/_3$ glossy,
    straw-coloured and horny . . . .   9. *S. vogelii*

5. Corolla limb white with faint mauve veins or with red to
 purple lines 6
 Corolla limb pink, blue or mauve to purple, usually with
 darker veins 7
6. Bracteoles 12–16 mm long; lower sepals 17–20 mm long,
 lateral 15–17 mm long, upper 20–26 mm long; corolla
 tube 7–10 mm long, with band of hairs just below
 insertion of stamens; seed with indistinct scales;
 montane forest . . . .   *4. S. obtusisepalus*
 Bracteoles 15–26 mm long; lower sepals 23–32 mm long,
 lateral 19–26 mm long, upper 26–38 mm long; corolla
 tube 11–13 mm long, with band of hairs 3–4 mm
 below insertion of stamens; seed with large distinct
 scales; lowland forest . . . .   *5. S. tanzaniensis*
7. Young branches glabrous; bracts, bracteoles and sepals
 with dense stalked glands (rarely without); upper
 sepal 18–27 mm long; corolla tube with band of hairs
 just below insertion of stamens . . . .   *6. S. glandulosissimus*
 Young branches glabrous, pubescent or sericeous;
 bracts, bracteoles and sepals without stalked glands;
 upper sepal 25–38 mm long; corolla tube with band
 of hairs 3–5 mm below insertion of stamens 8
8. Young branches pubescent; bracteoles 17–25 mm long;
 corolla tube 10–12 mm long, limb 19–23 mm long,
 lobes 4–6 mm long; filaments hairy at base; anthers
 with lateral thorn-like structure . . . .   *7. S. bequaertii*
 Young branches glabrous to sericeous; bracteoles
 10–17 mm long; corolla tube 13–18 mm long, limb
 10–17 mm long, lobes 9–12 mm long; filaments
 glabrous; anthers without lateral thorn-like structure   *8. S. uluguruensis*

1. **Sclerochiton kirkii** (*T.Anderson*) *C.B.Clarke* in F.T.A. 5: 110 (1899); S. Moore in
J.L.S. 40: 159 (1911); Napper in K.B. 24: 333 (1970); Drummond in Kirkia 10: 274
(1975); Vollesen in K.B. 46: 17 (1991); Lebrun & Stork, Enum. Pl. Afr. Trop. 4: 503
(1997); White *et al.*, For. Fl. Malawi: 118 (2001). Type: Mozambique, Moramballa,
Kirk s.n. (K!, holo.)

Shrub or small tree to 4(–6) m tall sometimes scandent; young branches glabrous.
Leaves slightly anisophyllous, sessile or petiole up to 1 cm long; lamina ovate to
elliptic or slightly obovate, largest 14–28 × 6.7–11.5 cm, apex acute to obtusely
acuminate, glabrous to sparsely puberulous. Cyme up to 6(–9) cm long; peduncle up
to 1.5(–2) cm long, glabrous; rachis glabrous; bracts clasping flowers, alternate ones
often sterile, coriaceous, dark green, glossy, broadly elliptic to obovate, (12–)15–24 ×
10–20 mm, apex truncate or obtuse, glabrous, margin scarious; pedicels ± 1 mm long;
bracteoles similar, clasping calyx. Sepals glabrous, basal $^2/_3$ thickened and straw-
coloured, apical part as bracts and bracteoles, oblong-lanceolate, 18–27 mm long,
dorsal 1–2 mm longer than the rest, 6–8 mm wide and acute to subacute, others
3–6 mm wide (ventral slightly wider) and acute to acuminate. Corolla limb deep blue
to bluish purple; tube white, 10–12 mm long, band of hairs just below insertion of
stamens; limb 28–38 mm long, horizontal; lobes 7–12 mm long. Filaments 7–12 mm
long, glabrous; anthers 6–7 mm long, densely bearded and hairy on one side.
Capsule 14–16 mm long. Seed 5–6 mm long, with large pectinate scales.

TANZANIA. Lindi District: Nachingwea, 4 Mar. 1953, *Anderson* 848!; Masasi District: 8 km on
 Chiwale–Masasi track, 13 March 1991, *Bidgood et al.* 1972!
DISTR. **T** 8; Malawi, Mozambique, Zimbabwe
HAB. Dry *Millettia stuhlmannii* forest on hillside, termite mound in woodland; 400–450 m

SYN. *Acanthus kirkii* T.Anderson in J.L.S. 7: 37 (1863); S. Moore in J.B. 18: 233 (1880); Lindau in P.O.A. C: 370 (1895)

NOTE. The two Tanzanian collections of *S. kirkii* differ from the southern African material in having sparsely puberulous leaves. In all other respects they are quite typical. Further collections are needed from Tanzania to show whether this is a constant difference. For the moment I have decided not to give this form any infraspecific rank.

2. **Sclerochiton insignis** (*Mildbr.*) *Vollesen* in K.B. 46: 20 (1991); Lebrun & Stork, Enum. Pl. Afr. Trop. 4: 503 (1997). Type: Tanzania, Lindi District: Lake Lutamba, *Schlieben* 5871 (B†, holo.; BM!, BR!, HBG!, K!, P!, LISC!, S!, SRGH!, iso.)

Shrub or small tree to 2(?–10) m tall; young branches glabrous. Leaves slightly anisophyllous, sessile or lower with petiole up to 2 mm long; lamina elliptic to obovate, largest ± 27 × 12.5 cm, apex subacute to rounded, base on some or all subamplexicaul, glabrous. Cyme up to 5 cm long; peduncle up to 1 cm long, glabrous; rachis glabrous; bracts as in *S. kirkii*, broadly elliptic to transversely elliptic, 15–25 mm long; flowers subsessile; bracteoles similar, 22–30 mm long, distinctly longer than bracts. Sepals as in *S. kirkii*, 37–43 mm long, subacute (dorsal) to acuminate, dorsal ± 10 mm wide, others 6–8 mm wide. Corolla limb bright blue; tube ± 13 mm long, band of hairs ± 5 mm below insertion of stamens; limb ± 5 cm long, horizontal; lobes 7–10 mm long. Filaments ± 2 cm long, glabrous; anthers ± 1 cm long, densely bearded and hairy on one side, sparsely hairy on the other side. Capsule and seed not seen.

TANZANIA. Rufiji District: Kibiti, 19 Dec. 1968, *Shabani* 266!; Lindi District: Lake Lutamba, 14 Jan. 1935, *Schlieben* 5871!
DISTR. **T** 6, 8; not known elsewhere
HAB. Coastal forest, riverine forest; 100–250 m

SYN. *Pseudoblepharis insignis* Mildbr. in N.B.G.B. 12: 718 (1935); T.T.C.L.: 14 (1949)
   Sclerochiton kirkii (T.Anderson) C.B.Clarke var. *insignis* (Mildbr.) Napper in K.B. 24: 34 (1970)

NOTE. Only known from these two collections. Napper (l.c.) argues that this taxon hardly differs from *S. kirkii*, and that it should only be treated as a variety of it. It differs in the subamplexicaul leaf-bases, the bracteoles being longer than the bracts, the much larger calyx and corolla, and in the longer filaments and anthers. It is – in my opinion – clearly a distinct species.

3. **Sclerochiton boivinii** (*Baill.*) *C.B.Clarke* in F.T.A. 5: 110 (1899); T.S.K.: 162 (1936); T.T.C.L.: 16 (1949); K.T.S.: 18 (1961); Iversen in Symb. Bot. Ups. 29(3): 162 (1991); Vollesen in K.B. 46: 23 (1991); K.T.S.L.: 608 (1994); Lebrun & Stork, Enum. Pl. Afr. Trop. 4: 503 (1997). Type: Kenya, Mombasa, *Boivin* s.n. (P!, holo.; K!, iso.)

Shrub to 5 m tall; young branches glabrous (rarely puberulous on uppermost node). Leaves slightly anisophyllous; petiole 0.5–7 cm long; lamina ovate to elliptic or slightly obovate, largest (8–)12–30 × (2.5–)4–11 cm, apex obtusely acuminate, glabrous or midrib hairy near base. Cyme up to 13 cm long; peduncle up to 1 cm long, puberulous; rachis puberulous; bracts, bracteoles and dorsal sepal pink to purplish brown or magenta (rarely yellowish green); bracts ovate to obovate or broadly so, 14–25 × 7–22 mm (uppermost down to 6 × 3 mm), acute to broadly rounded, glabrous to finely puberulous; pedicels 1–5(–7) mm long; bracteoles ovate to obovate (rarely narrowly so), (15–)20–27(–35) × 6–12 mm, subacute to acute. Sepals finely puberulous (rarely glabrous), larger veins and base thickened; dorsal obovate-spathulate, (25–)28–45 mm long, broadly rounded or 2–5-toothed at apex, ventral and lateral straw-coloured or purplish near tip, lanceolate to oblanceolate, acute to subacute, ventral 21–35 mm long, lateral 18–30 mm long. Corolla limb yellow to orange; tube 16–23 mm long, band of hairs ± 1 cm below insertion of

FIG. 13. *SCLEROCHITON BOIVINII* — **1**, habit, × ²/₃; **2**, form with short inflorescence, × ²/₃; **3**, bract, × 1; **4**, bracteole, × 1; **5**, sepals × 1; **6**, corolla, × 1.5; **7**, corolla opened up showing stamens, × 2; **8**, filament and anther, × 4; **9**, stigma, × 4; **10**, capsule, × ²/₃; **11**, seed, × 3. 1 & 3–5 from *Polhill & Robertson* 4795, 2 from *Koritschoner* 742, 6–9 from *Ruffo & Mmari* 1969; 10–11 from *Faulkner* 1350. Drawn by Eleanor Catherine. From K.B. 46: 24 (1991).

stamens; limb 13–18 mm long, deflexed; lobes 4–6 mm long. Filaments 6–8 mm long, glabrous; anthers 5–6 mm long, glabrous or bearded, sometimes also hairy on sides. Capsule 18–22 mm long. Seed 6–8 mm long, with large concentric pectinate scales. Fig. 13, p. 82.

KENYA. Kilifi District: N of Giriama, Adu, Jan. 1937, *Dale* 3664!; Kwale District: Shimba Hills, Mwele Mdogo Forest, 6 Feb. 1953, *Drummond & Hemsley* 1143! & Shimba Hills, Makadara Forest, 17 Sep. 1982, *Polhill & Robertson* 4795!
TANZANIA. Lushoto District: E Usambara Mts, Maramba, 18 Nov. 1936, *Greenway* 4748! & E Usambara Mts, Ndola, 17 Feb. 1954, *Faulkner* 1350! & W Usambara Mts, Dindira, 6 Aug. 1957, *Faulkner* 2032!
DISTR. **K** 7; **T** 3; not known elsewhere
HAB. Lowland evergreen or semi-evergreen forest (Kenya), lowland and medium altitude rainforest (Tanzania); 0–400 m in Kenya, (400–)800–1400(–1500) m in Tanzania.

SYN. *Pseudoblepharis boivinii* Baillon in Bull. Soc. Linn. Paris 2: 837 (1890); Lindau in E.J. 18: 63, t. 1, fig. 39 (1894) & in E. & P. Pf. IV, 3b: 319 (1895)
    P. heinsenii Lindau in E.J. 24: 320 (1897) & in E. & P. Pf., Nachtrag zu IV, 3b: 306 (1897). Type: Tanzania, E Usambara Mts, Nderema, *Heinsen* 4 (B†, holo.; K!, iso.)

NOTE. The differences in habitat and altitudinal range between the Kenyan and Tanzanian material are odd, but plants from the two areas are identical in all respects.
    An occasional form with yellowish green bracts, bracteoles and sepals has been collected both in Kenya (e.g. *Rawlins* 700) and in Tanzania (e.g. *Greenway* 5907). However some plants with only slightly coloured floral parts have also been seen, and I do not consider this form worthy of any taxonomic rank.
    The material from higher altitudes in the W Usambara Mts in Tanzania (e.g. *Lovett* 262) has few-flowered inflorescences or even some of the flowers solitary. But it has the coloured floral parts and yellow corolla of *S. boivinii*, and again I do not consider it sufficiently distinct to merit taxonomic recognition.

4. **Sclerochiton obtusisepalus** *C.B.Clarke* in F.T.A. 5: 111 (1899); Vollesen in K.B. 46: 30 (1991); Lebrun & Stork, Enum. Pl. Afr. Trop. 4: 503 (1997); White *et al.*, For. Fl. Malawi: 119 (2001). Type: Malawi, Masuku Plateau, *Whyte* 267 (K!, holo.)

Erect or scandent shrub or small tree to 5 m tall; young branches glabrous or sparsely hairy at nodes. Leaves slightly to strongly anisophyllous; petiole up to 2 cm long; lamina elliptic to slightly obovate, largest 9–17 × 3.5–6.5 cm, apex with an obtuse tip 1–1.5 cm long, glabrous. Cyme up to 5 cm long; peduncle 0–5 mm long, glabrous or sparsely puberulous; rachis puberulous or sparsely so; bracts and bracteoles green to brown, ovate, acute to acuminate with recurved tip, sometimes with a few large irregular teeth, glabrous to sparsely puberulous or sparsely pubescent; bracts 11–15 mm long; pedicels 1–4 mm long; bracteoles 12–16 mm long. Sepals straw-coloured to brownish or greenish or dorsal purplish, finely puberulous and finely ciliate, midrib thickened at base; dorsal oblong to slightly spathulate, 20–26 mm long, 3–5-toothed at apex; ventral and lateral narrowly oblong-obovate, acuminate to cuspidate, sometimes with a few large irregular teeth, ventral 17–20 mm long, lateral 15–17 mm long. Corolla limb white with faint mauve veins; tube 7–10 mm long, band of hairs just below insertion of stamens; limb 18–24 mm long; lobes 7–10 mm long. Filaments 9–11 mm long, sparsely hairy towards base; anthers 3–5 mm long, lower half bearded, two outer also hairy on outer side. Capsule 15–18 mm long, finely glandular. Seed 5–7 mm long, with concentric ridges of low indistinct scales.

TANZANIA. Rungwe District: Mt Rungwe, 24 Oct. 1956, *Richards* 6768!; Songea District: Matengo Hills, Lupembe Hill, 1 Apr. 1936, *Zerny* 564! & 20 May 1956, *Milne-Redhead & Taylor* 10388!
DISTR. **T** 7, 8; Malawi
HAB. Montane rainforest; (1400–)1850–2700 m

5. **Sclerochiton tanzaniensis** *Vollesen* in K.B. 46: 33 (1991); Lebrun & Stork, Enum. Pl. Afr. Trop. 4: 503 (1997). Type: Tanzania, Ulanga/Iringa District: Udzungwa Mts, Mwanihana Forest Reserve, Sanje Waterfall Camp, *Rodgers & Vollesen* 754 (K!, holo.; C!, DSM!, WAG!, iso.)

Spindly erect shrub (? rarely small tree) to 3 m tall; young branches glabrous (rarely densely pubescent). Leaves slightly anisophyllous; petiole up to 1.5(–4) cm long; lamina elliptic to slightly obovate (rarely broadly so), largest 10–19(–22) × 4–7.5(–14) cm, apex blunt or with an obtuse tip up to 1 cm long, glabrous or sparsely hairy on midrib above (rarely uniformly pubescent). Cyme up to 6 cm long; peduncle up to 5 mm long, glabrous; rachis glabrous or sparsely pubescent; bracts and bracteoles green or brownish, ovate to elliptic or oblong, acute to acuminate with straight tip, glabrous to sparsely pubescent or finely puberulous; bracts 9–20 mm long; pedicels 1–2 mm long, glabrous or sparsely puberulous; bracteoles 15–26 mm long. Sepals straw-coloured to greenish, tinged purplish in apical part, subglabrous to finely puberulous, midrib thickened at base; dorsal oblong, 26–38 mm long, 2–5-toothed at apex; ventral and lateral lanceolate to narrowly oblong, acute to cuspidate (rarely 2-toothed), ventral 23–32 mm long, lateral 19–26 mm long. Corolla limb white with red to purple guide lines; tube 11–13 mm long, band of hairs 3–4 mm below insertion of stamens; limb 18–23 mm long; lobes 5–6 mm long. Filaments 8–12 mm long, glabrous or finely glandular; anthers 5–6 mm long, lower half bearded, also with tuft of longer hairs at base. Capsule 17–21 mm long, finely glandular. Seed 6–7 mm long, with large distinct scales. Fig. 14: 1–11, p. 85.

Tanzania. Morogoro District: Kimboza Forest Reserve, 31 Mar. 1983, *Mwasumbi, Rodgers & Hall* 12433!; Ulanga/Iringa District: Udzungwa Mts, Mwanihana Forest Reserve, Sanje Waterfall Camp, 23 July 1983, *Polhill & Lovett* 5114!; Lindi District: 50 km W of Lindi, 4 May 1935, *Schlieben* 6479!
Distr. **T** 6–8; not known elsewhere
Hab. Lowland rainforest, riverine forest, dry coastal forest and thicket; 200–950(–1200) m

Note. *Bridson* 618 differs from the rest of the material in having pubescent stems and leaves, and also – quite extraordinarily – in having a long narrow spathulate upper lip to the corolla. I have not seen this phenomenon in any other of the numerous collections studied in this genus. Technically it puts the specimen out of *Sclerochiton* altogether, but I have considered it merely to be a monstrosity and have not considered it worthy of any taxonomic recognition. *Bridson* 591 – which was collected from the same locality –is perfectly normal S. *tanzaniensis*.
    Carmichael 568 differs in having dentate leaves and many axillary cymes scattered along the stems. It has been collected from a seemingly unbranched juvenile plant which probably explains its abnormal leaves and inflorescences.

6. **Sclerochiton glandulosissimus** *Vollesen* in K.B. 46: 35 (1991); Lebrun & Stork, Enum. Pl. Afr. Trop. 4: 503 (1997). Type: Tanzania, Morogoro District: Uluguru Mts, *Schlieben* 3935 (BM!, holo.; BR!, HBG!, LISC!, P!, S!, iso.)

Shrub or small tree to 5 m tall; young branches glabrous. Leaves slightly to strongly anisophyllous, drying blackish; petiole up to 3 cm long; lamina elliptic to obovate, largest 8–24 × 3.2–7.3 cm, apex with an obtuse tip up to 1 cm long, glabrous. Cymes terminal and axillary from upper leaf axils, up to 4 cm long; peduncle up to 8 mm long, glabrous or sparsely puberulous; rachis sericeous-puberulous and usually with scattered stalked glands; bracts and bracteoles yellowish brown to dark brown, ovate to elliptic, acute with recurved tip, densely covered with long-stalked capitate glands (rarely puberulous without glands); bracts 5–9 mm long; pedicels 1–4 mm long, sometimes with scattered capitate glands; bracteoles 9–18 mm long. Sepals strongly and uniformly thickened at base, colour and indumentum as bracts and bracteoles; dorsal oblong, 18–27 mm long, subacute or bifid at apex; ventral and lateral lanceolate, acuminate, ventral 15–23 mm long, lateral 12–20 mm long. Corolla limb blue to purple or lilac; tube 8–11 mm long, band of hairs just below insertion of

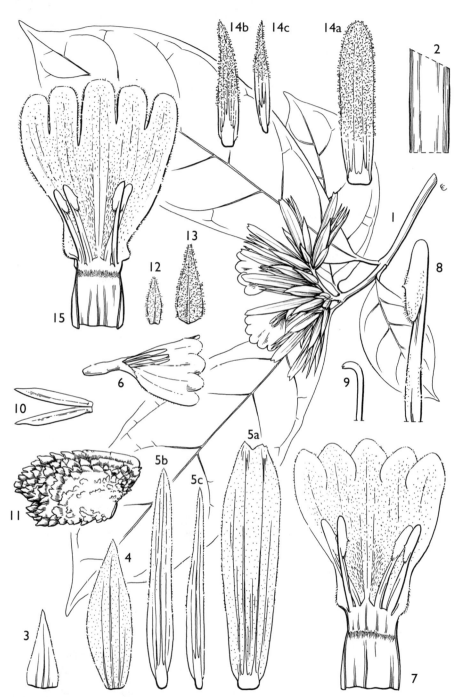

FIG. 14. *SCLEROCHITON TANZANIENSIS* — **1**, habit, × ²/₃; **2**, detail of stem, × 6; **3**, bract, × 2; **4**, bracteole, × 2; **5**, sepals × 2; **6**, corolla, × 1; **7**, corolla opened up showing stamens, × 2; **8**, filament and anther, × 4; **9**, stigma, × 4; **10**, capsule, × 1; **11**, seed, × 5. *S. GLANDULOSISSIMUS* — **12**, bract, × 2; **13**, bracteole, × 2; **14**, sepals × 2; **15**, corolla opened up showing stamens, × 2. 1–11 from *Bridson* 591, 12–15 from *D. Thomas* 3674. Drawn by Eleanor Catherine. From K.B. 46: 34 (1991).

stamens; limb 17–23 mm long; lobes 6–10 mm long. Filaments 7–10 mm long, hairy near base; anthers 3–4 mm long, bearded in lower half and with basal tuft of longer hairs. Capsule ± 2 cm long. Seed ± 6 mm long, with concentric ridges of large pectinate scales. Fig. 14: 12–15, p. 85.

TANZANIA. Ulanga District: Mahenge Plateau, Mt Muhulu, 12 May 1932, *Schlieben* 2187!; Ulanga/Iringa District: Udzungwa Mts, Mwanihana Forest Reserve, above Sanje, 13 Sep. 1984, *D.W. Thomas* 3674! & 17 June 1986, *Lovett et al.* 858!
DISTR. **T** 6, 7; not known elsewhere
HAB. Medium altitude rainforest; 1000–1700 m

NOTE. A very distinctive species, normally easily recognised by the densely glandular bracts, bracteoles and sepals. It also differs from *S. bequaertii* in drying blackish and in the shorter sepals.
    Schlieben 2187 differs from the rest of the material in having puberulous bracts, bracteoles and sepals instead of the usual dense glandular indumentum; but otherwise is quite typical. More material from the Mahenge area would be useful for deciding whether this form is worthy of taxonomic recognition.
    The type collection bears the label name *S. glischrocalyx*, but this name was apparently never published.

7. **Sclerochiton bequaertii** *De Wild.* in Rev. Zool. Afr. 8, suppl. Bot.: 35 (1920) & in Pl. Beq. 4: 24 (1926); Vollesen in K.B. 46: 36 (1991); Lebrun & Stork, Enum. Pl. Afr. Trop. 4: 503 (1997). Type: Congo-Kinshasa, Kivu, Kilo to Irumu, *Bequaert* 4878 (BR!, holo.; BR!, iso.)

Scandent or scrambling shrub to 5 m tall; young branches sparsely puberulous. Leaves strongly anisophyllous; petiole up to 1.2 cm long, puberulous to pubescent; lamina elliptic to slightly obovate, largest 6.5–12 × 3–5.3 cm, apex with an obtuse tip 1–1.5 cm long, base not decurrent, sparsely hairy along midrib and larger veins. Cyme up to 5 cm long; peduncle 0–8 mm long, sparsely puberulous; rachis similar; bracts and bracteoles yellowish brown to greenish or purplish tinged towards apex, ovate to elliptic or narrowly obovate, acute to acuminate, sparsely sericeous-puberulous; bracts 10–22 mm long; pedicels 1–3 mm long; bracteoles 17–25 mm long. Sepals straw-coloured or greenish to purple, puberulous to pubscent and finely ciliate, sometimes with scattered subsessile capitate glands, midrib strongly thickened at base; dorsal oblong to slightly spathulate, 25–32 mm long, subacute or 2–3-toothed at apex; ventral and lateral lanceolate, acuminate, ventral 19–28 mm long, lateral 15–21 mm long. Corolla limb blue to mauve or purple; tube 10–12 mm long, band of hairs ± 3 mm below insertion of stamens; limb 19–23 mm long; lobes 4–6 mm long. Filaments 9–11 mm long, hairy and with scattered glands towards base; anthers 3.5–5 mm long, bearded in middle part and with basal tufts. Capsule 17–21 mm long. Seeds ± 6 mm long, with concentric ridges or with indistinct scales.

UGANDA. Masaka District: Masaka road, June 1937, *Chandler* 1644!; Mengo District: Mawokota, Feb. 1905, *E. Brown* 174! & 25 km on Kampala-Mubende road, Oct. 1932, *Eggeling* 913!
DISTR. **U** 2, 4; Congo-Kinshasa, Rwanda
HAB. Medium altitude rainforest; 1100–1200 m

SYN. [*Pseudoblepharis obtusisepalus* sensu Lindau in Z.A.E.: 299 (1911), *non Sclerochiton obtusisepalus* C.B.Clarke (1899)]
    Sclerochiton sp. nr. bequaertii of Burtt-Davy, Check-lists Brit. Emp. 1, Uganda Prot.: 15 (1935)
    [*Sclerochiton obtusisepalus* sensu Raynal *et al.*, Fl. Med. Miss. Rwanda 1: 36 (1979); Troupin, Fl. Pl. Lign. Rwanda: 89, fig, 24, 2(1982); Champluvier in Fl. Rwanda 3: 485, fig. 148, 4 (1985), *non* C.B.Clarke (1899)]

NOTE. In recent years it has become general practice to unite *S. bequaertii* with *S. obtusisepalus*, but there are differences in habit, leaves, indumentum and corolla colour as well as a large disjunction in the distribution. They clearly deserve to be treated as distinct species.

8. **Sclerochiton uluguruensis** *Vollesen* in K.B. 46: 37 (1991); Lebrun & Stork, Enum. Pl. Afr. Trop. 4: 503 (1997). Type: Tanzania, Morogoro District: Uluguru Mts, Kinole road, *E.A. Bruce* 981 (K!, holo.; BM!, K!, iso.)

Shrub or small tree to 6 m tall; young branches glabrous or finely sericeous. Leaves strongly anisophyllous; petiole up to 1 cm long; lamina elliptic to slightly obovate, largest 6.5–12 × 2–5.5 cm, apex acute or with a short obtuse tip, base not decurrent, glabrous or sparsely sericeous along midrib. Cyme up to 3 cm long; peduncle 0–5 mm long, sericeous; rachis sericeous-puberulous; bracts and bracteoles brown or dark brown, ovate to elliptic, acute, slightly falcate, sparsely sericeous-puberulous; bracts 6–15 mm long; pedicels 3–5 mm long; bracteoles 10–17 mm long. Sepals falcate, brown or dark brown, dorsal tinged purplish, finely puberulous and finely ciliate, thickened at base; dorsal oblong, 30–38 mm long, acute or 2–5-toothed at apex; ventral and lateral lanceolate, acuminate, ventral 23–31 mm long, lateral 19–25 mm long. Corolla limb pink to mauve with red guide lines; tube 13–18 mm long, band of hairs ± 5 mm below insertion of stamens; limb 10–17 mm long; lobes 9–12 mm long. Filaments ± 11 mm long, glabrous, inserted ± 4 mm into corolla tube; anthers 5–6 mm long, glabrous or bearded, without ventral thorn-like structure. Capsule 16–20 mm long. Seed (immature) ± 6 mm long.

TANZANIA. Morogoro District: Uluguru Mts, Matombo–Tanana road, Feb. 1935, *E.A. Bruce* 813!; Iringa District: Udzungwa Mts, Ndunduru Forest Reserve, 7 Oct. 2000, *Luke et al.* 7046! & Udzungwa Mts National Park, 1 Oct. 2001, *Luke et al.* 8050!
DISTR. **T** 6, 7; not known elsewhere
HAB. Montane rainforest; 1600–1950 m

NOTE. A very distinct species which is only known from six collections. The shape of the upper sepal points to a relationship with *S. obtusisepalus*, and the hairy stem and pink to mauve corolla to a relationship with *S. bequaertii*. But otherwise it is quite distinct and differs from all other species in this group by the corolla tube being longer than the limb–a character which in our area is otherwise only seen in *S. boivinii*. It is also characterised by long pedicels, falcate sepals, long corolla lobes and anthers without ventral thorn-like structure.

9. **Sclerochiton vogelii** (*Nees*) *T.Anderson* in J.L.S. 7: 37 (1863); Lindau in E. & P. Pf. IV, 3b: 319 (1895); Clarke in F.T.A. 5: 111 (1899); F.W.T.A. 2: 258 (1931); F.W.T.A. (ed. 2) 2: 408 (1963); Heine in Fl. Gabon 13: 118 (1966); Vollesen in K.B. 46: 42 (1991); Lebrun & Stork, Enum. Pl. Afr. Trop. 4: 503 (1997). Type: Liberia, Cape Palmas, *Vogel* 54 (K!, lecto.)

Erect or scandent shrub to 3(–4) m tall; young branches puberulous or sericeous-puberulous or sparsely so (rarely glabrous). Leaves slightly to strongly anisophyllous, often glossy; petiole 0.2–2 cm long; lamina ovate to elliptic or slightly obovate, largest 5.5–22 × 2–8 cm, apex with an acute to obtuse tip up to 1.5 cm long, base attenuate or cuneate to truncate, glabrous or puberulous along midrib and larger veins (see also note after subsp. *holstii*). Cymes terminal and axillary, (1–)2–12-flowered, sessile; rachis up to 3.5 cm long, sericeous-puberulous; bracts lanceolate to ovate, 4–16 mm long, acuminate to acute, greenish to yellowish or dark brown, glabrous or sericeous-puberulous or sparsely so in apical part, finely ciliate; pedicels 0–3 mm long; bracteoles ovate-triangular or narrowly so, 9–20 mm long, acuminate to acute, basal part yellow and glossy, apical part greenish to dark brown, glabrous or finely puberulous in apical part, finely ciliate. Sepals with basal $^1/_5$–$^2/_3$ uniformly thickened, yellow and glossy, apical part as bracts, glabrous or finely ciliate, margin often inrolled, scarious or not; dorsal lanceolate, 18–32 mm long, acute to acuminate; ventral and lateral same shape but narrower, acuminate to cuspidate, ventral 16–28 mm long, lateral 14–26 mm long. Corolla limb pale mauve to mauve, pale to deep blue or bluish purple, with darker lines into throat; tube yellow, 8–12 mm long, band of hairs 2–3 mm below insertion of stamens; limb

15–26 mm long; lobes 3–7 mm long. Filaments 7–9 mm long, glabrous or finely glandular along upper edge; anthers 3–4.5 mm long, bearded in upper half and with tufts of longer hairs at base. Capsule 10–19 mm long. Seed 5–6 mm long, with large pectinate scales.

SYN. *Isacanthus vogelii* Nees in DC., Prodr. 11: 279 (1847); Bentham in Hooker, Niger Fl.: 48 (1849)

9a. subsp. **congolanus** (*De Wild.*) *Vollesen* in K.B. 46: 45 (1991); Lebrun & Stork, Enum. Pl. Afr. Trop. 4: 503 (1997). Type: Congo-Kinshasa, near Kimuenza, *Gillet* 1933 (BR!, holo.; BR! iso.)

Leaves below puberulous along all major veins, with 7–8 main veins each side; flowers sessile; basal $^1/_2$–$^2/_3$ of sepals thickened; corolla limb 20–26 mm long; capsule 15–19 mm long.

UGANDA. Toro District: Bwamba, Semliki Forest, July 1951, *Dale* 810! & Kibale Forest, 5 Sep. 1941, *Thomas* 3931! & Bwamba, Kidongo, Aug. 1937, *Eggeling* 3383!
TANZANIA. Kigoma District: Uvinza, W of Lugufu, 8 Feb. 1926, *Peter* 36509! & Kasye Forest, 20 Mar. 1994, *Bidgood et al.* 2843!; Ufipa District: Kalambo Falls, 29 Mar. 1955, *Exell et al.* 1318!
DISTR. **U** 2; **T** 4; Gabon, Congo (Brazzaville), Congo-Kinshasa, Rwanda, Angola, Zambia, Mozambique
HAB. Rainforest and riverine forest; 700–1500 m

SYN. *Butayea congolana* De Wild. in Ann. Mus. Congo, Bot., Ser. IV, Etud. Fl. Katanga 1: 150, pl. 42 (1903) & in Ann. Mus. Congo, Bot., Ser. V, 1: 216 (1906) & in Ann. Mus. Congo, Bot., Ser. V, 2: 202 (1907); Th. & H. Durand, Syll. Fl. Congo.: 425 (1909); De Wild. in Ann. Mus. Congo, Bot., Ser. V, 3: 480 (1912)
[*Sclerochiton holstii* sensu S. Moore in J.L.S. 37: 195 (1905); Burtt Davy, Check-lists Brit. Emp. 1. Uganda Prot.: 15 (1935), *non* (Lindau) C.B.Clarke (1899)]
S. albus De Wild. in Rev. Zool. Afr., Suppl. Bot. 8, 2: 33 (1920) & in Pl. Beq. 4: 24 (1926). Type: Congo-Kinshasa, Lubutu to Kirundu, *Bequaert* 6879 (BR!, holo.; BR!, iso.)
S. cyaneus De Wild. in Rev. Zool. Afr., Suppl. Bot. 8,2: 34 (1920) & in Pl. Beq. 4: 25 (1926). Type: Congo-Kinshasa, Penghe, *Bequaert* 2117 (BR!, holo.)
S. sousai Benoist in Bol. Soc. Brot., Ser. 2, 24: 24 (1950). Type: Angola, Luanda, Dando, *Carisso & Sousa* 28 (COI, holo.; P!, iso.)
[*S. obtusisepalus* sensu Burtt Davy, Check-lists Brit. Emp. 1, Uganda Prot.: 15 (1935), *non* C.B.Clarke (1899)]
S. sp. 1 of F. White in F.F.N.R.: 383 (1962)

9b. subsp. **holstii** (*Lindau*) *Napper* in K.B. 24: 333 (1970); Vollesen in Opera Bot. 59: 81 (1980); Iversen in Symb. Bot. Ups. 29(3): 162 (1991); Vollesen in K.B. 46: 46 (1991); K.T.S.L.: 608 (1994); Lebrun & Stork, Enum. Pl. Afr. Trop. 4: 503 (1997). Type: Tanzania, Tanga District: Amboni, *Holst* 2885 (B†, holo.; COI!, HBG!, K!, iso.)

Leaves below glabrous or with a few hairs along midrib, with 4–7 main veins each side; pedicels 0.5–3 mm long; basal $^1/_5$–$^1/_3$ of sepals thickened; corolla limb 15–21(–23) mm long; capsule 10–15 mm long

KENYA. Kwale District: 5 km S of Mazeras, Mwachi, 10 Sep. 1953, *Drummond & Hemsley* 4251!; Kilifi District: Kilifi-Kalobeni road, Cha Simba, 20 July 1969, *Adams* 26! & Kikuyuni to Lake Jilore, 26 Dec. 1954, *Verdcourt* 1185!
TANZANIA. Tanga District: Amboni, 29 June 1918, *Peter* K.793! & Mtimbwani, 6 Dec. 1935, *Greenway* 4228!; Uzaramo District: Banda Forest Reserve, 16 Aug. 1968, *Shabani* 187!
DISTR. **K** 7; **T** 3, 6, 8; not known elsewhere
HAB. Dry coastal *Brachylaena*, *Cynometra* and *Manilkara* forest and thicket, lowland evergreen rainforest; 0–300(–550) m

SYN. *Pseudoblepharis holstii* Lindau in E.J. 20: 35 (1894) & in P.O.A. C: 370 (1895) & in E. & P. Pf. IV, 3b: 319 (1895)
Sclerochiton holstii (Lindau) C.B.Clarke in F.T.A. 5: 111 (1899); T.T.C.L.: 16 (1949); K.T.S.: 18 (1961)
S. scissisepalus C.B.Clarke in F.T.A. 5: 111 (1899). Type: Kenya, near Mombasa, *Wakefield* s.n. (K!, holo.)
[*S. obtusisepalus* sensu T.S.K.: 162 (1936), *non* C.B.Clarke (1899)]
[*S. vogelii* sensu K.T.S.: 18 (1961), *non* (Nees) T.Anderson (1863)]

NOTE. In some of the flowers on the type of *S. scissisepalus* the upper sepal is bifid, but other flowers on the same branch have the usual entire upper sepal.

The material from the foothills of the E Usambara Mts from slightly higher altitudes (200–300 m) is completely glabrous, and leaves, bracts, bracteoles and flowers are all rather large for this subspecies. But large-leaved plants are also occasionally found on the coast as are glabrous plants. The pattern seen in the E Usambaras probably only reflects a slightly wetter and more shaded habitat.

Material from **T** 8 is from higher altitudes (400–550 m) than anything from further north. It differs in having scattered hairs all over the upper leaf-surface, and the flowers are mostly in 1-flowered inflorescences. But some inflorescences are usually 2–3-flowered and some collections from **T** 6 also have some 1-flowered inflorescences. Scattered collections from all parts of the distribution area have a few hairs on the upper leaf-surface. More material is needed from S Tanzania to decide whether this form should be recognised as a distinct taxon.

9. ACANTHUS

L., Sp. Pl.: 639 (1753) & Gen. Pl.: 286 (1754); Nees in DC., Prodr. 11: 270 (1847); G.P. 2: 1090 (1876); C.B. Clarke in F.T.A. 5: 105 (1899); Vollesen in K.B. 62: 233 (2007)

Cheilopsis Moq. in Ann. Sci. Nat., ser 1, 27: 230 (1832); Nees in DC., Prodr. 11: 272 (1847)

Perennial herbs, shrubs or small trees; cystoliths absent. Leaves opposite or in whorls of 4, usually leathery with pale yellow thickened midrib, veins and margins, alternate pair if present usually much reduced, with sessile scale-like glands. Flowers in terminal or axillary racemoid cymes; bracts large, leathery, with 3 prominent longitudinal veins, with spinose margin or entire; bracteoles 2, filiform to ovate, spinose or entire. Calyx divided to base into 4 glumaceous sepals which are thickened and horny at the base; dorsal sepal 1–3-veined, entire or with 2–3 small teeth, ventral 2-veined, entire or with 2 large triangular teeth or irregularly toothed, lateral 1-veined. Corolla tube short, white, thickened and spongy apically, with a dense band of thick retrorse hairs apically; limb split dorsally to give a 3–5-lobed lip, basally with a thickened callus, on both sides of and on callus with thick bulbous-based retrorse hairs; lobes broadly oblong, rounded or central emarginate. Stamens 4, exserted, inserted just inside throat, two posterior held above anterior; two pair of filaments similar, curved, thick and bony, laterally compressed; anthers 1-thecous, oblong, with brushes of stiff hairs along whole length ventrally. Ovary 2-locular with 2 ovules per locule; style linear, glabrous; stigma of 2 lanceolate, acute lobes. Capsule 2–4-seeded, woody, ellipsoid, glossy, sessile, flattened dorsiventrally and rounded apically. Seed discoid, glabrous or sericeous-puberulous.

About 25 species from southern Europe to the Pacific and south through Africa to Angola, Zambia and Malawi. The native species are occasionally cultivated and the Ethiopian *A. sennii* has also been cultivated in Kenya. It is immediately recognisable by its large dark red flowers.

1. Plant with persistent sharp terete straw-coloured
   interpetiolar spines; rachis, bracts and calyces with
   stalked capitate glands; anthers 6–9 mm long ......   *4. A. eminens*
   Plant without persistent sharp terete straw-coloured
   interpetiolar spines but often with small spiny stipule-
   like interpetiolar leaves; rachis, bracts and calyces
   without stalked capitate glands 2

2. Perennial herb with basal part of stem creeping and rooting, no small spiny stipule-like interpetiolar leaves and without scars at base of leaves; anthers 7–10 mm long; seed hairy .... 5. *A. ueleensis*
    Erect shrubby herbs or shrubs, with small spiny auricle-like leaves at base of normal leaves or with scars if caducous; anthers 3.5–5.5 mm long; seed glabrous 3
3. Bracts in middle of spike 1.4–2.8 cm long, longest bract teeth 3–5 mm long, small bract teeth inserted between large teeth; dorsal sepal 1.4–2.2 cm long, ventral 1.1–1.8 cm long, lateral 0.8–1.4 cm long .... 1. *A. polystachius*
    Bracts in middle of spike 2.5–3.7 cm long, longest bract teeth 7–12 mm long, small bract teeth inserted on base of large teeth; dorsal sepal 2.7–4 cm long, ventral 2.3–3.8 cm long, lateral 2.2– 2.7 cm long 4
4. Dorsal sepal 2.7–3.2 cm long, ventral 2.3–2.7 cm long; corolla pinkish mauve, thickened part of tube dorsally extended into a tooth; filaments 1.8–2 cm long .... 2. *A. kulalensis*
    Dorsal sepal 3.5–4 cm long, ventral 3.4–3.8 cm long; corolla deep purple, thickened part of tube not extended into a tooth; filaments 2.7–3.3 cm long ... 3. *A. austromontanus*

1. **Acanthus polystachius** *Delile*, Cent. Pl. Méroé: 72, t. 62, f. 2 (1826); Pichi Sermolli, Miss. Lago Tana 7,1: 140 (1951); E.P.A.: 953 (1964); Cufodontis in Senck. Biol. 46: 117 (1965); Lebrun & Stork, Énum. Pl. Afr. Trop. 4: 466 (1997); Ensermu in F.E.E. 5: 357 (2006); Vollesen in K.B. 62: 241 (2007). Type: Sudan, Singa [Singué], *Cailliaud* s.n. (P!, holo.)

Shrub to 4 m tall; young stems puberulous to tomentellous (rarely glabrous). Leaves with petiole 0.3–2.2(–2.5) cm long; lamina ovate to elliptic in outline, largest 11–35 × 6–16 cm, deeply lobed with large triangular spine-tipped lobes, each lobe with 1–2 spines on antrorse side or on both sides, apex acute to acuminate, spine-tipped, base truncate to cordate, beneath sparsely puberulous to tomentellous (rarely glabrous) on midrib and larger veins, glabrous to puberulous on lamina, above sparsely to densely pubescent on midrib and glabrous or with scattered long hairs on lamina (rarely uniformly puberulous); stipule-like interpetiolar leaves present, but often falling quickly, with lamina, up to 1.5 cm long, spiny. Cymes solitary or also 2–4 from upper axils, 4–25(–30) cm long; rachis puberulous to tomentose, with several pairs of sterile bracts at base; bracts pale green to straw-coloured, elliptic to slightly obovate, puberulous to densely pubescent (rarely almost glabrous but for ciliate margin), 1.4–2.8 cm long in middle of cyme, spine-tipped and with (5–)10–25 large and small teeth per side, longest 3–5 mm long, small teeth inserted between large, teeth hairy near base; bracteoles linear-lanceolate, 1–2.5 cm long, toothed in apical part. Calyx green or tinged purplish, glabrous to densely puberulous or pubescent, conspicuously ciliate; dorsal and ventral sepals ovate or broadly so, dorsal 1.4–2.2 cm long, obtuse or with 1–3-toothed tip and without or with 1(–4) weak lateral teeth per side, ventral 1.1–1.8 cm long, obtuse or with spiny tip or with 2 small additional apical teeth or with 2 triangular spine-tipped lobes, lateral ovate, 0.8–1.4 cm long, obtuse or acuminate. Corolla pale pink to pink or rose (more rarely white); tube 3–5 mm long below thickened rim which is 3–5 mm long; limb 5-lobed, 2.5–4 × 3–4 cm, below densely puberulous, above puberulous (rarely glabrous); callus with broad central groove and two narrow lateral ridges. Filaments 1.4–2.4 cm long, glabrous or 2 dorsal hairy on inside near base; anthers 3.5–5 mm long, glabrous or with long curly hairs on sides and back. Capsule 1.7–2.4 cm long. Seed ellipsoid to round, 7–9 mm long, glabrous. Fig. 15, p. 91.

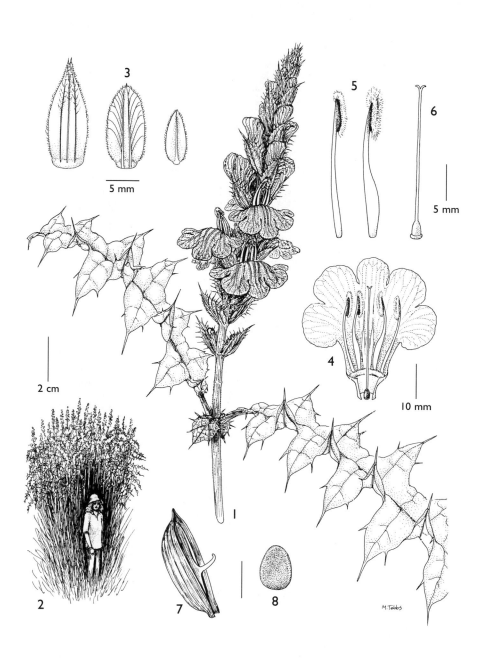

FIG. 15. *ACANTHUS POLYSTACHIUS* — **1**, habit; **2**, typical growth form; **3**, dorsal (left), ventral (middle) and lateral (left) sepals; **4**, corolla opened up; **5**, stamens; **6**, ovary, style and stigma; **7**, capsule; **8**, seed. 1 from *Drummond & Hemsley* 4759, 2 from private photo, 3 & 5–6 from *Eggeling* 3156, 4 from *Lugard* 233, 7 from *Tanner* 5372, 8 from *Ruffo* 401. Drawn by Margaret Tebbs.

UGANDA. Toro District: Busongora, Aug. 1936, *Eggeling* 3156!; Mengo District: Nakaziba Forest, 1 Dec. 1949, *Dawkins* 466! & Mabira Forest, 9 Nov. 1938, *Loveridge* 40!
KENYA. Uasin Gishu District: Turbo, Kipkarren, no date, *Brodhurst Hill* 698!; North Kavirondo District: Kakamega Forest, 15 Oct. 1953, *Drummond & Hemsley* 4759! & Mt Elgon, Nov. 1930, *Lugard* 233!
TANZANIA. Ngara District: Bushiri, Rubanga, 15 Oct. 1960, *Tanner* 5372!; Bukoba District: Ruiga Forest Reserve, 1 Sep. 1970, *Ruffo* 401!; Buha District: Kibondo, 7 Aug. 1950, *Bullock* 3078!
DISTR. U 2–4; K 3, 5; T 1, 4; Sudan, Ethiopia, Congo-Kinshasa, Rwanda, Burundi
HAB. Tall fireswept seasonally wet grassland, often forming dense stands at forest margins, rocky hills; 1100–2300 m

SYN. *Cheilopsis polystachius* (Delile) Moq. in Ann. Sci. Nat. Ser I, 27: 230 (1832); Nees in DC., Prodr. 11: 273 (1847); Richard, Tent. Fl. Abyss. 2: 151 (1850)
  Acanthus arboreus Forssk. var. *pubescens* Oliv. in Trans. Linn. Soc. 29: 129, t. 86 (1875). Type: Tanzania, Uzinzi, *Speke & Grant* 136 (K!, holo.)
  A. pubescens (Oliv.) Engl., Hochgebirgsfl. Trop. Afr.: 390 (1892); Lindau in P.O.A. C: 370 (1895); Turrill in K.B. 1913: 336 (1913); Bullock in K.B. 1933: 94 (1934); F.P.N.A. 2: 289 (1947); T.T.C.L.: 1 (1949); Meeuse in Fl. Pl. Afr. 32: t. 1268 (1958); Blundell, Wild Fl. E. Afr.: 385, fig. 701 (1987); U.K.W.F., ed. 2: 274 (1994); K.T.S.L.: 596 (1994); Ruffo *et al.*, Cat. Lushoto Herb. Tanzania: 1 (1996); Lebrun & Stork, Enum. Pl. Afr. Trop. 4: 466 (1997); Ensermu in F.E.E. 5: 356 (2006)
  [*A. arboreus* sensu C.B.Clarke in F.T.A. 5: 106 (1899); Lindau in Z.A.E.: 299 (1912) & in Wiss. Ergebn. Schwed. Rhod.-Kongo Exp., Bot. 1911–12: 306 (1916); De Wildeman, Contrib. Fl. Katanga: 202 (1921); F.P.S. 3: 165 (1956); K.T.S.: 16 (1961), *non* Forssk. (1775)]
  A. ugandensis C.B.Clarke in J.L.S. 37: 527 (1906). Type: Uganda, Masaka District: Buddu, *Dawe* 237 (K!, lecto.; selected by Vollesen in K.B. 62: 233 (2007)
  A. flamandii De Wild. in B.J.B.B. 4: 427 (1914) & in Pl. Beq. 4: 26 (1926). Type: Congo-Kinshasa, Kiamohanga, *Flamand* s.n. (BR!, holo.)
  A. arboreus Forssk. forma *albiflorus* Fiori in Nuov. Giorn. Bot. Ital. 47: 41 (1940). Type: Ethiopia, Agheresalam to Uondo, *Saccardo* s.n. (FT!, holo.)
  A. polystachius Delile var. *pseudopubescens* Cufod., Enum. Pl. Aeth.: 954 (1964), *nom. nud.*

NOTE. One of the most characteristic forest margin species in Uganda and NW Tanzania. It often forms a 5–10 m wide belt between the forest and the tall grassland, thus protecting the forest from the grassfires and eventually allowing the forest to expand.
  Iversen in Symb. Bot. Ups. 29, 3: 160 (1991) reports *A. pubescens* from both E and W Usambara Mts. No species of *Acanthus* has been collected in the Usambaras; it is unclear what Iversens records refer to.
  White-flowered forms are apparently much more common in Uganda than elsewhere. Occasionally these are cultivated in Kenya.

2. **Acanthus kulalensis** *Vollesen* in K.B. 62: 239 (2007). Type: Kenya, Northern Frontier District: Mt Kulal, *Luke* 10818 (K!, holo.; BR!, CAS!, EA!, K!, iso.)

Shrub to 2.5 m tall; young stems densely puberulous and with scattered longer hairs. Leaves with petiole to 7 mm long; lamina ovate to elliptic in outline, largest 10–17 × 4–7.5 cm, deeply lobed with large triangular spine-tipped lobes, each lobe with 1–2 spines on both sides, apex acuminate, spine-tipped, base cuneate to truncate, beneath puberulous or densely so on midrib and larger veins, glabrous to puberulous on lamina, above puberulous on midrib, otherwise glabrous or with a few scattered hairs; stipule-like interpetiolar leaves present, with lamina up to 5 mm long, spiny. Cymes terminal or also axillary from uppermost axils, 3–14 cm long; rachis densely sericeous-tomentose, with several pairs of sterile bracts at base; bracts leathery, with 3–5 prominent longitudinal veins, straw-coloured, obovate, densely sericeous in basal part, sparsely so upwards, 2.3–3 cm long in middle of cyme, spine-tipped and with 12–25 large and small teeth per side, longest ± 7 mm, small teeth inserted on larger near their base, teeth densely hairy throughout with longest hairs 1–2 mm long; bracteoles linear-lanceolate, 2.3–3.1 cm long, with 5–7 teeth per side, longest ± 7 mm long. Calyx green with purple tips, thickened part yellow, sericeous or densely so with hairs to 3 mm long, conspicuously ciliate; dorsal and ventral sepals ovate or ovate-elliptic, dorsal 1–3-veined, 2.7–3.2 cm long, obtuse to acute and spine-tipped or with acuminate 2-toothed tip and with 1–8 lateral teeth per side, longest tooth ± 4 mm long, ventral sepal 2-veined, 2.3–2.7 cm long, with 2 triangular spine-tipped lobes ± 5 mm long and with 1–6 lateral teeth per side, longest tooth ± 3 mm long, lateral sepals 1-veined, ovate, 2–2.2 cm long, acuminate, without or with 1–2 small teeth per side. Corolla tube white, limb pinkish-mauve; tube 2–3 mm long below thickened rim which is 4–7 mm long and extended dorsally into a 2–7 mm long tooth; limb 3.5–5 × 3–3.8 cm, below puberulous, above sparsely puberulous; callus with broad central groove and two narrow lateral ridges. Filaments 1.8–2 cm long, 2 dorsal hairy on inside near base; anthers 4–4.5 mm long, glabrous or with long curly hairs on sides. Capsule and seed not seen.

KENYA. Northern Frontier District: Mt Kulal, Oct. 1955, *Smart* 18! & 25 July 1976, *C.R. Field* 102! & 26 Dec. 2004, *Luke* 10818!
DISTR. **K** 1; not known elsewhere
HAB. Grassy glades in dry montane forest and on forest margins; 1950–2100 m

NOTE. Differs from *A. polystachius* in the long silky indumentum on bracts and calyx, the longer bract teeth, the longer sepals and in the long dorsal tooth on the thickened corolla rim. It also occurs a long way outside the distribution area of *A. polystachius*.

3. **Acanthus austromontanus** *Vollesen* in K.B. 62: 236 (2007). Type: Tanzania, Njombe District: Sunzi Mission, *Leedal* 7161 (K!, holo.; DSM!, K!, iso.)

Shrub to 3(–4) m tall; young stems puberulous or sparsely so. Leaves with petiole 0.5–1(–2) cm long; lamina elliptic or narrowly so in outline, largest 10–19 × 5.5–9 cm, deeply lobed with large triangular spine-tipped lobes, each lobe with a single spine on antrorse side, apex acute, spine-tipped, base truncate, beneath puberulous or sparsely so and with longer glossy hairs on midrib and larger veins, glabrous on lamina, above puberulous on midrib, otherwise glabrous; stipule-like interpetiolar leaves present (rarely absent), with lamina up to 7 mm long, spiny, soon caducous. Cymes 7–17(–20 in fruit) cm long; rachis sericeous-tomentose, with several pairs of sterile bracts at base; bracts with 3 prominent longitudinal veins, brown to purple, obovate, sparsely sericeous-puberulous or -pubescent, 2.7–3.7 cm long in middle of cyme, spine-tipped and with 15–28 large and small teeth per side, longest 8–12 mm, small teeth inserted on large, teeth with scattered hairs near base only, less than 1 mm long; bracteoles lanceolate to obovate, 3.2–3.8 cm long, toothed in apical $^1/_2$–$^3/_4$ with up to 25 teeth per side. Calyx green with purple apical part and teeth and yellow thickened part, sparsely to densely sericeous-pubescent; dorsal sepal 1–3-veined, ovate with long parallel-sided apical part, 3.5–4 cm long, acute, with 3–20 lateral teeth per side, longest tooth 3–5 mm long, ventral sepal 2-veined, obovate, 3.4–3.8 cm long, with 2 triangular lobes which are acute or 2–4-toothed and ± 8 mm long, with 6–17 lateral teeth per side, longest tooth ± 6 mm long, lateral sepals 1-veined, ovate-elliptic, 2.2–2.7 cm long, acuminate, without or with up to 12 lateral teeth per side (one side often with more than the other). Corolla tube white, limb deep purple; tube 6–7 mm long below thickened rim which is 8–10 mm long and deeply split dorsally; limb deflexed, 4–5.5 × 3–3.5 cm, finely and sparsely puberulous above and below; callus with broad central groove and two narrow lateral ridges. Filaments 2.7–3.3 cm long, glabrous; anthers, 4–5.5 mm long, with sparse long curly hairs on sides and sessile glands dorsally. Capsule ± 3 cm long. Seed (immature) discoid, glabrous, ± 8 × 6 mm.

TANZANIA. Njombe District: Ukinga, Sunzi Mission. 20 Aug. 1982, *Leedal* 7161! & Milo, 4 Sep. 1986, *P. Linder* 3861!; Njombe/Mbeya District: Itingule, Aug. 1936, *McGregor* 27!
DISTR. **T** 7; not known elsewhere
HAB. Montane grassland and forest margins, occasionally used as a hedge plant; 2000–2300 m

NOTE. Clearly related to *A. polystachius* but with a much larger differently coloured corolla and with different bracts and calyx.

4. **Acanthus eminens** *C.B.Clarke* in F.T.A. 5: 107 (1899); Bullock in K.B. 1933: 94 (1934); F.P.S. 3: 166 (1956); K.T.S.: 16 (1961); E.P.A.: 952 (1964); Blundell, Wild Fl. E. Afr.: 385, fig. 601 (1987); U.K.W.F., ed. 2: 274 (1994); K.T.S.L.: 596 (1994); Lebrun & Stork, Enum. Pl. Afr. Trop. 4: 466 (1997); Friis & Vollesen in Biol. Skr. 51(2): 435 (2005); Ensermu in F.E.E. 5: 355 (2006); Vollesen in K.B. 62: 239 (2007). Type: Kenya, Mau, *Scott Elliot* 6926 (K!, holo.)

Shrubby herb or shrub to 3 m tall; young stems glabrous. Leaves with petiole 0.5–2(–2.5) cm long; lamina narrowly elliptic to elliptic or obovate in outline, largest 13–37 × 5–15 cm, deeply lobed with large triangular spine-tipped lobes, each lobe with a spine on antrorse side, apex acuminate to cuspidate, spine-tipped, base attenuate to cuneate, glabrous; each node with one to several persistent sharp terete straw-coloured interpetiolar downwardly directed pungent pale yellow spines to 1.5 cm long, occasionally some with a small lamina. Cymes solitary or also 2 from upper leaf-pair, (5–)7–30 cm long; rachis upwards pubescent and with few to many short to long capitate glands, with several pairs of sterile bracts at base, sometimes merging into the vegetative leaves; bracts green to purplish, ovate to elliptic, finely puberulous to sparsely pubescent and upwards with few to many stalked capitate glands, 1.4–2.8 cm long in middle of cyme, spine-tipped and with 4–8 teeth per side (rarely entire); bracteoles subulate to linear, 5–11 mm long, entire. Calyx green to purplish, puberulous and with few to many stalked capitate glands; dorsal and ventral sepals longer than bract, oblong-obovate, similar or dorsal slightly longer, 2.1–4 cm long, dorsal entire or 2–3-toothed, ventral 2–3-toothed or irregularly toothed, lateral lanceolate, 1.3–2.5 cm long, acuminate. Corolla dark blue to dark purple (rarely pale blue to almost white); tube 4–5 mm long below thickened rim which is 3–4 mm long; limb 5-lobed, 3.5–5 × 3.5–5 cm, below puberulous and with capitate glands, above glabrous; callus with narrow central groove. Filaments 1.7–2.2 cm long, glabrous or with stalked glands near apex; anthers 6–9 mm long, densely glandular dorsally. Capsule 2.3–2.5 cm long. Seed ellipsoid to round, 9–10 mm long, glabrous.

UGANDA. Mbale District: Mt Elgon, Sipi, 16 Feb. 1924, *Snowden* 828! & 11 Dec. 1938, *A.S. Thomas* 2595! & 4 Feb. 2002, *Lye & Namaganda* 25513!
KENYA. Trans Nzoia District: 5 km SW of Suam Saw Mills, 22 Dec. 1967, *Mwangangi* 404!; Meru District: NW Mt Kenya, Marimba Forest, 14 Oct. 1960, *Verdcourt* 3000!; Kericho District: SW Mau Forest, Saosa Catchment, 14 Jan. 1959, *Kerfoot* 701!
DISTR. **U** 3; **K** 3–6; Sudan (Imatong Mts), Ethiopia
HAB. Primary and disturbed wet evergeen montane forest, forest margins, secondary grassland, often forming large stands, occasionally used as a hedge plant; (1200–)1500–2600 m

NOTE. Always easily recognised by the spinose nodes and large sepals combined with small bracts. This species must surely also occur in the N Ugandan mountains (**U** 1) as it is common in the Imatong Mts on the Sudan side.

5. **Acanthus ueleensis** *De Wild.* in Ann. Mus. Congo Bot., ser. 5, 3: 270 (1910); Napper in K.B. 24: 332 (1970); Lebrun & Stork, Enum. Pl. Afr. Trop. 4: 466 (1997); White *et al.*, For. Fl. Malawi: 113 (2001); Burrows, Pl. Nyika Plateau: 48 (2005); Vollesen in K.B. 62: 245 (2007). Type: Congo-Kinshasa, Uele River, *Seret* 272 (BR!, holo.; BR!, iso.)

Perennial herb to 1(–2) m tall with creeping rhizome; young stems somewhat fleshy, subglabrous to sparsely sericeous or sparsely puberulous. Leaves with petiole 0.5–2.2(–2.8) cm long; lamina elliptic to obovate in outline, largest 14–30 × 5–15 cm, varying (sometimes on the same stem) from unlobed with remotely denticulate margin to deeply lobed with oblong to triangular spine-tipped lobes, each lobe with 1–3 teeth on antrorse or on both sides, apex acuminate to cuspidate, spine-tipped, base attenuate to truncate with pair of large spines right at base and sometimes detached, sparsely puberulous (rarely pubescent) on midrib and larger veins, on lamina with scattered stiff erect bulbous-based hairs making surface rough to the touch, above sometimes without stiff hairs; alternate pair(s) of leaves absent. Cymes solitary, 3–15(–26) cm long (rarely with only 2–4 flowers apically on stem); rachis glabrous to puberulous, without or with 1-several pairs of sterile bracts at base; bracts green or purple-tipped or -mottled, elliptic-obovate, sparsely and minutely to densely puberulous, 2.5–4.7 cm long in middle of cyme of which the long triangular straight or recurved spine-tipped apical part 0.9–2.2(–3) cm, with 4–12 large and small teeth per side (sometimes small absent), longest 3–8 mm, small teeth inserted between large; bracteoles linear to linear-lanceolate, 1.4–2.8 cm long, entire or with 1–3 teeth per side. Calyx green or with purple apical part and sometimes veins, glabrous to puberulous; dorsal sepal ovate to elliptic, narrowing gradually or with parallel-sided apical part, 3.4–4.6 cm long, acute (rarely 2-toothed), no lateral teeth, ventral oblong-elliptic, narrowing gradually or with parallel-sided apical part, 2.6–3.8 cm long, bifid with 2 triangular teeth 1–4 mm long (these sometimes with 1–2 small additional teeth), without (very rarely with a single) lateral teeth, lateral lanceolate to narrowly ovate, 1.1–2.3 cm long, cuspidate, without (rarely with a single) lateral teeth per side. Corolla pale pink to pink or mauve (rarely white with pink veins); tube 3–6 mm long below thickened rim which is 3–6 mm long and with a broad shallow split dorsally without tooth; limb spreading, 5-lobed, 3–4.5 × 2.5–5 cm, below glabrous to sparsely puberulous, towards base with large glandular swellings, above glabrous; callus not grooved. Filaments 1.5–2.5 cm long, glabrous; anthers 7–10 mm long, without or with sparse long curly hairs on sides and without or with sessile glands dorsally. Capsule ± 2.5 cm long. Seed rhomboid, 7–8 mm long, sericeous.

UGANDA. Bunyoro District: Budongo Forest, Nov. 1935, *Eggeling* 2272!; Toro District: Bwamba Pass, 1925, *Maitland* 931! & 31 Jan. 1945, *Greenway & Eggeling* 7058!
TANZANIA. Kigoma District: Mt Livandabe [Lubalisi], 30 May 1997, *Bidgood et al.* 4199!; Ulanga District: 35 km S of Mahenge, Sali, 20 May 1932, *Schlieben* 2230!; Njombe District: Livingstone Mts, Madihani, 15 June 1992, *Gereau & Kayombo* 4747!
DISTR. U 2; T 4, 6, 7; Sudan, Congo-Kinshasa, Rwanda, Burundi, Malawi, Mozambique
HAB. Undergrowth in wet evergreen intermediate and montane forest, often along streams or in forest swamps, riverine forest; (700–)900–2500 m

SYN. [*Acanthus montanus* sensu C.B.Clarke in F.T.A. 5: 107 (1899) pro parte; Lindau in Z.A.E.: 299 (1912); De Wildeman, Contrib. Fl. Katanga: 202 (1921); F.P.N.A. 2: 292 (1947), *non* (Nees) T. Anderson (1863)]
    A. vandermeirenii De Wild. in B.J.B.B. 4: 428 (1914). Type: Congo-Kinshasa, between Lubile and Lukuya Rivers, *Vandermeiren* s.n. (BR!, holo.)
    A. vandermeirenii De Wild. var. *violaceo-punctatus* De Wild., Pl. Beq. 4: 27 (1926). Type: Congo-Kinshasa, Kifuku (Irumu), *Bequaert* 2684 (BR!, holo.)
    A. ueleensis De Wild. subsp. *mahaliensis* Napper in K.B. 24: 332 (1970); Lebrun & Stork, Enum. Pl. Afr. Trop. 4: 466 (1997). Type: Tanzania, Mpanda District: Mt Kungwe, Ntali River, *Newbould & Harley* 4594 (K!, holo.; K! iso.)

10. BLEPHARIS

A.L. Juss., Gen. Pl.: 103 (1789); Nees in DC., Prodr. 11: 265 (1847); G.P. 2: 1089 (1876); Lindau in E. & P. Pf. IV, 3b: 316 (1895); C.B. Clarke in F.T.A. 5: 94 (1899); Obermeijer in Ann. Transv. Mus. 19: 105 (1937); Vollesen, Blepharis: 30 (2000)

Acanthodium Delile, Fl. Aeg.: 97 & t. 33, fig. 2 (1813); Nees in DC., Prodr. 11: 273 (1847)

Annual or perennial herbs, subshrubs or shrubs. Leaves in whorls of 4 (very rarely opposite), one pair smaller than the other, sessile or subsessile, margin revolute, apex often with a pungent mucro. Flowers in spiciform cymes with numerous flowers, each subtended by a single bract (Subgen. *Acanthodium*) or spike reduced to a single terminal flower (Subgen. *Blepharis*); cyme then solitary or several together in axils of ordinary leaves each supported by (3–)4(–5) pairs of decussate bracts (Sect. *Blepharis*) or clustered in heads surrounded by floral leaves differing in shape and texture from ordinary leaves and with varying numbers of irregularly arranged bracts (Sect. *Scorpioideae*); bracts coriaceous, usually toothed (inner glumaceous in Sect. *Scorpioideae*); bracteoles 2 or absent, usually linear to lanceolate. Calyx divided to base into 4 glumaceous sepals, included in bracts or ± exserted, basal 1–2 mm thickened and horny; dorsal sepal longer than the others, 3(–5)-veined and 1–3-toothed apically, ± lateral teeth; ventral sepal 2(–5)-veined, bifid (rarely entire); lateral sepals lanceolate to ovate, 1-veined. Corolla with tube narrow basally, constricted below the expanded thickened and spongy upper 1–3 mm, 1–2 mm below insertion with band of upwardly directed hairs; limb horizontal, split dorsally to give a 3–5-lobed lower lip, with a thickened and spongy central callus which often terminates in 1–3 apical flanges, below puberulous to sericeous, above puberulous from sharply bent hairs (through 90°). Stamens 4, usually exserted (but hidden under dorsal sepal), inserted just inside tube on top of a thickened rim, two posterior held above anterior; filaments bony, posterior pair linear, curved, with knee at base, anterior pair slightly shorter, straight, broad and flattened, bifurcate at top into a short branch bearing the anther and an acute to obtuse ventral tooth-like appendage; anthers 1-thecous, elliptic to oblong, anterior pair slightly larger, fertile locule usually only $^1/_2$–$^2/_3$ of total length (whole length in Subgen. *Acanthodium*), densely bearded and with large apical tufts of long hairs, usually glandular. Ovary glabrous, 2-locular with 2 ovules per locule; disk large, annular; style usually glabrous, with two tufts of short glandular hairs at base; stigma subentire or with two distinct lanceolate subacute lobes. Capsule 2(–4)-seeded, ovoid-ellipsoid, sessile, glossy, flattened dorsi-ventrally, central wall thick and woody. Seed discoid, covered with long white branched hygroscopic hairs.

A genus of 129 species in the Old World tropics and subtropics. Especially numerous in eastern and southern Africa.

NOTE TO THE KEY. Despite the inflorescences clearly being of a cymose origin they are for reasons of brevity and general appearance referred to as "spikes" or "flowers" in the key and in the descriptions.

1. Flowers in many-flowered spikes, each subtended
   by a single bract; corolla limb 5-lobed (Subgen.
   Acanthodium) 2
   Flowers in 1-flowered spikes, subtended either by
   (3–)4(–5) pairs of decussate bracts or by
   several irregularly arranged bracts; corolla
   limb 3(–5)-lobed (Subgen. *Blepharis*) 3
2. Annual; petiole without teeth; leaf-margin entire
   or with spreading teeth; corolla-limb (9–)12–20
   (–23) mm long .... 1. *B. edulis* (p. 102)
   Perennial; petiole toothed almost to base; leaf-
   margin with forwardly directed teeth;
   corolla-limb (21–)23–31 mm long .... 2. *B. boranensis* (p. 104)
3. Indumentum on upper surface of corolla-limb of
   capitate glands (sometimes with simple hairs
   intermixed) apically and bent hairs basally;
   bracts in 4 decussate pairs with some
   or all teeth branched (Sect. *Inopinata*) ..... 15. *B. inopinata* (p. 114)
   Indumentum on upper surface of corolla-limb of
   bent hairs all over; bracts not with
   branched teeth 4

4. Flowers solitary, supported by 2 pairs of
    decussate bracts; bracts and sepals with stalked
    capitate glands 17
    Flowers solitary or clustered, each supported by
    (3–)4(–5) pairs of decussate bracts, or clustered
    in lateral elongated or head-like inflorescences
    surrounded by floral leaves and by irregularly
    arranged bracts; bracts and sepals with or
    without stalked capitate glands 5
5. Flowers solitary or clustered in axils of vegetative
    leaves; bracts in (3–)4(–5) decussate pairs,
    usually with scabrid and/or glochidiate bristles;
    bracts and sepals never glandular (Sect. *Blepharis*) 6
    Flowers clustered in lateral elongated or head-like
    often pedunculate inflorescences, surrounded
    by floral leaves differing from vegetative leaves
    and by a number of irregularly arranged bracts;
    bracts and sepals with or without stalked
    capitate glands (Sect. *Scorpioidea*) 17
6. Apical part of bracts sharply reflexed; stems
    quadrangular 7
    Apical part of bracts not reflexed; stems not
    quadrangular 8
7. Annual herb; largest leaf 4.5–8 × 0.4–1.5 cm, hairy
    at least on margins and midrib; reflexed tip of
    bract with 4–8 scabrid bristles on each side   . .   12. *B. involucrata* (p. 112)
    Perennial herb; largest leaf 2.5–4 × 0.3–0.5 cm,
    glabrous; reflexed tip of bract without or with
    1–2(–5) smooth or indistinctly scabrid bristles
    on each side . . . .   13. *B. refracta* (p. 113)
8. Bracts with entire margin or with 1 small smooth
    tooth per side 9
    Bracts with (1–)3–10 scabrid and/or glochidiate
    bristles on each side 12
9. Annual herb; inner bracts 15–21 mm long, with
    purple bands or blotches; spikes dispersed
    whole when mature . . . .   7. *B. glumacea* (p. 107)
    Shrubby herbs or shrubs; inner bracts 4–13 mm
    long, not with purple bands or blotches; spikes
    not dispersed whole when mature 10
10. Young branches, bracts and sepals glabrous; inner
    bracts ± 13 mm long; upper sepal ± 20 mm long,
    lower ± 18 mm long . . . .   10. *B. sp. A* (p. 109)
    Young branches, bracts and sepals hairy; inner
    bracts 4–8 mm long; upper sepal 7–13 mm
    long, lower 6–11 mm long 11
11. Branch-systems becoming divaricately branched
    and spiny; dorsal sepal 13–14 mm long;
    corolla 13–15 mm long . . . .   9. *B. turkanae* (p. 109)
    Branch-systems not becoming divaricately
    branched and not spiny; dorsal sepal
    7–9.5(–12) mm long; corolla 8–11 mm long   . .   8. *B. tanae* (p. 108)
12. Inner bracts 16–25 mm long; dorsal sepal
    23–33 mm long; corolla 27–44 mm long 13
    Inner bracts (4–)6–13(–15) mm long; dorsal
    sepal 9–25 mm long; corolla 10–32 mm long 14

13. Scendent subshrub to 2 m tall; inner bracts
    20–25 mm long; bracteoles 25–30 mm long;
    dorsal sepal 28–33 mm long; corolla 37–44 mm
    long; filaments 12–14 mm long; anthers
    5.5–7.5 mm long .... 5. *B. chrysotricha* (p. 106)
    Trailing perennial herb; inner bracts 16–20 mm
    long; bracteoles absent; dorsal sepal 23–26 mm
    long; corolla 27–33 mm long; filaments
    8–9 mm long; anthers 3.5–5 mm long ...... 14. *B. tanganyikensis* (p. 113)
14. Bracteoles absent; spikes solitary or in clusters,
    dispersed whole when mature; corolla white or
    cream with purple veins; procumbent or
    scrambling perennial (rarely annual) herb
    (rarely subshrub) .... 11. *B. maderaspatensis* (p. 109)
    Bracteoles present; spikes always solitary; corolla
    pale yellow to yellow, pale blue to deep blue or
    mauve to purple 15
15. Procumbent perennial herb; spikes dispersed
    whole when mature; corolla 10–18 mm long,
    blue to mauve or purple; filaments 3–5 mm
    long; anthers 1.5–2 mm long .... 6. *B. integrifolia* (p. 106)
    Erect or spreading subshrub or shrub; filaments
    4–10 mm long; anthers 2–5 mm long 16
16. Spikes not dispersed whole when mature; inner
    bracts 7–13 mm long; dorsal sepal (14–)
    17–25(–28) mm long, ventral 13–22 mm long;
    corolla 19–32 mm long, pale blue to deep
    blue or purple; capsule 9–11 mm long ...... 3. *B. hildebrandtii* (p. 104)
    Spikes dispersed whole when mature; inner bracts
    4–7 mm long; dorsal sepal 11–15 mm long,
    ventral 9–13 mm long; corolla 12–20 mm long,
    pale yellow to yellow with purple veins; capsule ±
    7 mm long .... 4. *B. ogadenensis* (p. 105)
17. Bracts and sepals without stalked capitate glands 18
    Bracts and sepals with stalked capitate glands,
    usually all over but in a few species only near
    base of bracts and/or sepals 27
18. Annual herbs 19
    Perennial herbs or subshrubs 23
19. Two of the four leaves in a whorl reduced to 3-fid
    spines; filaments ± 1 mm long, ± included
    in corolla tube .... 16. *B. trispina* (p. 115)
    All leaves with lamina, not reduced to 3-fid
    spines; filaments 4–9 mm long, exserted 20
20. Corolla 12–13 mm long; dorsal sepal 10–14 mm
    long, ventral 10–13 mm long; all heads with
    less than 10 flowers .... 17. *B. pusilla* (p. 115)
    Corolla 14–28 mm long; dorsal sepal 13–22 mm
    long, ventral 10–20 mm long; some or all
    heads with more than 10 flowers 21
21. Some peduncles usually over 1 cm long; floral
    leaves and bracts puberulous; dorsal sepal with
    ligulate apical part; ventral sepal 10–12 mm long 21. *B. kenyensis* (p. 118)
    Heads subsessile; floral leaves and bracts
    glabrous, finely sericeous or pilose from long
    shiny hairs; dorsal sepal narrowing gradually
    to apex; ventral sepal 13–20 mm long 22

22. Floral leaves, bracts and sepals glabrous to finely sericeous; corolla (15–)18–28 mm long; filaments 4–9 mm long . . . .   18. *B. longifolia* (p. 116)
 Floral leaves, bracts and sepals with long pilose hairs; corolla 14–18 mm long; filaments 3–4 mm long . . . .   19. *B. itigiensis* (p. 116)
23. Dorsal and ventral sepals with lateral teeth, dorsal 27–28 mm long, ventral 26–27 mm long, lateral ± 24 mm long; capsule ± 13 mm long   33. *B. petraea* (p. 127)
 Dorsal and ventral teeth without lateral teeth, dorsal 13–28 mm long, ventral 10–26 mm long, lateral 8–21 mm long; capsule (where known) 9–10 mm long 24
24. Dorsal sepal 16–28 mm long, ventral 15–26 mm long; heads subsessile (rarely with peduncle up to 1(–2) cm long) 25
 Dorsal sepal 13–17 mm long, ventral 10–15 mm long; some peduncles always over 1 cm long 26
25. Vegetative leaves coriaceous, glossy, larger pair ovate to elliptic or narrowly so, some or all with 3 or more teeth per side, largest 0.8–3.8 cm wide . . . .   23. *B. grandis* (p. 120)
 Vegetative leaves chartaceous, not glossy, larger pair linear-lancolate or narrowly elliptic, entire or some with 1–2 teeth near base, largest 0.5–1 cm wide . . . .   25. *B. torrei* (p. 121)
26. Largest vegetative leaf 9–18 mm wide; dorsal sepal 15–17 mm long, ventral 13–15 mm long; corolla 23–30 mm long . . . .   22. *B. pratensis* (p. 119)
 Largest vegetative leaf 2–7 mm wide; dorsal sepal 13–15 mm long, ventral 10–12 mm long; corolla 19–22 mm long . . . .   21. *B. kenyensis* (p. 118)
27. Annual herbs 28
 Perennial herbs or subshrubs 33
28. Stems finely sericeous with downwardly directed hairs . . . .   27. *B. affinis* (p. 122)
 Stems puberulous to pilose with spreading hairs 29
29. Dorsal sepal with ligulate-spathulate apical part, distinctly narrowed below apex, with lateral veins forming wide angles with main longitudinal veins and curving up towards apex 30
 Dorsal sepal gradually narrowed to apex or slightly narrowed below apex, lateral veins forming narrow angles with main longitudinal veins 31
30. Some or all vegetative leaves toothed in lower part; corolla 2–3 cm long, callus on limb with 3 equally strong ribs . . . .   28. *B. panduriformis* (p. 123)
 Leaves all entire (rarely some with a single tooth); corolla 3–3.5 cm long, callus on limb with 2 large lateral ribs and almost absent central rib . . . . .   27. *B. affinis* (p. 122)
31. Vegetative leaves entire; floral leaves without transverse ladder-like veins; callus on corolla limb with 2 large lateral ribs and almost absent central rib . . . .   30. *B. menocotyle* (p. 124)
 Some or all vegetative leaves toothed; floral leaves with or without transverse ladder-like veins; callus on corolla limb with 3 equally strong ribs 32

32. Vegetative leaves all subequal or one pair up to 2
    times longer than the other; heads sessile or
    peduncle up to 1(–2) cm long; dorsal sepal
    14–21 mm long; ventral 12–19 mm long,
    lateral 10–18 mm long; corolla limb distinctly
    3-lobed . . . .   29. *B. tenuiramea* (p. 124)
    Some or all whorls of vegetative leaves with one
    pair 5 or more times longer than the other;
    peduncle (0.5–)1–7(–12) cm long; dorsal sepal
    20–27 mm long; ventral 17–24 mm long, lateral
    18–26 mm long; corolla limb indistinctly lobed   31. *B. katangensis* (p. 125)
33. Largest vegetative leaf 1.5–4.8 × 0.9–1.9 cm;
    procumbent herb with coriaceous leaves . . . . .   24. *B. ilicifolia* (p. 120)
    Largest vegetative leaf (4–)5–23 cm long (if under
    5 cm long then also under 5 mm wide); erect or
    straggling herbs, leaves usually chartaceous 34
34. All vegetative leaves entire (rarely some with a
    single tooth) 35
    Some or all vegetative leaves toothed 40
35. Largest leaf pair (2–)4–8 times longer than
    smaller pair, 0.3–0.6(–1) cm wide . . . .   35. *B. uzondoensis* (p. 130)
    Largest leaf pair less than 2(–3) times longer
    than smaller pair, 0.3–3 cm wide 36
36. Plant with erect wiry densely pubescent stems
    0.75–1.25 m tall . . . .   33. *B. petraea* (p. 127)
    Plant with erect or decumbent stems, if more
    than 0.6 m long then decumbent 37
37. Floral leaves and bracts without capitate glands,
    conspicuously shaggy-ciliate; sepals with capitate
    glands only near base . . . .   20. *B. asteracantha* (p. 118)
    Floral leaves, bracts and sepals with long capitate
    glands all over 38
38. Callus on corolla limb with 3 equally strong ribs
    and 3 large apical flanges; stems puberulous to
    densely pubescent (rarely sericeous) . . . .   32. *B. stuhlmannii* (p. 126)
    Callus with 2 strong ribs (middle rib ± absent),
    two lateral apical flanges much larger than
    central; stems sericeous (rarely puberulous) 39
39. Young stems with spreading or downwardly
    directed hairs; largest vegetative leaf
    0.2–1.2(–2.2) cm wide; dorsal and ventral
    sepals with lateral teeth; anther appendage
    (1.5–)2–3 mm long . . . .   27. *B. affinis* (p. 122)
    Young stems with upwardly directed hairs; largest
    vegetative leaf 2–3.5 cm wide; dorsal and
    ventral sepals without lateral teeth; anther
    appendage ± 1 mm long . . . .   26. *B. tanzaniensis* (p. 121)
40. Floating aquatic herb with inflated spongy stems   20. *B. asteracantha* (p. 118)
    Terrestrial herbs or subshrubs 41
41. One pair of leaves in a whorl more than 5 times
    longer than the other pair (rarely down to 3
    times), smaller pair 3(–5)-veined from base
    and with 1(–2) large teeth per side 42
    Vegetative leaves all subequal or larger pair up to
    3 times longer than smaller pair, smaller pair
    not 3(–5)-veined from base, with more
    teeth per side 43

42. Stems 2–5 mm in diameter; largest leaf 0.7–2.5
     (–3) cm wide ....   34. *B. buchneri* (p. 128)
    Stems 1–1.5 mm in diameter; largest leaf 0.3–0.6
     (–1) cm wide ....   35. *B. uzondoensis* (p. 130)
43. Upper sepal ± 30 mm long.; rare form ....   34. *B. buchneri* (p. 128)
    Upper sepal 18–28 mm long 44
44. Callus on corolla limb with 3 equally strong ribs
     and 3 large apical flanges; stems puberulous
     to densely pubescent (rarely sericeous) .....   32. *B. stuhlmannii* (p. 126)
    Callus with 2 strong ribs (middle rib ± absent),
     two lateral apical flanges much larger than
     central; stems sericeous (rarely puberulous)   27. *B. affinis* (p. 122)

The East African species fall into four distinct groups:

Subgen. **Acanthodium** (*Delile*) *Oberm.* in Ann. Transv. Mus. 19: 109 (1937);
Vollesen, Blepharis: 32 (2000)
Flowers in many-flowered terminal (often seemingly lateral) spikes, each
subtended by a single bract. Corolla limb 5-lobed, outer lobes small almost tooth-like.
Tooth on anterior stamens as long or longer than anther, acute. Fertile theca as long
as the anther. Style hairy at base. Species 1–2.

Subgen. **Blepharis**
Flowers in 1-flowered spikes (many-flowered spikes reduced to one terminal
flower), each subtended either by (3–)4(–5) pairs of decussate bracts or by several
irregularly arranged bracts. Corolla limb 3-lobed. Tooth on anterior stamens
shorter than anther, broadly rounded. Fertile theca $\frac{1}{2}$–$\frac{2}{3}$ the length of the anther.
Style glabrous.

Sect. **Blepharis**, Vollesen, Blepharis: 159 (2000)
Spikes solitary or several clustered in axils of ordinary leaves, often dropping
whole when mature, narrowly ovoid to ovoid or ellipsoid, sessile. Bracts in
(3–)4(–5) decussate pairs, glumaceous, often with scabrid and/or glochidiate
bristles. Bracts and floral parts not with stalked capitate glands (very rarely with).
Bracteoles present or absent. Species 3–14.

Sect. **Inopinata** *Vollesen*, Blepharis: 231 (2000)
Spikes on a flattened axis in axils of ordinary leaves, persistent. Bracts in 4 decussate
pairs, with terminal and lateral bristles, some of the lateral bristles branched.
Bracteoles absent. Indumentum on lower surface of corolla of capitate glands only,
on upper surface of capitate glands apically and bent hairs basally. Species 15.

Sect. **Scorpioidea** *Vollesen*, Blepharis: 236 (2000)
Spikes on a flattened axis, clustered in lateral heads surrounded by floral leaves
differing in shape from the vegetative leaves (rarely with the spikes widely spaced
and supported by 2 pairs of decussate bracts, but then floral parts with dense
capitate glands). Floral leaves green, coriaceous, with toothed margin (rarely
entire) and pungent tip, teeth smooth (often hairy near base), never bristly or
glochidiate. Bracts irregularly arranged, outer like the floral leaves, inner
glumaceous, gradually smaller. Bracts and floral parts often with stalked capitate
glands; bracteoles absent. Species 16–35.

1. **Blepharis edulis** (*Forssk.*) *Pers.*, Syn. Pl. 2: 180 (1806); T. Anderson in J.L.S. 7: 36 (1863); Lindau in P.O.A. C: 369 (1895); C.B. Clarke in F.T.A. 5: 102 (1899) pro parte; Broun & Massey, Fl. Sudan: 341 (1929); Chiovenda, Fl. Somala: 251 (1929) & Fl. Somala 2: 350 (1932); Vollesen, Blepharis: 103 (2000); Friis & Vollesen in Biol. Skr. 51(2): 438 (2005); Ensermu in F.E.E. 5: 353 (2006); Thulin in Fl. Somalia 3: 380 (2006). Type: Yemen, Lohaje, *Forsskål* 905 (C!, holo.)

Erect or procumbent annual herb; stems up to 60 cm long, subglabrous to densely scabrid-puberulous when young. Leaves: lamina linear to lanceolate or narrowly elliptic, largest 2.5–12 × 0.3–1.2 cm, margin entire to sinuate-dentate, scabrid-puberulous or sparsely so, below also with longer hairs on midrib. Spikes terminal (in larger plants seemingly axillary due to development of lateral branches), 1–9(–12) cm long; peduncle 0–1(–2.5) cm long, with 1–2 pairs of small sterile bracts; fertile bracts elliptic to slightly obovate, 1.2–3.8 cm long, narrowing abruptly to a recurved pungent tip of $^1/_3$–$^1/_2$ the total length, each side with 3–8 straight teeth 1–7 mm long, pubescent to pilose or sparsely so (rarely puberulous) on veins, puberulous or sparsely so between; bracteoles 7–17 mm long, puberulous and with long pilose hairs. Sepals pubescent to pilose; dorsal ovate to elliptic with ligulate apical part, 10–18 mm long, 5-veined, apex truncate with 1–3 teeth; ventral elliptic or oblong, 8–12 mm long, 5-veined; lateral 5–9 mm long. Corolla bright blue or bluish mauve, with darker veins; tube 2–4 mm long; limb (9–)12–20(–23) mm long. Filaments 4–8 mm long, anterior pair hairy at base; appendage 3–5 mm long, as long or longer than the 2–4 mm long anthers. Capsule 7–9 mm long. Seed ± 5 × 5 mm. Fig. 16, p. 103.

UGANDA. Karamoja District: near Moroto, 28 Oct. 1939, *A.S. Thomas* 3078! & Karasuk, Amudat, Sep. 1959, *Tweedie* 1904! & Karita, 10 Oct. 1964, *Leippert* 5091!
KENYA. Turkana District: Lake Turkana [Rudolph], 17 May 1953, *Padwa* 157A!; Garissa District: Garissa–Jara Jila road, 26 Nov. 1978, *Brenan et al.* 14747!; Masai District: Nairobi–Magadi road, 11 Aug. 1951, *Bally* 8016!
TANZANIA. Masai District: road to Engaruka, 27 Feb. 1970, *Richards* 25550!; Lushoto District: 8 km SE of Mkomazi, 1 May 1953, *Drummond & Hemsley* 2324!; Iringa District: 90 km N of Iringa, 17 July 1956, *Milne-Redhead & Taylor* 11176!
DISTR. U 1; K 1–4, 6, 7; T 2, 3, 6, 7; West Africa to Ethiopia and Somalia, north to Egypt and through Arabia and Iran to NW India
HAB. Dry *Acacia-Commiphora* bushland, *Acacia* bushland and grassland, semidesert scrub, lava outcrops, hardpans, often on alkaline or saline soils; 150–1450 m

SYN. *Acanthus edulis* Forssk., Fl. Aegypt-Arab.: 114 (1775); Vahl, Symb. Bot. 1: 48 (1790)
   Acanthodium spicatum Del., Fl. Aegypt.: 97 & tab. 33, fig. 2 (1813); Nees in DC., Prodr. 11: 274 (1847); Solms-Laub. in Schweinf., Beitr. Fl. Aeth.: 102 (1867). Type: Egypt, Suez, *Delile* s.n. (P, holo.; K!, iso.)
   Acanthus tetragonus R. Br. in Salt, Voy. Abyss., App. IV: 65 (1814), *nom. nud.*
   Blepharis hirta (Nees) Martelli var. *latifolia* Martelli, Fl. Bogos: 65 (1886). Type: Eritrea, Keren, *Beccari* s.n. (FT!, holo.)
   B. edulis (Forssk.) Pers. var. *oblongata* Terrac. in Bull. Soc. Bot. Ital. 1892: 424 (1892). Type: Somalia, El Anot, *Baudi & Candeo* s.n. (FT!, holo.)
   [*B. linariifolia* sensu C.B.Clarke in F.T.A. 5: 100 (1899) pro parte; Chiovenda, Fl. Somala: 251 (1929) & Fl. Somala 2: 340 (1932); Napper, Israel J. Bot. 21: 166 (1972); U.K.W.F.: 579 (1974); Blundell, Wild Fl. E. Afr. 389 & fig. 848 (1987); Iversen in Symb. Bot. Ups. 29(3): 160 (1991); Kokwaro, Med. Pl. East Afr. (ed. 2): 22 (1993), *non* Pers. (1806)]
   [*B. hirta* sensu Chiovenda, Nuov. Giorn. Bot. Ital., N. S. 26: 101 (1919) & Atti Ist. Bot. Pavia, ser. 4, 7. 146 (1936), *non* (Nees) Martelli (1886)]
   B. obovata Chiov., Agric. Colon. 20: 48 (1926). Type: Somalia, *Gorini* 64 (FT!, holo.)
   B. edulis (Forssk.) Pers. forma *minima* Chiov. in Fl. Somala 2: 350 (1932). Type: Somalia, Mogadishu, *Senni* 563 (FT!, holo.)
   [*B. persica* sensu F.P.S. 3: 171 (1956); E.P.A.: 952 (1964), *non* (Burm. f.) Kuntze (1891)]

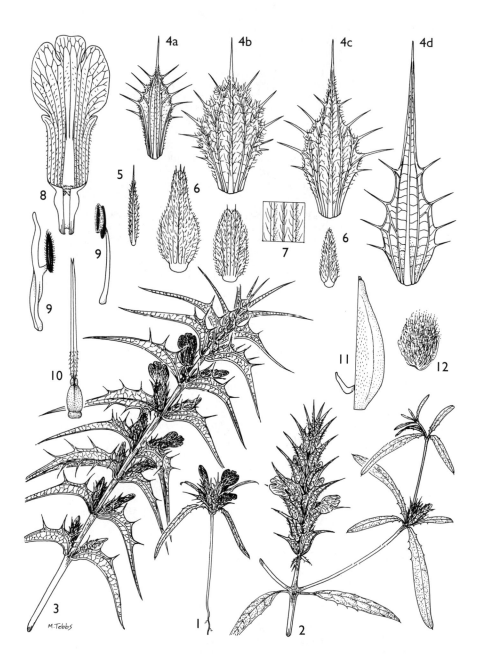

FIG. 16. *BLEPHARIS EDULIS* — **1**, habit (small plant), × ²/₃; **2**, habit (large plant), × ²/₃; **3**, habit (large plant, open spike), × ²/₃; **4a–d**, series of bracts, variation in shape and indumentum, × 2; **5**, bracteole, × 2; **6**, sepals, × 2; **7**, detail of surface of dorsal sepal, × 6; **8**, corolla (stamens removed), × 3; **9**, stamens, × 3; **10**, ovary, style and stigma, × 4; **11**, capsule, × 4; **12**, seed, × 2. 1 & 4a from *Mungai* 172/83, 2 from *Modha* 11, 3 from *Hepper* 5712, 4b from *Mathenge* 79, 4c from *Vessey & Coates* 16, 4d & 8–10 from *Wood* 3426, 5–7 from *Bally* 9767, 11–12 from *Goyder* 4004. Drawn by Margaret Tebbs. From Blepharis, a taxonomic revision: 106 (2000).

[*B. ciliaris* sensu Collenette, Fl. Saudi Arabia: 29 (1985); Miller, Pl. Dhofar: 8 (1988); Lebrun *et al.*, Cat. Pl. Vasc. Djibouti: 212 (1989); Hepper & Friis, Pl. Forssk. Fl. Aeg.-Arab.: 64 (1994); Audru *et al.*, Pl. Vasc. Djibouti, Fl. Ill. 2,2: 720 (1994); U.K.W.F., ed. 2: 274 (1994); Lebrun & Stork, Enum. Pl. Afr. Trop. 4: 473 (1997); Wood, Handb. Yemen Fl.: 268 (1997), *non* (L.) B. L. Burtt (1956)]

NOTE. *Blepharis edulis* shows an enormous plasticity. Plants from the beginning of the rainy season (or in dry years) are small and often only have a single spike. Through the rainy season (and into the dry where moisture prevails) the plants continue to grow and become increasingly branched, in the end producing large procumbent plants with thick semi-woody stems up to 60 cm long.

2. **Blepharis boranensis** *Vollesen*, Blepharis: 110 (2000); Ensermu in F.E.E. 5: 360 (2006); Thulin in Fl. Somalia 3: 380 (2006). Type: Ethiopia, 5 km S of Harar, Achim, *Burger* 3560 (K!, holo.; EA!, ETH!, FT!, US!, iso.)

Perennial herb with prostrate stems from woody rootstock; stems to 35 cm long, subglabrous to densely puberulous when young. Leaves sometimes purple tinged; lamina elliptic or narrowly so (rarely lanceolate), largest (3–)4–12 × (0.3–)0.5–1.5(–2) cm; petiole of some or all on each side with 1–2 teeth within 2 mm of base, margin on each side with (10–)20-over 50 (usually some over 20) fine forwardly directed teeth, glabrous to sparsely puberulous and with sparse to dense broad glossy hairs on midrib beneath. Spikes 2–7.5(–12) cm long; peduncle 0.5–1 cm long, with 4–6 pairs of sterile bracts; fertile bracts elliptic-obovate or broadly so, 2.2–3.5 × 0.8–1.6 cm of which the straight to slightly recurved narrowly triangular spinose tip 0.5–1 cm, each side with (5–)7–12 straight or slightly reflexed teeth of which the longest 4–7 mm, glabrous to minutely puberulous and usually with scattered to dense puberulous to pilose hairs, often only on veins; bracteoles 14–21(–23) mm long, minutely puberulous or sparsely so and distinctly puberulous-ciliate. Sepals sparsely to densely minutely puberulous or sericeous-puberulous, distinctly pilose-ciliate; dorsal ovate-elliptic or broadly so, 17–23 (–25) mm long, 5(–7)-veined from base, narrowing abruptly to a subligulate apical part which narrows gradually to an often irregularly 3-toothed apex (often appearing irregularly toothed because apical part becomes scarious and disintegrates); ventral broadly ovate-elliptic, 12–17(–20) mm long, 5(–7)-veined from base, narrowing gradually to an apex with 2 triangular and mucronate teeth; lateral 8–10 mm long. Corolla bright blue to dark blue, (25–)27–35 mm long of which the tube 4–7 mm. Filaments 7–10 mm long, glabrous to sparsely glandular, all four hairy towards base; appendage 4–6.5 mm long, from same length to longer than anther; anthers 4–6 mm long. Capsule 10–11 mm long. Seed 6–7 × ± 5 mm.

KENYA. Northern Frontier District: 6 km S of Funya Nyata, Koloba Hill, no date, *Powys* 927!
DISTR. **K** 1; Ethiopia, Djibouti, Somalia
HAB. *Acacia-Commiphora* bushland on rocky hills; no altitude given

NOTE. *Blepharis boranensis* is widespread in SE Ethiopia where it almost exclusively occurs on crystalline limestone. There is no indication of the substrate of the single Kenyan collection. The species is almost certainly more widespread in NE Kenya. It differs most conspicuously from *B. edulis* in the perennial habit, more numerous forwardly directed leaf-teeth, spiny petiole and larger corolla.

3. **Blepharis hildebrandtii** *Lindau* in E.J. 20: 29 (1894) & in P.O.A. C: 369 (1895) & in E. & P. Pf. IV, 3b: 317 (1895); Blundell, Wild Fl. E. Afr.: 388 & pl. 847 (1987); U.K.W.F. ed. 2: 274 (1994); Vollesen, Blepharis: 163 (2000); Ensermu in F.E.E. 5: 361 (2006). Type: Kenya, Voi River, *Hildebrandt* 2490 (B†, holo.). Neotype: Kenya, Voi, *E. Polhill* 459 (K!, neo, selected by Vollesen, l.c.).

Erect to spreading or procumbent subshrub or shrub, often forming cushions; stems up to 75 cm long (occasionally longer if growing through other plants), puberulous or sericeous-puberulous or sparsely so, sometimes only in a broad band. Leaves: lamina linear-lanceolate to elliptic, narrowly ovate or narrowly obovate, largest 0.8–4 × 0.2–0.7(–1.2) cm, often only partly developed at time of flowering, margin entire, glabrous to puberulous, scabrid, usually with dense white dots (bulbous bases of hairs that have rubbed off). Spikes solitary, persistent; bracts oblong or oblong-elliptic, puberulous or sparsely so, pilose along margin near apex, terminating in a recurved bristle and on each side with (1–)3–5(–6) finely scabrid and minutely glochidiate bristles 0.5–6 mm long; lower pair 2–5 mm long, second pair 4–7.5 mm long, third pair 5–10 mm long, inner pair 7–13 mm long; bracteoles (9–)10–17 mm long, with a terminal bristle 2–6 mm long. Sepals puberulous or sparsely so, veins sometimes pubescent, ciliate; dorsal ovate-elliptic or narrowly so, (14–)17–25(–28) mm long, with slightly ligulate apical part or narrowing gradually to the cuspidate to subacute apex (rarely with 1–2 small lateral teeth), apical part purplish to brownish; ventral similar, 13–22 mm long, with two long apical teeth; lateral 9–18 mm long. Corolla pale blue to deep blue or bluish purple, with darker veins, 19–32 mm long of which the tube 4–7 mm; callus with 3 apical flanges. Filaments 6–10 mm long, glabrous; appendage ± 1 mm long, broadly rounded; anthers 3–5 mm long. Capsule 9–11 mm long. Seed ± 6 × 5 mm, ovoid.

KENYA. Masai District: 40 km on Nairobi–Magadi road, 27 Mar. 1959, *Greenway* 9567!; Nairobi District: Nairobi, 17 Jan. 1951, *Verdcourt* 416!; Teita District: near Manyani, May 1971, *Tweedie* 4015!
TANZANIA. Masai District: Ol Donyo Sambu, 27 June 1941, *Hornby* 2112!; Lushoto District: Mkomazi, opposite Lassa, 30 Nov. 1935, *Burtt* 5332! & 8 km SE of Mkomazi, 1 May 1953, *Drummond & Hemsley* 2357!
DISTR. **K** 3, 4, 6, 7; **T** 2, 3; Ethiopia
HAB. *Acacia-Commiphora* bushland and grassland, usually on rocky slopes but also on alluvial soils in flat areas; 250–2150 m

SYN. *Blepharis fruticulosa* C.B.Clarke in F.T.A. 5: 99 (1899); U.K.W.F.: 579 (1971). Type: Kenya, Kedong Valley, *Scott Elliot* 6612 (K!, holo.)
     B. ruwenzoriensis C.B.Clarke in F.T.A. 5: 99 (1899); Ruffo *et al.*, Cat. Lushoto Herb. Tanzania: 3 (1996); Lebrun & Stork, Enum. Pl. Afr. Trop. 4: 474 (1997). Type: Kenya, Machakos District: Lanjoro, *Scott Elliot* 6379 (K!, holo.; BM!, iso.)
     [*B. setosa* sensu C.B.Clarke in F.T.A. 5: 98 (1899), quoad *Hildebrandt* 2490 et pro syn., *non* Nees (1847)]

NOTE. Even though the original collection is no longer extant, there is no doubt that *Blepharis hildebrandtii* is the correct name for what has for a long time been called *B. fruticulosa*. This is the only species occurring around Voi which fits the original description (subshrub, linear white-dotted leaves, large solitary flowers). The neotype fits the diagnosis in all respects except for slightly wider leaves.

4. **Blepharis ogadenensis** *Vollesen*, Blepharis: 167 (2000); Ensermu in F.E.E. 5: 361 (2006); Thulin in Fl. Somalia 3: 382 (2006). Type: Somalia, Hiraan, Bulo Burti, Halgen, *Kuchar* 16669 (K!, holo.)

Erect shrub to 1 m tall; young branches whitish puberulous or densely so, soon glabrous with greyish to brownish bark. Leaves: lamina oblong-elliptic to obovate, largest 0.5–1.5 × 0.2–0.4 cm, margin entire, strongly recurved, sparsely sericeous-puberulous beneath (densest on veins), above with dense minute bulbous-based hairs (covered in small white dots when older). Spikes solitary, dispersed whole when mature; bracts greenish to yellowish brown, elliptic or oblong, finely sericeous-puberulous or densely so, terminating in a recurved bristle and on each side with 1–3 retrorsely barbed and finely glochidiate bristles 1–7 mm long; lower pair 1–2 mm long, second pair 2–4 mm long, third pair 3–5 mm long, inner pair 4–7 mm long; bracteoles elliptic-obovate, 7–13 mm long, of which about half is a long terminal bristle. Sepals greenish to brownish or purplish mottled or uniformly purplish towards apex, puberulous or sparsely so, ciliate; dorsal ovate-elliptic, 11–15 mm long, narrowing gradually to the cuspidate apex (rarely slightly ligulate); ventral similar, 9–13 mm long, with two ± 1 mm long mucros; lateral 7–11 mm long. Corolla pale yellow or yellow with purple veins, 12–20 mm long of which the tube 3–6 mm; callus with 3 weak or distinct apical ribs and 3 apical flanges. Filaments 4–7 mm long, with scattered hairs and glands near base; appendage 0.5–1 mm long, broadly rounded; anthers 2–3 mm long. Capsule ± 7 mm long. Seed not seen.

KENYA. Northern Frontier District: Losai National Reserve, 10 km S of Laisamis, 7 Feb. 1979, *Jonsell & Moberg* 4615!
DISTR. **K** 1; Ethiopia, Somalia
HAB. Open *Acacia-Commiphora* bushland on rocky hills; ± 725 m

5. **Blepharis chrysotricha** *Lindau* in E.J. 20: 32 (1894) & in P.O.A. C: 370 (1895); C.B. Clarke in F.T.A. 5: 101 (1899); Lebrun & Stork, Enum. Pl. d'Afr, Trop. 4: 473 (1997); Vollesen, Blepharis: 171 (2000). Type: Tanzania, Mpwapwa District: Usagara, Mlali, *Stuhlmann* 199 (B†, holo.). Neotype: Tanzania, Mpwapwa, *Greenway* 2401 (K!, neo; BR!, EA!, iso., selected by Vollesen, l.c.).

Scandent or scrambling subshrub to 2 m tall; young stems sparsely sericeous-puberulous, hispid. Leaves: lamina ovate to elliptic, largest 7–9.5 × 3–4.2 cm, margin entire or with distant mucronate teeth, slightly scabrid, subglabrous to sparsely pubescent or pilose, densest on vein. Spikes solitary or more usually clustered towards end of branches and then subtended by small bract-like leaves, persistent; bracts whitish with conspicuous green veins, elliptic to obovate, puberulous to pubescent or sparsely so along veins and margins, terminating in a recurved bristle and on each side with 6–8 scabrid bristles 3–7 mm long; outer pair 7–12 mm long, second pair (7–)10–16 mm long, third pair (12–)15–19 mm long, inner pair 20–25 mm long; bracteoles 25–30 mm long. Sepals finely sericeous-puberulous or pubescent towards base (rarely also on veins), greenish or brownish towards apex; dorsal ovate-elliptic, 28–33 mm long, with long ligulate apical part, acute; ventral similar, 22–25 mm long; lateral 16–20 mm long. Corolla blue (rarely white or pale magenta); tube 12–14 mm long; limb 25–30 mm long, callus with 1 or 3 apical flanges. Filaments 12–14 mm long, glabrous; appendage ± 0.5 mm long, broadly rounded; anthers 5.5–7.5 mm long. Capsule ± 15 mm long, 4-seeded. Seed ± 6 × 5 mm.

TANZANIA. Mpwapwa District: Mpwapwa, 18 Aug. 1930, *Greenway* 2401!; Rufiji District: Beho Beho, 12 June 1977, *Vollesen* MRC 4660!; Iringa District: Mtandika Village, Lukose River, 18 May 1986, *Congdon* 73!
DISTR. **T** 5–7; not known elsewhere
HAB. Dry coastal forest and thicket, *Acacia-Commiphora* bushland, dry riverine forest, on sandy soil or on rocky slopes; (100–)700–1200 m

SYN. *B. hornbyae* Milne-Redh. in Hooker, Ic. Pl.: t. 3266 (1935); T.T.C.L.: 5 (1949); Vollesen in Opera Bot. 59: 79 (1980); Ruffo *et al.*, Cat. Lushoto Herb. Tanzania: 2 (1996); Lebrun & Stork, Enum. Pl. Afr. Trop. 4: 473 (1997). Type: Tanzania, Mpwapwa, *Hornby* 1 (K!, holo.; EA!, K!, P!, iso.)

NOTE. Even though the type of *Blepharis chrysotricha* was destroyed in Berlin, there can be no doubt as to its identity. No other Tanzanian species in sect. *Blepharis* has the large bracts and corolla described combined with long anthers with a very short tooth.

6. **Blepharis integrifolia** (*L.f.*) *Schinz* in Viert. Nat. Ges. Zürich 60: 416 (1915); Obermeyer in Ann. Transv. Mus. 19: 130 (1937); F.P.N.A. 2: 289 (1947); E.P.A.: 951 (1964); U.K.W.F.: 579 (1974); Champluvier in Fl. Rwanda 3: 441 (1985); U.K.W.F. ed. 2: 274 (1994); Lebrun & Stork, Enum. Pl. Afr. Trop. 4: 473 (1997); Vollesen, Blepharis: 173 (2000); Friis & Vollesen in Biol. Skr. 51(2): 438 (2005); Ensermu in F.E.E. 5: 361 (2006). Type: South Africa, Cape Province, *Thunberg* s.n. (UPS, holo.; K!, microfiche).

Procumbent mat-forming perennial herb with woody rootstock; stems up to 35 cm long, puberulous or sparsely so or sparsely to densely finely sericeous (rarely subglabrous). Leaves: lamina linear to elliptic or slightly obovate, largest 1–4 × 0.2–1.2 cm, margin entire, subglabrous with scabrid edges or with scattered (rarely dense) long pilose hairs. Spikes solitary, dispersed whole when mature; bracts straw-coloured to greenish, elliptic-oblong to obovate (rarely ovate), uniformly finely puberulous (rarely also glandular), usually puberulous along midrib and at base of bristles, terminating in a recurved bristle and on each side with 4–5(–6) finely scabrid and glochidiate bristles 1–2(–3) mm long; lower pair 2–3 mm long, second pair 3–6 mm long, third pair 5–8 mm long, inner pair 7–12 mm long; bracteoles 8–12(–15) mm long, with a terminal glochidiate bristle. Sepals finely puberulous (very rarely also glandular) and pilose-ciliate; dorsal oblong, 10–14(–19) mm long, with slightly ligulate apical part or narrowing gradually to the acute to truncate apex, sometimes with two small lateral teeth; ventral similar, 10–12(–16) mm long, with two broad apical teeth; lateral 7–11 mm long. Corolla blue to mauve or purple with darker veins, 10–18 mm long of which the tube 2–3 mm long; callus with central apical flange. Filaments 3–5 mm long, glabrous; appendage ± 0.5 mm long, broadly rounded; anthers 1.5–2 mm long. Capsule 7–9 mm long. Seed ± 6 × 4 mm, ovoid.

UGANDA. Karamoja District: Kangole, July 1957, *J. Wilson* 367!; Toro District: Lake Edward Plains, Nov. 1950, *Purseglove* 3508!; Kigezi District: Queen Elizabeth National Park, Ishasha River Camp, 15 May 1961, *Symes* 721!
KENYA. Northern Frontier District: Moyale, 31 July 1952, *Gillett* 13663!; West Suk District: near Moropus, Marich Pass, June 1966, *Tweedie* 3297!; Thika District: Thika River, 14 Falls, 17 June 1951, *Verdcourt* 521!
TANZANIA. Mwanza District: Kwimba, June 1932, *Rounce* 53!; Mbulu District: Tarangire National Park, 13 Mar. 1969, *Richards* 24355! & 14 Feb. 1970, *Richards* 25418!
DISTR. **U** 1, 2; **K** 1–4, 7; **T** 1–3; Sudan, Ethiopia, Congo-Kinshasa, Rwanda, Angola, Mozambique, Botswana, Namibia, Swaziland, South Africa; also in India
HAB. Grassland and open *Acacia* bushland, usually on clayey or loamy soil but occasionally on gravelly or stony soil; (200–)700–1900 m

SYN. *Acanthus integrifolius* L.f.; Suppl. Pl.: 294 (1782); Thunb., Fl. Cap. (ed. 2): 456 (1823)
   Blepharis saturejifolia Pers., Syn. Pl. 2: 180 (1807); Nees in DC., Prodr. 11: 265 (1847); T. Anderson in J.L.S. 7: 34 (1863); Lindau in P.O.A. C: 369 (1895) & in E. & P. Pf. IV, 3b: 317 (1895); De Wild., Pl. Beq. 4: 26 (1926). Type: as for *Acanthus integrifolius*.
   B. molluginifolia Pers. Syn. Pl. 2: 180 (1807); T. Anderson in J.L.S. 9: 500 (1867); C.B. Clarke in F.T.A. 5: 98 (1899). Type: India Orientalis, *Herb. Vahl* (C!, holo.)
   B. setosa Nees in DC., Prodr. 11: 265 (1847); C.B. Clarke in F.T.A. 5: 98 (1899), excl. syn.; Dinter in F.R. 15: 350 (1918). Type: South Africa, near Pretoria, Aapies River, *Burke* s.n. (K!, holo.)
   B. rupicola Engl., Hochgebirgsfl. Trop. Afr.: 389 (1892). Type: Ethiopia, Addi Dschoa, *Schimper* 521 (B†, holo.; BM!, K!; iso.)

NOTE. The Uganda material usually has appressed indumentum or is subglabrous and usually has no long pilose hairs on the leaves. The Ethiopia, Kenya and Tanzania material has spreading indumentum and long pilose hairs on the leaves. But in the material from southern Africa other combinations of these characters occur, and at the moment it seems wisest not to recognise any infraspecific taxa.

   Gillett 13663 from **K** 1, Moyale differs from all other East African collections in having scattered stalked capitate glands on bracts and sepals and also has very long bracteoles and sepals (maximum dimensions in the description), It is also said to grow in "montane scrub". More material is needed to decide if this collection represents a distinct taxon. Large bracts and sepals have been seen in material from southern Africa, but the capitate glands are unique in Sect. *Blepharis*, and have never been seen in any other collection.

7. **Blepharis glumacea** *S.Moore* in J.B. 18: 232 (1880); Engl., Hochgebirgsfl. Trop. Afr.: 390 (1892); Lindau in E. & P. Pf. IV, 3b: 317 (1895); C.B. Clarke in F.T.A. 5: 97 (1899); Heine in F.W.T.A. ed. 2, 2: 410 (1963); Binns, Checklist Herb. Fl. Malawi: 12 (1968); Björnstad in Serengeti Res. Inst. Publ. No. 215: 25 (1976); Lebrun & Stork, Enum. Pl. Afr. Trop. 4: 473 (1997); Vollesen, Blepharis: 190 (2000). Type: Angola, Huilla, between Catumba and Ohay, *Welwitsch* 5052 (BM!, holo.; C!, K!, P!, iso.)

Erect or procumbent annual herb; stems up to 30(–50) cm long, pubescent to pilose or sparsely so, usually only in a single band. Leaves: lamina linear to narrowly ovate or narrowly elliptic (widest near base of stems), largest 5–12 × 0.3–1.7(–2.5) cm, margin entire, scabrid, pilose or sparsely so above and on margins, below subglabrous or sparsely pilose along midrib. Spikes solitary, terminal on a lateral sericeous branch up to 5 cm long, surrounded by reduced leaves, dispersed whole when mature; bracts coriaceous, glabrous and glossy, straw-coloured with purplish bands and blotches, ovate to elliptic, acute to acuminate with straight tip, margin entire; lower pair 6–9 mm long, second pair 9–13 mm long, third pair 13–19 mm long, inner pair 15–21 mm long; bracteoles absent. Sepals coloured as bracts, very finely ciliate, otherwise glabrous; dorsal ovate-elliptic, 14–20 mm long, narrowing gradually to the acute or 2-toothed apex; ventral similar, 12–16 mm long, acuminate or 2-toothed at apex; lateral 9–14 mm long. Corolla white, pale blue or pale mauve; tube 3–5 mm long; limb 11–19 mm long, callus with 3 apical flanges, central usually smaller. Filaments 2–6 mm long, glabrous, often purple spotted; appendage 1–4 mm long, broadly rounded; anthers 1.5–4.5 mm long. Capsule 8–9 mm long. Seed ± 6 × 6 mm, ovoid.

TANZANIA. Dodoma District: 40 km S of Itigi, 20 Apr. 1964, *Greenway & Polhill* 11645!; Sumbawanga District: 11 km from Kawimbe, 4 May 1961, *Richards* 15102!; Songea District: 139 km E of Songea, 4 June 1956, *Milne-Redhead & Taylor* 10494!
DISTR. T 4, 5, 7, 8; Mali, Benin, Nigeria, Cameroon, Central African Republic, Chad, Congo-Kinshasa, Angola, Zambia, Malawi, Zimbabwe
HAB. *Brachystegia* woodland on grey to orange sandy to loamy soil, rocky outcrops; 900–1700 m

SYN. *Blepharis kassneri* S.Moore in J.B. 51: 213 (1913); De Wildeman, Etud. Fl. Katanga 2: 147 (1913) & Contrib. Fl. Katanga: 201 (1921); Lebrun & Stork, Enum. Pl. Afr. Trop. 4: 473 (1997). Type: Congo-Kinshasa, *Kässner* 3003 (BM!, holo.; BR!, E!, HBG!, K!, P!, Z!, iso.)

8. **Blephasis tanae** *Napper* in K.B. 24: 328 (1970); Kuchar, Pl. Somalia (CRDP Techn. Rep. Ser., no. 16): 245 (1986); Lebrun & Stork, Enum. Pl. Afr. Trop. 4: 474 (1997); Vollesen, Blepharis: 194 (2000); Thulin in Fl. Somalia 3: 383 (2006). Type: Kenya, 15 km S of Garissa, near Tana River, *Lucas* 21 (K!, holo.; EA!, iso.)

Densely branched cushion-forming subshrub; stems up to 20(–30) cm long, finely sericeous-puberulous in a single band (rarely uniformly), soon glabrous. Leaves: lamina linear to narrowly elliptic or narrowly obovate, largest 1.5–3.5(–5) × 0.2–0.8 cm, margin entire, glabrous to sparsely hispid-pubescent, densest along margins. Spikes solitary, persistent; bracts whitish to yellowish brown, ovate to elliptic, glabrous to sparsely sericeous-puberulent (rarely puberulous along midrib), terminating in a short recurved mucro and on each side entire or with one smooth lateral tooth 0.5–1 mm long; lower pair 1–2 mm long (sometimes absent), second pair 2–3.5 mm long, third pair 3–6 mm long, inner pair 4–7 mm long; bracteoles absent. Sepals coloured as bracts glabrous to finely and sparsely sericeous-puberulous; dorsal elliptic, 7–9.5(–12) mm long, with ligulate apical part, subacute and mucronate; ventral similar, 6–8(–10) mm long, with two short broad obtuse teeth (sometimes almost obsolete); lateral 3.5–6(–7) mm long. Corolla pale blue to deep blue or bluish purple, 8–11 mm long of which the tube 3–4 mm; callus without flanges or with a single indistinct one. Filaments 1–2 mm long, ± included in tube; appendage ± 0.5 mm long, broadly rounded; anthers 0.5–2 mm long. Capsule 4–5 mm long. Seed 3–4 × 3 mm.

KENYA. Garissa District: Garissa–El Lein road, 10 May 1974, *Gillett & Gachati* 20571!; Kitui District: turning to Sosoma on Garissa road, 24 Jan. 1954, *Bally* 9483!; Tana River District: Thika-Garissa road, 10 km from Tana River, 10 June 1974, *Faden* 74/788!
DISTR. K 1, 4, 7; Somalia
HAB. *Acacia-Commiphora* bushland on sandy to loamy alkaline to saline soils in places liable to seasonal flooding; 50–250(–600) m

9. **B. turkanae** *Vollesen*, Blepharis: 195 (2000). Type: Kenya, Turkana, Lokori–Sigor road, *Mathew* 6422 (K!, holo.; EA!, iso.)

Densely branched dwarf-shrub to 25 cm tall, branch-systems becoming divaricately branched and the single branches eventually becoming spine tipped; branches finely uniformly sericeous-puberulous, soon glabrous with smooth white bark. Leaves: lamina narrowly elliptic-obovate, largest 1–2.1 × 0.3–0.6 cm, margin entire, finely and sparsely puberulous. Spikes solitary, persistent; bracts greenish with white margins to yellowish brown, ovate-elliptic, finely sericeous, acuminate to acute with straight tip and entire margins; lower pair 2–3 mm long, second pair 3–4 mm long, third pair 6–7 mm long, inner pair 8–9 mm long; bracteoles absent. Sepals green, brownish towards margins, finely and uniformly sericeous; dorsal elliptic 13–14 mm long, with long ligulate apical part, acuminate; ventral similar, 11–12 mm long, with two acute teeth; lateral 8–9 mm long. Corolla cream with blue veins, 13–15 mm long of which the tube 4–5 mm; callus without apical flanges. Filaments 4–5 mm long, exserted, glabrous; appendage ± 0.5 mm long, broadly rounded; anthers 2–3 mm long. Capsule ± 6 mm long. Seed not seen.

KENYA. Northern Frontier District: Kalacha to Marsabit, 30 Dec. 2004, *Luke* 10868!; Turkana District: 15 km on Lokori–Sigor road, 28 May 1970, *Mathew* 6422! & near Lokori, 12 Aug. 1968, *Mwangangi & Gwynne* 1128! &
DISTR. **K** 1, 2; not known elsewhere
HAB. *Acacia* bushland on rocky lava hills; 350–1050 m

NOTE. Only known from four collections. A very distinct species recognised by the development of spiny branches, a character to my knowledge unique in *Blepharis*. It is related to *B. tanae* from which it also differs in the indumentum, in the larger bracts, sepals and corolla and in the different habitat.

10. **Blepharis sp. A**

Subshrub; stems up to 20 cm long, ± straggling, all parts glabrous. Leaves glaucous; lamina elliptic to obovate, largest ± 3 × 0.8 cm, margin entire. Spikes solitary, persistent; bracts whitish to straw-coloured, ovate-elliptic, acuminate to cuspidate with straight tip and entire margin; lower pair ± 2 mm long, second pair ± 6 mm long, third pair ± 10 mm long, inner pair ± 13 mm long; bracteoles absent. Sepals straw-coloured; dorsal elliptic, 19–21 mm long, with long ligulate apical part, acute; ventral similar, ± 18 mm long, with two short obtuse teeth; lateral ± 10 mm long. Corolla, capsule and seed not seen.

KENYA. Kitui/Tana River District: E of Ukamba, Nthua [Thua] River, 1911–12, *Lindblom* s.n.!
DISTR. **K** 4/7; not known elsewhere
HAB. No notes but probably similar to *B. tanae*.

SYN. *Blepharis sp. aff. tanae* of Vollesen, Blepharis: 195 (2000)

NOTE. A very distinct species which is only known from a single incomplete specimen collected over 90 years ago. It is related to *B. tanae* but differs in being completely glabrous and in the much larger bracts and sepals.

11. **Blepharis maderaspatensis** (*L.*) *Roth*, Nov. Pl. Sp. Ind. Or.: 320 (1821); Obermeyer in Ann. Transv. Mus. 19: 132 (1937); F.P.N.A. 2: 289 (1947); T.T.C.L.: 5 (1949); F.P.S. 3: 171 (1956); F.P.U.: 142 (1962); Heine in F.W.T.A. ed. 2, 2: 410 (1963); E.P.A.: 952 (1964), excl. syn. *Ruellia ciliaris*; Napper in K.B. 24: 323 (1970); U.K.W.F.: 579 (1974); Björnstad in Serengeti Res. Inst. Publ. No. 215: 25 (1976); Champluvier in Fl. Rwanda 3: 441 (1985); Blundell, Wild Fl. E. Afr.: 389, pl. 602 (1987); Iversen in Symb. Bot. Ups. 29(3): 160 (1991); Ruffo *et al.*, Cat. Lushoto Herb. Tanzania: 3 (1996); Lebrun & Stork, Enum. Pl. Afr. Trop. 4: 473 (1997); Vollesen, Blepharis: 198 (2000); Friis & Vollesen in Biol. Skr. 51(2): 438 (2005); Ensermu in F.E.E. 5: 363 (2006); Thulin in Fl. Somalia 3: 384 (2006). Type: India (LINN, holo.; K, microfiche)

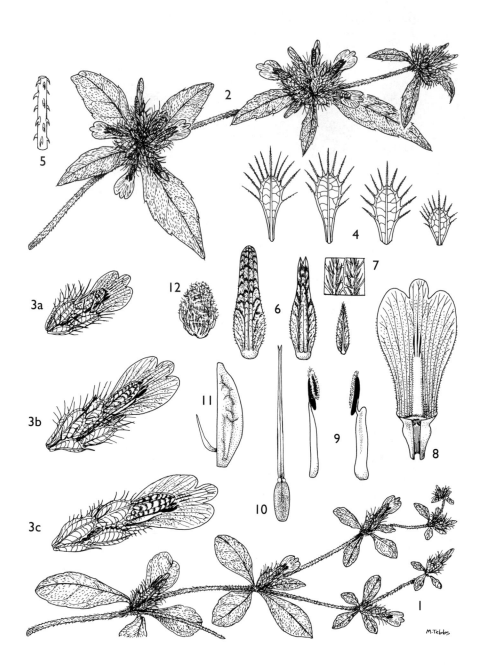

Fig. 17. *BLEPHARIS MADERASPATENSIS* — **1**, habit (plant with single flowers), × ²/₃; **2**, habit (plant with clustered flowers), × ²/₃; **3a–c**, three spikes showing variation in bract and corolla size, × 2; **4**, series of bracts, × 2; **5**, detail of part of bristle, × 16; **6**, sepals, × 2; **7**, detail of surface of dorsal sepal, × 6; **8**, corolla opened up (stamens removed), × 3; **9**, stamens, × 3; **10**, ovary and style, × 4; **11**, capsule, × 4; **12**, seed, × 2. 1 from *Bidgood et al.* 1063, 2, 3b, 4–7 & 10 from *Richards* 5531, 3a from *Burger* 989, 3c from *Mooney* 6828, 8–9 from *Richards* 5826, 11–12 from *Burtt* 6098. Drawn by Margaret Tebbs. From Blepharis, a taxonomic revision: 202 (2000).

Suberect, procumbent or scrambling perennial (rarely annual) herb (rarely a suberect or trailing subshrub); stems up to 2 m long, sometimes rooting at nodes, pubescent to pilose or sparsely so or sericeous, sometimes only in a single band (rarely subglabrous). Leaves often paler beneath; lamina elliptic (rarely ovate or slightly obovate), largest 2–9(–12.5) × 0.8–3.5(–5) cm, margin entire or distantly toothed, hispid-puberulous to pubescent or sparsely so, often densest on veins (rarely subglabrous). Spikes solitary or in clusters of up to 11, dispersed whole when mature; bracts greenish to yellowish brown, sometimes with purple veins, elliptic-obovate to orbicular, subglabrous to puberulous or pubescent (rarely densely so), terminating in a recurved bristle and on each side with 4–10 retrorsely barbed and finely glochidiate bristles 1–5 mm long; lower pair 2–8 mm long, second pair 4–12 mm long, third pair 5–12(–14) mm long, inner pair 6–13(–16) mm long; bracteoles absent. Sepals coloured as bracts, puberulous or sparsely so (rarely pubescent or sparsely pilose) and ciliate; dorsal ovate-elliptic, 9–22(–25) mm long, narrowing gradually to the acute to rounded apex; ventral similar, 8–20 mm long, with two small apical teeth; lateral 5–10(–12) mm long. Corolla white or cream with purple veins, 12–27(–30) mm long of which the tube 4–8 mm, callus with one (rarely 3) inconspicuous (rarely distinct) apical flange; hairs on upper surface of limb strongly bent. Filaments 2–9 mm long, sparsely glandular (rarely hairy); appendage 0.5–1 mm long, broadly rounded; anthers 2–4 mm long. Capsule 5–7 mm long. Seed ± 4.5 × 3 mm. Fig. 17, p. 110.

UGANDA. Karamoja District: Lodoketemit, July 1958, *Kerfoot* 310!; Acholi District: Agoro, Oct. 1947, *Dale* 481!; Busoga District: 12 km W of Kidera, 16 Oct. 1952, *G.H S. Wood* 414!
KENYA. Laikipia District: 6 km S of Suguta Marmar, 25 Oct. 1978, *Gilbert et al.* 5109!; Tana River District: 50 km S of Garsen, Kurawa, 24 Sep. 1961, *Polhill & Paulo* 553!; Lamu District: Takwa, 31 Aug. 1956, *Rawlins* 50!
TANZANIA. Mpanda District: Mahali Mts, Lumbye River Mouth, 20 Sep. 1958, *Jefford et al.* 2520!; Dodoma District: 4 km on Kilimatinde-Manyoni road, 14 Apr. 1988, *Bidgood et al.* 1063!; Mbeya District: Usangu Flats, 10 km beyond Nyamakuyu, 5 Mar. 1987, *Lovett et al.* 1949!; Pemba, near Ole, 26 Oct. 1929, *Vaughan* 896!
DISTR. U 1–3; K 1–7; T 1–8; Z; P; widespread in tropical Africa and through Arabia to India and Thailand
HAB. A wide range of grassland, bushland, woodland and forest habitats, also commonly in disturbed and secondary vegetation and ruderal; 0–1850 m

SYN. *Acanthus maderaspatensis* L., Sp. Pl. 2: 639 (1753) & Syst. Nat. (ed.12) 2: 427 (1767); Burm. f., Fl. Ind.: 139 (1768), as *maderaspatanus*
    Blepharis boerhaviifolia Pers., Syn. Pl. 2: 180 (1806); Nees in DC., Prodr. 11: 266 (1847); Richard, Tent. Fl. Abyss. 2: 150 (1850); T. Anderson in J.L.S. 7: 34 (1863); Engler, Hochgebirgsfl. Trop. Afr.: 389 (1892); Lindau in P.O.A. C: 369 (1895); C.B. Clarke in F.T.A. 5: 96 (1899); Hiern, Cat. Welw. Pl. Afr. 4: 811 (1900); De Wild., Not. Fl. Katanga 3: 24 (1914) & Contrib. Fl. Katanga: 201 (1921) & Suppl. 4: 92 (1932). Type: as for *Acanthus maderaspatensis*
    B. rubiifolia Schumach., Beskr. Guinea Pl.: 292 (1829). Type: Guinea, *Thonning* (C!, holo.)
    B. abyssinica Hochst. in Schimper, Iter Abyss. III.1492 (1844). Type: Ethiopia, near Axum, *Schimper* III.1492 (FI!, FI-W!, FT!, K!, P!, iso.)
    B. boerhaviifolia Pers. var. *maderaspatensis* (L.) Nees in DC., Prodr. 11: 267 (1847)
    B. togodelia Solms-Laub. in Schweinf., Beitr. Fl. Aeth.: 108 (1867); Lindau in P.O.A. C: 369 (1895). Type: Eritrea, Wadi Togodele, *Ehrenberg* s.n. (BR!, K!, P!, iso.)
    B. boerhaviifolia Pers. var. *nigrovenulosa* De Wild. & T.Durand in Ann. Mus. Congo, ser. 2, 1: 46 (1899); De Wild., Contrib. Fl. Katanga: 201 (1921). Type: Congo-Kinshasa, Moanda, *Vanderyst* s.n. (BR!, holo.)
    B. breviciliata Fiori in Bull. Soc. Bot. Ital. 1911: 61 (1911) & Boschi e Piante Legn. Eritrea: 346 (1912); E.P.A.: 951 (1964). Type: Eritrea, Mt Lesa, *Fiori* 716 (FT!, holo.; FI!, iso.)
    B. maderaspatensis (L.) Roth var. *abyssinica* (Hochst.) Fiori, Boschi e Piante Legn. Eritrea: 346 (1912); Pichi Sermolli, Miss. Stud. Lago Tana 7: 257 (1951); Napper in K.B. 24: 325 (1970); Champluvier in Fl. Rwanda 3: 442 (1985)
    B. maderaspatensis (L.) Roth var. *boerhaviifolia* (Pers.) Fiori, Boschi e Piante Legn. Eritrea: 346 (1912)

B. *maderaspatensis* (L.) Roth subsp. *rubiifolia* (Schumach.) Napper in K.B. 24: 325 (1970); Vollesen in Opera Bot. 59: 79 (1980); Champluvier in Fl. Rwanda 3: 442 (1985); Iversen in Symb. Bot. Ups. 29(3): 160 (1991)

NOTE. Napper in K.B. 24: 323 (1970) divides the East African material of this widespread, often semi-weedy species, into two subspecies and one of these into two varieties: subsp. *maderaspatensis* var. *maderaspatensis* with small (12–18 mm long) corolla and upper sepal twice as long as longest bracts occurs along the coast while var. *abyssinica* with large (20–27 mm long) corolla occurs inland in Uganda and W Tanzania. Subsp. *rubiifolia* with included or slightly exserted upper sepal and small (11–18 mm long) corolla occurs inland all over East Africa.

If we also consider the Indian material the picture becomes much less clear. Most of the collections from here are intermediate between all the above taxa with a large variation in corolla length combined with an upper sepal which is usually $^1/_2$–$^2/_3$ longer than the longest bracts.

Closer examination of the East African material also shows that there are intermediate collections. Most material from T 5 & 7 – which ought to be subsp. *rubiifolia* – have somewhat exserted upper sepal and corolla up to 22 mm long (e.g. *Bidgood et al.* 1063, *Greenway & Polhill* 11521, *Lovett et al.* 1949). Some specimens from upland K 4 are also intermediate between the two subspecies and between the varieties of subsp. *maderaspatensis* (e.g. *Hucks* 203, *Mungai et al.* 291A).

Examination of the material from southern tropical Africa shows a large variation in length of the upper sepal with most material ± intermediate between the two subspecies. See also Vollesen, l.c. for a detailed discussion of the variation in this species.

Adamson 590 from Kenya, Northern Frontier District: Melka Ioni is a trailing subshrub with very large sepals and corolla (maximum dimensions in text). *Gillett* 12827 from Moyale has similarly large flowers and also very large inner bracts (maximum dimension in text). These collections seem to be but extreme forms.

12. **Blepharis involucrata** Solms in Schweinf., Beitr. Fl. Aeth.: 107 (1867); Björnstad in Serengeti Res. Inst. Publ. No. 215: 25 (1976); Vollesen in Opera Bot. 59: 79 (1980); Vollesen, Blepharis: 207 (2000). Type: Sudan, Kordofan, near Obeid, Mulbes, *Cienkowski* 360 (LE, holo.; W!, iso.)

Erect to procumbent annual herb; stems up to 35 cm long, quadrangular, sparsely to densely hispid-pubescent (rarely sericeous). Leaves: lamina linear to narrowly ovate or narrowly elliptic, largest 4.5–8 × 0.4–1.5 cm, margin entire, subglabrous to sparsely hispid-pubescent, densest on veins. Spikes solitary, dispersed whole when mature; bracts greenish with purple-brown veins and bristles, ovate to elliptic, puberulous to pubescent and ciliate, with a reflexed tip about half the length of the bract, this tip ending in a bristle and on each side with 4–8 scabrid bristles 1–4 mm long; lower pair (excl. tip) 3–5 mm long, second pair 4–6 mm long, third pair 6–9 mm long, inner pair 7–9 mm long; bracteoles absent. Sepals coloured as bracts, puberulous to pubescent and ciliate; dorsal elliptic, 11–14 mm long, narrowing gradually to the acute apex; ventral ovate-elliptic, 10–13 mm long; lateral 9–12 mm long. Corolla white to pale blue or pale mauve; tube 5–6 mm long; limb 6–8 mm long, callus without or with inconspicuous flanges. Filaments 1–2 mm long, ± included in tube, glabrous; appendage broadly rounded, almost as long as the 1.5–2 mm long anther. Capsule ± 7 mm long. Seed ± 5 × 4 mm.

TANZANIA. Mpanda District: Rukwa Valley, Tumba, 12 Mar. 1952, *Siame* 156!; Dodoma District: 10 km N of Manyoni, 15 Apr. 1988, *Bidgood et al.* 1127!; Morogoro District: 40 km NNW of Morogoro, 11 Sep. 1972, *Mwasumbi* DSM 2696!
DISTR. T 1, 4–8; Cameroon, Chad, Sudan, Zambia, Malawi, Mozambique, Zimbabwe
HAB. Seasonally waterlogged grassland and *Acacia* bushland on alkaline hardpans, *Acacia drepanolobium* bushland and grassland on greyish to black clay, old rice fields, roadsides; 100–1400 m

SYN. *Blepharis pinguior* C.B.Clarke in F.T.A. 5: 97 (1899); T.T.C.L.: 5 (1949); Binns, Checklist Herb. Fl. Malawi: 12 (1968); Lebrun & Stork, Enum. Pl. Afr. Trop. 4: 473 (1997). Type: Malawi, *Buchanan* 914 (K!, holo.)

NOTE. This and the following species do not seem to be close to any other species. They are immediately recognisable by the sharply reflexed bract-tip.

13. **Blepharis refracta** *Mildbr.* in N.B.G.B. 11: 821 (1933); Lebrun & Stork, Enum. Pl. Afr. Trop. 4: 473 (1997); Vollesen, Blepharis: 210 (2000). Type: Tanzania, Ulanga District: Rupia, *Schlieben* 2382 pro parte (B†, holo.; BM!, BR!, G!, MO!, P!, Z!, iso.)

Procumbent perennial herb; stems up to 20 cm long, with a narrow band of finely sericeous hairs. Leaves: lamina lanceolate, largest 2.5–4 × 0.3–0.5 cm, margin entire, glabrous. Spikes solitary, dispersed whole when mature; bracts purplish brown, ovate to elliptic, puberulous and ciliate, with a reflexed tip about half the length of the bract, this tip ending in a bristle and on each side without or with 1–2(–5) smooth or indistinctly scabrid bristles 0.5–1.5 mm long; lower pair (excl. tip) 3–4 mm long, second pair 5–6 mm long, third pair 7–8 mm long, inner pair 8–10 mm long; bracteoles absent. Sepal colour and indumentum as bracts; dorsal ovate-elliptic, 10–12 mm long, narrowing gradually to the acute apex; ventral similar, 9–11 mm long, with two subulate teeth; lateral 9–10 mm long. Corolla mauve, 11–13 mm long of which the tube 4–5 mm, callus on limb without apical flanges. Filaments ± 1 mm long, ± included in tube, glabrous; appendage broadly rounded, almost as long as the 2–2.5 mm long anthers. Capsule (fide Mildbraed) ± 7 mm long. Seed (fide Mildbraed) ± 6 × 5 mm.

TANZANIA. Ulanga District: Selous Game Reserve, Rupia, 14 June 1932, *Schlieben* 2382!
DISTR. T 6; not known elsewhere
HAB. *Brachystegia* woodland on sandy soil; 400 m

SYN. *Blepharis refracta* Mildbr. var. *pinnatispina* Mildbr. in N.B.G.B. 11: 821 (1933). Type: Tanzania, Ulanga District: Rupia, *Schlieben* 2382 pro parte (B†, holo.; BM!, iso.)

NOTE. Only known from this collection. The isotype of var. *pinnatispina* shows all transitions from no lateral bristles on the recurved bract-tip (var. *refracta*) to having 1–2(–5) bristles per side (var. *pinnatispina*).

    Closely related to *Blepharis involucrata* from which it differs in being a perennial herb with smaller glabrous leaves and in the recurved bract-tip without or with 1–2(–5) short, smooth or indistinctly scabrid bristles per side.

14. **Blepharis tanganyikensis** (*Napper*) *Vollesen*, Blepharis: 217 (2000). Type: Zambia, Mbala District: 8 km on Kawimbe-Old Sumbawanga road, *Richards* 15228 (K!, holo.; LISC!, S!, SRGH!, iso.)

Trailing or straggling perennial herb with several stems from a woody rootstock; stems up to 35 cm long, puberulous to sericeous. Leaves: lamina lanceolate to narrowly ovate or narrowly elliptic, largest 5–7 × 0.5–1.5 cm, margin entire, glabrous to sparsely scabrid-pubescent. Spikes in dense terminal clusters of 2–8(–16) or some solitary, persistent, each subtended by a whorl of small bract-like leaves; bracts straw-coloured to dark brown, obovate or narrowly so, puberulous to pubescent, terminating in a recurved bristle and on each side with 5–7 finely retrorsely scabrid (pilose near base) and finely glochidiate bristles (2–)3–7 mm long; outer pair 6–9 mm long, second pair 8–13 mm long, third pair 12–16 mm long, inner pair 16–20 mm long; bracteoles absent. Sepals coloured as bracts, puberulous and ciliate; dorsal elliptic, 23–26 mm long, apical part somewhat ligulate, apex subacute to rounded with 1–3(–5) small teeth; ventral ovate-elliptic, 19–23 mm long, with two acute teeth 2–7 mm long; lateral 12–16 mm long. Corolla creamy white to pale violet, 27–33 mm long of which the tube 6–8 mm; callus 3-ribbed apically and with 3(–4) distinct apical flanges. Filaments 8–9 mm long, sparsely glandular near apex; appendage 1–2 mm long broadly rounded; anthers 3.5–5 mm long. Capsule and seed not seen.

TANZANIA. Mbeya District: Mbozi, Apr. 1935, *Horsbrugh-Porter* s.n.! & May 1935, *Horsbrugh-Porter* s.n.!; Rungwe District: Mlale, 10 July 1972, *Leedal* 1198!
DISTR. **T** 7; Congo-Kinshasa, Zambia, Malawi
HAB. *Brachystegia* woodland on sandy to stony soil and on rocky outcrops; 1500–1800 m

SYN. *B. cuanzensis* S.Moore subsp. *tanganyikensis* Napper in K.B. 24: 326 (1970); Lebrun & Stork, Enum. Pl. Afr. Trop. 4: 473 (1997)

NOTE. Napper (l.c.) says that this taxon is only slightly different from *Blepharis cuanzensis* from W Angola, Zambia and SW Congo. But there are in actual fact so many differences that the two must be treated as distinct species. In *B. tanganyikensis* the flowers are in dense clusters, the sepals are short, the corolla is white to pale violet with the limb uniformly hairy all over the upper surface. In *B. cuanzensis* the corolla is bright blue or dark purple and the limb is only hairy in the central part.

15. **Blepharis inopinata** *Vollesen*, Blepharis: 233 (2000). Type: Tanzania, Mpanda District: 72 km on Chala-Mpanda road, *Bidgood et al.* 3883 (K!, holo.; BR!, C!, CAS!, DSM!, EA!, K!, NHT!, P!, UPS!, WAG!, iso.)

Erect unbranched annual herb (rarely with a single branch); stems up to 45 cm long, sericeous-pubescent or pilose when young. Leaves subequal or smaller down to ²/₃ of larger; lamina elliptic to obovate, largest 4–11 × 1.2–3.2 cm, margin entire, sparsely scabrid-sericeous, scabrid on margins. Flowers in 1-flowered persistent spikes subtended by 4 pairs of decussate bracts and with a number of reduced bract-like leaves at base, the spikes in dense sessile clusters of 2–5 on a flattened sparsely puberulous axis in axils of lower leaf-whorl just above the cotyledons (rarely also higher up); bracts green with brownish midrib and bristles, obovate, sparsely puberulous or sericeous-puberulous, terminating in a recurved bristle and on each side with 3–5 bristles 1–4 mm long, scabrid at base, upper bristles branched and sometimes with secondary bristles at base; lower pair 7–11 mm long, second pair 9–14 mm long, third pair 12–18 mm long, inner pair 15–20 mm long; bracteoles absent. Sepals green with brown veins or dorsal uniformly brown towards apex, puberulous and with scattered capitate glands, ciliate; dorsal ovate-elliptic, 22–26 mm long, narrowing gradually to the acute apex, withour or with 1–2 small lateral teeth per side, with numerous strong parallel secondary veins; ventral elliptic-obovate, 20–22 mm long, with 2 teeth 3–5 mm long and with a small recurved central tooth; lateral 18–20 mm long. Corolla yellow, 18–23 mm long of which the tube ± 6 mm, callus not ribbed, with 3 indistinct apical flanges; limb 8–12 mm wide, distinctly 3-lobed, indumentum beneath exclusively of capitate glands or with simple hairs intermixed, above of straight capitate glands and sometimes simple hairs apically and of bent hairs basally. Filaments 2–3 mm long, sparsely glandular; appendage ± 3 mm long, tapering gradually to an obtuse tip; anthers 2–3 mm long. Capsule and seed not seen.

TANZANIA. Mpanda District: Katavi National Park, Sitalike, 19 March 1973, *Ludanga* MRC1475! & 72 km on Chala–Mpanda road, 11 May 1997, *Bidgood et al.* 3883! & 5 km on Uruwira–Inyonga road, 3 June 2000, *Bidgood et al.* 4599!
DISTR. **T** 4; Zambia
HAB. *Brachystegia, Pterocarpus-Burkea* and *Isoberlinia* woodland on white sandy soil or grey clayey loam; 900–1100 m

NOTE. An extraordinary species which combines characters from Sect. *Blepharis* (flowers in 1-flowered spikes subtended by 4 pairs of decussate bracts with bristles) and Sect. *Scorpioidea* (spikes condensed into head-like inflorescences, glandular indumentum, ventral sepal with long teeth and central tooth, long curved anther tooth).
    But it also has unique characters: bristles branched, corolla-indumentum beneath of glandular hairs only (or with simple hairs intermixed) and above of straight capitate glands and sometimes simple hairs apically and of bent hairs basally.

16. **Blepharis trispina** *Napper* in K.B. 24: 328 (1970); Lebrun & Stork, Enum. Pl. Afr. Trop. 4: 474 (1997); Vollesen, Blepharis: 242 (2000). Type: Tanzania, Mbulu District: Mugugu, *Welch* 566 (K!, holo.; BR!, EA!, iso.)

Erect or procumbent annual herb; stems up to 35 cm long, finely sericeous on one side. Leaves very dissimilar: the laminate pair of each whorl with lamina linear-lanceolate to narrowly elliptic, largest 4–7.5 × 0.4–1.3 cm, margin entire, glabrous below but for a few hairs along midrib, above with puberulous midrib, scabrid along margins; the alternate pair reduced to robust very pungent 3-fid spines, central tooth 3–10 mm long, lateral shorter. Heads from near base to top of stems; floral leaves and outer bracts lanceolate, up to 3.5(–5) cm long, very finely sericeous, no capitate glands; inner bracts 1–2 cm long, linear-lanceolate. Sepals with indumentum as bracts, finely ciliate; dorsal ovate-elliptic, 11–14 mm long, narrowing gradually to the cuspidate apex, no lateral teeth; ventral similar, 10–13 mm long, with two acuminate teeth; lateral 9–11 mm long. Corolla pale blue to bright blue (rarely white), 9–14 mm long of which the tube 2.5–5 mm; callus with 3 indistinct apical flanges. Filaments ± 1 mm long, ± included in tube, glabrous; appendage 1–1.5 mm long, rounded; anthers 1–1.5 mm long. Capsule 6–7 mm long. Seed ovoid, ± 5 × 3 mm.

TANZANIA. Shinyanga District: Seseku, 10 June 1931, *Burtt* 2520!; Dodoma District: 8 km on Manyoni–Kilimatinde road, 14 Apr. 1988, *Bidgood et al.* 1102!; Mbeya District: Usangu Flats, Mbarali Estate, 19 Mar. 1986, *Pocs* 8632/G!
DISTR. **T** 1, 2, 4, 5, 7; not known elsewhere
HAB. *Acacia* bushland on seasonally waterlogged loamy to clayey alkaline hardpans; 700–1400 m

17. **Blepharis pusilla** *Vollesen*, Blepharis: 243 (2000). Type: Tanzania, Mbulu District: 15 km on Katesh–Singida road, *Bidgood et al.* 4395 (K!, holo.; C!, CAS!, NHT!, iso.)

Erect or decumbent annual herb; stems up to 5 cm long, sparsely puberulous to sparsely sericeous. Leaves subequal or smaller pair ³/₄ of larger; lamina lanceolate to narrowly oblong, largest 2.5–5.2 × 0.3–0.5(–0.8) cm, margin entire or some with a single tooth near base, sparsely puberulous on both sides, densest on midrib and margins. Heads slightly elongated, with less than 10 flowers, sometimes reduced to 2(–1) flowers, at base of stem or in axils of cotyledons, subsessile; axes finely sericeous; floral leaves linear to lanceolate, up to 2.2 × 0.3 cm, straight, glabrous to finely sericeous, margin with hard spines up to 1.5 mm long; inner bracts linear-lanceolate, up to 1.5 cm long. Sepals green or tinged purplish, with dark green veins, straw-coloured towards base, glabrous to finely sericeous, ciliate towards apex; dorsal elliptic, 10–14 mm long, narrowing gradually to an acute apex, no lateral teeth; ventral elliptic, 10–13 mm long, with 2 cuspidate teeth 1–3 mm long, no lateral teeth; lateral 9–12 mm long. Corolla pale to dark blue, 12–13 mm long of which the tube ± 3 mm; limb obovate, ± 5 mm wide, distinctly 3-lobed, central lobe longer, subacute; callus not ribbed. Filaments 3–4 mm long, exserted, glabrous; appendage ± 0.5 mm long, broadly rounded, strongly curved. Capsule ± 7 mm long. Seed ± 5 × 4 mm.

TANZANIA. Mbulu District: 15 km on Katesh–Singida road, 16 June 1997, *Bidgood et al.* 4395!; Manyoni District: 32 km on Singida–Itigi road, 24 May 2006, *Bidgood et al.* 6143! & 34 km on Itigi–Singida road, 24 May 2006, *Bidgood et al.* 6167!
DISTR. **T** 2, 5; not known elsewhere
HAB. *Acacia* bushland on grey sandy soil and on grey crumbling clay, degraded Itigi Thicket; 1250–1750 m

NOTE. Known only from these collections. One of the few species in sect. *Scorpioidea* which occurs outside the Zambesian *Brachystegia* associations. It is closest to the following species and to the southern African *Blepharis leendertziae*, another species which interestingly enough also mainly occurs outside the Zambesian Domain.

18. **Blepharis longifolia** *Lindau* in E.J. 20: 32 (1894) & in P.O.A. C: 370 (1895) & in E. & P. Pf. IV, 3b: 318 (1895); C.B. Clarke in F.T.A. 5: 104 (1899) pro parte; Vollesen in Opera Bot. 59: 79 (1980); Lebrun & Stork, Enum. Pl. Afr. Trop. 4: 473 (1997); Vollesen, Blepharis: 249 (2000). Type: Tanzania, Wala [Oalla] River, *Stuhlmann* 48 (B†, holo.) Neotype: Tanzania, Shinyanga District: Samui Hills, *B.D. Burtt* 5656 (K!, neo; BM!, BR!, COI!, EA!, FI!, G!, K!, P!, PRE!, S!, Z!, iso.)

Erect or procumbent annual herb; stems up to 50 cm long, glabrous or pilose with long many-celled hairs or sericeous. Leaves usually subequal; lamina linear-lanceolate to narrowly ovate, largest 6–17 × 0.4–1(–1.7) cm, margin of some or all with a few teeth in basal half, subglabrous to pilose, finely scabrid-ciliate on margins. Cotyledons usually not present at time of flowering, if present only rarely supporting heads. Heads from near base to top of plant, subsessile; floral leaves and outer bracts lanceolate to narrowly obovate, up to 4 cm long, glabrous to finely sericeous, no capitate glands; inner bracts 8–18 mm long, linear-oblong. Sepals with similar indumentum, finely ciliate (rarely with long cilia); dorsal ovate-elliptic, 13–22 mm long, narrowing gradually to the cuspidate apex, no lateral teeth; ventral similar, 13–20 mm long, with two acuminate teeth; lateral 12–19 mm long. Corolla blue or bright blue, (15–)18–28 mm long of which the tube 3.5–5 mm, callus with 3 distinct apical flanges. Filaments 4–9 mm long, glabrous; appendage ± 1 mm long, rounded, tip curved in towards anther; anthers 2.5–4.5 mm long. Capsule ± 9 mm long. Seed ± 5 × 3.5 mm. Fig. 18, 1–10, p. 117.

TANZANIA. Shinyanga District: near Shinyanga, June 1933, *Bax* 311!; Singida District: Iwumba Mbuga, 10 Mar. 1928, *Burtt* 1468!; Njombe District: 17 km on Makumbako-Mbeya road, 23 Mar. 1988, *Bidgood et al.* 632!
DISTR. **T** 1, 2, 4–8; not known elsewhere
HAB. Seasonally waterlogged *Acacia* bushland and grassland on grey to black clay; (200–)1000–1500 m

NOTE. The type is no longer extant, but this is the only species from anywhere near the type-locality which fits the original description.
    Blepharis longifolia normally flowers in the rainy season as an erect ± unbranched herb. But where local conditions allow it to continue growing into the dry season more robust plants often develop, usually ± branched and often procumbent. *Bax* 311 is a good example of this robust late season form.

19. **Blepharis itigiensis** *Vollesen*, Blepharis: 251 (2000). Type: Tanzania, Dodoma District: 41 km on Itigi–Rungwa road, *Greenway & Polhill* 11671 (K!, holo.; B!, LISC!, P!, S!, iso.)

Erect to decumbent unbranched annual herb; stems up to 25 cm long. Leaves subequal, pilose with long many-celled shiny hairs; lamina lanceolate, largest 4–14 × 0.4–1 cm, margin of all entire or a few with a single tooth near base, sparsely pilose with many-celled hairs, finely scabrid on margins. Cotyledons present at time of flowering. Heads all in axils of cotyledons (rarely also in axils of first leaf-pair), subsessile; axis sericeous; floral leaves and outer bracts lanceolate or narrowly oblong, up to 3 cm long, sparsely to densely pilose with long many-celled hairs, no capitate glands; inner bracts 14–17 mm long, narrowly oblong. Sepals with similar indumentum, finely ciliate; dorsal ovate, 15–18 mm long, narrowing gradually to the cuspidate apex, no lateral teeth; ventral similar, 15–17 mm long, with 2 acuminate teeth; lateral 13–15 mm long. Corolla mauve to blue or dark blue, 14–18 mm long of which the tube 3–5 mm; limb ± 7 mm wide, distinctly 3-lobed, callus with 3 distinct apical flanges. Filaments 3–4 mm long, glabrous; tooth ± 1 mm long, rounded, tip curved in towards anther; anthers 2.5–3 mm long. Immature capsule ± 7 mm long. Seed not seen. Fig. 18, 11–17, p. 117.

TANZANIA. Dodoma District: 41 km on Itigi–Rungwa road, 21 Apr. 1964, *Greenway & Polhill* 11671! & 37 km on Itigi–Rungwa road, 25 Mar. 1965, *Richards* 19858!; Manyoni District: 34 km on Itigi–Singida road, 24 May 2006, *Bidgood et al.* 6165!

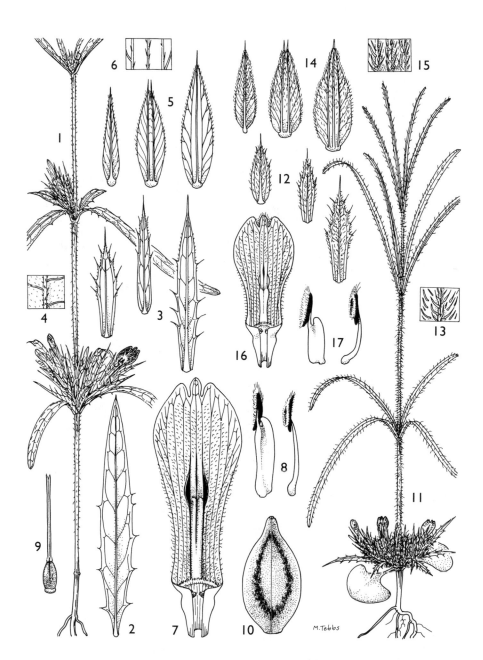

FIG. 18. *BLEPHARIS LONGIFOLIA* — **1**, habit, × ²/₃; **2**, large leaf, × ²/₃; **3**, series of floral leaves and bracts, × 2; **4**, detail of surface of floral leaf, × 6; **5**, sepals, × 2; **6**, detail of surface of dorsal sepal, × 6; **7**, corolla opened up (stamens removed), × 3; **8**, stamens, × 3; **9**, ovary, style and stigma, × 4; **10**, capsule, × 4. *B. ITIGIENSIS* — **11**, habit, × ²/₃; **12**, series of floral leaves and bracts, × 2; **13**, detail of surface of floral leaf, × 6; **14**, sepals, × 2; **15**, detail of surface of dorsal sepal, × 6; **16**, corolla opened up (stamens removed), × 3; **17**, stamens, × 3.5. 1 & 5–9 from *Burtt* 5656, 2 from *Bidgood et al.* 632, 3–4 from *Bax* 310, 10 from *Bax* 311, 11–17 from *Greenway & Polhill* 11671. Drawn by Margaret Tebbs. From Blepharis, a taxonomic revision: 250 (2000).

DISTR. **T** 5; not known elsewhere

HAB. Open *Acacia* bushland and grassland on damp pale orange sandy loam or on grey crumbling clay; 1150–1400 m

NOTE. A very striking species related to *Blepharis longifolia* but no doubt distinct. It grows mainly in the transition zone between ridgetop *Brachystegia* woodland and *Acacia* valley bottom grassland. This is quite a widespread micro-habitat in central Tanzania, and the species has probably been overlooked. It could be interpreted as a neotenous form of *B. longifolia* with the heads produced in the axils of the cotyledons, but it also differs from *B. longifolia* in the spreading pilose indumentum and in the smaller flowers.

20. **Blepharis asteracantha** *C.B.Clarke* in F.T.A. 5: 100 (1899); Lebrun & Stork, Enum. Pl. Afr. Trop. 4: 473 (1997); Vollesen, Blepharis: 257 (2000). Type: Zambia, S of Lake Tanganyika, Fwambo, *Nutt* s.n. (K!, holo.)

Erect to procumbent terrestrial perennial herb with creeping rootstock or aquatic herb with floating (or trailing on mud) spongy stems; stems up to 60 cm long, distinctly angular, glabrous or with sparse pilose or appressed hairs, soon with glossy yellowish brown bark. Leaves subequal, lamina linear-lanceolate to elliptic, largest 7–23 × 0.3–1.5(–2) cm, margin entire (rarely some with 1(–2) teeth), glabrous to sparsely scabrid-puberulous. Heads globose, on upper part of stems; peduncle (1.5–)3–14 cm long, sericeous-puberulous or sparsely so, with or without capitate glands; axes sericeous-puberulous, with or without capitate glands; floral leaves and outer bracts lanceolate to narrowly ovate-elliptic, 3–9 cm long, puberulous or sparsely so and conspicuously shaggy-ciliate with long white hairs (some over 1 mm long), no capitate glands; inner bracts similar, up to 2.5 cm long, with or without capitate glands. Sepals with similar indumentum, with capitate glands towards base; dorsal elliptic, 20–29 mm long, with short to long ligulate apical part or narrowing gradually to apex, apex acute to truncate, 1–3-toothed, lateral teeth absent; ventral ovate-elliptic, 17–26 mm long, not ligulate, shortly to deeply bifid, without (very rarely with) central tooth, without lateral teeth; lateral 15–25 mm long. Corolla bright or deep blue with darker veins, 25–35 mm long of which the tube 4–7 mm; limb distinctly 3-lobed, below with few to many capitate glands, callus without or with 2 indistinct ribs, 2 lateral apical flanges larger than central. Filaments 5–9 mm long, glabrous; appendage 1–2 mm long, broadly rounded; anthers 3–5 mm long. Capsule and seed not seen.

TANZANIA. Kigoma District: 8 km S of Katare, Malagarasi Swamps, 26 Aug. 1952, *Lowe* 553a!; Ufipa District: 60 km on Mbala [Abercorn]–Mwasye road, 16 Mar. 1950, *Bullock* 2643!; Mbeya District: Mbosi, 5 Apr. 1932, *Davies* 487!

DISTR. **T** 4, 7; Congo-Kinshasa, Zambia

HAB. Forming floating swards in lakes and swamps, trailing on drying mud, floodplains, wet grassland, weed in maize fields, on grey to black clayey or peaty soil, rarely on sand; 1050–1850 m

SYN. *Blepharis bequaertii* De Wild. in F.R. 13: 146 (1914) & Not. Fl. Katanga 4: 83 (1914) & Contrib. Fl. Katanga: 200 (1921); Lebrun & Stork, Enum. Pl. Afr. Trop. 4: 473 (1997). Type: Congo-Kinshasa, Bukama, *Bequaert* 123 (BR! holo.; BR!, iso.)

NOTE. The aquatic habit of some populations in W Tanzania and N Zambia (e.g. *Bally* 7562, *Lowe* 553) is unique to *Blepharis*, but apart from the spongy floating or creeping stems, there are no characters to separate them from terrestrial forms, and some erect terrestrial specimens from Congo (e.g. *Malaisse* 6358) have the same spongy stems.

21. **Blepharis kenyensis** *Vollesen*, Blepharis: 263 (2000). Type: Kenya, Kilifi, *Graham* 1630 (K!, holo.; EA!, FHO!, K!, iso.)

Erect to procumbent annual or short-lived perennial herb; stems up to 25 cm long, puberulous to pubescent or densely so. Leaves with the smaller $\frac{1}{2}$–$\frac{2}{3}$ of larger; lamina linear-lanceolate, largest 4–8 × 0.2–0.7 cm, margin entire (rarely with a single tooth near base), uniformly puberulous or sparsely so on both sides. Heads on upper part of stems only; peduncle up to 3.5 cm long, indumentum as stems; axes puberulous; floral leaves and outer bracts lanceolate, up to 3 cm long, puberulous and finely ciliate, no capitate glands; inner bracts lanceolate to ovate-elliptic, up to 18 mm long. Sepals sericeous-puberulous, no capitate glands; dorsal ovate-elliptic, 13–15 mm long, apical part ligulate, apex acuminate, no lateral teeth; ventral similar, 10–12 mm long, with 2 acute teeth up to 1.5 mm long, no lateral teeth; lateral 8–11 mm long. Corolla bright blue, 19–22 mm long of which the tube 3–5 mm; limb distinctly 3-lobed, callus not ribbed, with 3 distinct apical flanges. Filaments 4.5–6 mm long, glabrous; appendage 1–1.5 mm long, broadly rounded, not curved; anthers 3.5–4 mm long. Capsule and seed not seen.

KENYA. Kilifi District: Kilifi, *Graham* 1630! & 45 km N of Malindi, 3 Oct. 1958, *Moomaw* 1004!; Lamu District: Witu Mudirate, Mukunumbe, Dec. 1956, *Rawlins* 298!
DISTR. **K** 7; not known elsewhere
HAB. On sandy to clayey soil in shallow damp depressions in coastal bushland, roadsides; 0–50 m

NOTE. This grows together with *Blepharis pratensis*, and in the herbaria the two have been confused. It differs in usually being annual, in the linear leaves, spreading indumentum and smaller flowers.

22. **Blepharis pratensis** *S.Moore* in J.B. 15: 294 (1877); Lindau in P.O.A. C: 369 (1895); C.B. Clarke in F.T.A. 5: 99 (1899), quoad specim. ex Kenya; Lebrun & Stork, Enum. Pl. Afr. Trop. 4: 474 (1997); Vollesen, Blepharis: 265 (2000). Type: Kenya, Lamu, *Hildebrandt* 1906 (BM!, holo.; G!, K!, L!, M!, P!, W!, WU!, iso.)

Erect to procumbent much branched perennial or shrubby herb with creeping branched rootstock; stems up to 1 m long, sparsely to densely and finely sericeous or puberulous. Leaves with the smaller $^1/_2$–$^2/_3$ of larger; lamina narrowly elliptic or narrowly oblong, largest 4–8 × 0.9–1.8 cm, margin entire or some with 1–2 indistinct teeth near base, subglabrous to sparsely sericeous or uniformly puberulous, densest along midrib. Heads on upper part of stems only; peduncle up to 4 cm long, indumentum as branches; axes puberulous; floral leaves and outer bracts lanceolate to narrowly ovate, up to 2(–3) cm long, subglabrous to puberulous, no capitate glands; inner bracts similar, up to 19 mm long. Sepals puberulous (sericeous-pubescent along veins), no capitate glands; dorsal elliptic-obovate, 15–17 mm long, apical part ligulate with acute apex (rarely with truncate to retuse apex with 3 small teeth or with central vein forking near tip to give two terminal teeth); ventral ovate-elliptic, 13–15 mm long, with slightly ligulate apical part, with 2 acute teeth up to 2 mm long; lateral 10–12 mm long. Corolla bright blue or mauvish blue, 23–30 mm long of which the tube 3–6 mm; limb distinctly 3-lobed, callus ribbed and with 3 very distinct apical flanges. Filaments 5–7 mm long; appendage ± 1 mm long, broadly rounded, not curved; anthers 4–5 mm long. Capsule ± 9 mm long. Seed not seen.

KENYA. Kitui District: Kitui to Mutomo, 15 Mar. 1969, *Napper* 1936!; Kilifi District: 6 km N of Malindi, Sabaki, 2 Nov. 1961, *Polhill & Paulo* 690!; Kwale District: near Kikoneni, Kitoni Hot Springs, July 1967, *Makin* 416!
DISTR. **K** 4, 7; not known elsewhere
HAB. Damp shallow depressions on sandy soil in coastal bushland, hot springs, dry *Acacia-Commiphora* bushland; 0–50(–1000) m

SYN. [*Blepharis panduriformis* sensu C.B.Clarke in F.T.A. 5: 103, quoad *Wakefield* s.n., *non* Lindau (1877)]

NOTE. *Napper* 1936 has the central vein in the dorsal sepal forking near the tip to give two apical teeth, but this is probably best considered an abnormality. In *Makin* 416 the dorsal sepal is truncate or retuse with two small lateral teeth.

23. **Blepharis grandis** *C.B.Clarke* in F.T.A. 5: 104 (1899); Binns, Checklist Herb. Fl. Malawi: 12 (1968); Moriarty, Wild Fl. Malawi: 84, pl. 42, 2 (1975); Björnstad in Serengeti Res. Inst. Publ. No. 215: 25 (1976); Lebrun & Stork, Enum. Pl. Afr. Trop. 4: 473 (1997); Vollesen, Blepharis: 266 (2000). Type: Malawi, Mpata to Tanganyika Plateau, *Whyte* s.n. (K!, lecto., selected by Vollesen, l.c.)

Erect (rarely straggling or procumbent) perennial herb with several usually unbranched stems from a woody rootstock; stems up to 1 m long, glabrous to densely pubescent (rarely sericeous). Leaves stiff and leathery, glossy, greyish beneath, one pair usually only $\frac{1}{8}$–$\frac{2}{3}$ of the other; lamina narrowly ovate to ovate or narrowly elliptic to elliptic (rarely lanceolate), largest 5–14 × 0.8–3.8 cm, margin toothed (rarely only towards base or some entire), glabrous to sparsely pubescent below, above finely scabrid-puberulous (rarely glabrous or puberulous to pubescent), scabrid on margins. Heads on upper part of stems only, subsessile; axis glabrous to pubescent or pilose; floral leaves and outer bracts lanceolate to elliptic, up to 5 cm long, glabrous to pilose, no capitate glands; inner bracts linear-lanceolate to ovate or obovate, up to 2.5(–3) cm long. Sepals subglabrous to pubescent, ciliate, no capitate glands; dorsal elliptic-obovate, (16–)20–28 mm long, narrowing gradually to the acuminate to acute apex (rarely with 1–2 small teeth at base of terminal); ventral similar, 16–26 mm long, with 2 acuminate teeth (very rarely also with a central tooth); lateral 13–21 mm long. Corolla pale blue to bright dark blue or pale to dark bluish purple (very rarely pure white), (21–)25–35 mm long of which the tube 6–10(–12) mm; limb distinctly 3-lobed, callus 3-ribbed and with 3 distinct apical flanges. Filaments 6–10 mm long, glabrous; appendage 1–2.5 mm long, broadly rounded, straight; anthers 4–6 mm long. Capsule 9–10 mm long. Seed not seen.

TANZANIA. Shinyanga District: near Shinyanga, Apr. 1933, *Bax* 308!; Kondoa District: Jagose Hills, 18 May 1929, *Burtt* 2079!; Mbeya District: Songwe Valley, 2 km N of Mbeya–Tunduma road, 25 Mar. 1988, *Bidgood et al.* 685!
DISTR. **T** 1, 4, 5, 7, 8; Zambia, Malawi, Mozambique
HAB. *Brachystegia* woodland, *Combretum* and *Uapaca* bushland, grassland, on a wide range of soils from black clay to loamy or sandy soil or on rocky limestone slopes; 900–2000 m

SYN. *Blepharis carduacea* Lindau in E.J. 30: 410 (1901); T.T.C.L.: 5 (1949); Lebrun & Stork, Enum. Pl. Afr. Trop. 4: 473 (1997). Type: Tanzania, Mbeya District: Usafu, Punguluma [Bunguluma] Mts, *Goetze* 1085 (B†, holo.; BR!, E!, G!, L!, P!, iso.)
  B. frutescens Gilli in Ann. Naturhist. Mus. Wien 77: 47 (1973); Lebrun & Stork, Enum. Pl. Afr. Trop. 4: 473 (1997). Type: Tanzania, Njombe District: Madunda, *Gilli* 517 (W!, holo.)
  [*B. acanthodioides* sensu Gilli in Ann. Naturhist. Mus. Wien 77: 47 (1973), *non* Klotzsch (1861)]
  [*B. pratensis* sensu Ruffo *et al.*, Cat. Lushoto Herb. Tanzania: 3 (1996), quoad *Carmichael* 182, *non* S.Moore (1877)]

NOTE. Plants from N Tanzania (**T** 1, 4) are much less hairy and have wider bracts and sepals than plants from S Tanzania. But the transition seems to be quite gradual, and I find it impossible to separate out any infraspecific taxa. The material from Malawi seem to indicate that here subglabrous and hairy plants can be found in ± the same locality.
  Blepharis grandis is normally strictly erect, but when it grows in dense vegetation or continues to grow into the dry season straggling or even decumbent plants develop.

24. **Blepharis ilicifolia** *Napper* in K.B. 24: 330 (1970); Lebrun & Stork, Enum. Pl. Afr. Trop. 4: 473 (1997); Vollesen, Blepharis: 269 (2000). Type: Tanzania, Njombe District: 3 km NW of Njombe, *Milne-Redhead & Taylor* 11060 (K!, holo.; BR!, EA!, iso.)

Procumbent perennial herb with several stems from woody rootstock; stems up to 30 cm long, pubescent. Leaves leathery, glossy, smaller ± $\frac{2}{3}$ of larger; lamina ovate to elliptic, largest 1.5–4.8 × 0.7–1.9 cm, margin entire or some with 1–2 small teeth per side, pubescent or sparsely so beneath (densest on midrib), subglabrous to sparsely pubescent above, ciliate. Heads on upper part of stems, globose, subsessile; axes finely puberulous and with sparse to dense capitate glands; floral leaves and outer bracts lanceolate to elliptic, up to 18 mm long, indumentum similar; inner bracts similar, up to 32 mm long. Sepals purplish towards tip, puberulous and with scattered capitate glands (dense towards base), ciliate; dorsal ovate-elliptic, 18–29 mm long (mucro 1–2 mm), narrowing gradually to the acuminate apex, no lateral teeth; ventral similar, 18–26 mm long, with slightly ligulate apical part, with 2(–3) acuminate teeth 2–4 mm long, no lateral teeth; lateral 15–21 mm long. Corolla pale blue to bright blue, (22–)26–33 mm long of which the tube 4–8 mm; limb obovate-spathulate, 14–22 mm wide, distinctly 3-lobed, hairy all over upper surface; callus with 3 equally strong ribs. Filaments 5–9 mm long, glabrous or sparsely glandular; appendage 0.5–1.5 mm long, broadly rounded; anthers 3.5–5 mm long. Capsule and seed not seen.

TANZANIA. Iringa District: Mufindi, Itulituli Ridge, 22 June 1986, *Congdon* 118!; Njombe District: Njombe, Aug. 1931, *Hornby* 22 & 8 Nov. 1936, *Hornby* 22A
DISTR. **T** 7; not known elsewhere
HAB. Montane grassland, lawns; 1750–1950 m

NOTE. Only known from nine collections from two small areas in the Southern Highlands of Tanzania. Immediately recognisable by the combination of procumbent habit, broad leathery leaves and the very fine glandular indumentum on bracts and sepals.
　　Its habit has enabled it to survive and even prosper on the fairways of the golf-course at Mufindi. This is the only area where it has been collected recently and a large healthy population is still found here.

25. **Blepharis torrei** *Vollesen*, Blepharis: 269 (2000). Type: Tanzania, Songea District: 1.5 km S of Gumbiro, *Milne-Redhead & Taylor* 10122 (K!, holo.; B!, BR!, EA!, K!, LISC!, SRGH!, iso.)

Erect to procumbent perennial herb from woody rootstock; stems up to 60 cm long, pilose with long many-celled glossy hairs. Leaves subequal or smaller ³/₄ of larger; lamina narrowly elliptic (lower) to linear-lanceolate, largest 8.5–11 × 0.5–1 cm, margin entire or some with 1–2 teeth per side in lower half, sparsely pilose on both sides. Heads on upper part of stems or along the whole length, subsessile or peduncle up to 1(–2) cm long, pilose; axis sparsely pilose; floral leaves and outer bracts lanceolate to narrowly elliptic, up to 3 cm long, sparsely to densely pilose with long many-celled hairs, no capitate glands; inner bracts similar, tinged purplish towards apex, 13–18 mm long. Sepals purplish on veins and towards apex, sparsely pilose and long-ciliate; dorsal ovate-elliptic, 17–25 mm long, narrowing gradually to the acute apex or with ligulate apical part and then subacute to rounded, no lateral teeth; ventral similar, with slightly ligulate apical part, 15–22 mm long, with 2 cuspidate teeth, no lateral teeth; lateral 13–19 mm long. Corolla lavender, 15–30 mm long of which the tube 3–7 mm; limb distinctly 3-lobed, callus ribbed apically and with 3 distinct apical flanges. Filaments 6–10 mm long, with scattered glands near top; appendage + 2 mm long, broadly rounded, slightly curved; anthers ± 4 mm long. Capsule and seed not seen.

TANZANIA. Songea District: 1.5 km S of Gumbiro, 8 May 1956, *Milne-Redhead & Taylor* 10122!
DISTR. **T** 8; Mozambique
HAB. *Acacia-Brachystegia boehmi* wooded grassland on concrete-like clayey hardpan; 875 m

NOTE. Known only from this collection and two from N Mozambique. Easily recognised by the indumentum of long pilose hairs.

26. **Blepharis tanzaniensis** *Vollesen*, Blepharis: 273 (2000). Type: Tanzania, Uzaramo District: Bagamoyo to Dar es Salaam, *Bally* 16921 (K!, holo.; EA!, iso.)

Erect shrubby perennial herb; stems up to 50 cm tall, sparsely and finely sericeous when young with antrorse hairs. Leaves drying blackish, subequal; lamina ovate to elliptic, largest 8–12 × 2–3.5 cm, margin entire, glabrous or with a few appressed hairs along midrib. Heads elongated, on upper part of stems; peduncle up to 2 cm long, puberulous or finely sericeous; axes puberulous, no capitate glands; floral leaves and outer bracts narrowly ovate or narrowly elliptic, up to 2.5 cm long, puberulous, with few to many stalked capitate glands; inner bracts similar or elliptic, up to 2 cm long. Sepals puberulous or sericeous-puberulous, with scattered capitate glands, ciliate; dorsal ovate-elliptic 20–24 mm long, with ligulate apical part, apex acute, 1-toothed, no lateral teeth; ventral elliptic, 18–20 mm long, not ligulate, shallowly bifid, central and lateral teeth absent; lateral ± 17 mm long. Corolla pale blue, 28–35 mm long of which the tube 6–8 mm; limb distinctly 3-lobed, below with dense brown capitate glands, callus as in *B. affinis*. Filaments 8–10 mm long, glabrous; appendage 1–1.5 mm long, broadly rounded, slightly curved; anthers ± 5 mm long. Capsule and seed not seen.

TANZANIA. Uzaramo District: Bagamoyo to Dar es Salaam, 6 Sep. 1975, *Bally* 16921!; Kilwa District: 25 km N of Kilwa Masoko, 24 Sep. 1966, *B.J. Harris* 371!
DISTR. **T** 6, 8; not known elsewhere
HAB. Probably coastal bushland or grassland; 0–25 m

NOTE. Related to *Blepharis affinis* but clearly distinct. The stems have upwardly directed hairs, the leaves are wider, the peduncle and inflorescence axis are puberulous, the dorsal and ventral sepals lack lateral teeth and the appendage on the anterior stamens is shorter.

27. **Blepharis affinis** *Lindau* in P.O.A. C: 369 (1895) & in E.J. 24: 319 (1897); C.B. Clarke in F.T.A. 5: 97 (1899); T.T.C.L.: 4 (1949); Vollesen in Opera Bot. 59: 79 (1980); Ruffo *et al.*, Cat. Lushoto Herb. Tanzania: 2 (1996); Lebrun & Stork, Enum. Pl. Afr. Trop. 4: 473 (1997); Vollesen, Blepharis: 273 (2000). Type: Tanzania, Uzaramo District: Bagamoyo, *Stuhlmann* 7259 (K!, lecto.; selected by Vollesen, l.c.)

Erect or procumbent annual, perennial or shrubby herb, unbranched to strongly branched; stems up to 1 m long, finely sericeous or sparsely so with retrorse hairs (more rarely subglabrous or pubescent to pilose). Leaves drying blackish, subequal or smaller down to ½ of larger; lamina linear to narrowly elliptic (rarely elliptic near base), largest 5–13.5 × 0.2–1.2(–2.2) cm, margin entire or some with 1–3 teeth per side in basal part, glabrous or sparsely to densely sericeous-puberulous (rarely pilose). Heads globose or elongated, on upper part of stems (rarely at base); sessile or peduncle up to 1 cm long, sericeous (rarely puberulous); axes sericeous or sericeous-puberulous, without or with sparse (rarely dense in **T** 8) stalked capitate glands; floral leaves and outer bracts lanceolate to narrowly ovate or narrowly elliptic, up to 2(–4) cm long, sericeous-puberulous and with few to many capitate glands; inner bracts similar or elliptic, up to 1.8 cm long. Sepals puberulous to pubescent, with few to many glands, ciliate; dorsal ovate-elliptic, 18–26 mm long, with long ligulate apical part (in annual plants slightly spathulate and narrowed below apex), apex acute to truncate, 3-toothed and sometimes with 1(–2) lateral teeth per side; ventral elliptic, 15–22 mm long, not ligulate, usually deeply bifid, usually with central tooth and always with 1–2 lateral teeth per side; lateral 14–22 mm long. Corolla pale blue to bright blue or mauve, (22–)25–37 mm long of which the tube (4–)6–9 mm; limb distinctly 3-lobed, below with dense brown capitate glands, callus with two large lateral ribs, central ± absent, two lateral apical flanges much larger than central. Filaments (5–)6–10 mm long, glabrous; appendage (1.5–)2–3 mm long, obtuse or tapering to an obtuse tip, curved; anthers 4–6 mm long. Capsule ± 1 cm long. Seed not seen.

TANZANIA. Kilosa District: Mikumi National Park, Mugira track, 16 June 1973, *Greenway & Kanuri* 15167!; Uzaramo District: Dar es Salaam University Campus, *Mwasumbi & Mhoro* 11103!; Songea District: Gumbiro, 29 June 1956, *Milne-Redhead & Taylor* 10924!
DISTR. **T** 3, 6, 8; Mozambique
HAB. *Brachystegia* woodland, usually on loamy soil, coastal bushland and grassland, riverbanks; near sea level to 500(–900) m

SYN. *Blepharis hirsuta* Mildbr. in N.B.G.B. 13: 285 (1936); Lebrun & Stork, Enum. Pl. Afr. Trop.
4: 473 (1997). Type: Tanzania, Ulanga District: Mahenge to Shuguri Falls, Saidi Ngwega,
Schlieben 2280 (B†, holo.; BM!, BR!, G!, HBG!, M!, MO!, P!, Z!, iso.)
[*B. panduriformis* sensu Vollesen in Opera Bot. 59: 79 (1980), *non* Lindau (1894)]
[*B. pratensis* sensu Ruffo *et al.*, Cat. Lushoto Herb. Tanzania: 3 (1996), quoad *Paulo* 107,
non S.Moore (1877)]

NOTE. In most of its area *Blepharis affinis* is easily distinguished by the perennial habit,
appressed indumentum, leaves drying blackish and the inflorescences towards tip of the
stems. But in **T** 8 some specimens become ± intermediate with *B. panduriformis*. The
indumentum here is often spreading, the annual habit is more common as are inflorescences
towards the base of the stems, and the upper sepal is often slightly spathulate. These
specimens differ mainly from *B. panduriformis* in the larger corolla and usually entire leaves,
but there are also differences in the shape of the ribs and flanges of the callus on the corolla
limb (see descriptions). An extreme form from **T** 8 is represented by *Issa* 116 and *Migeod* 20:
a small ephemeral herb with inflorescences in the axils of the cotyledons, but with large
corolla and appressed indumentum.
   Bidgood et al. 430 from the foothills of the Nguru Mts is from much higher altitude (900 m)
than the rest of the material. It has very long floral leaves, and in some respects approaches
Blepharis stuhlmannii. But it dries blackish and has the corolla callus typical of *B. affinis*.

28. **Blepharis panduriformis** *Lindau* in E.J. 20: 30 (1894) & in P.O.A. C: 369 (1895)
& in E. & P. Pf. IV, 3b: 318 (1895); C.B. Clarke in F.T.A. 5: 103 (1899) pro parte;
T.T.C.L.: 5 (1949); Ruffo *et al.*, Cat. Lushoto Herb. Tanzania: 3 (1996) pro parte;
Lebrun & Stork, Enum. Pl. Afr. Trop. 4: 474 (1997); Vollesen, Blepharis: 276 (2000).
Type: Tanzania, Kondoa District: W of Irangi, *Stuhlmann* 4226 (B†, holo.). Neotype:
Tanzania, Mpwapwa District: 22 km on Mpwapwa road from Morogoro–Dodoma
road, *Bidgood et al.* 939 (K!, neo; BR!, C!, CAS!, DSM!, EA!, K!, NHT!, UPS!, WAG!,
iso.; selected by Vollesen, l.c.)

Erect to procumbent annual herb (most often erect and single-stemmed); stems up
to 45(–75) cm long, pilose or sparsely so with shiny many-celled hairs and with a band
of puberulous hairs. Leaves subequal or smaller down to $^2/_3$ of larger; lamina linear to
narrowly ovate or narrowly elliptic, largest 4–11 × 0.3–1.1 cm, margin on some or all
with 1–4 teeth per side in basal half, subglabrous to sparsely pilose, densest above and
along veins beneath, scabrid above. Heads usually elongated, usually only near base of
stems; peduncle up to 1.5(–3) cm long, puberulous to pubescent; axes finely
sericeous-puberulous, usually with stalked capitate glands; floral leaves and outer
bracts lanceolate to narrowly elliptic, up to 4(–7) cm long, indumentum similar, with
capitate glands; inner bracts similar or elliptic, up to 1.4 cm long, with capitate glands.
Sepals with similar indumentum, distinctly ciliate; upper ovate-elliptic, 16–23 mm
long, with a long ligulate-spathulate apical part distinctly narrowed dorsal apex, apex
subacute to truncate or retuse, 1–3-toothed; ventral ovate-elliptic, 12–17 mm long, not
ligulate, bifid, usually also with central tooth and a single lateral tooth per side; lateral
13–18 mm long. Corolla pale blue to bright blue or mauve (rarely white), 20–30 mm
long of which the tube 4–6 mm; limb distinctly 3-lobed, below with many long-stalked
capitate glands, callus with 3 equally strong ribs and 3 large apical flanges. Filaments
4–7 mm long, glabrous; tooth 2.5–4 mm long, tapering to an obtuse tip, curved;
anthers 3.5–5 mm long. Capsule 7–9 mm long. Seed ± 6 × 4 mm.

TANZANIA. Dodoma District: 11 km E of Itigi, 8 Apr. 1964, *Greenway & Polhill* 11443!; Mpwapwa
   District: 22 km on Mpwapwa road from Morogoro–Dodoma road, 8 Apr. 1988, *Bidgood et al.*
   939!; Mbeya District: Igawa, 31 Mar. 1962, *Polhill & Paulo* 1959!
DISTR. **T** 4–7; not known elsewhere
HAB. *Acacia-Commiphora* bushland on sandy to loamy or clayey soil, grassland on black cotton
   soil, roadsides; 600–1500 m

SYN. [*Blepharis affinis* sensu Björnstad in Serengeti Res. Inst. Publ. 215: 25 (1976), *non*
   Lindau (1895)]

NOTE. In its typical form *Blepharis panduriformis* is an unbranched or sparsely branched erect herb, but occasionally decumbent plants occur, sometimes mixed with the normal erect plants. Plants which continue to grow into the dry season get larger, more branched and eventually get almost woody stems. The more vigorous of these get a strong superficial resemblance to *B. stuhlmannii*, but are always immediately recognisable by the characteristic dorsal sepal. See also note after *B. affinis*.

A common and often dominant element of the early dry season flora in the *Acacia* bushlands of Central Tanzania where it can sometimes be so plentiful that it suffuses the ground with a pale blue colour.

29. **Blepharis tenuiramea** *S.Moore* in J.B. 38: 205 (1900); C.B. Clarke in F.T.A. 5: 512 (1900); Binns, Checklist Herb. Fl. Malawi: 12 (1968); Lebrun & Stork, Enum. Pl. Afr. Trop. 4: 474 (1997); Vollesen, Blepharis: 278 (2000). Type: Malawi, *Buchanan* 387 (BM!, holo.; E!, iso.)

Erect to procumbent single-stemmed unbranched or slightly branched annual herb; stems up to 40 cm long, hispid-pubescent or sparsely so with shiny many-celled hairs and sometimes with a band of puberulous hairs. Leaves subequal or smaller pair $^2/_3$ of larger; lamina linear to narrowly ovate-elliptic, largest 4–15 × 0.2–1.5 cm, margin on or some or all with 1–3 small teeth per side in basal half, subglabrous to hispid-pubescent along midrib, hispid-puberulous above. Heads dense or elongated and up to 7 cm long, near base or towards tip of stems, sessile (rarely with peduncle up to 1(–2) cm long); axes sericeous-puberulous, with scattered stalked capitate glands; floral leaves and outer bracts lanceolate to ovate-elliptic, up to 7 cm long, finely sericeous-puberulous, with few to many capitate glands; inner bracts similar, up to 2.5 cm long. Sepals puberulous or sparsely so, often also with pilose hairs, with capitate glands; dorsal ovate-elliptic or narrowly so, 14–21 mm long, narrowing gradually or with slightly ligulate apical part, apex acute, with 1–3 lateral teeth per side; ventral elliptic-obovate, 12–19 mm long, not or slightly ligulate, deeply bifid, with or without central and with 1–3 lateral teeth per side; lateral 10–18 mm long. Corolla pale to dark blue to mauve, 15–26 mm long of which the tube 2–6 mm; limb distinctly 3-lobed, below with scattered capitate glands, callus with 3 equally strong ribs, with 2 strong lateral apical flanges, central weak. Filaments 2–6 mm long, broad pair or all with long hairs on inside, all sparsely glandular; tooth 1–3 mm long, tapering to a rounded tip, curved; anthers 2.5–4 mm long. Capsule 8–13 mm long. Seed ± 7 × 5.5 mm.

TANZANIA. Ufipa District: 60 km on Chala–Mpanda road, 11 May 1997, *Bidgood et al.* 3869!; Manyoni District: 32 km on Itigi–Manyoni road, 28 May 2006, *Bidgood et al.* 6272!; Mbeya District: Songwe Valley, 2 km N of Mbeya–Tunduma road, 25 Mar. 1988, *Bidgood et al.* 692! DISTR. **T** 4, 5, 7; Congo-Kinshasa, Zambia, Malawi, Mozambique, Zimbabwe, Botswana, Namibia HAB. *Combretum* bushland on rocky limestone slope, *Brachystegia* woodland on grey clayey loam, *Acacia* bushland on sandy-loamy soil, old rice fields, roadside ditches; 900–1350 m

SYN. *Blepharis caloneura* S.Moore var. *angustifolia* Oberm. in Ann. Transv. Mus. 19: 135 (1937); Binns, Checklist Herb. Fl. Malawi: 12 (1968); Lebrun & Stork, Enum. Pl. Afr. Trop. 4: 473 (1997). Type: Zimbabwe, Victoria Falls, *Flanagan* 3245 (BOL!, lecto., selected by Vollesen, l.c.)

30. **Blepharis menocotyle** *Milne-Redh.* in Hook., Ic. Pl. 32: t. 3198 (1933); Lebrun & Stork, Enum. Pl. Afr. Trop. 4: 473 (1997); Vollesen, Blepharis: 281 (2000). Type: Zambia, Solwezi District, Mutanda Bridge, *Milne-Redhead* 537 (K!, holo.)

Erect unbranched (rarely branched) single-stemmed annual herb; stems up to 20 cm long, hispid-pubescent or sparsely so when young and with a band of puberulous hairs (rarely subglabrous). Leaves subequal or smaller down to $^2/_3$ of larger; lamina linear-lanceolate to narrowly ovate or narrowly elliptic, largest 5–19(–23) ×

0.2–1.5(–2) cm, margin entire, glabrous to sparsely hispid-puberulous or -pubescent, densest along midrib, scabrid on margins. Heads globose, in axils of cotyledons or towards tip of stems, more rarely along whole length, subsessile; axes sericeous or sparsely so; floral leaves and outer bracts lanceolate to narrowly ovate, up to 2 cm long, with up to 4 spreading teeth per side, subglabrous to finely sericeous (sometimes pubescent to pilose along veins) and with few to many capitate glands, ciliate, with central and 2 lateral veins but without transverse ladder-like veins; inner bracts similar, up to 1.5 cm long. Sepals uniformly sericeous-puberulous or sparsely so, with usually dense long capitate glands; dorsal ovate or narrowly so, 15–24 mm long, narrowing gradually to apex or with slightly ligulate apical part, apex acute to truncate, with large central tooth and without or with 2(–4) small subterminal teeth, no lateral teeth, secondary veins conspicuously raised, leaving under straight narrow angles; ventral elliptic-obovate, 13–19 mm long, not ligulate, deeply bifid (teeth 5–7 mm long), with 1–2 central teeth and 1–3 lateral teeth per side, veins similar; lateral 13–20 mm long, with 3–7 small apical teeth. Corolla blue or mauve, 12–16 mm long of which the tube 3–5 mm; limb obovate-spathulate, 5–7 mm wide, deeply 3-lobed; callus with 2 strong lateral ribs, central almost absent. Filaments 2–4 mm long, sparsely glandular; appendage ± 1 mm long, rounded, strongly curved; anthers 2–3 mm long. Capsule 8–11 mm long. Seed 6–8 × 4–6 mm.

TANZANIA. Ufipa District: 34 km on Namanyere–Kipili road, 5 May 1997, *Bidgood et al.* 3752!; Mpanda District: 96 km on Mpanda–Uvinza road, 19 May 1997, *Bidgood et al.* 4061! & 20 km on Mwese road from Mpanda–Uvinza road, 6 June 2000, *Bidgood et al.* 1604!
DISTR. **T** 4; Congo-Kinshasa, Angola, Zambia, Malawi
HAB. In tall mature *Brachystegia* woodland on sandy to loamy soil; 950–1350 m

NOTE. In Shaba province of Congo *Blepharis menocotyle* has been recorded as an indicator-species for soils with high contents of cobalt, copper and manganese.

31. **Blepharis katangensis** *De Wild.* in Ann. Mus. Congo, Bot., ser. 4, 1: 146 (1903) & in Contrib. Fl. Katanga: 201 (1921) & Suppl. 4: 92 (1932) & Suppl. 5: 54 (1933); Lebrun & Stork, Enum. Pl. Afr. Trop. 4: 473 (1997); Vollesen, Blepharis: 284 (2000). Type: Congo-Kinshasa, Katanga, Lukafu, *Verdick* 438 (BR!, holo.)

Erect (rarely procumbent) single-stemmed unbranched (rarely branched) annual herb; stems up to 60 cm long, sparsely to densely pilose with shiny many-celled hairs, sometimes also with a band of puberulous hairs. Leaves: some or all whorls with one pair more than 5 times longer than the other, but some often subequal; lamina narrowly ovate or narrowly elliptic, largest 6.5–22 × 0.5–1.9 cm, margin of larger with 3–13 widely spaced teeth per side, of smaller with 3–5 larger teeth per side, sparsely pubescent (or pilose along midrib), hispid-puberulous above. Heads dense, near top of stems; peduncle (0.5–)1–7(–12) cm long, usually ± deflexed, indumentum as stems; axes puberulous to pubescent, with capitate glands; floral leaves and outer bracts narrowly ovate to narrowly elliptic, up to 7(–9) cm long, pubescent to pilose, ± capitate glands; inner bracts similar, up to 23 mm long, with capitate glands. Sepals puberulous to pubescent (often also with long shiny many-celled hairs), with capitate glands; dorsal ovate, 20–27 mm long, narrowing gradually to apex or with slightly ligulate apical part, apex acute, lateral teeth absent; ventral narrowly ovate-elliptic, 17–24 mm long, not ligulate, bifid, usually also with 1(–2) central teeth and with 1–3 lateral teeth per side; lateral 18–26 mm long. Corolla blue to dark blue or mauve, 18–26 mm long of which the tube 5–8 mm; limb very indistinctly lobed, without capitate glands below, callus with 3 equally strong ribs and 3 large apical flanges (or central weaker). Filaments 4–7 mm long, sparsely glandular; tooth 1–1.5 mm long, rounded, curved or straight; anthers 3–4 mm long. Capsule ± 13 mm long. Seed ± 11 × 7 mm.

TANZANIA. Mbulu District: Dareda to Babati, Mangati, 12 Aug. 1926, *Peter* 44087!; Ufipa District: Ntatanda, 7 Apr. 1970, *Sanane* 1108!; Songea District: Gumbiro, 8 May 1956, *Milne-Redhead & Taylor* 10120!
DISTR. **T** 2, 4, 5, 7, 8; Congo-Kinshasa, Angola, Zambia, Malawi
HAB. *Acacia-Brachystegia boehmii* wooded grassland on clay, grassland on black cotton soil, on wet mud at edge of pools, roadsides; 850–1750 m

SYN. [*Blepharis buchneri* sensu C.B.Clarke in F.T.A. 5: 101 (1899), quoad *Whyte* s.n.; Benoist, Bol.
     Soc. Brot., Ser. 2, 24: 23 (1950), quoad *Exell & Mendonca* 316; Binns, Checklist Herb. Fl.
     Malawi: 12 (1968), *non* Lindau (1894)]
     [*B. linariifolia* sensu De Wild., Ann. Mus. Congo Belge, Ser. 4, 2: 147 (1913) & Contrib. Fl.
     Katanga: 201 (1921), *non* Pers. (1806)]

NOTE. This species has very large fruits and seeds for a species in sect. *Scorpioidea*. It also has a very indistinctly lobed corolla limb with the two lateral lobes almost absent.

32. **Blepharis stuhlmannii** *Lindau* in E.J. 20: 31 (1894) & in P.O.A. C: 370 (1895) & in E. & P. Pf. IV, 3b: 318 (1895); U.K.W.F.: 579 (1974); Raynal *et al.*, Fl. Med. Miss. Rwanda 1: 31 (1980); U.K.W.F. ed. 2: 274 (1994); Ruffo *et al.*, Cat. Lushoto Herb. Tanzania: 3 (1996); Vollesen, Blepharis: 290 (2000). Type: Tanzania, Mwanza, *Stuhlmann* 4596 (B†, holo.). Neotype: Tanzania, Musoma District: Serengeti National Park, Seronera River, *Greenway & Kanuri* 13624 (K!, neo; EA!, MO!, P!, iso.; selected by Vollesen, l.c.)

Erect to procumbent perennial herb with several stems from woody rootstock; stems up to 0.5(–1) m long, puberulous (sometimes only in a single band) to pubescent or densely so (rarely sericeous). Leaves subequal or smaller down to $^1/_2(-^1/_3)$ of larger; lamina linear-lanceolate to narrowly ovate or narrowly elliptic (rarely elliptic), largest 2.5–13(–16) × 0.4–2(–3.5) cm, margin on some or all with 1–5(–8) teeth per side (very rarely all entire), glabrous to puberulous or pubescent, scabrid above and along margins. Heads on upper part of stems (rarely at base), usually dense but sometimes much elongated and repeatedly bifurcate with up to 2 cm long internodes between the flowers, occasionally the whole plant one large inflorescence with no vegetative leaves; peduncle (in plants with dense heads) 0.5–5(–6) cm long, puberulous to pubescent (rarely densely so), with or without capitate glands; axes puberulous or sericeous-puberulous, with few to many capitate glands; floral leaves and outer bracts lanceolate to narrowly ovate or narrowly elliptic, 1–8 cm long, indumentum as axes; inner bracts similar or elliptic, up to 2.5 cm long. Sepals puberulous to pubescent and with usually dense capitate glands, distinctly ciliate; dorsal ovate-elliptic, 17–28(–32) mm long, with short or long ligulate apical part (rarely slightly spathulate below tip or narrowing gradually to tip), apex acute to truncate, with 1–3 apical teeth and below these sometimes with 1–2(–4) lateral teeth per side; ventral elliptic, 15–25(–30) mm long, not or slightly ligulate, bifurcate (usually deeply) at tip, with or without central tooth and without or with 1–2(–4) lateral teeth per side; lateral 15–25 mm long. Corolla white to pale blue, blue or mauve, pale greyish pink to salmon pink, 23–33(–39) mm long of which the tube 4–9(–13) mm; limb distinctly 3-lobed, below with dense capitate glands, callus with 3 equally strong ribs and 3 large apical flanges. Filaments (4–)5–9 mm long, glabrous or sparsely glandular; appendage 2–4 mm long, triangular in outline, flattened towards the anther, tapering to an obtuse tip, curved; anthers 3.5–6 mm long. Capsule 9–13 mm long. Seed 6–8 × 4–5 mm.

UGANDA. Ankole District: Ankole, 2 Apr. 1940, *Thomas* 3364!; Kigezi District: Kamwezi, Feb. 1948, *Purseglove* 2581!
KENYA. Masai District: Nguruman Hills, Lenyora, 27 Sep. 1944, *Bally* 3831! & Narok, Olodunguru, 13 July 1961, *Glover et al.* 2106! & 44 km on Narok–Masai Mara road, 12 July 1979, *Gilbert* 5749!
TANZANIA. Mwanza District: Geita, Karumo, 2 Mar. 1953, *Tanner* 1247!; Mbulu District: Magugu, 16 Apr. 1964, *Welch* 559!; Mpanda District: Kapapa, 12 Sep. 1970, *Richards* 25865!

DISTR. **U** 2; **K** 6; **T** 1–5, 7; Rwanda, Congo-Kinshasa, Zambia

HAB. In northern part of area in *Acacia* wooded grassland and in grassland, in southern part (**T** 4–7) in *Brachystegia* woodland and wooded grassland, sometimes persisting in cultivated areas, on a wide range of soils from rocky slopes through sandy to loamy soil or even black clay; (450–)850–2200 m

SYN. *Blepharis trinervis* C.B.Clarke in F.T.A. 5: 105 (1899); De Wildeman in Ann. Mus. Congo, Ser. 4, 1: 147 (1903); Th. & H. Durand, Syll. Fl. Cong.: 423 (1909); De Wildeman, Contrib. Fl. Katanga: 202 (1921). Type: Congo-Kinshasa, *Cornet* 84 (BR!, holo.)

　[*B. panduriformis* sensu C.B.Clarke in F.T.A. 5: 103 (1899), quoad *Scott Elliot* 8190; Ruffo *et al.*, Cat. Lushoto Herb. Tanzania: 3 (1996), quoad *Carmichael* 1488, *non* Lindau (1894)]

　[*B. pungens* sensu C.B.Clarke in F.T.A. 5: 104 (1899), quoad specim. ex Tanzania; T.T.C.L.: 5 (1949), *non* Klotzsch (1861)]

　B. verdickii De Wild. in Ann. Mus. Congo, Ser. 4, 1: 147 (1903); Th. & H. Durand, Syll. Fl. Cong.: 423 (1909); De Wildeman, Contrib. Fl. Katanga: 202 (1921); Lebrun & Stork, Enum. Pl. Afr. Trop. 4: 474 (1997). Types: Congo-Kinshasa, *Verdick* 166 (BR!, syn.) & 595 (BR!, syn.)

　B. cristata S.Moore in J.L.S. 37: 194 (1905); Champluvier in Fl. Rwanda 3: 441 (1985); Lebrun & Stork, Enum. Pl. Afr. Trop. 4: 473 (1997). Type: Uganda, Ankole District: Rufua River, *Bagshawe* 506 (BM!, holo.; K!, iso.)

　B. evansii Turrill in K.B. 1912: 331 (1912).Type: Kenya, Masai District: Guaso Nyiro, *Evans* 754 (K!, holo.)

　B. sp. aff. affinis of Björnstad in Serengeti Res. Inst. Publ. No. 215: 25 (1976)

NOTE. Even though the type is no longer extant, there is no doubt as to the identity of this species. It is the only species of this section occurring anywhere near Mwanza which fits the original description. Lindau's indication of a plant with quite widely spaced flowers is particularly revealing.

　Blepharis stuhlmannii is by far the most variable species in Sect. *Scorpioideae*, but there seems to be no satisfactory ways of subdividing the material. Collections from **K** 6; **T** 1, 2, 5 and partly **T** 4 are uniform apart from a large variation in the contraction of the heads. But when the species gets into the *Brachystegia* woodlands of **T** 4 & 7 it suddenly exhibits an enormous variation. Here there are pyrophytic forms without vegetative leaves, either with dense heads (*Bally* 7493), with more open heads but almost glabrous (*Boaler* 653) or with large open dichotomously branched inflorescences and very glandular (*Richards* 25865). Forms with large elliptic leaves and with elliptic floral leaves (*Semsei* FH 2516) also occur here, as well as a rare form with entire leaves (*Milne-Redhead & Taylor* 11075).

　The material from Uganda, Rwanda and neighbouring areas of Congo (*Blepharis cristata*, sensu stricto) is normally eglandular and has small leaves (2.5–6(–8.5) cm long) and short sepals (dorsal 17–23 mm versus 20–28 mm in Kenya and Tanzania and ventral 14–21 mm versus (16–)18–25 mm). In all other respects this material is identical with material from Kenya and N Tanzania, and a number of collections from Rwanda are more or less intermediate with glandular bracts and sepals and with either short or long sepals (e.g. *Lewalle* 2625, *Becquet* 564 and *Raynal* 20718).

　The type of *Blepharis evansii* is exceptionally hairy, but other Kenyan specimens and some from northern Tanzania are intermediate to the normal sparser indumentum of the Tanzanian material.

　The only collection from **T** 3 (*Faulkner* 1381) is from a much lower altitude (450 m) than usual, and shows some resemblance with *Blepharis affinis*. But it has a spreading puberulous indumentum combined with dentate leaves, a combination never seen in *B. affinis*. It also lacks the outer lateral teeth on the lower sepal and has a callus with 3 equally strong ribs.

33. **Blepharis petraea** *Vollesen* in K.B. 57: 454 (2002). Type: Tanzania, Mpanda District: 42 km on Uvinza–Mpanda road, *Bidgood, Leliyo & Vollesen* 4737 (K!, holo.; BR!, CAS!, DSM!, EA!, K!, NHT!, iso.)

Stiffly erect perennial herb with 1(–2) stems from creeping or suberect woody rootstock; stems 0.7–1.3 m tall, wiry, glabrous to densely pubescent when young, glabrescent. Leaves: large pair 1.5–3 times longer than small pair; lamina linear-lanceolate to lanceolate, large pair 11–21 × 0.8–2.5 cm, with entire margin or with a single tooth near base, beneath glabrous or with a few curly hairs on midrib, above glabrous to uniformly sparsely crisped-puberulous, scabrid on margins and above.

Heads on upper part of stem (rarely also lower down), dense and globose; peduncle 0.5–4 cm long, sparsely to densely sericeous-puberulous; axes densely sericeous-puberulous; floral leaves and outer bracts sericeous-puberulous (pubescent on margins), narrowly ovate-elliptic, 2–4 cm long, with distinct submarginal and transverse ladder-like veins, with up to 6 teeth per side; inner bracts similar, up to 1.5 cm long. Sepals uniformly puberulous (hairs slightly longer on veins), without or towards base with a few stalked capitate glands; dorsal ovate-elliptic, 27–28 mm long (terminal mucro ± 3 mm), tapering gradually or with a long slightly ligulate apical part, apex acute, on each side with 1–2 lateral teeth; ventral ovate-elliptic, 26–27 mm long, tapering gradually to apex, bifurcate with two ± 5 mm long teeth, on each side with one lateral tooth; lateral ± 24 mm long. Corolla pale mauve to purple with white tube, 30–35 mm long of which the tube 7–10 mm; limb obovate, 15–20 mm wide, distinctly 3-lobed, callus with three equally strong ribs; filaments 7–8 mm long, sparsely glandular towards base, appendage ± 3 mm long, triangular in outline, flattened towards the anther; anthers 5–7 mm long. Capsule ± 13 mm long. Seed ± 7 × 4.5 mm.

TANZANIA. Mpanda District: 42 km on Uvinza–Mpanda road, 24 June 2000, *Bidgood et al.* 4737! & 12 km N of Uzondo Camp on Uvinza road, 2 May 2006, *Bidgood et al.* 5776!
DISTR. **T** 4; not known elsewhere
HAB. *Brachystegia* woodland on grey loamy soil, large rocky outcrops with *Brachystegia microphylla* thicket on gravelly-stony soil; 1450–1600 m

34. **Blepharis buchneri** *Lindau* in E.J. 20: 30 (1894); C.B. Clarke in F.T.A. 5: 101 (1899); Lindau in in Fries, Rhod.-Kongo Exp. 1911–12, 1, Bot.: 306 (1916); De Wild., Contrib. Fl. Katanga, Suppl. 4: 92 (1932) & Contrib. Fl. Katanga, Suppl. 5: 54 (1933); Champluvier in Fl. Rwanda 3: 441 (1985); Ruffo *et al.*, Cat. Lushoto Herb. Tanzania: 2 (1996); Lebrun & Stork, Enum. Pl. Afr. Trop. 4: 473 (1997); Vollesen, Blepharis: 295 (2000). Type: Angola, Pungo Andongo, *von Mechow* 109 (BR!, lecto.; K!, Z!, iso.; selected by Vollesen, l.c.).

Erect to straggling (rarely procumbent) perennial or shrubby herb; stems up to 1(–1.5) m long, glabrous to finely sericeous-puberulous, sometimes with longer hairs intermixed (rarely pubescent). Leaves: larger pair more than 5(–3) times longer than smaller (but see note); lamina of larger lanceolate to narrowly ovate or narrowly elliptic, largest 7–21 × 0.7–2.5(–3) cm, margin on some or all toothed in lower $^{1}/_{2}$–$^{2}/_{3}$; smaller pair rhomboid-triangular (rarely elliptic) in outline, 1–3.5 cm long, 3(–7)-veined from base, with one large triangular tooth on each side, sparsely scabrid-pubescent on both sides (rarely glabrous). Heads globose or elongated, on upper part of stems; peduncle up to 8 cm long, puberulous or sericeous-puberulous or densely so with ordinary hairs and stalked capitate glands; axes sericeous-puberulous, with scattered capitate glands; floral leaves and outer bracts narrowly ovate-elliptic, 2–6.5 cm long, puberulous to sericeous or sparsely so, with few to many capitate glands; inner bracts similar, up to 2.5 cm long, with dense capitate glands. Sepals with similar indumentum, ciliate; dorsal ovate-elliptic or narrowly so, (20–)26–38 mm long, with long (rarely short) ligulate apical part, apex acute (rarely truncate), 3-toothed, lateral teeth absent (rarely 1–2 per side); ventral similar, (18–)20–29 mm long, not ligulate, shortly to deeply (teeth up to 8 mm long) bifid, without (rarely with) central tooth, without or with 1–4(–6) lateral teeth per side; lateral (18–)20–32 mm long. Corolla bright blue or purple, (27–)33–50 mm long of which the tube 5–10 mm; limb distinctly 3-lobed, below with few to many capitate glands, callus with 3 equally strong ribs or central weaker and with 3 large apical flanges. Filaments (5–)6–14 mm long, glabrous; appendage 4–6 mm long, triangular in outline and flattened towards the anther, tapering to an obtuse curved tip; anthers (4–)5–7.5 mm long. Capsule 10–13 mm long. Seed ± 7 × 4 mm, ovoid. Fig. 19, p. 129.

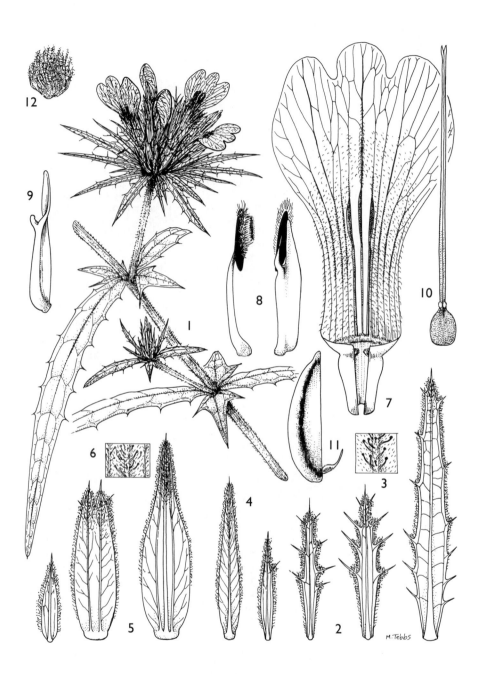

FIG. 19. *BLEPHARIS BUCHNERI* — **1**, habit, × ²/₃; **2**, series of floral leaves and bracts, × 2; **3**, detail of surface of floral leaf, × 6; **4**, inner floral leaf, × 2; **5**, sepals, × 2; **6**, detail of surface of dorsal sepal, × 6; **7**, corolla opened up (stamens removed), × 3; **8**, stamens, × 3; **9**, anterior stamen with anther removed, × 3; **10**, ovary, style and stigma, × 4; **11**, capsule, × 4; **12**, seed, × 2. 1 from *Bullock* 3886, 2–4 from *Juniper* 132, 5–10 from *Newbould* 4414, 11–12 from *Bally* 7507. Drawn by Margaret Tebbs. From Blepharis, a taxonomic revision: 297 (2000).

TANZANIA. Kigoma District: 16 km N of Kigoma, Kakombe, 6 July 1959, *Newbould & Harley* 4218!; Mpanda District: Mahali Mts, Lubugwe River, 14 July 1958, *Jefford et al.* 132!; Ufipa District: Milepa, 24 May 1951, *Bullock* 3886!

DISTR. **T** 4, 7; Congo-Kinshasa, Burundi, Angola, Zambia

HAB. Grassland on black clay, riverine woodland and bushland, *Brachystegia* woodland on sandy soil and rocky outcrops, lakeshores; 750–2150 m

SYN. [*Blepharis acanthodioides* sensu S.Moore in J.B. 18: 229 (1880), *non* Klotzsch (1861)]
    [*B. longifolia* sensu C.B.Clarke in F.T.A. 5: 104 (1899), quoad *Scott Elliot* 8354; De Wild., Contrib. Fl. Katanga: 201 (1921), *non* Lindau (1894)]
    B. buchneri Lindau var. *major* De Wild. in Ann. Mus. Congo Bot., ser. 4, 1: 146 (1903) & Contrib. Fl. Katanga: 201 (1921) & Suppl. 4: 92 (1932) & Suppl. 5: 54 (1933). Type: Tanzania, Mpanda District: Karema, *Storms s.n.* (BR!, lecto.; BR!, iso.)

NOTE. *Batty* 719 from Iringa and *Geilinger* 3157 from Mbeya District – the only collections seen from **T** 7 – have short (22 mm long) dorsal sepal and small (30 mm long) corolla. The ventral sepal also has a central tooth, but this a variable character in other species and no great emphasis can be put on it. More collections are needed from **T** 7 to decide whether this is a distinct taxon.
    Richards 7254 from Ufipa District: Mt Chala also seems to be but an abnormal form of *Blepharis buchneri*. It has short (10–15 cm long) procumbent stems and most leaves are ± subequal. But in some whorls one pair is 2–3 times longer than the other, and the collection has the large calyx and corolla of *B. buchneri*.
    Bally 7507 from Ugalla River has the smaller leaf pair about ¹/₃ of the larger and some of the smaller elliptic and 7-veined. The dorsal sepal is only 20 mm long, ventral and lateral 18 mm long and the corolla 27 mm long. There is the possibility that this is a hybrid between *B. buchneri* and *B. stuhlmannii*, but specimens with similarly small sepals and corolla are known from Zambia as well.

35. **Blepharis uzondoensis** *Vollesen*, Blepharis: 299 (2000). Type: Tanzania, Mpanda District: 56 km on Uvinza–Mpanda road, Uzondo Plateau, *Bidgood et al.* 4117 (K!, holo.; C!, CAS!, NHT!, iso.)

Procumbent perennial herb with short creeping sometimes branched rootstock and fleshy roots; stems up to 35(–75) cm long, slender (1–1.5 mm in diameter), glabrous to uniformly puberulous. Leaves with largest pair (of middle and upper leaves) (2–)4–8 times longer than smaller pair; lamina of larger pair linear-lanceolate, largest 6.5–11 × 0.3–0.6(–1) cm, margin entire or some with up to 7 inconspicuous teeth per side, lamina of smaller pair narrowly ovate-elliptic, 0.8–2(–5) cm long, entire or (usually) with one large tooth per side, glabrous to sparsely puberulous, beneath dense along veins, above uniformly so. Heads globose, on upper part of stems only; peduncle 2–5(–6) cm long, subglabrous to puberulous with eglandular hairs; axes sericeous-puberulous or densely so, without capitate glands; floral leaves and outer bracts with strong submarginal vein and strong transverse veins, narrowly elliptic, 2–4 cm long, subglabrous to puberulous, without capitate glands; inner bracts similar, up to 1.5 cm long. Sepals puberulous (pubescent along veins), with scattered stalked capitate glands towards base, densely crisped-ciliate with hairs to 2 mm long; dorsal ovate-elliptic, 20–24 mm long (mucro ± 1 mm), tapering gradually to an acute apex, with 1–3 lateral teeth per side; ventral similar, 19–22 mm long, apical part not ligulate, deeply bifid with teeth 3–5 mm long, with or without central tooth, with 1–2 lateral teeth per side; lateral 19–21 mm long. Corolla purple or bluish purple, pale on the outside, callus and tube yellowish, 29–32 mm long of which the tube 5–6 mm; limb broadly obovate, 19–22 mm wide, distinctly 3-lobed, middle lobe slightly shorter to slightly longer, narrower than lateral and narrowed at base; callus with 3 equally strong ribs. Filaments 5–7 mm long, glandular; appendage 2–3 mm long, triangular in outline and flattened towards the anther, tapering to an obtuse curved tip; anthers 4–5 mm long. Capsule and seed not seen.

TANZANIA. Mpanda District: 61 km on Uvinza–Mpanda road, Uzondo Plateau, 20 May 1997, *Bidgood et al.* 4103! & 56 km on Uvinza–Mpanda road, Uzondo Plateau, 21 May 1997, *Bidgood et al.* 4117! & 29 May 2000, *Bidgood et al.* 4497!

DISTR. **T** 4; not known elsewhere

HAB. Large areas of seepage grassland on shallow sandy soil overlying ironstone pans; 1400–1600 m

NOTE. *Blepharis uzondoensis* differs from *B. buchneri* in the thin procumbent stems, the linear-lanceolate leaves and ventral and lateral sepals which are of equal length.

11. **STREPTOSIPHON**

Mildbr. in N.B.G.B. 12: 719 (1935); Napper in K.B. 24: 323 (1970); Vollesen in K.B. 45: 504 (1990) & in K.B. 49: 405 (1994); Lebrun & Stork, Enum. Pl. Afr. Trop. 4: 504 (1997)

Shrub. Leaves opposite, entire. Flowers one per bract, in dense terminal spiciform cymes; bracts imbricate, foliaceous, papery, in 4 rows; bracteoles 2, glumaceous. Calyx glumaceous, divided to the base in 5 subequal 1-toothed sepals. Corolla finely sericeous-puberulent outside; basal part of tube straight, middle part twisted through 180°, apical part straight, split dorsally to give a small entire or 3-lobed lip which due to the twisting of the tube is held dorsally. Stamens 4, included, inserted just above twisted part of tube on a thickened flange; anthers 1-thecous, oblong, rounded at both ends, finely bearded on one side. Ovary 2-locular, with 2 ovules per locule, glabrous; style glabrous; stigma entire, dilated, truncate. Capsule sessile, ovoid, glabrous. Seed with large pectinate scales.

Monotypic genus.

Streptosiphon hirsutus *Mildbr.* in N.B.G.B. 12: 719 (1935); T.T.C.L.: 16 (1949); Vollesen in K.B. 49: 405 (1994); Lebrun & Stork, Enum. Pl. Afr. Trop. 4: 504 (1997). Type: Tanzania, Lindi District: 40 km W of Lindi, Noto Plateau, *Schlieben* 6098 (B†, holo.; BM!, BR!, K!, P! iso.)

Erect shrub to 1 m tall, often single-stemmed and unbranched; young branches yellowish hirsute to villose. Leaves sometimes clustered towards end of branches; petiole up to 2.5 cm long, villose; lamina elliptic to obovate, largest 14–25 × 5–10 cm, apex rounded or drawn out into a short acute tip, base attenuate to truncate, pilose or sparsely so on both sides, densely so along midrib and larger veins. Cyme solitary, 4–11 cm long, sessile; rachis pilose; bracts yellowish green or slightly purplish tinged, ovate-elliptic or broadly so, 18–25 × 8–17 mm, subacute to obtuse and apiculate, uniformly puberulous, margin entire (rarely dentate); bracteoles thickened at base, narrowly ovate, 12–19 mm long, cuspidate, puberulent on back, puberulous-ciliate and sometimes with puberulous midrib. Sepals thickened at base, narrowly ovate (dorsal slightly wider), 16–20 mm long, cuspidate, puberulent towards tip and finely ciliate. Corolla-limb pale lilac to purple; tube white, cylindric, basal part ± 5 mm long, twisted part ± 6 mm long, upper part ± 5 mm long; limb 5–7 mm long, slightly hooded. Filaments ± 1.5 mm long; anthers ± 3 mm long. Style 12–14 mm long. Capsule 18–22 mm long. Seed (immature) ± 5 × 4 mm. Fig. 20, p. 132.

TANZANIA. Lindi District: Rondo Plateau, Mchinjiri, Mar. 1952, *Semsei* 671! & Rondo Forest Reserve, 7 Feb. 1991, *Bidgood et al.* 1391! & Rondo Plateau, St. Cyprians College, 21 Feb. 1991, *Bidgood et al.* 1687!

DISTR. **T** 8; not known elsewhere

HAB. Semi-deciduous lowland forest; 200–700 m

NOTE. The absence of cystoliths, sepals which are thickened and horny at base and the 1-lipped corolla leaves no doubt as to the correctness of placing *Streptosiphon* in *Acanthoideae*. It differs from all other genera here in its peculiar corolla. Within *Acanthoideae* it must be considered closest to the other genera with included stamens: *Crossandra*, *Crossandrella* and *Stenandrium*.

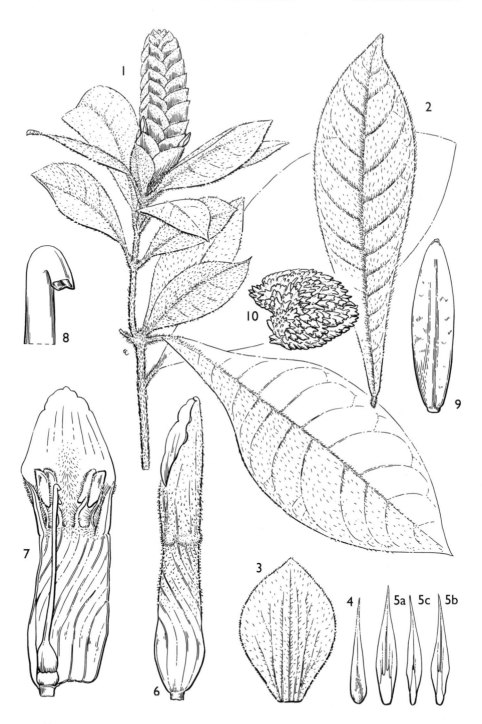

FIG. 20. *STREPTOSIPHON HIRSUTUS* — **1**, habit, × ½; **2**, leaf, × ½; **3**, bract, × 1.5; **4**, bracteole, × 1.5; **5**, sepals × 1.5; **6** corolla, × 3; **7**, corolla opened up, × 3; **8**, stigma, × 20; **9**, capsule, × 2; **10**, seed, × 6. 1 & 3–8 from *Bidgood et al.* 1568, 2 & 9–10 from *Bidgood et al.* 1687. Drawn by Eleanor Catherine. From K.B. 49: 402 (1994).

Streptosiphon is probably closest to *Crossandra* with which it shares the 1-lipped corolla and seeds with large pectinate scales. But the 5 subequal 1-toothed sepals point more towards *Stenandrium* which, however, has a subactinomorphic 5-lobed corolla. A tendency for the corolla limb to be only obscurely lobed is seen in *Crossandrella*, but this genus has a quite different calyx.

Streptosiphon differs from all three genera in having the stamens inserted on a thickened flange, a character which point at a relation with the genera with exserted stamens: *Acanthus*, *Blepharis* and *Sclerochiton*.

12. STENANDRIUM

Nees in Lindl., Intr. Nat. Syst. Bot., ed. 2: 444 (1836), *nom. conserv.* & in DC., Prodr. 11: 28 (1847); G.P. 2: 1096 (1876); Lindau in E. & P. Pf. IV, 3b: 320 (1895); Wasshausen in Lundell, Fl. Texas 1: 257 (1966); Long in Journ. Arn. Arbor. 51: 281 (1970); Daniel in Ann. Miss. Bot. Gard. 71: 1028 (1984); Vollesen in K.B. 47: 176 (1992)

Stenandriopsis S.Moore in J.B. 44: 153, t. 478B (1906); Lindau in E. & P. Pf. IV, 3b, Nachtr. 4: 286 (1914); Benoist in Bull. Mus. Nat. Hist. Nat., ser. 2, 15: 231 (1943); Bremek. in Acta Bot. Neerl. 4: 644 (1955) & in Bull. Bot. Surv. India 7: 25 (1965)

Perennial herbs or (on Madagascar) shrubs. Leaves opposite or in whorls of 4. Flowers single, sessile, in dense or lax sessile or pedunculate racemiform cymes; bracts small or large and foliaceous, lower sterile; bracteoles 2, linear to lanceolate, acuminate. Calyx glumaceous, divided to the base in 5 1-toothed sepals which are thickened and horny at base, dorsal slightly longer and wider than the rest. Corolla glabrous on the outside (hairy in Madagascar); tube linear, widened above stamens; limb subequally 5-lobed, 2 upper lobes erect and slightly narrower, 3 lower spreading. Stamens 4, inserted in tube, subsessile; anthers 1-thecous, included, curved, oblong with apiculate tip, finely bearded. Ovary with 2 ovules per locule, glabrous; style glabrous or sparsely hairy; stigma with 2 broad subequal lobes or the upper larger. Capsule 2–4-seeded, oblong-ellipsoid to triangular in outline, beaked, glumaceous, glabrous and shiny. Seed discoid, tuberculate.

About 50 species from southern USA to Argentina and Chile, 10 species in tropical Africa and 10 species in Madagascar.

1. Bracts 14–18 mm long, longer than calyx, with serrate
    margin, imbricate; spikes sessile; corolla pale mauve,
    pale blue or white with purple markings; tube 2.5–3 cm
    long . . . .   1. *S. guineense*
   Bracts 4–10 mm long, shorter than calyx, with entire
    margin, not imbricate; spikes pedunculate; corolla white,
    tube 0.9–2.1 cm long 2
2. Sepals 1–1.5 cm long; corolla tube 1.7–2.1 cm long, lobes
    in lower lip (1–)1.3–1.7 cm long; capsule ± 1.3 cm
    long; seed 4–5 mm . . . .   4. *S. grandiflorum*
   Sepals 0.7–1 cm long; corolla tube 0.9–1.4 cm long, lobes
    in lower lip 0.7–0.9 cm long; capsule 0.8–1.1 cm long;
    seed ± 2.5 mm long (unknown in *S. afromontanum*) 3
3. Bracts 7–10 mm long, almost as long as the calyx; bracteoles
    6–8.5 mm long; flowers subopposite . . . .   2. *S. warneckei*
   Bracts 4–7 mm long, about half as long as calyx;
    bracteoles 4–6 mm long; flowers alternate . . . .   3. *S. afromontanum*

1. **Stenandrium guineense** (*Nees*) *Vollesen* in K.B. 47: 182 (1992); Lebrun & Stork, Enum. Pl. Afr. Trop. 4: 504 (1997); Friis & Vollesen in Biol. Skr. 51(2): 454 (2005). Type: "Guinea Coast", *Herb. Hooker*, without collector or date (K!, holo.)

Perennial herb, basal part of stem creeping and rooting, apical part erect, to 20 cm tall, tomentose from long many-celled downwardly directed hairs. Leaves opposite, beneath paler with dark venation; petiole 0.3–2 cm long; lamina obovate, largest 8–14.5 × 4–7 cm, apex subacute to rounded, base subcordate, subglabrous to densely pubescent, densest on veins. Spike terminal, solitary, sessile, 3–8(–12) cm long; axis puberulous; flowers subopposite; fertile bracts imbricate, greyish or straw-coloured to green, lanceolate to narrowly ovate, 14–18 × 3–6 mm, acute and terminating in a straight mucro up to 1 mm long, glabrous to finely puberulous, serrulate in upper part, with many parallel veins; bracteoles linear-lanceolate, ± 1 cm long, finely puberulous. Sepals lanceolate, 5.5–7 mm long, acute, finely serrulate near tip, glabrous or puberulous near tip. Corolla pale blue, pale violet, pale mauve or white with purple markings in throat; tube 2.5–3 cm long; lobes in upper lip ± 1 cm long, in lower lip 1.2–1.5 cm long. Style with 2 broad fimbriate lobes, the upper longer. Capsule 4-seeded, 9–11 mm long. Seed 3–3.5 mm long, irregularly triangular in outline, tuberculate. Fig. 21, 1–3, p. 135.

UGANDA. Bunyoro District: Budongo Forest, July 1935, *Eggeling* 2130!; Mengo District: Sezibwa Falls, May 1915, *Dümmer* 2617! & 4 Aug. 1938, *Thomas* 2333!
DISTR. U 2, 4; West Africa, Cameroon, Gabon, Congo-Kinshasa, Rwanda, Sudan, Angola
HAB. Rainforest ground floor; 900–1250 m

SYN. *Crossandra guineënsis* Nees in DC., Prodr. 11: 281 (1847); Hooker in Bot. Mag. 104: t. 6346 (1878); Lindau in P.O.A. C: 370 (1895); C.B. Clarke in F.T.A. 5: 117 (1899); S. Moore in J.L.S. 37: 195 (1905); Lindau in Z.A.E.: 300 (1911); De Wild., Pl. Beq. 4: 27 (1926); Milne-Redhead in K.B. 1935: 281, fig. 2 (1935); Heine in F.W.T.A. (ed. 2) 2: 409 (1963)
Stenandriopsis guineënsis (Nees) Benoist in Bull. Mus. Hist. Nat. Paris, ser. 2, 15: 235 (1943) & in Bol. Soc. Brot., ser 2, 24: 25 (1950); Aké Assi, Contrib. Etud. Fl. Côte d'Ivoire 1: 209 (1961); Heine in Fl. Gabon 13: 102 (1966); Burkill, Useful Pl. W. Trop. Afr. 1: 27 (1985)

2. **Stenandrium warneckei** (*S.Moore*) *Vollesen* in K.B. 47: 188 (1992); Lebrun & Stork, Enum. Pl. Afr. Trop. 4: 504 (1997). Type: Tanzania, Lushoto District: Amani, *Warnecke* 230 (BM!, holo.; EA!, iso.)

Perennial herb, basal part of stem creeping and rooting, apical part erect, to 25 cm tall, tomentose with long many-celled downwardly directed purplish hairs. Leaves opposite, paler beneath; petiole 1–5 cm long; lamina oblong to slightly obovate, largest 4.5–8(–9) × 2–4.2(–5) cm, apex broadly rounded, base cordate, subglabrous to densely pubescent, densest on veins. Spike terminal, solitary, 2–9 cm long; peduncle (0.5–)1–3(–4) cm long, pubescent or densely so with downwardly directed purplish hairs; axis sparsely puberulous; flowers subopposite (or 1 developed with a sterile bract opposite); fertile bracts glumaceous, not imbricate, straw-coloured to brown, narrowly ovate-triangular, 7–10 × ± 2 mm, acuminate, entire, glabrous, with many parallel veins; bracteoles similar, 6–8.5 mm long. Sepals similar, 7.5–10 mm long, many-veined. Corolla white; tube 9–14 mm long, curved; lobes 4–7 mm long, subequal, rounded. Capsule 4-seeded, 8–11 mm long, beak recurved when ripe. Seed ± 2.5 mm long, with dense long tubercles. Fig. 21, 4–9, p. 135.

TANZANIA. Lushoto District: Derema to Ngambo, 15 Feb. 1918, *Peter* 60642! & Nderema, 24 Dec. 1956, *Verdcourt* 1723! & Amani to Marvera, 5 Aug. 1969, *Magogo* 1268!
DISTR. T 3; not known elsewhere
HAB. Rainforest ground floor; 650–1100 m

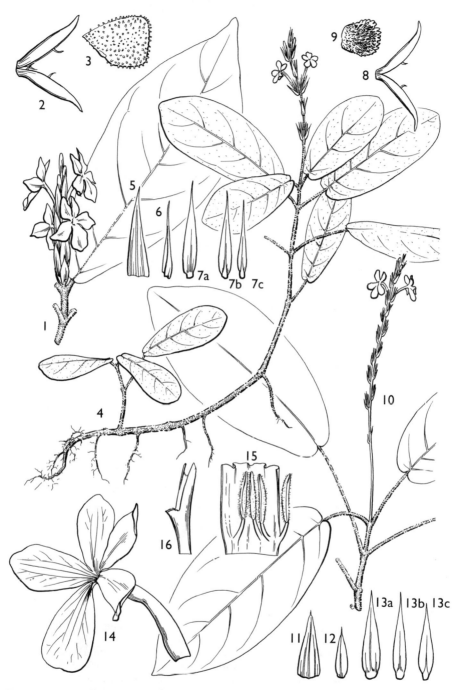

FIG. 21. *STENANDRIUM GUINEENSE* — **1**, inflorescence, × ²/₃; **2**, capsule, × 2; **3**, seed, × 5. *S. WARNECKEI* — **4**, habit, × ²/₃; **5**, bract × 3; **6**, bracteole, × 3; **7**, sepals, × 3; **8**, capsule, × 2; **9**, seed, × 5. *S. AFROMONTANUM* — **10**, flowering branch, × ²/₃; **11**, bract × 3; **12**, bracteole, × 3; **13**, sepals, × 3; **14**, corolla, × 3; **15**, corolla tube opened up with stamens, × 5; **16**, stigma, × 20. 1 from *Thomas* 2333, 2–3 from *Dümmer* 2617, 4–7 from *Greenway* 7918, 8–9 from *Peter* 60139, 10–13 from *Lovett* 274, 14–16 from *Drummond & Hemsley* 1847. Drawn by Eleanor Catherine. From K.B. 47: 184 (1992).

Syn. *Crossandra warneckei* S.Moore in J. B. 51: 214 (1913) & in J. B. 67: 271 (1929)
   C. usambarensis Mildbr. in N.B.G.B. 11: 63 (1930). Type: as for *C. warneckei* (B†, holo.; BM!, EA!, iso.)
   Stenandriopsis warneckei (S.Moore) Napper in K.B. 24: 342 (1970); Iversen in Symb. Bot. Ups. 29(3): 162 (1991)

3. **Stenandrium afromontanum** (*Mildbr.*) *Vollesen* in K.B. 47: 189 (1992); Lebrun & Stork, Enum. Pl. Afr. Trop. 4: 504 (1997). Type: Tanzania, Ulanga District: Sali, *Schlieben* 1955 (B†, holo.; BM!, BR!, HBG!, K!, LISC!, P!, S!, iso.)

Habit and indumentum as *S. warneckei*; stems to 15 cm tall. Leaves oblong or elliptic to obovate, largest 5–11.5 × 2.3–6 cm, apex subacute to broadly rounded, base cordate. Spike terminal, solitary or 2–3 together, 2–12 cm long; peduncle 1–5 cm long, glabrous to pubescent; axis glabrous to pubescent; flowers alternate; fertile bracts glumaceous, not imbricate, brown to purplish, ovate-triangular, 4–7 × 2–3.5 mm, acuminate, entire, glabrous or pubescent near base; bracteoles lanceolate to narrowly ovate, 4–6 mm long, glabrous. Sepals similar, 7–10 mm long, many-veined. Corolla white; tube 1–1.2 cm long, curved; lobes in upper lip 5–7 mm long, in lower lip 7–9 mm long. Capsule 8–10 mm long. Immature seed with dense long tubercles. Fig. 21, 10–16, p. 135.

Tanzania. Morogoro District: Nguru Mts, Manyangu Forest, 27 Mar. 1953, *Drummond & Hemsley* 1847! & Uluguru Mts, Kitundu, 25 Mar. 1935, *E.A. Bruce* 934!; Iringa District: Udzungwa Mts, Sanje, 16 Apr. 1984, *Lovett* 274!
Distr. **T** 6, 7; not known elsewhere
Hab. Rainforest ground floor; (600–)900–1400 m

Syn. *Crossandra afromontana* Mildbr. in N.B.G.B. 11: 822 (1933)
   Stenandriopsis afromontana (Mildbr.) Benoist in Bull. Mus. Hist. Nat. Paris, ser. 2, 15: 236 (1943)

Note. *S. afromontana* is closely related to *S. warneckei*, and the two could with some justification be treated as two subspecies. But the separating characters used here are constant and I have seen no intermediates.

4. **Stenandrium grandiflorum** *Vollesen* in K.B. 55: 967 (2000). Type: Tanzania, Iringa District: Udzungwa Mts, Kihansi River Gorge, *Lovett* 5053 (K!, holo.; DSM!, NHT!, iso.)

Perennial herb, basal part of stem creeping and rooting, apical part erect, to 10 cm tall, tomentose with long many-celled downwardly directed purplish hairs. Leaves opposite on stems but often in a whorl of 4 just below inflorescences, above dark green with paler central area to almost white along veins, paler beneath; petiole 1–4 cm long; lamina oblong to obovate, largest 7–15 × 2.5–7.5 cm, apex rounded, base cordate, glabrous above, beneath sericeous along the strongly raised veins. Spikes 4–18 cm long, solitary or up to 4 from axils of upper leaves, sometimes with a pair of bracts near base of peduncle, these bracts sometimes with a single flower or developing lateral spikes; peduncle 1.5–8.5 cm long, with 1–3 pairs of sterile bracts, sericeous with many-celled purple hairs or glabrous upwards; axis with scattered hairs near base, glabrous upwards; flowers alternate or opposite; fertile bracts glumaceous, not imbricate, dark brown, ovate-triangular or narrowly so, 8–10 × 2–3 mm, acuminate, entire, glabrous or lower sparsely puberulous; bracteoles narrower, 7–10 mm long. Sepals dark brown to straw-coloured, lanceolate, 10–15 mm long, cuspidate. Corolla white; tube 17–21 mm long, straight or gently curved; lobes obovate, (10–)13–17 mm long, longer in lower lip, rounded. Capsule 4-seeded, ± 13 mm long, beak recurved when ripe. Seed triangular in outline, 4–5 mm long, with dense elongated tubercles.

Tanzania. Iringa District: Udzungwa Escarpment, Kihansi River Gorge, 22 June 1995, *Lovett* 5053! & 21 March 1997, *Lovett* 5066!
Distr. **T** 7; not known elsewhere
Hab. Wet evergreen lowland forest; 650–900 m

13. **CROSSANDRA**

Salisbury, Parad. Lond.: t. 12 (1805); Nees in DC., Prodr. 11: 280 (1847); G.P. 2: 1094 (1876); C.B. Clarke in F.T.A. 5: 112 (1899); Napper in K.B. 24: 334 (1970); Vollesen in K.B. 45: 504 (1990)

Perennial (rarely annual) herbs, subshrubs or shrubs. Leaves opposite or in whorls of 4, from below middle usually gradually narrowed into a long-decurrent base. Flowers single, in dense sessile or pedunculate, terminal or axillary spiciform cymes; bracts imbricate, in 4 rows, persistent, papery to coriaceous, with straight or recurved often pungent tip or 3-lobed at apex, margin entire, toothed, spinulose or setose; lower sterile and smaller; bracteoles 2, usually linear to lanceolate and acuminate to cuspidate, usually herbaceous. Calyx glumaceous, divided to base in 5 unequal sepals which are thickened and horny at base; dorsal sepal 2-veined and 2-toothed (in *C. cephalostachya* and *leucodonta* 1-veined and 1-toothed), broader than and often enveloping the rest, ventral and lateral 1-veined and 1-toothed, lateral smaller than ventral. Corolla puberulous outside; tube linear, usually widened above stamens, with a band of hairs at insertion of stamens; limb and top of tube split posteriorly to give a single 5-lobed lower lip, lobes oblong, obtuse, middle broadest and often emarginate. Stamens 4, included, usually inserted ± $\frac{1}{3}$ down the tube, subsessile; anthers 1-thecous, oblong, usually rounded to finely apiculate, finely bearded. Ovary 2-locular with 2 ovules per locule, glabrous or minutely hairy near top; style usually glabrous, gibbous or not below the obliquely trumpet-shaped or obscurely 2-lobed stigma which is at the same height as anthers. Capsule 4-seeded, ellipsoid, sessile, glumaceous, glabrous and shiny. Seed discoid, obliquely ovoid, covered by large appressed laciniate scales.

52 species in tropical Africa and Madagascar with 1 species extending to India and 1 species endemic in Arabia.

1. Bracts deeply 3-lobed; corolla limb white or pale lilac (rarely yellow) 2
   Bracts entire or with dentate, spinulose or setose margin, never deeply 3-lobed; corolla limb yellow to orange or light to bright red 5
2. Dorsal sepal 1-veined from base, terminating in a single long awn-like tooth 3
   Dorsal sepal with 2 strong veins from base and terminating in 2 teeth 4
3. Largest leaf 4–6.5 cm long; basal part of bracts dark green, apical part (teeth) white, finely puberulous apically; corolla white or with a faint lilac tinge, tube ± 10 mm long, limb ± 9 mm long   14. *C. leucodonta* (p. 150)
   Largest leaf 10.5–20 cm long; bracts not bi-coloured, densely pubescent and with stalked capitate glands; corolla limb pale yellow or yellow, tube ± 6 mm long and limb ± 5 mm long   13. *C. cephalostachya* (p. 149)
4. Leaves gradually narrowed below middle, with (7–)9–17 main lateral veins each side, apex acuminate to acute, concave below tip; corolla limb white ....   11. *C. tridentata* (p. 148)
   Leaves abruptly narrowed below middle, with 5–7 main lateral veins each side, apex subacute to rounded, not concave below tip; corolla limb pale lilac ....   12. *C. friesiorum* (p. 149)
5. Bracts with broad white margin, becoming frayed with age; bracteoles glumaceous, elliptic; style hairy......................   8. *C. puberula* (p. 145)
   Bracts without white margin, not frayed with age; bracteoles herbaceous, linear-lanceolate; style glabrous 6

6. Bracts with spinulose or setose margin; spikes sessile or with peduncle up to 5 mm long 7

Bracts with entire margin, but often conspicuously ciliate; some or all peduncles over 1 cm long (except *C. spinosa*) 8

7. Annual herb, stem usually unbranched; leaves in a terminal rosette; bract margin spinulose with 4–6 prickles per side, each 3–6 mm long; bracteoles 11–14 mm long; longest sepal 12–15 mm long .... 9. *C. stenostachya* (p. 147)

Perennial or shrubby herb; leaves not in a terminal rosette; bract margin setose with numerous bristles per side, each up to 1 mm long; bracteoles 2–3 mm long; longest sepal 6–7 mm long .... 10. *C. pungens* (p. 148)

8. Bracts ending in a long straight spine; corolla yellow (rarely apricot) 9

Bracts ending in a short recurved spine or with mucronate to acute non-spiny tip; corolla orange to red 10

9. Peduncle 2–10.5 cm long; bracts 10–13 mm long; bracteoles 13–15 mm long; leaves subglabrous or sparsely sericeous beneath; erect shrub or shrubby herb to 1 m tall .... 4. *C. baccarinii* (p. 141)

Spikes sessile or peduncle up to 5 mm long; bracts 13–20 mm long; bracteoles 15–22 mm long; leaves sparsely to densely uniformly puberulous; erect or procumbent herb to 40 cm tall ...... 5. *C. spinosa* (p. 142)

10. Bracts ending in a short recurved spine, concave below tip; sepals pilose, not with capitate glands; stamens inserted near middle of corolla tube; style shorter than bract; corolla tube bent at insertion of stamens 11

Bracts mucronate or acute, gradually narrowed to the straight tip; sepals glabrous or pilose, but then also with capitate glands; stamens inserted in upper $^1/_3$ of corolla tube; style longer than bract 12

11. Shrubby herb with elongated leafy stems; bracts 11–21 × 5–12(–14) mm, length/width-ratio 1.4–2.2(–3), if over 2.2 then bract less than 15 mm long .... 6. *C. mucronata* (p. 142)

Pyrophytic usually stem-less rosette herb; bracts 15–26 × 6–10 mm, length/width-ratio 2.4–3.3 (–3.7) .... 7. *C. subacaulis* (p. 144)

12. Plant with fine appressed silvery indumentum; bracts coriaceous, without reticulate secondary veins; corolla with capitate glands, tube bent at insertion of stamens .... 3. *C. infundibuliformis* (p. 140)

Plant without appressed silvery indumentum; bracts papery, with reticulate secondary veins; corolla without capitate glands, tube straight 13

13. Bracts, bracteoles and sepals with long capitate glands; sepals puberulous on back, dorsal 8–17 mm long .... 1. *C. nilotica* (p. 139)

Bracts, bracteoles and sepals without long capitate glands; sepals glabrous or sparsely puberulous on back, dorsal 5–8 mm long ... 2. *C. massaica* (p. 140)

1. **Crossandra nilotica** *Oliver* in Trans. Linn. Soc. 29: 128, t. 85 (1875); Lindau in P.O.A. C: 370 (1895); C.B. Clarke in F.T.A. 5: 115 (1899), excl. syn.; De Wild., Contrib. Fl. Katanga 1: 202 (1921) & Pl. Beq. 4: 28 (1926); F.P.N.A. 2: 294 (1947); Napper in K.B. 24: 338 (1970); Champluvier in Fl. Rwanda 3: 450 (1985); Vollesen in K.B. 45: 121 & 508 (1990); Lebrun & Stork, Enum. Pl. Afr. Trop. 4: 476 (1997); Friis & Vollesen in Biol. Skr. 51(2): 440 (2005); Ensermu in F.E.E. 5: 364 (2006). Types: Uganda, West Nile District: Madi, *Grant* 685 (K!, lecto.); Tanzania, Biharamulo District: Usui, *Grant* 135 (K!, syn.)

Erect or straggling to scrambling perennial or shrubby herb to 1.5 m tall; young stems glabrous to sericeous-puberulous. Leaves in whorls of 4; sessile or petiole up to 1.5(–4) cm long; lamina ovate to elliptic, largest 6.5–18(–23) × 2–7 cm, apex subacuminate to acute (rarely rounded), the actual tip rounded, subglabrous to puberulous (rarely pubescent), densest on veins. Spikes 2–8.5 cm long; peduncles 2–18 cm long, puberulous or sparsely so (rarely pubescent); fertile bracts ovate to elliptic or broadly so (rarely obovate or narrowly ovate), 12–33 × 4–16 mm, length/width-ratio 1.3–2.2(–3), narrowing gradually to an acute tip with a straight mucro up to 0.5 mm long, finely puberulous or sparsely so and with usually dense stalked capitate glands, ciliate on edges (sometimes also on veins) from long many-celled hairs, with 3–5 longitudinal veins and raised reticulum; bracteoles 8–18 mm long, finely puberulous, with long capitate glands, ciliate. Sepals acuminate to cuspidate, finely puberulous and with sparse to dense capitate glands towards tip, dorsal broadly elliptic, 8–17 mm long, ventral ovate-elliptic, 8–17 mm long, lateral 6–14 mm long. Corolla light red to bright red or scarlet (? sometimes orange); tube 20–32 mm long, straight, not infundibuliform apically; limb 10–16 mm long. Capsule ± 14 mm long. Seed ± 3 × 3 mm.

UGANDA. Karamoja District: Moroto, Oct. 1952, *Dale* 813!; Bunyoro District: Budongo Forest, 26 Nov. 1971, *Synnott* 774!; Mbale District: Bungulilo, 1 Mar. 1951, *G.H. Wood* 107!
KENYA. Turkana District: 32 km NE of Amodet, Sep. 1968, *Carr* 470!
TANZANIA. Bukoba District: Buhamira, Oct. 1931, *Haarer* 2248!; Mwanza District: Uzinza, Geita, 3 June 1937, *Burtt* 6543!; Kigoma District: 15 km N of Kigoma, Kakombe, 7 July 1959, *Newbould & Harley* 4278!
DISTR. U 1–4; K 2; T 1, 4; Sudan, Eritrea, Ethiopia, Congo-Kinshasa, Rwanda, Burundi, Angola, Zambia, Malawi
HAB. Dry forest, riverine forest, secondary forest, riverine scrub, termite mounds, shaded places in woodland and wooded grassland; 750–1550 m

SYN. *Crossandra smithii* S.Moore in J. B. 38: 462 (1900). Type: Sudan, Magois (Msai), *Donaldson Smith* s.n. (BM!, holo.)
    C.. nilotica Oliver var. *acuminata* Lindau in Ann. Ist. Bot. Roma 6: 76 (1896), *nom. nud.*; C.B. Clarke in F.T.A. 5: 115 (1899), excl. spec. ex Somalia; Fiori, Boschi e piante legn. Eritrea: 346 (1912); Jex-Blake, Gard. E. Afr. (ed. 4): 337, pl. 12, 2 (1957)
    C. nilotica Oliver subsp. *acuminata* (C.B.Clarke) Napper in K.B. 24: 338 (1970). Type: Eritrea, near Ghinda, Donkollo, *Schweinfurth & Riva* 303 (K!, lecto.; FT!, iso.; selected by Napper (l.c.))
    C. rhynchocarpa (Klotzsch) Cuf., E.P.A.: 955 (1964), *non Barleria rhynchocarpa* Klotzsch (1861)

NOTE. The description of a 4-lobed calyx with 2 large and 2 small decussate lobes and 2 stamens plus 2 staminodes clearly indicates that *Barleria rhynchocarpa* is a true *Barleria*, and that it was erroneously synonymised by Clarke (l.c.). This mistake unfortunately led Cufodontis to publish the combination *Crossandra rhynchocarpa*. Napper (l.c.) has explained the intricate synonymy of *C. nilotica* in detail.
    There are gradual morphological changes from south to north in the distribution area. Southern plants (Tanzania, Angola, Zambia) have wide bracts (length/width-ratio 1.3–2(–2.2)) and long sepals (12–17 mm), while northern plants (Sudan, Ethiopia) have narrow bracts (length/width-ratio 1.7–3) and short sepals (8–13 mm). But the variation is quite gradual with most Ugandan material falling in between, and there is no justification for maintaining subsp. *acuminata* for the northern form. The lectotype of *C. nilotica* from N Uganda has sepals 13 mm long and bracts exactly twice as long as wide. The corresponding measurements for the lectotype of subsp. *acuminata* are exactly the same.

2. **Crossandra massaica** *Mildbr.* in J. Arn. Arb. 11: 54 (1930); F.P.N.A. 2: 294 (1947); T.T.C.L.: 6 (1949); Turrill in Bot. Mag. (n. s.) 174: t. 404 (1962); Heine in F.W.T.A. (ed. 2) 2: 409 (1963); Vollesen in K.B. 45: 511 (1990); Lebrun & Stork, Enum. Pl. Afr. Trop. 4: 476 (1997); Friis & Vollesen in Biol. Skr. 51(2): 440 (2005); Ensermu in F.E.E. 5: 365 (2006). Type: Kenya, Kikemu (probably misprint for Kisumu), *Linder* 2639 (B†, holo.; K!, iso.)

Erect or straggling to scandent shrubby herb to 1.3 m tall; young stems glabrous to finely sericeous-puberulous. Leaves in whorls of 4; sessile or petiole up to 1 cm long; lamina ovate to elliptic or narrowly so, largest 4–13 × 1–4(–5) cm, apex acute to acuminate, the actual tip sometimes rounded, subglabrous to sparsely puberulous (rarely puberulous), densest on veins. Spikes 1–5(–7) cm long; peduncles (1–)2–16(–21) cm long, finely puberulous-sericeous or sparsely so; fertile bracts elliptic or narrowly so, 10–20 × 3–9 mm, usually 2–3 times as long as wide, narrowing gradually to an acute tip with a straight mucro less than 0.5 mm long, finely puberulous or sparsely so and sparsely ciliate on edges, no stalked capitate glands, with 3–5 strong longitudinal veins and raised reticulum; bracteoles 9–16 mm long, finely puberulous and ciliate towards tip. Sepals acuminate, pilose-ciliate towards tip, glabrous or sparsely puberulous towards tip, no capitate glands, dorsal broadly elliptic, 5–8 mm long, ventral ovate-elliptic, 7–10 mm long, lateral 3–6 mm long. Corolla pale red to vermillion, crimson or orange-red; tube 17–25 mm long, straight, not infundibuliform apically; limb 8–15 mm long. Capsule 10–13 mm long. Seed ± 3.5 × 3 mm.

UGANDA. Karamoja District: Kidepo National Park, 15 May 1972, *Synnott* 1052!; Busoga District: Buswale, Siavona Hill, 26 Mar. 1953, *G.H. Wood* 652! & Lake Victoria, Berkeley Bay, Jan. 1894, *Scott Elliot* 7105!
KENYA. Samburu District: Ngare-Narok, 21 Dec. 1958, *Newbould* 3324!; Laikipia District: Laikipia, 12 May 1923, *Battiscombe* 1130!; Masai District: Mara Masai Reserve, Teleke River, 25 Feb. 1950, *Bally* 7751!
TANZANIA. Musoma District: Banagi, Mugungu River, 3 Apr. 1961, *Greenway* 9967! & Zanaki, 20 June 1959, *Tanner* 4370! & Lobo Hill, 20 Dec. 1969, *Ole Sayalel* 11!
DISTR. U 1, 3; **K** 1–6; **T** 1; Ghana, Sudan, Ethiopia
HAB. Dry *Olea-Euclea* forest and -scrub, *Euclea-Acokanthera* upland bushland, *Grewia* bushland, *Acacia-Euphorbia* woodland, often on rocky hills or along dry rocky riverbeds, dry riverine forest; 1100–2100 m

SYN. [*Crossandra undulaefolia* sensu Rendle in J.B. 34: 398 (1896), *non* Salisbury (1805)]
    C. puberula Klotzsch var. *smithii* C.B.Clarke in F.T.A. 5: 117 (1899). Types: Kenya, Northern Frontier District: Kulal to Marsabit, *Donaldson Smith* 401 (BM!, syn.) & Laikipia, *Gregory* s.n. (BM!, syn.)
    [*C. nilotica* sensu C.H. Wright in Johnstone, Uganda Protec.1: 341 (1902); Jex-Blake, Wild Fl. Kenya: 107 (1948); Blundell, Wild Fl. Kenya: 104, pl. 27 (1982); F. White, Veg. Afr.: 176 (1983); Blundell, Wild Fl. E. Afr.: 390, pl. 528 (1987), quoad distrib. Kenya, *non* Oliver (1875)]
    [*C. mucronata* sensu C.H. Wright in Johnstone, Uganda Protec.1: 341 (1902), *non* Lindau (1894)]
    C. nilotica Oliver subsp. *massaica* (Mildbr.) Napper in K. B. 24: 339 (1970); U.K.W.F.: 581 (1974); Burkill, Useful Pl. W. Trop. Afr. 1: 9 (1985)

NOTE. *C. massaica* is closely related to *C. nilotica*, and the two were treated as subspecies by Napper (l.c.). But the separating characters are quite constant, and I have seen no intermediate specimens. It is true that the two have almost allopatric distributions in Tanzania and SW Kenya, but in Uganda, Sudan, NW Kenya and Ethiopia there is a considerable overlap. They also seem to have more or less identical habitat preferences, and I find it preferable to treat them as two distinct species.

3. **Crossandra infundibuliformis** (*L.*) *Nees* in Wall., Pl. As. Rar. 3: 98 (1832). Type: "India" (LINN!, holo.)

Shrub to 2 m tall; young stems densely silvery sericeous. Leaves in whorls of 4, sessile or with petiole up to 3 mm long; lamina elliptic, largest 4.5–6.5 × 1.3–2 cm, apex subacute to rounded, with spreading or silvery sericeous indumentum. Spikes 2–4.5 cm long; peduncle 11–18 cm long, subglabrous to sparsely sericeous. Sterile bracts silvery sericeous; fertile coriaceous, elliptic, 9–15 × 4–5 mm, narrowing gradually to an acute tip with a straight pungent mucro up to 0.5 mm long, puberulous and with dense capitate glands, ciliate on edges and on lower part of veins, with 3–5 strong longitudinal veins, no reticulum but a few veins branching off from main veins under very narrow angles; bracteoles 8–13 mm long, puberulous and with dense capitate glands, densely ciliate. Sepals cuspidate, sparsely puberulous and with scattered capitate glands, glabrous towards base, densely ciliate, dorsal broadly elliptic, 11–12 mm long, ventral elliptic 11–15 mm long, lateral 8–10 mm long. Corolla light red to bright red, with scattered capitate glands; tube 18–22 mm long, bent near insertion of stamens, upper part infundibuliform; limb 15–20 mm long. Capsule 11–12 mm long, puberulous near tip. Seed ± 3 × 3 mm (immature).

subsp. **boranensis** *Vollesen* in K.B. 45: 125 & 514 (1990); Lebrun & Stork, Enum. Pl. Afr. Trop. 4: 476 (1997); Ensermu in F.E.E. 5: 366 (2006). Type: Ethiopia, Sidamo Region, El Siro Wells, *Friis, Mesfin & Vollesen* 3216 (K!, holo.; C!, ETH, UPS, iso.)

Indumentum silvery-sericeous.

KENYA. Northern Frontier District: Dawa River, Murri, 30 June 1951, *Kirrika* 111!
DISTR. **K** 1; Ethiopia
HAB. *Acacia-Commiphora* woodland and bushland on sandy soil derived from limestone or on rocky limestone slopes; 650 m (900–1300 m in Ethiopia).

NOTE. Only known from 4 collections from adjacent areas of NE Kenya and SE Ethiopia It differs from the other subspecies of the variable *C. infundibuliformis* in the appressed silvery indumentum on stems and leaves and in the small leaves (see Vollesen in K.B. 45: 122 (1990)).
   C. infundibuliformis subsp. *infundibuliformis* – a native of India – is occasionally cultivated in Tanzania and has also been recorded as an escaped weed in the Sudan. It differs in having a spreading non-silvery indumentum and larger leaves.

4. **Crossandra baccarinii** *Fiori* in Bull. Soc. Bot. Ital. 1915: 53 (1915) & in Miss. Stefanini-Paoli 1. (Bot.): 138 (1916); E.P.A.: 954 (1964); Kuchar, Pl. Somalia (CRDP Techn. Rep. Ser., no. 16): 245 (1986); Vollesen in K.B. 45: 516 (1990); Lebrun & Stork, Enum. Pl. Afr. Trop. 4: 476 (1997); Ensermu in F.E.E. 5: 367 (2006); Thulin in Fl. Somalia 3: 386 (2006). Type: Somalia, between Mansur and Availe, *Paoli* 582 (FT!, holo.)

Erect shrubby herb or shrub to 1 m tall; young stems finely greyish sericeous. Leaves in whorls of 4, sessile; lamina ovate to elliptic or narrowly so, largest 3–6 × 0.8–2.2 cm, apex subacute to rounded, subglabrous to finely sericeous beneath, densest on veins. Spikes 2–5 cm long; peduncle 2–10.5 cm long, finely greyish sericeous; fertile bracts narrowly ovate, 10–13 × 2–4 mm, with a long cuspidate straight or slightly recurved pungent tip, sericeous-puberulous and densely pilose from long many-celled hairs, with strong central and two lateral longitudinal veins; bracteoles 13–15 mm long, with a long cuspidate straight or slightly recurved tip, finely puberulous and pilose. Sepals cuspidate, pilose-ciliate towards tip, dorsal broadly ovate, 8–10 mm long, ventral ovate, 8–10 mm long., lateral 6–8 mm long. Corolla yellow; tube 1.5–2 cm long; limb 1.3–1.7 cm long. Capsule ± 9 mm long (immature). Seed not seen.

KENYA. Northern Frontier District: Chandlers Falls, 26 Jan. 1956, *J. Adamson* 587! & Solberawara, 27 Dec. 1976, *Powys* 332!
DISTR. **K** 1; Ethiopia, Somalia
HAB. Dry open *Acacia-Commiphora* bushland on sandy soil; ± 600 m

SYN. [*Crossandra spinosa* sensu Jex-Blake, Gard. E. Afr. (ed. 4): 230, pl.12, 1 1957), *non* G.Beck (1888)]

5. **Crossandra spinosa** *G.Beck* in Paulitschke, Harar: 459, fig. 13 & 14 (1888); C.B. Clarke in F.T.A. 5: 114 (1899); S. Moore in J.B. 39: 302 (1901); Hutch. & E.A. Bruce in K.B. 1941: 169 (1941); E.P.A.: 955 (1964); Kuchar, Pl. Somalia (CRDP Techn. Rep. Ser., no. 16): 245 (1986); Vollesen in K.B. 45: 516 (1990); Lebrun & Stork, Enum. Pl. Afr. Trop. 4: 476 (1997); Ensermu in F.E.E. 5: 367 (2006); Thulin in Fl. Somalia 3: 386 (2006). Type: Ethiopia, near Harar, Mt Haqim, *Hardegger* s.n. (W holo.; K!, fragment & photo).

Erect shrubby herb to 40 cm tall or perennial or shrubby herb with prostrate branches up to 20 cm long; young stems finely sericeous to puberulous. Leaves in whorls of 4, sessile or petiole up to 5 mm long; lamina ovate, elliptic or obovate or narrowly so, largest (1.5–)3–6.5(–8) × 0.6–1.5(–3.2) cm, apex subacute to broadly rounded, sparsely to densely uniformly puberulous. Spikes 1–3.5 cm long, sessile or peduncle up to 5 mm long, sericeous to puberulous; fertile bracts narrowly ovate, 13–20 × 2–4 mm, with a long cuspidate straight or slightly recurved pungent tip, sericeous-puberulous and pilose or densely so from long many-celled hairs, with strong central and two lateral longitudinal veins; bracteoles 15–22 mm long, with a long cuspidate straight or slightly recurved tip, sericeous-puberulous and pilose. Sepals cuspidate, puberulous and pilose-ciliate towards tip, dorsal broadly ovate, 7–10 mm long, ventral ovate, 8–11 mm long, lateral 6–8 mm long. Corolla yellow (rarely apricot); tube 1.5–2 cm long; limb 1–1.5 cm long. Capsule ± 13 mm long, hairy at tip. Seed ± 3 × 3 mm.

KENYA. Northern Frontier District: Dawa River, Murri, 3 July 1951, *Kirrika* 127!
DISTR. **K** 1; Ethiopia, Djibouti, Somalia
HAB. *Acacia-Commiphora* bushland on rocky limestone slopes; ± 600 m

SYN. *Crossandra parvifolia* Lindau in E.J. 20: 37 (1895); Lindau in E. & P. Pf. IV,3b: 319 (1895); E.P.A.: 955 (1964). Type: Somalia, Meid, *Hildebrandt* 1404 (B†, holo.)
   C. parviflora Lindau in Schweinf. & Volkens, Liste Pl. Ghika-Comanesti: 17 (1897); Glover, Check-list Brit. and Ital. Somal.: 63 (1947), *nom. nud.*

NOTE. The Kenyan collection has appressed indumentum and is erect. Most of the Ethiopian collections are perennial herbs with trailing stems and spreading indumentum. The Somalian collections are often more or less intermediate in habit as well as in indumentum, and at the moment I prefer to treat all as one variable species. At least some of the erect forms have apricot-coloured corollas, and if it could be confirmed that erect plants always have apricot flowers, a case might be made for recognising two taxa. But relying solely on corolla colour has proved unreliable in other parts of the genus, e.g. in *C. spinescens* from Zambia and Zimbabwe.

6. **Crossandra mucronata** *Lindau* in E.J. 20: 35 (1894); Lindau in P.O.A. C: 370 (1895); Rendle in J.B. 34: 398 (1896); Clarke in F.T.A. 5: 116 (1899); S. Moore in J. Bot. 39: 302 (1901); Chiov., Fl. Somala 2: 350 (1932); E.P.A.: 955 (1964); Napper in K.B. 24: 336 (1970); U.K.W.F.: 581 (1974); Blundell, Wild Fl. Kenya: 105 (1982) & Wild Fl. E. Afr.: 390, pl. 527 (1987); Vollesen in K.B. 45: 518 (1990); Iversen in Symb. Bot. Ups. 29(3): 160 (1991); U.K.W.F. ed. 2: 274 (1994); Lebrun & Stork, Enum. Pl. Afr. Trop. 4: 476 (1997); Ensermu in F.E.E. 5: 368 (2006); Thulin in Fl. Somalia 3: 386 (2006). Type: Tanzania, Lushoto District: Usambara Mts, Nyika, *Holst* 571 (B†, holo.). Neotype: Kenya, Teita District: Voi, *Napier* 890 (K!, neo.; EA!, iso.; selected by Napper)

Shrubby herb or shrub to 1(–1.5) m tall, erect or somewhat scandent; young stems subglabrous to puberulous or pubescent. Leaves in whorls of 4, sessile or with petiole up to 2(–3) cm long; lamina ovate to elliptic, largest (3.5–)5.5–14(–18) × (1.5–)2–5(–7.2) cm, apex acute to broadly rounded, glabrous to puberulous or pubescent. Spikes 2–12.5 cm long; peduncle 1–16 cm long, subglabrous to finely sericeous or puberulous; fertile bracts obovate, 11–21 × 5–12(–14) mm, length/width-ratio 1.4–2.2(–3), concave (with distinct "shoulders") below a recurved

FIG. 22. *CROSSANDRA MUCRONATA* — **1**, habit, × ²⁄₃; **2**, bract, × 2; **3**, bracteole, × 2; **4**, sepals, × 2; **5**, corolla × 1.5; **6**, corolla opened up with stamens, × 10; **7**, fruiting inflorescence, × ²⁄₃; **8**, capsule, × 3; **9**, seed, × 6. 1 from *Robertson* 5062, 2–4 from *Bally* 8045, 5–6 from *Archer* 736, 7–9 from *Gilbert & Thulin* 1716. Drawn by Eleanor Catherine.

pungent mucro 0.5–3 mm long, puberulous and ciliate or densely so on edges and lower part of back, usually also with stalked capitate glands (see note), with 3–5 strong longitudinal veins and raised reticulum; bracteoles (7–)10–20 mm long, puberulous and densely ciliate. Sepals cuspidate, sparsely puberulous and sparsely to densely pilose on back (sometimes only on midrib), glabrous towards base, ciliate or densely so, dorsal broadly elliptic, 8–13 mm long, ventral ovate-elliptic, 9–15 mm long, lateral 6–11 mm long. Corolla pale orange to orange or salmon pink to bright red; tube yellow, 15–24 mm long, bent near insertion of stamens, upper part infundibuliform; limb 10–18 mm long. Stamens inserted near middle of tube. Capsule 9–14 mm long. Seed 3–4 × 2.5–3.5 mm. Fig. 22, p. 143.

Uganda. Karamoja District: 5–10 km N of Lokapel, 16 Nov. 1968, *Lye* 441! & 40–80 km N of Kacheliba, 9 May 1953, *Padwa* 107! & Amudat, 11 June 1959, *Symes* 537!
Kenya. Northern Frontier District: Ol Lolokwe [Ol Donyo Sabachi], 26 Mar. 1978, *Gilbert* 5036!; Machakos District: Mtito Andei, 16 Jan. 1961, *Greenway* 9743!; Teita District: Buchuma to Voi, Maunga, 10 Dec. 1961, *Polhill & Paulo* 942!
Tanzania. Pare District: Mgigile, May 1928, *Haarer* 1369! & Buiko, 22 July 1915, *Peter* 60244! & Mamba Ndungu, 15 Apr. 1971, *Sangiwa* 121!
Distr. **U** 1; **K** 1–4, 6, 7; **T** 2, 3; Ethiopia, Somalia, Zimbabwe, South Africa
Hab. *Acacia-Commiphora* woodland and bushland on sandy to loamy or stony soil or on rocky slopes and outcrops, dry upland forest and dry riverine forest and thicket; 0–1500(–2200) m

Syn. *Crossandra subacaulis* C.B.Clarke in F.T.A. 5: 116 (1899) quoad *Fischer* 253 (HBG!, syn.); *Gregory* s.n., Nyika Country (BM!, syn.); *Gregory* s.n., Sabaki River (BM!, syn.); *Volkens* 1669 (BM!, syn.), non sensu str.

Note. Specimens from the northern part of our area (**U** 1, **K** 1, 2, 3 and part of **K** 4) have many capitate glands on the bracts, while specimens from Tanzania and the southern part of **K** 7 are usually without glands. Specimens from central Kenya vary between these two. This is a useful additional character for separating *C. mucronata* from *C. subacaulis* in Uganda where some collections are otherwise more or less intermediate. The bracts in *C. subacaulis* have no or only a few scattered glands.

7. **Crossandra subacaulis** *C.B.Clarke* in F.T.A. 5: 116 (1899); C.H. Wright in Johnstone, Uganda Protec. 1: 341 (1902); Mildbr. in N.B.G.B. 9: 498 (1926); Stapf in Bot Mag. 156: t. 9336 (1933); Jex-Blake, Wild Fl. Kenya: 107, fig. 86 (1948); F.P.S. 3: 172 (1956); Napper in K.B. 24: 337 (1970); U.K.W.F.: 581 (1974); Blundell, Wild Fl. Kenya: 104, pl. 32 (1982); Vollesen in K.B. 45: 520 (1990); U.K.W.F. ed. 2: 274 (1994); Lebrun & Stork, Enum. Pl. Afr. Trop. 4: 476 (1997); Friis & Vollesen in Biol. Skr. 51(2): 440 (2005). Type: Kenya, Kitui, *Hildebrandt* 2716 (K!, lecto.; BM!, P!, iso., selected by Napper).

Acaulescent perennial herb from woody rootstock (rarely with stems up to 5 cm long). Leaves in a rosette, usually pressed to the ground, sessile or with petiole up to 1(–5) cm long; lamina ovate, elliptic or obovate, largest 4–20 × 1.5–6.5 cm, apex acute to broadly rounded (rarely retuse), subglabrous to densely puberulous. Spikes 2–12 cm long; peduncle 1–20 cm long, puberulous to pubescent or sparsely so; fertile bracts oblong-obovate or narrowly so, 16–26 × 6–10 mm, length/width-ratio 2.4–3.3(–3.7), concave (with distinct "shoulders") below a recurved pungent mucro 0.5–5 mm long, finely sericeous-puberulous or sparsely so and ciliate or densely so on edges and back, sometimes with a few capitate glands, with 3–5 strong longitudinal veins and slightly raised reticulum; bracteoles 13–24(–27) mm long, puberulous and densely pilose-ciliate. Sepals cuspidate, sparsely puberulous and sparsely to densely pilose on back (sometimes only on midrib), glabrous towards base, ciliate or densely so, dorsal elliptic or broadly so, 9–14 mm long, ventral ovate-elliptic, 10–17 mm long, lateral 7–12 mm long. Corolla pale orange to orange or salmon pink to bright red; tube yellow, 15–26 mm long, bent near insertion of stamens, upper part infundibuliform; limb 12–22 mm long. Stamens inserted near middle of tube. Capsule 9–14 mm long. Seed ± 4 × 3.5 mm.

UGANDA. West Nile District: Moyo, Sep. 1940, *Purseglove* 1012!; Karamoja District: Timuarea, Apr. 1960, *Wilson* 882! & Kidepo National Park, 13 May 1972, *Synnott* 940!
KENYA. Machakos/Kitui District: Yatta Plains, 28 Jan. 1938, *Edwards* 90!; Masai District: Namanga to Kajiado, 17 Dec. 1961, *Polhill & Paulo* 1009! & 125 km on Nairobi–Namanga road, 15 Dec. 1959, *Verdcourt* 2593!
TANZANIA. Mwanza District: Manano Island, 21 Nov. 1951, *Tanner* 486!; Moshi District: 26 km on Moshi–Same road, 7 Jan. 1967, *Richards* 21907!; Morogoro District: 25 km E of Morogoro, 26 Nov. 1955, *Milne-Redhead & Taylor* 7433!
DISTR. U 1, 4; **K** 4–7; **T** 1–4, 6, 8; Central African Republic, Congo-Kinshasa, Sudan, ?Mozambique
HAB. Fireswept grassland, *Acacia* wooded grassland and bushland, *Combretum-Terminalia* wooded grassland and bushland, *Acacia-Commiphora* bushland, *Brachystegia* woodland; 450–2000 m

SYN. *Crossandra nilotica* Oliv. var. *acuminata* Oliv. in Trans. Linn. Soc., ser. 2, 2: 345 (1887); Lindau in P.O.A.C: 370 (1895), *nom. nud.*, *non C. nilotica* Oliv. var. *acuminata* C.B.Clarke in F.T.A. 5: 115 (1899)
[*Crossandra nilotica* sensu Sacleux in Bull. Mus. Hist. Nat. Paris 16: 400 (1910), *non* Oliv. (1875)]

NOTE. Clarke (l.c.) cites a specimen from Mozambique, Boror, *Peters* s.n. which is no longer extant. This is the only record from Mozambique. But four of the other collections cited by Clarke as *C. subacaulis* are actually stunted specimens of *C. mucronata* (see p. 144), and similar stunted specimens of *C. mucronata* are known from Zimbabwe. It is therefore possible that the *Peters*-specimen – from an area not very far from Zimbabwe – was a similar specimen. On the other hand, proper *C. subacaulis* does occur as far south as Lindi in **T** 8, and it most likely occurs at least in N Mozambique as well.

C. subacaulis differs from *C. mucronata* mainly in the acaulescent habit and in the narrower bracts, and they might with some justification be treated as subspecies. The two have at least partly differing ecological preferences with *C. subacaulis* preferring higher rainfall areas, but there is a large overlap and the distribution areas also show a large overlap in Kenya, Uganda and Tanzania, seemingly indicating that they are better treated as two distinct species.

8. **Crossandra puberula** *Klotzsch* in Peters, Reise Mossamb. Bot. 1: 214 (1861); Lindau in P.O.A. C: 370 (1895); Clarke in F.T.A. 5: 117 (1899); E.A. Bruce in Fl. Pl. Afr. 28: pl. 1098 (1951); Vollesen in Opera Bot. 59: 80 (1980) & in K.B. 45: 521 (1990); Lebrun & Stork, Enum. Pl. Afr. Trop. 4: 476 (1997). Type: Mozambique, Rios de Sena, *Peters* s.n. (B†, holo.). Neotype: Mozambique, Mocuba District: Namagoa, *Faulkner* P.317 (K!, neo.; BR!, COI!, EA!, K!, P!, PRE!, S!, SRGH!, iso.)

Erect perennial or shrubby herb to 40 cm tall, from creeping woody rhizome; young stems puberulous or sparsely so. Leaves in pseudo-whorls of 4, sessile or petiole up to 1 cm long; lamina usually abruptly narrowed (lyrate) below middle, more rarely gradually narrowed, upper part ovate to elliptic, largest 11.5–21 × 4.5–10 cm, entire or slightly crenate, apex subacute to broadly rounded, sparsely puberulous, densest on veins. Spikes 4–25 cm long; peduncle 1–17 cm long, sparsely to densely puberulous; fertile bracts 15–21 mm long, narrowly ovate, glumaceous, green with a broad white margin when young, becoming frayed with age, glabrous to finely puberulous, with several parallel longitudinal veins, no reticulum; bracteoles glumaceous, elliptic, 9–14 mm long, with a fine terminal bristle 1–2 mm long, finely puberulous. Sepals ovate, terminating in a bristle 1–2 mm long, puberulous towards tip, dorsal and ventral 6–9 mm long, lateral ± 2 mm shorter. Corolla pale peach or salmon pink; tube 2–3 cm long, slightly widened above insertion of stamens, slightly bent; limb 1.3–1.8 cm long. Style hairy. Capsule 9–11 mm long. Seed 3–4 × 3–4 mm. Fig. 23, p. 146.

TANZANIA. Morogoro District: Mikumi National Park, 12 Apr. 1970, *Batty* 1049!; Masasi District: Masasi, 13 Mar. 1960, *Hay* 88!; Tunduru District: 32 km on Tunduru–Songea road, 4 Mar. 1963, *Richards* 17727!
DISTR. **T** 6, 8; Malawi, Mozambique, Zimbabwe

Fig. 23. *CROSSANDRA PUBERULA* — **1**, habit, × ²/₃; **2**, bract, × 4; **3**, bracteole, × 4; **4**, sepals, × 4; **5**, corolla × 1.5; **6**, corolla opened up with stamens, × 8; **7**, fruiting inflorescence, × ²/₃; **8**, capsule, × 2; **9**, seed, × 6. 1–4 from *Chase* 4457, 5–6 from *Chase* 4456, 7–9 from *Meller* s.n. Drawn by Eleanor Catherine.

Hab. Woodland and wooded grassland, often on termite mounds, usually in shade; 200–900 m

Syn. *Crossandra pubescens* Klotzsch in Peters, Reise Mossamb. Bot. 1: 213 (1861); Lindau in
  P.O.A. C: 370 (1895) & in E. & P. Pf. IV, 3b: 319 (1895). Type: Mozambique, Boror, *Peters*
  s.n. (B†, holo.)
  Crossandra jashi Lindau in E. J. 38: 70 (1905); Mildbraed in N.B.G.B.11: 823 (1933);
  T.T.C.L.: 6 (1949). Types: Tanzania, Lindi District: Maua, Kipindimbi, *Busse* 2655 (BM!,
  BR!, EA!, HBG!, isosyn.) & Seliman-Mamba, *Busse* 2666 (BM!, BR!, EA!, HBG!, isosyn.)

Note. *C. pubescens* is no more than a form of *C. puberula* with a slightly denser indumentum
  than usual, especially on peduncles and bracts. It is also a pyrophytic rosette herb, but both
  subglabrous pyrophytic plants and hairy subshrubs have been seen, and – as no other
  differences have been found – it seems best only to recognise one taxon. *Faulkner* P.317
  from Mozambique has been particularly useful in illustrating the variation possible within
  a single population.

9. **Crossandra stenostachya** (*Lindau*) *C.B.Clarke* in F.T.A. 5: 113 (1899); E.P.A.: 955
(1964); U.K.W.F.: 581 (1974); Blundell, Wild Fl. Kenya: 105 (1982) & Wild Fl. E. Afr.:
390 (1987); Vollesen in K.B. 45: 529 (1990); U.K.W.F. ed. 2: 274 (1994); Lebrun &
Stork, Enum. Pl. Afr. Trop. 4: 476 (1997); Ensermu in F.E.E. 5: 369 (2006); Thulin in
Fl. Somalia 3: 386 (2006). Type: Kenya, Machakos/Kitui District: Ukamba,
Hildebrandt 2720 (B†, holo.; COR, iso.; K!, fragment & photo).

Erect annual herb to 35 cm tall, unbranched with an apical rosette of leaves and
1–4 sessile spikes or with up to 4 branches up to 10 cm long from the rosette and
these each ending in a rosette with a single sessile spike; young stems glandular-
puberulous to pilose. Leaves subsessile; lamina obovate or narrowly so, largest 5–15
× 1–5 cm, apex subacute to broadly rounded, pubescent or sparsely so, usually only
on veins when mature. Spikes 2–12 cm long; fertile bracts broadly obovate to
orbicular or transversely elliptic, 12–16 × 10–17 mm, ending in a straight spine
2–6 mm long and on each side with 4–6 prickles 3–6 mm long (sometimes also with
a few short ones between these), basal part pilose, apical part puberulous, venation
reticulate, very prominently raised; bracteoles 11–14 mm long, pilose. Sepals
lanceolate with long cuspidate-spinulose tip, pilose near tip, dorsal 12–15 mm long,
ventral 10–13 mm long, lateral 9–11 mm long. Corolla bright yellow; tube up to 15 mm
long; limb 8–13 mm long. Anthers with a small apical horn. Capsule 8–10 mm long.
Seed 3–4 × 2.5–3 mm.

Uganda. Karamoja District: Kangole, 22 May 1940, *A.S. Thomas* 3493! & *J. Wilson* 379! &
  Bokora, 5 km S of Kautaku, 12 June 1970, *Lye* 5661!
Kenya. Northern Frontier District: Mt Marsabit, 19 Feb. 1956, *J.G. Williams* EA 11036!; Kitui
  District: 80 km S of Kitui, 31 Jan. 1957, *Bogdan* 4378!; Tana River District: 6 km on
  Tula–Garsen road, 7 July 1974, *Faden* 74/986!
Tanzania. Pare District: Lake Jipe, 31 Oct. 1915, *Peter* 60338! & 17 Aug. 1974, *Mhoro & Backeus*
  2190! & Kisiwani, 27 June 1942, *Greenway* 6496!
Distr. U 1; K 1, 3, 4, 6, 7; T 3; Ethiopia, Somalia
Hab. Grassland and *Acacia* bushland on black cotton soil, wet depressions in *Acacia-Commiphora*
  bushland, riverine forest and riverbanks; (150–)300–1200 m

Syn. *Sclerochiton stenostachyus* Lindau in E.J. 20: 27 (1894); Lindau in P.O.A. C: 369 (1895) &
  Lindau in E. & P. Pf. IV, 3b: 316 (1895)
  Crossandra. stenostachya (Lindau) C.B.Clarke var. *somalensis* Fiori in Miss. Stefanini-Paoli 1.
  Bot.: 137 (1916); E.P.A.: 956 (1964); Kuchar, Pl. Somalia (CRDP Techn. Rep. Ser., no.
  16): 245 (1986). Type: Somalia, Matagoi to El Bar, *Paoli* 694 (FT!, holo.)

Note. The pollen grains of this species are the longest ever reported in terrestrial
  angiosperms. They are rod-shaped and ± 0.5 mm long (visible with the naked eye). See
  Brummitt, Ferguson & Poole in Pollen et Spores 22: 12 (1980).
  The peculiar small apical horns on the anthers have not been seen in any other species.
  They are probably a rudiment of the second locule.

10. **Crossandra pungens** *Lindau* in E.J. 20: 36 (1894) & in P.O.A. C: 370 (1895); Clarke in F.T.A. 5: 114 (1899); S. Moore in J.B. 39: 302 (1901); Vollesen in Opera Bot. 59: 80 (1980) & in K.B. 45: 530 (1990); Iversen in Symb. Bot. Ups. 29(3): 160 (1991); Lebrun & Stork, Enum. Pl. Afr. Trop. 4: 476 (1997). Types: Kenya, near Mombasa, *Wakefield* s.n. (K!, lecto.); Tanzania, Lushoto District: E Usambara Mts, Duga, *Holst* 3215 (B†, syn.)

Perennial or shrubby herb to 1 m tall; young stems crisped-puberulous or sparsely so. Leaves opposite, dark green with pale green areas along veins, sessile or petiole up to 1(–3.5) cm long; lamina elliptic or narrowly so, largest 7–26.5 × 2–6 cm, entire to sinuate, apex subacute to broadly rounded, subglabrous to sparsely crisped-puberulous, densest on veins. Spikes 2–10(–16) cm long, sessile or peduncle up to 5 mm long; fertile bracts broadly elliptic to subcircular, 15–21 × 12–17 mm, ending in a straight mucro 1–2 mm long and on each side with numerous bristles up to 1 mm long, sparsely (rarely densely) pilose on veins, otherwise uniformly puberulous, venation prominently raised, reticulate; bracteoles 2–3 mm long, acute to obtuse, glabrous. Sepals acute, pilose near tip, dorsal broadly elliptic, 6–7 mm long, ventral narrowly ovate, 6–7 mm long, lateral 5–6 mm long. Corolla yellow to golden yellow or orange yellow, glandular-puberulous; tube 2–3 cm long; limb 8–13 mm long. Capsule 11–13 mm long. Seed 3–4 × ± 3 mm.

KENYA. Kwale District: Mombasa–Tanga road, 1 km from border, 14 Aug. 1953, *Drummond & Hemsley* 3755A!; Lamu District: Boni Forest, 6 Sep. 1961, *Gillespie* 312! & Witu Forest, *Rawlins* 71!
TANZANIA. Tanga District: Kivindana, 11 Jan. 1959, *Faulkner* 2219! & 3 Aug. 1965, *Faulkner* 3608!; Bagamoyo District: 30 km W of Bagamoyo Ferry, 5 July 1960, *Leach & Brunton* 10179!
DISTR. **K** 7; **T** 3, 6; not known elsewhere
HAB. Undergrowth in evergreen or semi-evergreen coastal forest and thicket, riverine forest, termite mounds; 0–500 m

11. **Crossandra tridentata** *Lindau* in P.O.A. C: 370 (1895); Lindau in E.J. 24: 320 (1897); Clarke in F.T.A. 5: 114 (1899); T.T.C.L.: 6 (1949); U.K.W.F.: 581 (1974); Blundell, Wild Fl. Kenya: 105 (1982); Vollesen in K.B. 45: 531 (1990); Iversen in Symb. Bot. Ups. 29(3): 160 (1991); U.K.W.F. ed. 2: 274 (1994); Lebrun & Stork, Enum. Pl. Afr. Trop. 4: 476 (1997). Type: Tanzania, Kilimanjaro, Useri, *Volkens* 1994 (B†, holo.; BM!, K!, iso.)

Erect shrubby herb to 1.5 m tall (rarely basal part of stem creeping and rooting); young stems conspicuously swollen for 0.5–3 cm below nodes, purplish, subglabrous to finely sericeous. Leaves opposite or in whorls of 4, dark green above, pale beneath; petiole 0.5–4 cm long; lamina elliptic to obovate, largest (6.5–)8.5–22 × (2.5–)3–7.8 cm, apex acute to acuminate, concave below the blunt tip; entire to slightly crenate; with (7–)9–17 main lateral veins each side, subglabrous to finely and sparsely sericeous beneath, densest on veins. Spikes 1–3.5(–4) cm long; peduncle 1–10.5 cm long, finely sericeous; fertile bracts elliptic or broadly so, 10–18 × 5–8 mm, often purplish tinged towards apex, subglabrous to puberulous (rarely also sparsely pilose) and ciliate, with 3 strong longitudinal veins, no reticulum, apically deeply 3-lobed, lobes triangular, acute to acuminate; bracteoles (4–)5–10 mm long, ciliate towards tip. Sepals acute, glabrous or puberulous to finely ciliate near tip, dorsal broadly elliptic, 6–8.5 mm long, ventral oblong, 5.5–8 mm long, lateral 4–6.5 mm long. Corolla white with yellow patch in throat; tube 8–13 mm long, narrowed above insertion of stamens; limb 8–13 mm long. Capsule 8–11 mm long, hairy at apex. Seed 3–4 × 3–4 mm.

KENYA. Nyeri District: E Aberdare Mts, *Dale* 1862!; Teita District: Ngangao Forest, 6 Feb. 1953, *Bally* 8753! & 15 Sep. 1953, *Drummond & Hemsley* 4351!

TANZANIA. Arusha District: Mt Meru, 9 Aug. 1960, *Greenway* 9721!; Lushoto District: E Usambara Mts, Ngomeni, 14 Sep. 1915, *Peter* 60328!; Morogoro District: Nguru Mts, Turiani, 20 Aug. 1971, *Schlieben* 12270!
DISTR. **K** 4, 6, 7; **T** 2, 3, 6; not known elsewhere
HAB. Ground layer in lowland and upland rainforest, in deep shade; 700–2400 m

12. **Crossandra friesiorum** *Mildbr.* in N.B.G.B. 9: 498 (1926); Hepper in Bot. Mag. (n.s.) 179: t. 647 (1973); U.K.W.F.: 581 (1974); Blundell, Wild Fl. Kenya: 105 (1982) & Wild Fl. E. Afr.: 389, pl. 123 (1987); Vollesen in K.B. 45: 532 (1990); U.K.W.F. ed. 2: 275 (1994); Lebrun & Stork, Enum. Pl. Afr. Trop. 4: 476 (1997). Type: Kenya, Fort Hall District: Thika, *R.E. & Th.C.E. Fries* 4 (UPS, holo.; S!, iso.)

Perennial herb, basal part of stems creeping and rooting, apical part ascending, up to 25 cm tall; young stems not conspicuously swollen below nodes, finely sericeous to puberulous. Leaves sometimes variegated along midrib; petiole 0.5–5.5 cm long; lamina abruptly narrowed below middle (lyrate), lower part often sinuate, upper part ovate, largest 5–14.5(–16) × 2–5.5 cm, apex subacute to rounded, not concave below tip; apical part entire (rarely sinuate); with 5–7 main lateral veins each side, subglabrous to puberulous, densest on veins. Spikes 1–3 cm long; peduncle 5–17 cm long, puberulous and with long pilose hairs; fertile bracts elliptic to obovate in outline, 8–12 mm long, purplish tinged towards apex, puberulous and pilose or densely so, conspicuously ciliate, veins as in *C. tridentata*, deeply 3-lobed (lowermost pair often entire), lobes narrowly triangular, acute to acuminate; bracteoles 3–6 mm long, ciliate towards tip. Calyx as in *C. tridentata*, longest lobe 4–7 mm long. Corolla pale lilac with yellow throat; tube 9–12 mm long, narrowed above insertion of stamens; limb 10–15 mm long. Capsule 7–10 mm long, hairy at apex. Seed ± 3 × 3 mm.

KENYA. Fort Hall District: Thika Falls, 14 Aug. 1938, *Bally* 7462! & 12 Apr. 1968, *Faden* 68/132!; Teita District: Maungu Hills, 31 May 1970, *Faden* 70/173!
TANZANIA. Moshi District: Ngulu, Jan. 1928, *Haarer* 985!; Same District: Pare Mts, Chome, 20 Mar. 1980, *Kibuwa* 5236! & Mkomazi Game Reserve, Maji Kununua Ridge, 7 June 1996, *Abdallah et al.* 96/81!
DISTR. **K** 4, 7; **T** 2, 3; not known elsewhere
HAB. Dry lowland and medium altitude *Drypetes-Syzygium* and *Croton megalocarpus* forest, riverine *Newtonia-Filicium* forest; 900–1600 m

NOTE. *Polhill & Verdcourt* 316 from Meru Forest (**K** 4) is somewhat intermediate between this species and *C. tridentata*. It has less lyrate leaves than typical *C. friesiorum* and is much less hairy. It is from a wetter forest type than the rest of the material of this species, and is probably no more than an extreme shade form. The area where it was collected is however isolated from the rest of the range of *C. friesiorum*, and further collections from the Meru area are needed to clarify its status.

13. **Crossandra cephalostachya** *Mildbr.* in N.B.G.B. 11: 1083 (1934). Type: Tanzania, Morogoro District: Uluguru Mts, Kinole, *Schlieben* 2862 (B†, holo.; BM!, BR!, EA!, HBG!, P!, S!, iso.)

Shrubby herb to 1.2 m tall, basal part of stem creeping; young stems only slightly swollen below nodes, finely sericeous to puberulous. Leaves opposite or in whorls of 3–4 towards apex of stems, dark green above, pale beneath; petiole 1–6 cm long; lamina elliptic to slightly obovate, largest 10.5–20 × 3.5–6.5 cm, apex subacute, concave below the blunt tip, base attenuate, narrowing gradually into petiole; entire to crenate; with 12–15 main lateral veins each side, sparsely sericeous on veins and minutely ciliate. Spikes quadrangular, 2–4 cm long; peduncle 2.5–5 cm long, sericeous-puberulous; fertile bracts broadly obovate in outline, 8–10 × 8–12 mm, densely pubescent and with long stalked capitate glands, with 3 strong longitudinal veins, no reticulum, apically deeply 3-lobed, lobes triangular, acute to obtuse, sometimes reflexed; bracteoles 5–8 mm long, densely pubescent and with capitate glands in apical $^2/_3$. Sepals acute (lateral obtuse), puberulous towards apex and ciliate, dorsal broadly elliptic, 5–7 mm long, ventral elliptic, 4.5–6.5 mm long, lateral 2.5–3.5 mm long. Corolla pale yellow or yellow with darker orange-yellow centre; tube ± 6 mm long, narrowed above insertion of stamens; limb ± 5 mm long. Capsule 6–8 mm long, hairy at apex. Seed ± 3 × 2.5 mm.

TANZANIA. Lushoto District: Mtai Forest Reserve, 20 Jan. 1987, *Ruffo & Mmari* 2033!; Morogoro District: Mkungwe Forest Reserve, 25 Jan. 2001, *Jannerup & Mhoro* 285!; Iringa District: Udzungwa Mts National Park, 5 Oct. 2001, *Luke et al.* 8161!
DISTR. **T** 3, 6, 7; not known elsewhere
HAB. Lowland to montane evergreen forest; 750–1450 m

NOTE. This was originally thought to be synonymous with *C. tridentata*, but recent good collections from the Udzungwa Mts with good notes have shown it to be quite distinct. The densely hairy bracts, smaller yellow flowers and smaller fruit separate it. Most significantly the dorsal sepal is 1-veined and 1-toothed pointing to it being closer to *C. leucodonta* than to *C. tridentata*.

14. **Crossandra leucodonta** *Vollesen* in K.B. 55: 965 (2000). Type: Tanzania, Iringa District: Udzungwa Escarpment, Luhega Forest Reserve, *Frimodt-Møller et al.* NG092 (K!, holo.; C!, iso.)

Perennial herb, basal part of stems creeping and rooting, apical part erect, to 25 cm tall; young stems not conspicuously swollen below nodes, sericeous-tomentose with long curly many-celled hairs. Leaves opposite or in whorls of 4, dark green above, paler beneath; petiole 0.5–2 cm long; lamina gradually narrowed below middle, ovate to elliptic, largest 4–10 × 2–4 cm, apex subacute to acuminate, with a more or less well-defined tip, base attenuate, margin slightly crenate, sparsely sericeous-pubescent along veins beneath, above with scattered hairs on veins and lamina; with 3–5 main lateral veins each side. Spikes 1–2 cm long; peduncle 1.5–3.5 cm long, sericeous-tomentose with long many-celled curly hairs; fertile bracts obovate in outline, dark green with white lobes, 14–18 × 5–6 mm, basal part with scattered long many-celled curly hairs, apical part of lobes finely puberulous, deeply 3-lobed and lower down also with a single tooth per side, lobes narrowly triangular, 7–9 × 1–2 mm, tapering gradually to an acuminate tip; bracteoles 3–5 mm long, linear-lanceolate, with a long cuspidate tip, indumentum as bracts. Sepals glabrous, dorsal elliptic, 8–9 mm long, with a long cuspidate awn-like tip, 1-toothed, 1-veined from base, ventral oblong, ± 7 mm long, acuminate, lateral lanceolate, ± 5 mm long. Corolla white or with a faint pale lilac tinge to limb; tube ± 10 mm long; limb ± 9 mm long. Capsule 8–9 mm long. Seed 2–3 mm long.

TANZANIA. Iringa District: Udzungwa Escarpment, Luhega Forest Reserve, 22 Feb. 1996, *Frimodt-Møller et al.* NG092! & 25 Feb. 1996, *Hørlyck & Jøker* TZ459! & Ilutila Forest, 17 Feb. 2000, *Ndangalasi* 429!
DISTR. **T** 7; not known elsewhere
HAB. Wet evergreen montane forest with *Parinari excelsa*, *Strombosia scheffleri*, *Syzygium guineense*, *Myrianthus holstii* and *Ficalhoa laurifolia*; 1500–1850 m

IMPERFECTLY KNOWN SPECIES

Crossandra leikipiensis Schweinf. in von Höhnel, Zum Rudolph See, Sonderabdr.: 7 (1892); Engl., Hochgebirgsfl. Trop. Afr.: 391 (1892) (as *likipiensis*); Lindau in P.O.A. C: 370 (1895); Clarke in F.T.A. 5: 115 (1899). Type: Kenya, Laikipia [Leikipia], *von Höhnel* 75 (B†, holo.)

NOTE. *C. leikipiensis* must remain an imperfectly known species. The original description is very short and does not allow any conclusions as to its identity. Clarke (l.c.) – who examined the type material and prepared a longer description from it – obviously thought it different from another specimen (*Thomson* s.n.) also from Laikipia. All material from this area examined by me (including *Thomson* s.n.) is referable to *C. massaica* but differs in several respects from the description given by Clarke for *C. leikipiensis*. The bracts and sepals are e.g. described as having long silky hairs, a character which would fit *C. nilotica* better. But *C. nilotica* does not occur anywhere near Laikipia.

14. CROSSANDRELLA

C.B.Clarke in K.B. 1906: 251 (1906); Lindau in E. & P. Pf. IV, 3b, Nachtr. 4: 285 (1914); Heine in Adansonia, Ser. 2, 11: 641 (1971)

Shrubby herbs or shrubs. Leaves opposite. Flowers single, sessile, in loose terminal spikes; bracts small, green; bracteoles 2, larger, green. Calyx divided almost to base in 4 green sepals which are thickened and horny at base, dorsal and ventral broad, ventral 2-fid, lateral narrower. Corolla glabrous outside, tube widened around stamens, limb split dorsally to give an obscurely 5-lobed or crenate lower lip. Stamens 4, inserted in middle of tube, not on a thickened flange, subsessile; anthers 1-thecous, included, oblong, rounded at both ends, finely bearded. Ovary 2-locular with 2 ovules per locule; style glabrous; stigma entire, bent. Capsule 2-seeded, woody, sessile, beaked, cracking transversely when mature. Seed discoid, tuberculate.

2 species, 1 in Liberia and 1 from Nigeria eastwards to Uganda and NW Tanzania.

Crossandrella dusenii (*Lindau*) *S.Moore*, Cat. Pl. Nig. Talbot: 74 (1913); T.T.C.L.: 6 (1949); Heine in F.W.T.A. (ed. 2) 2: 412 (1963) & in Fl. Gabon 13: 110 & pl. 22, fig. 11–20 (1966) & in Adansonia, Ser. 2, 11: 647 (1971); Lebrun & Stork, Enum. Pl. Afr. Trop. 4: 476 (1997). Type: Cameroon, *Dusen* 348A (B†, holo.; BM!, fragment and sketch of holotype)

Shrubby herb or shrub to 1 m tall, basal part of stem sometimes creeping and rooting; stems fleshy, with 2 longitudinal bands of hairs when young, otherwise glabrous. Leaves with petiole up to 2 cm long; lamina elliptic or narrowly so, long-decurrent on the petiole, largest 13–26 × 3.2–5.5 cm, apex acuminate, glabrous. Spikes 5–17 cm long; peduncle 1–2 cm long, glabrous to puberulous; rachis puberulous, flattened; flowers opposite; bracts lanceolate to narrowly ovate, 5–8 mm long, glabrous, reflexed; bracteoles 8–12 mm long, elliptic, obtuse, finely puberulous along edges, 3–5-veined. Dorsal and ventral sepals 8–11 mm long, glabrous, with numerous strong veins, dorsal ovate, ventral broadly so or elliptic, lateral sepals linear, 6–8 mm long, 1-veined. Corolla blue to purple or white, glandular hairy at base of lip and in a ventral band in upper half of tube; tube 8–11 mm long; limb 5–7 mm long, deflexed and curved backwards. Anthers ± 3.5 mm long. Capsule 10–11 mm long, glabrous. Seed 4–5 mm long, irregularly ellipsoid- circular, sparsely tuberculate. Fig. 24, p. 152.

UGANDA. Bunyoro District: Budongo Forest, Sep. 1933, *Eggeling* 1410!; Mengo District: Entebbe, 1909, *Fyffe* 65!; Masaka District: Minziro Forest, Oct. 1925, *Maitland* 1053!
TANZANIA. Bukoba District: Kikuru Forest Reserve, 16 Sep. 1934, *Gillman* 159! & Minziro Forest Reserve, 6 July 2000, *Bidgood et al.* 4884!
DISTR. **U** 2, 4; **T** 1; Nigeria, Cameroon, Gabon, Equatorial Guinea, Congo-Kinshasa
HAB. Ground layer in evergreen lowland rainforest and swamp forest; 850–1200 m

SYN. *Pseudoblepharis dusenii* Lindau in E.J. 20: 34 (1894); Lindau in E. & P. Pf. IV, 3b: 319 (1895) & in Z.A.E.: 299 (1911)
  Acanthus dusenii (Lindau) C.B.Clarke in F.T.A. 5: 108 (1899)
  Crossandrella laxispicata C.B.Clarke in K.B. 1906: 251 (1906). Type: Uganda, Mengo District: Mawokota, *E. Brown* 210 (K!, holo.; BM!, iso.)

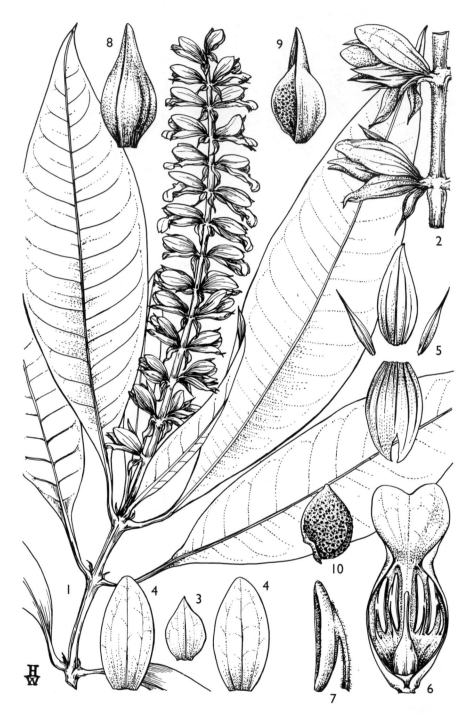

FIG. 24. *CROSSANDRELLA DUSENII* — **1**, habit, × 1; **2**, detail of fruiting inflorescence, × 2; **3**, bract, × 3; **4**, bracteoles, × 3; **5**, sepals, × 3; **6**, corolla, tube opened up, × 5; **7**, anther, × 15; **8**, capsule, × 3; **9**, valve of capsule with seed, × 3; **10**, seed, × 3. 1–7 from *Gillman* 159, 8–10 from *Eggeling* 1410. Drawn by Heather Wood.

15. BRILLANTAISIA

P. Beauv., Fl. Owar. 2: 67, t. 100, fig. 2 (1818); Burkill in F.T.A. 5: 37 (1899); Cramer in K.B. 46: 335 (1991); Sidwell in Bull. Nat. Hist. Mus. London 28: 67 (1998)

Belantheria Nees in DC., Prodr. 11: 96 (1847)

Leucoraphis Nees in DC., Prodr. 11: 97 (1847)

Ruelliola Baillon in Bull. Soc. Linn. Paris 2: 852 (1890) & in Hist. Pl. 10: 427 (1891); Lindau in E. & P. Pf. IV, 3b: 307 (1895)

Perennial (rarely annual) or shrubby herbs or soft wooded shrubs with conspicuous cystoliths, usually drying blackish; stems quadrangular with conspicuously swollen nodes. Leaves opposite, usually with two decurrent often toothed flanges on the petiole. Flowers in large open or contracted panicles, rarely in racemoid cymes; bracts foliaceous; bracteoles absent. Calyx divided almost to the base into 5 sepals, these linear to ± spathulate, similar or dorsal larger. Corolla distinctly 2-lipped, lips ± of equal length; tube short and broad, with a transverse fold just above insertion of stamens, slightly to distinctly constricted in upper part; upper lip 2-lobed, strongly hooded and laterally compressed; lower lip 3-lobed, ± flat or sides bent down, with a distinct "hinge" at base and with distinct "herring bone" pattern. Stamens 2 (dorsal), held in the upper lip, ventral reduced to 2 usually conspicuous staminodes; filaments long, inserted in tube and fused with it for some distance, broad and flattened, with a forwardly directed tooth; basal part hairy; anthers long, linear to narrowly oblong, 2-thecous, muticous, thecae parallel, at same height. Ovary 2-locular, with numerous ovules; disk large, annular; style linear, hairy, held under upper lip or exserted; stigma with linear ventral lobe usually wavy along the edge, dorsal lobe reduced to a small tooth. Capsule linear, tetragonous with valves deeply grooved on the back; retinaculae indurated, curved. Seeds 6–15 per locule, discoid, ovoid-cordiform, with dense hygroscopic hairs.

13 species in tropical Africa, mainly in the Guineo-Congolian forests, 2 species extending to Madagascar.

1. Ovary and capsule glabrous or with a few long non-
   glandular hairs at the apex . . . .   4. *B. lamium*
   Ovary and capsule densely hairy and/or glandular, if
   only in apical half then distinctly glandular 2
2. Flowers in dense racemoid cymes without discernible
   lateral branches . . . .   9. *B. madagascariensis*
   Flowers in open to contracted panicles with well-
   developed lateral branches 3
3. Anthers 1–5 mm long; capsule 10–23 mm long 4
   Anthers 5–11 mm long; capsule (17–)20–45 mm long 7
4. Corolla with long stiff hairs on lower lip; young stems
   and inflorescences with long thin pilose hairs; whole
   plant strongly aromatic 5
   Corolla without long stiff hairs on lower lip; young
   stems and inflorescences with or without thin pilose
   hairs; whole plant not strongly aromatic 6
5. Corolla 12–17(–20) mm long; anthers 1–2.5 mm long;
   pedicels 0.5–1 mm long; capsule 10–15 mm long . . .   1. *B. pubescens*
   Corolla 32–40 mm long; anthers 4–5 mm long;
   pedicels 0.5–4 mm long; capsule 13–18 mm long . . .   2. *B. riparia*

6. Corolla 28–40 mm long with tube 9–13 mm long;
   longest pedicel 1–4 mm long; inflorescences axes
   with dense white pilose hairs; largest leaf 17–34 cm
   long; capsule 22–30 mm long, the valves ± 1.5 mm
   wide . . . .   3. *B. stenopteris*
   Corolla 17–27 mm long with tube 5–7 mm long;
   longest pedicel 5–12 mm long; inflorescence axes
   without or with scattered usually purple pilose hairs;
   largest leaf (5–)7–14 cm long; capsule 17–23 mm
   long, the valves 2–3 mm wide . . . .   6. *B. vogeliana*
7. All sepals linear-lanceolate to slightly spathulate, dorsal
   less than twice as wide as the rest; capsule valves
   2–3 mm wide; seed ± 1.5 mm long . . . .   5. *B. owariensis*
   Dorsal sepal oblong, 2–4 times wider than the rest;
   capsule valves 3–6 mm wide; seed 3–4 mm long 8
8. Inflorescence axes and branches puberulous or
   densely so, all hairs less than 0.5 mm long, no stalked
   capitate glands; capsule 25–45 mm long, the valves
   3–5 mm wide . . . .   7. *B. cicatricosa*
   Inflorescence axes and branches with long purple
   eglandular hairs up to 7 mm long (some always over
   3 mm) and with stalked capitate glands; capsule
   37–42 mm long, the valves ± 6 mm wide . . . .   8. *B. richardsiae*

1. **Brillantaisia pubescens** *Oliv.* in Trans. Linn. Soc. 29: 125, pl. 125 (1875); Burkill in F.T.A. 5: 38 (1899); Benoist, Cat. Pl. Madag. Acanthacées: 13 (1939); Vollesen in Opera Bot. 59: 79 (1980); Vollesen & Brummitt in K.B. 36: 570 (1981); Lebrun & Stork, Enum. Pl. Afr. Trop. 4: 475 (1997); Sidwell in Bull. Nat. Hist. Mus. London 28: 84, fig. 8 (1998). Type: Tanzania, Morogoro District: Ukutu [Khutu], Kilengwe [Kirengwe], *Speke & Grant* s.n. (K!, holo.)

Much branched viscid aromatic annual (? sometimes short-lived perennial) herb; stems erect, trailing or scrambling, up to 1 m tall, puberulent and with usually dense thin glossy white pilose hairs up to 4 mm long, with scattered stalked capitate glands. Leaves with petiole up to 7 cm long; lamina broadly ovate-cordiform, largest 8–21(–30) × 3.5–14 cm, apex subacute to subacuminate with an obtuse tip, base truncate to cuneate, margin entire to slightly crenate, beneath puberulous with long thin hairs and with scattered broad hairs, above subglabrous. Panicle open, 10–40 cm long; indumentum as stems but with denser capitate glands; primary bracts elliptic or broadly so, from leaf-size down to 5 × 4 mm near apex, indumentum as leaves or denser and with capitate glands; secondary bracts elliptic to orbicular, 2–10 × 2–6 mm; pedicels 0.5–1 mm long, puberulous and with capitate glands. Calyx puberulous, with long pilose hairs and with usually numerous stalked glands; lobes linear, tapering or spathulate, ventral and lateral 8–10(–13) mm long, dorsal 12–14 mm long. Corolla with upper lip white and lower pale mauve to violet or purple, 12–17(–20) mm long, finely puberulous outside; tube 5–10 mm long; lower lip 7–12(–15) mm long, with long stiff hairs in central part, lobes ovate, 1–3 mm long; anthers 1–2.5 mm long, glabrous; staminodes minute. Capsule 10–15 mm long, the valves 2–3 mm wide, finely glandular-puberulous in apical half and ± scattered non-glandular hairs. Seed ± 1.5 mm long.

TANZANIA. Rufiji District: Utete, 28 June 1956, *Anderson* 1121!; Lindi District: Lake Lutamba, 12
   Sep. 1934, *Schlieben* 5323! & Litipo Forest Reserve, 4 Nov. 1984, *Mwasumbi & Mponda* 12650!
DISTR. **T** 3, 6, 8; Congo-Kinshasa, Zambia, Malawi, Mozambique, Zimbabwe, Madagascar
HAB. Riverine forest and thicket, riverbanks, *Acacia* woodland on alluvial clay; 25–450(–800) m

SYN. *Brillantaisia anomala* Lindau in P.O.A. C: 366 (1895). Type: Mozambique, Villa Gouveio, *de Carvalho* s.n. (COI!, holo.)
    Hygrophila pubescens (Oliv.) Benoist in Fl. Madag., Acanthacées 1: 36 (1967), *non* Nees (1847)

2. **Brillantaisia riparia** (*Vollesen & Brummitt*) *Sidwell* in Bull. Nat. Hist. Mus. London 28: 86, fig. 9 (1998). Type: Malawi, 20 km N of Kasungu, Dwangwa River, *Pawek* 3908 (K!, holo.; SRGH!, iso.)

Much branched viscid aromatic annual (? sometimes short-lived perennial) herb; stems erect, trailing or scrambling, up to 1 m tall, finely puberulous and with usually dense thin glossy white pilose hairs up to 4 mm long, with scattered stalked capitate glands. Leaves with petiole 1.4–2.2 cm long; lamina ovate-oblong, largest 2.5–5.2 × 4.5–9.5 cm, apex acute with an obtuse tip, base cuneate, margin entire to slightly crenate, beneath puberulous with long thin hairs and with scattered broad hairs, above subglabrous. Panicle rather narrow, 10–20 cm long, occasionally with vegetative growth continuing at apex; indumentum as stems; primary bracts as leaves, from leaf-size down to 8 × 6 mm near apex, indumentum as leaves or denser and with capitate glands; secondary bracts elliptic to orbicular, 2–10 × 2–6 mm; pedicels 0.5–4 mm long, puberulous and with capitate glands. Calyx puberulous, with long pilose hairs and with usually numerous stalked glands; lobes linear, tapering or spathulate, ventral and lateral 9–14 mm long, dorsal 14–18 mm long. Corolla blue to purple, 32–40 mm long, finely puberulous outside; tube 13–15 mm long; lower lip 20–25 mm long, with long stiff hairs in central part, lobes ovate, 3–5 mm long; anthers 4–5 mm long, glabrous; staminodes 3–4 mm long. Capsule 13–18 mm long, the valves 2–3 mm wide, finely glandular-puberulous in apical half and ± scattered non-glandular hairs. Seed ± 1.5 mm long.

TANZANIA. Ulanga District: Selous Game Reserve, Lukula, along Luwegu River, 18 Nov. 1976, *Vollesen* MRC4123!; Lindi District: Tendaguru, Nubemkuru River, 22 Aug. 1926, *Migeod* 291!
DISTR. T 6, 8; Malawi, Mozambique
HAB. Riverine forest, riverbanks; ± 450 m

SYN. *Brillantaisia pubescens* Oliv. var. *riparia* [Vollesen in Opera Bot. 59: 80 (1980), *nom. nud.*]; Vollesen & Brummitt in K.B. 36: 571 (1981); Lebrun & Stork, Enum. Pl. Afr. Trop. 4: 475 (1997)

3. **Brillantaisia stenopteris** *Sidwell* in Bull. Nat. Hist. Mus. London 28: 89, fig. 11 (1998). Type: Tanzania, Morogoro, Kombola, *Schlieben* 4068 (LISC!, holo.; BR!, iso.)

Perennial herb; stems erect or basal part rooting, to 1 m tall, puberulous on uppermost node and with scattered thin pilose hairs and capitate glands. Leaves often purplish tinged; petiole up to 4 cm long; lamina ovate-cordiform or broadly so, largest 17–34 × 9–19 cm, apex drawn out into an acuminate apex with rounded tip, base truncate to cuneate, margin subentire to dentate, beneath with long and short thin hairs and scattered broad hairs along veins, above subglabrous. Panicle open, 5–15 cm long, puberulous and with thin glossy pilose hairs up to 3 mm long and with dense stalked capitate glands; primary bracts green or purple, ovate-elliptic, from leaf-size down to 5 × 4 mm near apex, indumentum as leaves but denser and with capitate glands; secondary bracts elliptic, 3–9 × 1–3 mm; pedicels 1–4 mm long. Calyx purplish, puberulous and with long broad pilose hairs and with dense capitate glands; ventral and lateral lobes linear, tapering or slightly spathulate, 7–12 mm long, dorsal linear-oblong, distinctly spathulate, more than twice as broad as the rest, 8–12 mm long. Corolla with upper lip yellow-green flushed with purple and lower lip white or pale blue to mauve with white base, 28–40 mm long, crisped-puberulous and with long capitate glands; tube 9–13 mm long; lower lip 18–23 mm long, without long stiff hairs, lobes ovate, 3–7 mm long; anthers 3–4 mm long, glabrous; staminodes ± 2 mm long, glabrous. Capsule 2.2–3 cm long, the valves ± 1.5 mm wide, with scattered minute glands. Seed ± 1 mm long.

TANZANIA. Morogoro District: Kanga Mt, 4 July 1983, *Polhill & Lovett* 4954! & Kimboza Forest
Reserve, July 1983, *Rodgers et al.* 2532!; Kilosa District: Mamboya Forest Reserve, 13 Nov. 1987,
Pócs et al. 87224/D!
DISTR. **T** 6; not known elsewhere
HAB. Dry evergreen or semi-evergreen lowland forest extending to lower montane evergreen
forest; 200–750(–1300) m

SYN. [*Brillantaisia pubescens* Oliv. var. *riparia* sensu Rodgers *et al.*, Conserv. val. & stat. Kimboza
F.R.: 51 (1983), *non* Vollesen & Brummitt]

4. **Brillantaisia lamium** (*Nees*) *Benth.* in Hook., Niger Fl.: 477 (1849); Burkill in
F.T.A. 5: 38 (1899); Benoist in Bull. Soc. Bot. France 60: 334 (1913); F.W.T.A. 2: 254
(1931); Exell, Cat. Vasc. Pl. Sao Tomé: 260 (1944); Heine in F.W.T.A. (Ed. 2) 2: 406,
fig. 300 (1963) & in Fl. Gabon 13: 88, pl. 17, 4–8 (1966); U.K.W.F. ed. 2: 267 (1994);
Lebrun & Stork, Enum. Pl. Afr. Trop. 4: 475 (1997); Sidwell in Bull. Nat. Hist. Mus.
London 28: 97, fig. 14 (1998); Friis & Vollesen in Biol. Skr. 51(2): 439 (2005);
Ensermu in F.E.E. 5: 371 (2006). Type: Liberia, Cape Palmas, *Ansell* s.n. (K!, lecto.;
P!, iso.; selected by Sidwell, l.c.)

Coarse perennial herb; stems up to 1.5(–2) m long, creeping and rooting at the
base, then erect or ± scandent, subglabrous to puberulous and then also with broad
pilose many-celled glossy hairs. Leaves with petiole up to 6(–9) cm long; lamina
cordiform or broadly so (rarely ovate), largest (6–)10–17 × (3.5–)5–10.5 cm, apex
subacute to acuminate, usually drawn out into an obtuse tip, base subcordate to
cordate (rarely rounded to subcuneate), usually without decurrent flanges on
petiole, margin entire to crenate, pubescent with broad glossy hairs, sometimes only
along veins with subglabrous lamina. Panicle very loose, 10–25(–40) cm long,
glabrous to puberulous and with sparse to dense long broad glandular hairs; primary
bracts elliptic-obovate or broadly so, from leaf-size down to 5 × 2 mm near apex,
glabrous to puberulous and with broad glossy hairs; pedicels 0.5–4 mm long. Calyx
glabrous to puberulous and with long broad glossy hairs; lobes linear-lanceolate,
8–14 mm long, tapering gradually, similar or dorsal slightly longer and wider. Corolla
tube whitish, lower lip blue to dark blue or purple to dark brillant purple, upper lip
greenish tinged bluish to dark purple, 30–55 mm long, with dark purple glandular
hairs; tube (5–)7–11 mm long; lower lip 27–45 mm long, lobes triangular-ovate, 4–7 mm
long; anthers 6–9 mm long, glabrous or hairy; staminodes filiform. Ovary glabrous.
Capsule 25–45 mm long, the valves 3–5 mm wide, glabrous except sometimes with
long a few non-glandular hairs near apex. Seed 1.5–2 mm long.

UGANDA. West Nile District: Paidha, 29 Aug. 1953, *Chancellor* 206!; Mengo District: Kampala,
Port Bell, Apr. 1939, *Chandler* 2788!; Masaka District: Katera, Malabigambo Forest, 2 Oct.
1953, *Drummond & Hemsley* 4584!
KENYA. Nairobi District: N of Nairobi, 12 Oct. 1972, *Hansen* 713! & NW Nairobi, Kyuna Estate,
25 May 1980, *Gillett* 22810!; Masai District: Ololua Forest, 26 Feb. 1987, *Mungai & Ndiangui*
40/87!
TANZANIA. Bukoba District: Bugandika, Jan. 1932, *Haarer* 2483! & Minziro Forest, March 1953,
Procter 147!; Mwanza District: Geita, South Nzera, 5 Jan. 1973, *Ebbels* 8842!
DISTR. **U** 1, 2, 4; **K** 4 (see note); **T** 1; widespread in the forest regions of West Africa to S
Sudan, W Ethiopia and N Angola
HAB. Swamps and continuously waterlogged grassland areas in forest, swamp forests, lake
shores, streams sides; 1100–1600 m

SYN. *Leucoraphis lamium* Nees in DC., Prodr. 11: 97 (1847)
    [*Brillantaisia owariensis* sensu Hook. in Bot. Mag.: t.4717, fig. 3 (1853); T. Anderson in J.L.S.
        7: 21 (1863); Engl. in E.J. 7: 339 (1886), *non* P. Beauv. (1818)]
    B. eminii Lindau in E.J. 17: 103 (1893); Lindau in P.O.A. C: 366 (1895); Burkill in F.T.A. 5: 38
        (1899); Jex-Blake, Some Wild Fl. Kenya: 103, fig. 84 (1948); T.T.C.L.: 5 (1949); F.P.S. 3: 172
        (1956); Synnott, Checklist Budongo Forest Reserve (C.F.I. Occ. Papers 27): 68 (1985).
        Types: Tanzania, Bukoba, *Stuhlmann* 3664 (B†, syn.; K!, iso.) & *Stuhlmann* 3995 (B†, syn.)

B. subcordata De Wild. & T.Durand in Compt. Rend. Soc. Bot. Belg. 38: 44 (1899) & Contrib. Fl. Congo 1,2 : 47 (1900); C.B. Clarke in F.T.A. 5: 510 (1900). Type: Congo-Kinshasa, Bokakata, *Dewevre* 802 (BR!, holo.)

B. subcordata De Wild. & T.Durand var. *macrophylla* De Wild. & Th. Dur. in Contrib. Fl. Congo 1, 2: 47 (1900); C.B. Clarke in F.T.A. 5: 510 (1900). Type: Congo-Kinshasa, *Duchesne* 13 (BR!, holo.)

NOTE. Typical *B. lamium* from West Africa is much more slender and subglabrous and with considerably smaller flowers than the East African plants. But eastwards through Nigeria, Cameroon and Congo the plants become gradually coarser and more hairy and the flowers become larger. Over the whole range there is no way to separate the large-flowered East African material (*B. eminii*, sensu stricto) from the typical West African material. A similar pattern where eastern plants from higher altitudes have larger flowers than western lowland plants is known in other *Acanthaceae*, e.g. from *Anisosepalum alboviolaceum*.

All specimens seen from Kenya, apart from the one cited above from Masai District are escaped and naturalised from original garden plantings of Ugandan material. The species is now quite common in swamps around Nairobi.

5. **Brillantaisia owariensis** *P.Beauv.* in Fl. Owar. 2: 68, t. 100, fig. 2 (1818); Bentham in Hook., Niger Fl.: 477 (1849); Hooker, Bot. Mag.: t. 4717 (1853), excl. fruit; Engler in E.J. 8: 65 (1887); Th. Durand & Schinz, Etud. Fl. Congo 1: 217 (1896); Burkill in F.T.A. 5: 40 (1899); Benoist in Bull. Soc. Bot. France 60: 335 (1913); F.W.T.A. 2: 254 (1931); Heine in F.W.T.A. ed. 2, 2: 406 (1963); U.K.W.F. ed. 2: 267 (1994); Lebrun & Stork, Enum. Pl. Afr. Trop. 4: 475 (1997); Sidwell in Bull. Nat. Hist. Mus. London 28: 90 (1998); Friis & Vollesen in Biol. Skr. 51(2): 439 (2005). Type: Nigeria, Benin, Agathon, *Palisot de Beauvois* s.n. (P, holo.)

Perennial or shrubby herb, usually glandular and sticky; stems up to 2 m tall, erect or rooting at base, glabrous to crisped-puberulous, sometimes also with long pilose hairs. Leaves with petiole up to 14(–17) cm long; lamina ovate-cordiform or broadly so, largest 11–28 × 6–20 cm, apex acuminate, sometimes drawn out into a short obtuse tip, base cuneate to cordate, margin regularly to irregularly and sometimes grossly crenate-dentate (rarely subentire), glabrous to densely crisped-puberulous on veins, glabrous to sparsely puberulous on lamina. Panicle from very lax to narrow and contracted, (5–)10–45 cm long, finely puberulous and with usually dense broad glandular hairs, often also with long pilose hairs; primary bracts green to purplish, ovate to elliptic or obovate, from leaf-size to 10 × 3 mm near apex, finely puberulous and with long pilose hairs, towards apex with stalked capitate glands; secondary bracts lanceolate to elliptic, 5–18 × 2–5 mm; pedicels 0.5–7(–11) mm long. Calyx dark purple, with dense capitate glands, ± puberulous hairs and ± scattered pilose hairs; lobes linear-lanceolate, tapering gradually to distinctly spathulate, ventral and lateral (8–)12–19 mm long, dorsal (10–)14–21 mm long. Corolla pale to deep mauve, pale to deep purple or blue to dark blue with white throat, (25–)30–53 mm long, with purple to dark purple glandular hairs; tube 7–14 mm long; lower lip 20–38 mm long, lobes 2–5 mm long, ovate, subacute (lateral) or rounded. Filaments and style white; anthers black, (5–)6–9(–11) mm long, hairy along connective; staminodes filiform, 10–15 mm long, widened apically. Capsule (17–)20–35 mm long, the valves 2–3 mm wide, glandular-puberulous and usually with longer non-glandular hairs. Seed ± 1.5 mm long.

UGANDA. West Nile District: Agoro, *Eggeling* 798!; Mengo District: Entebbe, Nov. 1930, *Snowden* 1847! & Kampala, Oct. 1904, *E. Brown* 103!

KENYA. Uasin Gishu District: Kipkarren, Dec. 1933, *Dale* 3208!; North Kavirondo District: W of Kimilili, Sep. 1958, *Tweedie* 1707! & Malaba Forest, March 1967, *Tweedie* 3413!

TANZANIA. Bukoba District: Minziro Forest, 3 July 1958, *Willan* 331!; Mpanda District: W of Kasangazi, Lubugwe River, 26 July 1958, *Juniper et al.* 256! & Kungwe-Mahali Peninsula, NW slopes of Musenabantu, 12 Aug. 1959, *Harley* 9302!

DISTR. **U** 1–4; **K** 3, 5; **T** 1, 4; widespread in the forest regions of West and Central Africa, east to SE Sudan and south to N Angola

HAB. Wet lowland and montane forest, often along streams or in damp places, riverine and swamp forest, riverbanks; (750–)900–1850 m

SYN. *Belantheria belvisiana* Nees in DC., Prodr. 11: 96 (1847). Type: as for *B. owariensis*
  Brillantaisia patula T.Anderson in J.L.S. 7: 21 (1863); Burkill in F.T.A. 5: 41 (1899); Benoist in Bull. Soc. Bot. France 60: 335 (1913); F.W.T.A. 2: 254 (1931); Exell, Cat. Vasc. Pl. Sao Tomé: 260 (1944); F.P.N.A. 2: 269 (1947); Heine in F.W.T.A. ed. 2, 2: 406 (1963) & in Fl. Gabon 13: 94, pl. 19, fig.1–6 (1966). Type: "Congo", *Smith* s.n. (K!, holo.; P!, iso.)
  B. alata Oliv. in Trans. Linn. Soc. 29: 125, t. 124 (1875); S. Moore in J.B. 18: 197 (1880); Th. Dur. & Schinz, Etud. Fl. Congo 1: 216 (1896); Durand & De Wild. in Comp. Soc. Bot. Belg. 36: 83 (1897).Type: Uganda, Bunyoro [Unyoro], *Speke & Grant* 583 (K!, holo.)
  B. nitens Lindau in E.J. 17: 102 (1893); Burkill in F.T.A. 5: 41 (1899); Lindau in Z.A.E.: 292 (1911); Benoist in Bull. Soc. Bot. France 60: 336 (1913); F.W.T.A. 2: 254 (1931); F.P.N.A. 2: 270 (1947); Heine in F.W.T.A., ed. 2, 2: 406 (1963); Champluvier in Fl. Rwanda 3: 448 (1985); Blundell, Wild Fl. E. Afr.: 389, pl. 849 (1987). Type: Cameroon, W of Buea, *Preuss* 847 (B†, holo.; K!, iso.)
  B. dewevrei De Wild. & T.Durand in Compt. Rend. Soc. Bot. Belg. 38: 45 (1899); C.B. Clarke in F.T.A. 5: 510 (1900). Type: Congo-Kinshasa, between Loukolela and N'Gombi, *Dewevre* 751 (BR!, holo.; BR!, iso.)
  B. nyanzarum Burkill in F.T.A. 5: 39 (1899); F.P.N.A. 2: 272 (1947); F.P.S. 3: 171 (1956); U.K.W.F.: 583 & 584 (1974); Synnott, Checklist Budongo Forest Reserve (C.F.I. Occ. Papers 27): 68 (1985). Type: Kenya, 'Kavirondo', *Scott Elliot* 6999 (K!, holo.)
  B. mahonii C.B.Clarke in K.B. 1906: 251 (1906); S. Moore in J.B. 45: 89 (1907); Lind & Tallantire, Common Fl. Pl. Uganda: 140, fig. 82 (1962). Type: Uganda, Mengo District: Entebbe, *Mahon* s.n. (K!, holo.)

NOTE. There seems no way of separating the multitude of forms which have been united under this species. This is the most widespread species in *Brillantaisia* and also by far the most variable. It especially seems to have an enormous range of leaf-shapes and -dentations, and these have been widely used to separate species. But over the total range of the species they all merge together.
  Forms with very open inflorescences and very large flowers have traditionally been separated out as *B. patula*, but again they merge into typical forms. See also note under *B. lamium*.
  B. owariensis has been grown as an ornamental in Nairobi and occasionally within its natural range in Uganda.

6. **Brillantaisia vogeliana** (*Nees*) *Benth.* in Hooker, Niger Fl.: 477 (1849); T. Anderson in J.L.S. 7: 21 (1863); Burkill in F.T.A. 5: 40 (1899); Benoist in Bull. Soc. Bot. France 60: 335 (1913); F.W.T.A. 2: 254 (1931); Exell, Cat. Vasc. Pl. Sao Tomé: 260 (1943); F.P.N.A. 2: 272 (1947); Heine in F.W.T.A., ed. 2, 2: 406 (1963) & in Fl. Gabon 13: 92, pl. 18, fig. 1–6 (1966); U.K.W.F. ed. 2: 267 (1994); Lebrun & Stork, Enum. Pl. Afr. Trop. 4: 475 (1997); Sidwell in Bull. Nat. Hist. Mus. London 28: 95, fig. 13 (1998); Friis & Vollesen in Biol. Skr. 51(2): 440 (2005). Type: Bioko [Fernando Po], *Vogel* 179 (K!, holo.; K!, iso.)

Annual or short-lived perennial herb; stems up to 1.5 m tall, basal part creeping and rooting, apical part erect, glabrous or puberulous on upper nodes. Leaves with petiole up to 14 cm long; lamina ovate-cordiform or broadly so, largest (5–)7–14(–35) × 5–10(–15) cm, apex acute to acuminate or drawn out into a short obtuse tip, base rounded to cordate, margin subentire or crenate to sharply dentate, with scattered broad glossy hairs, densest along veins. Panicle open, 5–20(–35) cm long, with dense purple (rarely white) broad glandular hairs, finely puberulous and with long pilose hairs; primary bracts green to purple, ovate to elliptic, from leaf-size down to 5 × 3 mm near apex, with long broad glossy purplish (rarely white) hairs, no capitate glands; secondary bracts purple, 3–8 × 1–3 mm; pedicels of central flowers 1–4 mm long, of lateral flowers 5–12 mm long. Calyx dark purple (rarely green), glandular-puberulous, finely puberulous and with long pilose hairs; lobes tapering or slightly spathulate, linear-lanceolate, ventral and lateral 7–15 mm long, dorsal 9–16 mm long. Corolla blue or purple with white tube, 15–27 mm long, with long scattered glandular hairs; tube 5–7 mm long; lower lip 10–18 mm long, lobes ± 2 mm long, ovate; anthers 3–4.5 mm long; staminodes ± 5 mm long, puberulous, widened apically. Capsule 17–23 mm long, the valves 2–3 mm wide, glandular-puberulous and with scattered non-glandular hairs. Seed ± 1.5 mm long.

UGANDA. Bunyoro District: Budongo Forest, Sonso River, Nov. 1935, *Eggeling* 2290! & Budongo Forest, Dec. 1935, *Eggeling* 3323!; Mengo District: Mabira Forest, 8 Nov. 1938, *Loveridge* 25!
KENYA. North Kavirondo District: Kakamega Forest, 11 Dec. 1956, *Verdcourt* 1683!
DISTR. **U** 2, 4; **K** 5; Ghana, Nigeria, Cameroon, Bioko, Sao Tomé, Gabon, Central African Republic, Congo-Kinshasa, Sudan
HAB. Wet montane forest, usually along paths or in clearings; 1050–1700 m

SYN. *Leucoraphis vogeliana* Nees in DC., Prodr. 11: 97 (1847)

NOTE. Normally one of the more distinctive species in the group around *B. owariensis*, characterised by by its small slender stature and small flowers with small anthers. But at the upper end of its variation it approaches *B. owariensis*.

7. **Brillantaisia cicatricosa** *Lindau* in E.J. 20: 4 (1894) & in P.O.A. C: 366 (1895); Burkill in F.T.A. 5: 39 (1899); F.P.N.A. 2: 270 (1947); Troupin, Fl. Pl. Lign. Rwanda: 84 (1982); Champluvier in Fl. Rwanda 3: 446 (1985); U.K.W.F. ed. 2: 267 (1994); Lebrun & Stork, Enum. Pl. Afr. Trop. 4: 475 (1997); Friis & Vollesen in Biol. Skr. 51(2): 439 (2005). Type: Congo-Kinshasa, W slope of Ruwenzori Mts, Butahu Valley, *Stuhlmann* 2301 (B†, holo.)

Coarse often strongly aromatic shrubby perennial herb or shrub to 5 m tall, erect (rarely scandent); stems glabrous to densely puberulous (rarely velutinous). Leaves sometimes purplish tinged; petiole up to 17 cm long, upper often subsessile; lamina ovate to cordiform or broadly so, largest 14–43 × 11–32 cm, apex drawn out into an acuminate or cuspidate tip, base cuneate to cordate, margin grossly and irregularly dentate with larger and smaller teeth, the larger teeth up to 2 cm long and often toothed again, glabrous to densely puberulous along veins, glabrous or subglabrous on lamina. Panicle ± contracted to rather open, purplish, 10–40 cm long, puberulous to pubescent, without capitate glands; primary bracts green or purple, ovate-elliptic or broadly so, from leaf size down to 6 × 3 mm near apex, entire or lower toothed, puberulous or sparsely so along veins, no glands; secondary bracts elliptic, 2–20 × 1–7 mm; pedicels 1–9(–12) mm long, puberulous. Calyx purplish, puberulous or densely so and with scattered to dense short-stalked capitate glands (rarely without), sometimes also with long pilose hairs; ventral and lateral lobes linear-lanceolate, 9–20 mm long, tapering gradually or slightly spathulate, dorsal narrowly oblong, 10–22 mm long, 2–4 times wider and slightly longer than the rest. Corolla pale blue to brilliant royal blue, mauve to violet or purple, upper lip usually paler and often with dark streaks, 25–52 mm long, with purplish glandular hairs all over; tube 5–12 mm long; lower lip 17–40 mm long, lobes ovate, 4–13 mm long. Filaments, style and stigma white to mauve; anthers grey to black, 4–9 mm long, glabrous or hairy; staminodes up to 1 cm long, filiform, puberulous. Capsule 25–45 mm long, the valves 3–5 mm wide, glandular-puberulous or densely so, ± scattered non-glandular hairs. Seed 3–4 mm long. Fig. 25, p. 160.

UGANDA. Kigezi District: Kinkizi, March 1946, *Purseglove* 2022! & Kayonza Forest Reserve, 4 Aug. 1960, *Paulo* 646!; Toro District: Kanyawara, 17 Nov. 1982, *Struhsaker* 370!
KENYA. North Kavirondo District: Kakamega Forest, 15 Oct. 1953, *Drummond & Hemsley* 4768! & 14 Apr. 1973, *Hansen* 937!; Kericho District: Changna Tea Estate, 2 Dec. 1967, *Perdue & Kibuwa* 9242!
TANZANIA. Morogoro District: Uluguru Mts, 26 Oct. 1932, *Wallace* 399! & near Bunduki, Ngeta River, 17 Aug. 1951, *Greenway & Eggeling* 8594!; Rungwe District: Kyimbila, 4 June 1907, *Stolz* 152!
HAB. Wet montane evergreen forest, bamboo forest, usually in wet places along streams, forest edges and clearings; 1200–2600 m
DISTR. **U** 2–4; **K** 3, 5; **T** 6–8; Congo-Kinshasa, Rwanda, Burundi, Zambia, Malawi, Mozambique, Zimbabwe

FIG. 25. *BRILLANTAISIA CICATRICOSA* — **1**, inflorescence, × ²/₃; **2**, basal leaf, × ¹/₃; **3**, calyx, × 6; **4**, upper lip of corolla with stamens, staminodes and attachment of stamens, × 3; **5**, anther, × 6; **6**, style and stigma, × 3; **7**, whole capsule, × 3; **8**, single capsule valve, × 6; **9**, seed, × 6. 1 & 3–6 from *Univ. College* 116, 2 from *Verdcourt* 1674, 7–9 from *Pawek* 2780. Drawn by Margaret Tebbs.

SYN. *Brillantaisia ulugurica* Lindau in E.J. 22: 112 (1895); Burkill in F.T.A. 5: 43 (1899); T.T.C.L.:
6 (1949); Lebrun & Stork, Enum. Pl. Afr. Trop. 4: 475 (1997). Types: Tanzania,
Morogoro District: Uluguru Mts, *Stuhlmann* 8850 (B†, syn.) & *Stuhlmann* 9224 (B†, syn.)

B. *kirungae* Lindau in Götzen, Durch Afrika von Ost nach West, Sonderabdr.: 9 (1896);
Burkill in F.T.A. 5: 42 (1899); Th. & H. Durand, Syll. Fl. Congo: 417 (1909); F.P.N.A. 2:
269 (1947); Sidwell in Bull. Nat. Hist. Mus. London 28: 93, fig. 12 (1998); White *et al.*,
For. Fl. Malawi: 115 (2001). Type: Congo-Kinshasa, N of Lake Kivu, Mt Kirunga, *von
Götzen* 48 (B†, holo.)

B. *subulugurica* Burkill in F.T.A. 5: 42 (1899); Binns, Checklist Herb. Fl. Malawi: 12 (1968);
Lebrun & Stork, Enum. Pl. Afr. Trop. 4: 475 (1997). Type: Mozambique, Makua, Namuli
Hills, *Last* s.n. (K!, holo.)

B. *grandidentata* S.Moore in J.B. 45: 331 (1907); Lebrun & Stork, Enum. Pl. Afr. Trop. 4:
475 (1997). Type: Uganda, Toro District: Fort Portal, *Bagshawe* 1270 (BM!, holo.)

B. *cicatricosa* Lindau var. *kivuensis* Mildbr. in B.J.B.B. 17: 86 (1943); F.P.N.A. 2: 270 (1947).
Type: Congo-Kinshasa, Lake Magera, *de Witte* 1434 (BR!, holo.; BR!, iso.)

[B. *nitens* sensu U.K.W.F.: 583 (1974), *non* Lindau (1893)]

NOTE. No plant fitting the original description of *B. cicatricosa* has been recollected near the
type locality. Mildbraed presumably saw the type in Berlin prior to describing var. *kivuensis*,
but the possibility that the two taxa are not the same species clearly exists. Points in favour of
them being conspecific are the short calyx and hairy – but not glandular – capsule.

Sidwell (l.c.) uses *B. kirungae* as the valid name for this species and treats *B. cicatricosa* as an
insufficiently known species because the description of the calyx indumentum (subglabrous)
doesn't fit with the material seen by her. But among the East Congo (Kivu) material at BR
there are several collections with a rather sparsely hairy calyx.

There are several reports of mass-flowering of this species, but even in years where most
plants do not flower occasional flowering plants can usually be found.

8. **Brillantaisia richardsiae** *Vollesen* **sp. nov.** a *B. cicatricosa* panicula atque calyce
pilos longos purpureos atque glandulas stipitatas capitatas gerentibus et capsule
maiora differt. Type: Tanzania, Ufipa District: Sumbawanga, Nsanga Forest, *Richards*
12983 (K!, holo.; K!, iso.)

Strongly unpleasantly aromatic shrubby herb to 3 m tall; stems erect, glabrous to
sparsely puberulous. Leaves with petiole up to 9 cm long, upper leaves sessile; lamina
ovate to cordiform, largest 15–23 × 11–14 cm, apex drawn out into a cuspidate tip,
base cuneate to cordate, often very unequal, margin very irregularly and coarsely
crenate-dentate with small and large teeth, upper subentire, finely puberulous along
veins, glabrous on lamina. Panicle rather open, 20–30 cm long, with dense long
purple eglandular hairs up to 7 mm long (some always over 3 mm long), also with
purple puberulous hairs and scattered to dense stalked capitate glands; primary
bracts green or purplish, ovate-cordiform or broadly so, from leaf-size to 10 × 7 mm
near apex, puberulous and with scattered glands and long pilose hairs towards apex;
secondary bracts greenish to purple, indumentum as panicle; pedicels 2–6 mm long.
Calyx dark purple, with dense purple glands and long eglandular hairs up to 4 mm
long; ventral and lateral lobes linear-lanceolate, tapering gradually, 12–15 mm long,
dorsal oblong, 3–4 times wider than the rest, tapering, 16–19 mm long. Corolla rich
bluish purple to brilliant purple or royal blue, 40–50 mm long, with purple hairs; tube
10–12 mm long; lower lip 30–40 mm long, lobes ovate, rounded, 4–7 mm long.
Filaments and style orange; anthers 8–9 mm long, hairy; staminodes ± 1 cm long,
puberulous. Capsule 37–42 mm long, the valves ± 6 mm wide, densely glandular-
puberulous and with scattered eglandular hairs. Seed ± 4 mm long.

TANZANIA. Ufipa District: Sumbawanga, Mbizi Forest, 9 July 1957, *Whellan* 1363! & 17 June
1960, *Leach & Brunton* 10069! & Nsanga Forest, 6 Aug. 1960, *Richards* 12983!
DISTR. **T** 4; not known elsewhere
HAB. Shaded places in dense montane forest; 1800–2200 m

NOTE. Sidwell (l.c., p. 95) claims to have seen several intermediates between this and *B. cicatricosa*.
To me the specimens she cites are all typical *B. cicatricosa*.

9. **Brillantaisia madagascariensis** *Lindau* in E.J. 17: 103 (1893); Burkill in F.T.A. 5: 43 (1899); F.W.T.A. 2: 254 (1931); T.T.C.L.: 5 (1949); Heine in F.W.T.A. ed. 2, 2: 406 (1963); E.P.A.: 930 (1964); Benoist in Fl. Madag., Acanthacées 1: 29 (1967); U.K.W.F.: 583 (1974); Synnott, Checklist Budongo Forest Reserve (C. F. I. Occ. Papers 27): 68 (1985); Champluvier in Fl. Rwanda 3: 445 (1985); Iversen in Symb. Bot. Ups. 29(3): 160 (1991); U.K.W.F. ed. 2: 267 (1994); Lebrun & Stork, Enum. Pl. Afr. Trop. 4: 475 (1997); Sidwell in Bull. Nat. Hist. Mus. London 28: 104 (1998); Friis & Vollesen in Biol. Skr. 51(2): 439 (2005); Ensermu in F.E.E. 5: 372 (2006). Type: Madagascar, Betsileo, Nandihizana, *Hildebrandt* 3901 (B†, holo.; BM!, P!, iso.)

Perennial herb to 1.5(–2) m tall, but often flowering from 20 cm tall; stems often creeping and rooting in basal part, apical part erect, puberulous to pubescent or sparsely so. Leaves with petiole up to 14 cm long, upper often sessile; lamina ovate or broadly so (rarely elliptic or cordiform), largest 7–27 × 3–15 cm, apex acute to acuminate or drawn out into an acute or obtuse tip, base cuneate to truncate, margin entire to crenate-dentate, subglabrous to pubescent with broad glossy hairs, densest on veins and lamina often glabrous. Inflorescence a spiciform raceme, 4–19(–28) cm long, interrupted basally, confluent towards apex, axes puberulous to densely pubescent with broad curly many-celled hairs, towards apex with stalked glands; primary bracts green or purplish, ovate-elliptic, from leaf-size to 8 × 5 mm near apex, glabrous to puberulous and pilose-ciliate; secondary bracts ovate to elliptic or obovate or broadly so, up to 17 × 12 mm; pedicels 0.5–3 mm long, puberulous. Calyx green or purplish, puberulous, with scattered pilose hairs and with sparse to dense stalked capitate glands; lobes linear-lanceolate, tapering or dorsal slightly spathulate, ventral and lateral 9–23 mm long, dorsal 11–27 mm long. Corolla blue to dark blue, mauve to purple or lilac, upper lip often paler and lower darker towards base, 25–48 mm long, with many-celled purple hairs; tube white to pale violet, 7–13 mm long; lower lip 15–35 mm long, lobes ovate, 3–8 mm long, middle widest. Filaments and style white, stigma purple; anthers pale grey to black, 3–5 mm long; staminodes filiform, puberulous. Capsule (15–)18–28 mm long, the valves 3–5 mm wide, glandular-puberulous and ± scattered long eglandular hairs. Seed 2–3 mm long.

Uganda. Ankole District: Kalinzu Forest, 24 May 1961, *Symes* 749!; Masaka District: SSW of Katera, Malabigambo Forest, 2 Oct. 1953, *Drummond & Hemsley* 4556!; Mengo District: Kawanda, 4 Oct. 1969, *Lye & Morrison* 4359!

Kenya. North Kavirondo District: Kakamega Forest, 9 Dec. 1956, *Verdcourt* 1656! & May 1971, *Tweedie* 3961! & 8 km S of Rondo, 2 Dec. 1974, *Vuyk* 330!

Tanzania. Arusha District: Ngurdoto Crater, Sage, 11 Oct. 1965, *Greenway & Kanuri* 12107!; Lushoto District: E Usambara Mts, Ngambo to Derema, 26 Sep. 1917, *Peter* 60588!; Morogoro District: Nguru Mts, Minangya, Liwale River, 21 Aug. 1951, *Greenway & Farquhar* 8634!

Distr. **U** 2–4; **K** 5; **T** 1–4, 6, 7; Guinea, Liberia, Bioko, Cameroon, Gabon, Congo, Congo-Kinshasa, Sudan, Ethiopia, Rwanda, Burundi, Zambia, Madagascar

Hab. Wet montane forest, often along paths, in clearings or along streams, swamp forest, streambanks; 850–1800 m

Syn. *Brillantaisia spicata* Lindau in E.J. 20: 4 (1894) & in Z.A.E.: 293 (1911); Lebrun & Stork, Enum. Pl. Afr. Trop. 4: 475 (1997). Types: Tanzania, Lushoto District: Usambara Mts, Bangarra, Lutindi, *Holst* 3316 (B†, syn.; COI!, K!, iso.) & Gonja, Handei, *Holst* 4216 (B†, syn.; COI!, K!, P!, iso.)

   B. bagshawei S.Moore in J.B. 46: 312 (1908). Type: Uganda, Bunyoro District: Bugoma Forest, *Bagshawe* 1387 (BM!, holo.)

16. **HYGROPHILA**

R. Brown, Prodr. Fl. Nov. Holl. 1: 479 (1810); Nees in DC., Prodr. 11: 85 (1847); G.P. 2: 1075 (1876); Lindau in E. & P. Pf. IV, 3b: 296 (1895); Burkill in F.T.A. 5: 30 (1899); Heine in K.B. 16: 171 (1962) & in Adansonia 11: 656 (1971); Cramer in Nord. J. Bot. 9: 261 (1989)

Asteracantha Nees in Wallich, Pl. As. Rar. 3: 75 (1832) & in DC., Prodr. 11: 247 (1847); Lindau in E. & P. Pf. IV, 3b: 297 (1895)

Polyechma Hochst. in Flora 24: 376 (1841); Nees in DC., Prodr. 11: 82 (1847)

Cardanthera Bentham in G.P. 2: 1074 (1876)

Synnema Bentham in DC., Prodr. 10: 538 (1846); Burkill in F.T.A. 5: 29 (1899)

Hemigraphis sensu C.B.Clarke in F.T.A. 5: 58 (1899) *et auct. plur. seq. quoad spec. Afr., non sensu stricto*

Erect to procumbent annual or perennial herbs, with usually conspicuous linear cystoliths. Leaves opposite, sessile or subsessile, usually entire. Flowers in lax axillary cymes, these often congested into ± dense axillary clusters and these again into compound spike-like inflorescences; bracts various; bracteoles 2, linear to lanceolate, or absent. Calyx ± deeply divided into 5 sepals (rarely 4 by fusion of the two ventral sepals); sepals similar or dorsal often larger. Corolla contorted in bud, 2-lipped (rarely subactinomorphic); tube usually cylindric in basal part and widened into an apical pouch; upper lip 2-lobed, shorter than lower, forming a ± distinct hood; lower lip 3-lobed, plicate-rugose ("herring-boned") at base, usually with two large orange spots and usually with long stiff hairs. Stamens 4 (rarely 2 with dorsal reduced to staminodes or 2 without trace of staminodes), didynamous (dorsal pair smaller), held under upper lip (rarely included in tube); filaments of dorsal and ventral stamen of each pair united at base; anthers 2-thecous, oblong, muticous, those of ventral pair slightly larger, usually hairy on connective. Ovary 2-locular, with 4 to over 40 ovules per locule (rarely 2); disk inconspicuous; style filiform, usually hairy throughout, dorsal lobe of stigma minute, ventral linear (rarely both lobes developed). Capsule linear-oblong, rounded or flattened, valves sulcate dorsally; retinaculae slender, curved (rarely conical and straight). Seeds (4–)8–over 40, discoid, obliquely ovoid or orbicular, with dense hygroscopic hairs.

About 100 species in all tropical regions. Usually in wet habitats with several truly aquatic species, and also with some rheophytic species.

1. Flowers in dense heads, surrounded by hardened spines; two ventral sepals fused almost to apex, other three only at base 2
   Inflorescences various, but never with hardened spines; all sepals usually fused to the same height 3
2. Annual herb; corolla 22–40 mm long of which the expanded part of tube and limb 13–27 mm, lobes in lower lip 4–8 mm long, all lobes rounded; spines usually 6, stout, up to 5 cm long (rarely absent in some inflorescences) . .   14. *H. schulli* (p. 176)
   Perennial herb; corolla 40–44 mm long of which the expanded part of tube and limb 27–31 mm, lobes in lower lip 9–12 mm long, middle lobe retuse; spines usually less than 6, slender, up to 2 cm long . . . .   15. *H. richardsiae* (p. 178)

3. Corolla 4–8(–10) mm long, subactinomorphic with
    5 subequal lobes, without long stiff hairs; stamens
    included in corolla-tube; style ± 1 mm long, with
    two well-developed flattened stigma lobes . . . . .   12. *H. abyssinica* (p. 174)
    Corolla 7–44 mm long, distinctly 2-lipped with a 2-
    lobed upper lip and a 3-lobed lower lip usually
    with long stiff hairs; stamens held under the
    upper lip; style over 5 mm long, with one filiform
    stigma lobe 4
4. Corolla 7–12 mm long, without orange spots on
    lower lip, with 2 fertile stamens and with or
    without 2 staminodes 5
    Corolla 9–44 mm long, usually with 2 large orange
    spots on lower lip, with 4 fertile stamens (rarely
    some flowers with 2 stamens and 2 staminodes) 6
5. Flowers in an open panicle; corolla 7–9 mm
    long, with long stiff hairs on lower lip, with 2
    fertile stamens and 2 staminodes . . . .   8. *H. linearis* (p. 171)
    Flowers in a narrow spiciform panicle; corolla ±
    12 mm long, without long stiff hairs on lower
    lip, with 2 fertile stamens and no staminodes   4. *H.* sp. *A* (p. 167)
6. Stalked capitate glands absent from all
    vegetative parts, if present only on corolla 7
    Stalked capitate glands present on either stems,
    leaves or bracts, plant often ± viscid 11
7. Corolla 27–44 mm long; inflorescence a dense head 8
    Corolla 11–15 mm long; inflorescence racemiform
    or paniculate; all sepals fused to same height 9
8. Corolla ± 27 mm long, bright blue, linear tube
    longer than throat plus lobes, lower lip with
    curly hairs; all sepals fused to ± 8 mm from
    base; filaments ± 4 mm long . . . .   16. *H.* sp. *C* (p. 179)
    Corolla 40–44 mm long, white, linear tube
    much shorter than throat plus lobes, lower lip
    with straight hairs; two ventral sepals fused
    almost to apex, the rest for 2–3 mm at base;
    filaments 6–16 mm long . . . .   15. *H. richardsiae* (p. 178)
9. All flowers solitary; sepals fused for 0.5–1 mm
    at base; corolla with clavate hairs on lower
    lip; capsule 5–9 mm long, with over 20 seeds
    per locule . . . .   13. *H. hippuroides* (p. 175)
    Some or all bracts usually with more than one
    flower; sepals fused for 3–6 mm; corolla with
    tapering hairs on lower lip; capsule (where
    known) 12–14 mm long, with 14–16 seeds
    per locule 10
10. Stems, inflorescence axes and calyx glabrous or
    sparsely hairy at nodes (stems) or edges (calyx);
    largest leaf 4.5–8 cm long; basal fused part of
    calyx 4–6 mm long; corolla 12–14 mm long . . .   6. *H. pobeguinii* (p. 169)
    Stems, inflorescence axes and calyx uniformly
    strigose pubescent or densely so; largest leaf ±
    2.5 cm long; basal fused part of calyx ± 3 mm
    long; corolla ± 15 mm long . . . .   7. *H.* sp. *B* (p. 171)
11. Capsule 16–35 mm long, glabrous; sepals fused
    for 2–5 mm at base 12
    Capsule 5–13 mm long, finely puberulous and
    often also with capitate glands; sepals fused for
    0.5–2(–3) mm at base 13

12. Corolla 30–40 mm long; flowers in a large open
   panicle; capsule (18–)22–35 mm long .... 1. *H. didynama* (p. 165)
   Corolla 22–26 mm long; flowers in a contracted
   spiciform to racemiform panicle; capsule
   16–22 mm long .... 2. *H. uliginosa* (p. 166)
13. Corolla 9–17 mm long; capsule 5–9 mm long
   with 4–14 seeds per locule 14
   Corolla 17–37 mm long; capsule 8–13 mm long
   with 12–28 seeds per locule 15
14. Leaves glabrous below, glabrous or sparsely
   puberulous above, without capitate glands;
   filaments with thick blunt retrorse hairs;
   capsule with 4–12 seeds per locule .... 3. *H. cataractae* (p. 167)
   Leaves on both sides puberulous or sparsely so
   and with few to many capitate glands;
   filaments glabrous or with thin acute hairs;
   capsule with 10–14 seeds per locule .... 5. *H. pilosa* (p. 168)
15. Flowers in pedunculate heads; basal part of
   bracts white; corolla 33–37 mm long; capsule
   11–13 mm long, with 6–8 seeds per locule;
   seed 1.5–2 mm long .... 11. *H. albobracteata* (p. 173)
   Flowers in ± interrupted spiciform inflorescences;
   basal part of bracts usually not white; corolla
   17–33 mm long; capsule 8–12 mm long, with
   12–28 seeds per locule; seed 0.5–1 mm long 16
16. Sepals all linear, all ± same length; corolla 17–20
   (–23) mm long, apical part of lobes with long
   curly hairs; capsule with 24–28 seeds per locule 10. *H. asteracanthoides* (p. 172)
   Upper sepal oblanceolate, distinctly longer than
   the lower; corolla 20–33 mm long, apical part
   of lobes puberulous; capsule with 12–14 seeds
   per locule .... 9. *H. spiciformis* (p. 172)

1. **Hygrophila didynama** (*Lindau*) *Heine* in K.B. 16: 178 (1962); Lebrun & Stork, Enum. Pl. Afr. Trop. 4: 482 (1997). Type: Malawi, Lufiro, *Descamps* s.n. (BR!, holo.)

Erect or scrambling many-stemmed perennial herb with creeping rhizome; stems up to 1.5(–2) m long, quadrangular, glabrous to puberulous and then also with usually scattered stalked capitate glands. Leaves lanceolate to narrowly elliptic (rarely elliptic), largest 4–16 × 0.7–2.5(–3) cm, apex acute to rounded, base attenuate to rounded, subglabrous to hirsute-strigose. Inflorescence an open panicle up to 75 cm long; axes finely puberulous or sparsely so and with few to many capitate glands; floral leaves ovate with subcordate or cordate base, finely puberulous and with capitate glands; flowers in pedunculate dichasial cymes with up to 12 flowers but usually solitary towards apex; peduncles up to 5 cm long; bracts as floral leaves; pedicels 0.5–2(–3) mm long. Calyx finely puberulous or sparsely so and with usually dense capitate glands, 10–21 mm long of which basal fused part 2–5 mm, ventral and lateral lobes linear-lanceolate, dorsal similar or slightly wider, 1–5 mm longer than the rest. Corolla pale mauve to mauve or purple, "herring-bones" conspicuous, orange-veined, widened part of tube almost saccate, palate bullate ± closing throat, puberulous and with capitate glands, 30–40 mm long of which linear tube 8–11 mm, expanded tube 8–11 mm and limb 13–18 mm; upper lip ovate, lobes truncate to retuse, 2–4 mm long; lower lip with slightly clavate hairs, broadly elliptic, lobes rounded to retuse, 3–5 mm long (middle wider than lateral). Filaments glabrous or with short stubby hairs near apex, upper 9–13 mm long and lower 13–20 mm; anthers glabrous, purple edged, 2–3 mm long. Stigmatic lobe ± 2 mm long. Capsule glabrous, (18–)22–35 mm long, with 10–20 seeds per locule. Seed ± 2 mm long.

TANZANIA. Mbeya District: Unyika, Mlowo, Chitete River, 8 July 1976, *Leedal* 3725b!
DISTR. **T** 7; Congo-Kinshasa, Zambia, Malawi, Zimbabwe
HAB. Not recorded, but in the rest of its area on riverbanks and in riverbeds, sometimes rheophytic; no altitude recorded

SYN. *Brillantaisia didynama* Lindau in E.J. 24: 313 (1897); Burkill in F.T.A. 5: 36 (1899)
   Hygrophila gigas Burkill in F.T.A. 5: 36 (1899); S. Moore in J.B. 58: 46 (1920). Type: Zambia, Fort Young, *Nicholson* s.n. (K!, holo.)
   Hygrophila gilletii De Wild. in Ann. Mus. Congo, Ser. 5, 1: 314, pl. 50 (1906); Lindau in Schwed. Rhod.-Kongo-Exp. 1: 305 (1916); Lebrun & Stork, Enum. Pl. Afr. Trop. 4: 482 (1997). Type: Congo-Kinshasa, Leopoldville, *Gillet* 2703 (BR!, holo.; BR! iso.)

2. **Hygrophila uliginosa** *S.Moore* in J.B. 18: 197 (1880); Burkill in F.T.A. 5: 32 (1899); F.W.T.A. 2: 608 (1931); F.P.S. 3: 176 (1956); Heine in F.W.T.A. (ed. 2) 2: 396 (1963); U.K.W.F. ed. 2: 268 (1994); Lebrun & Stork, Enum. Pl. Afr. Trop. 4: 482 (1997). Type: Angola, Lombe River and near Bumba, *Welwitsch* 5106 (BM!, holo.; K!, iso.)

Erect perennial herb from stout rootstock, with fleshy fusiform roots; stems up to 2 m tall, quadrangular, glabrous to hispid-puberulous. Leaves lanceolate to ovate or elliptic, largest 8–12 × 1–3.7 cm (smaller and narrower towards apex), apex tapering to an acute tip, base decurrent, sparsely hispid-sericeous or hispid-puberulous along veins below and on edges. Inflorescence elongated, up to 50 cm long; axes towards apex puberulous or hispid-puberulous and usually with minute stalked capitate glands; flowers in axillary condensed cymes which are many-flowered and pedunculate near base and few-flowered or of solitary flowers and sessile near apex; peduncle 2–10 mm long, finely puberulous and with minute capitate glands; bracts up to 2.5 cm long; pedicels 1–4 mm long. Calyx finely puberulous and with minute capitate glands, 8–17 mm long of which basal fused part 3–5 mm; lobes linear, dorsal slightly longer than the rest. Corolla pale to deep mauve or pale to deep purple, with white "herring-bones", puberulous, no capitate glands, 22–26 mm long of which linear part of tube 8–10 mm and expanded part of tube and lobes 14–16 mm; upper lip ovate; lower elliptic, without well-defined orange spots, with slightly clavate hairs, lobes oblong, 1–3 mm long. Filaments finely hairy towards base, upper 6–8 mm long and lower 9–12 mm long; anthers 2–3 mm long. Stigmatic lobe ± 2 mm long. Capsule glabrous, 16–22 mm long, with 14–16 seeds per locule. Seed ± 1 mm long.

UGANDA. Bunyoro District: Budongo Forest, 30 Nov. 1938, *Loveridge* 140!; Busoga District: July 1905, *E. Brown* 283!; Mengo District: Namulonge, 26 Feb. 1952, *Norman* 79!
KENYA. Central Kavirondo District: Gem Kalanyo, 1 Jan. 1969, *Kokwaro* 1770!; North Kavirondo District: Bungoma, Nov. 1971, *Tweedie* 4140!
TANZANIA. Bukoba District: Minziro Forest Reserve, Nyakabanga, 16 Aug. 1999, *Sitoni et al.* 769!; Kigoma District: Lubalisi Village to Ntakatta, 8 June 2000, *Bidgood et al.* 4618!; Mpanda District: Kalaya, 22 Aug. 1959, *Harley* 9388!
DISTR. **U** 1–4; **K** 5; **T** 1, 4; Nigeria, Cameroon, Chad, Central African Republic, Congo-Kinshasa, Burundi, Sudan, Angola, Zambia
HAB. Seasonally inundated grass-swamps, papyrus swamps, riverbanks, old ricefields; 750–1500(?–2100) m

SYN. *Hygrophila acutisepala* Burkill in F.T.A. 5: 34 (1899); F.P.S. 3: 176 (1956); F.P.N.A. 2: 273 (1947); Lebrun & Stork, Enum. Pl. Afr. Trop. 4: 482 (1997). Type: Sudan, *Petherick* s.n. (K!, holo.)
   H. seretii De Wild. in Ann. Mus. Congo, Ser. 5, 3: 264 (1910). Type: Congo-Kinshasa, Dungu to Faradje, *Seret* 704 (BR!, holo.)
   H. thonneri De Wild., Etud. Fl. Bangala and Ubangi: 250 (1911); Lebrun & Stork, Enum. Pl. Afr. Trop. 4: 482 (1997). Type: Congo-Kinshasa, Mogbogoma, *Thonner* 184 (BR!, holo.)
   H. sp. A of U.K.W.F.: 583 (1974); Blundell, Wild Fl. E. Afr.: 392 (1987)

NOTE. A single collection (*E. Brown* 283) has been seen where the two longer (lower) stamens are completely absent. In species where reductions in the stamens occur, it is normally the two short (upper) stamens which are reduced. I consider this collection to be no more than an aberrant specimen.

Agnew (U.K.W.F., l.c.) in the description of his *sp. A* states that the calyx is glabrous. But the sole collection (*Kokwaro* 1770) cited by him has a finely puberulous calyx just like the rest of the material.

3. **Hygrophila cataractae** *S.Moore* in J.L.S. 37: 459 (1906); Eyles in Trans. Roy. Soc. S. Afr. 5: 481 (1916); Milne-Redhead in Mem. NY Bot. Gard. 9, 1: 20 (1954); Binns, Checklist Herb. Fl. Malawi: 13 (1968); U.K.W.F. ed. 2: 268 (1994); Lebrun & Stork, Enum. Pl. Afr. Trop. 4: 482 (1997). Type: Zimbabwe, Victoria Falls, Livingstone Island, *Gibbs* 159 (BM!, holo.; K!, iso.)

Erect, ascending or procumbent perennial (? rarely annual) herb; stems quadrangular, up to 75 cm long, glabrous or finely puberulous on upper nodes, occasionally with longer hairs intermixed, ± stalked capitate glands. Leaves elliptic or narrowly so, largest 4–7 × 0.3–2 cm, apex acute to subacute (rarely rounded), base subamplexicaul, glabrous below, glabrous or sparsely puberulous above. Inflorescence spiciform to ± paniculate, interrupted or confluent towards apex, up to 25 cm long; axes finely puberulous or densely so, ± longer hairs and ± capitate glands (sometimes glabrous in Uganda); cymes up to 12-flowered (fewer or solitary towards apex), sessile or basal with peduncle up to 2 cm long; bracts lanceolate or narrowly ovate, up to 1.5 cm long; pedicels 0.5–2 mm long; bracteoles linear-lanceolate, 4–6 mm long. Calyx finely puberulous and with longer hairs on edges, with (rarely without) stalked capitate glands, 7–13 mm long of which fused basal part 0.5–1.5 mm; ventral and lateral lobes linear, dorsal linear-oblanceolate and 2–4 mm longer than the rest. Corolla white or very pale yellow, "herring-bones" indistinct, bluish to purplish veined, sparsely puberulous and with scattered capitate glands, 11–14 mm long of which the linear tube 4–6 mm and expanded tube plus limb 7–8 mm; upper lip ovate, lobes truncate; lower lip broadly obovate, with long tapering red-tipped hairs; lobes 2–3 mm long, broadly rounded. Filaments with thick blunt retrorse hairs, upper 1–3 mm long, lower 3–5 mm long; anthers 0.5–1.5 mm long. Capsule finely puberulous and with a few capitate glands near apex, 5–9 mm long, with 4–12 seeds per locule. Seed ± 1 mm long.

UGANDA. Teso District: Serere, Dec. 1931, *Chandler* 278!; Busoga District: Bugaya, 30 Nov. 1949, *Jameson* 79!; Mengo District: Kakoge, 21 Dec. 1955, *Langdale-Brown* 1799!
KENYA. North Kavirondo District: Bungoma, Aug. 1964, *Tweedie* 2881!
TANZANIA. Buha District: 150 km on Kibondo–Kasulu road, 15 July 1960, *Verdcourt* 2829!; Mpanda District: Mt Kungwe, Kasoje, 17 July 1959, *Newbould & Harley* 4425!; Iringa District: Ndembera Valley, 29 Sep. 1971, *Perdue* 11585!
DISTR. **U** 3, 4; **K** 5; **T** 4, 7; Congo-Kinshasa, Zambia, Malawi, Mozambique, Zimbabwe
HAB. Grass swamps, riverbanks, riverbeds, waterholes; 750–1400 m

NOTE. Plants from Uganda and Kenya tend to be less hairy than those from Tanzania and further south in Africa but are otherwise identical. Some specimens from Tanzania (*Geilinger* 3117, *Richards* 13402) have the lower cymes very long-peduncled and are indeed close to forms of *H. pilosa*; but are easily distinguished from this species by the different leaf-indumentum and different hairs on the filaments.

4. **Hygrophila sp. A**

Annual (? short-lived perennial) herb; stems creeping and rooting, ascending apically, spongy and inflated basally, up to 40 cm long, crisped-pubescent. Leaves narrowly elliptic, largest ± 5 × 0.8 cm, apex acute, on both sides pubescent with broad glossy hairs. Inflorescence spiciform, interrupted basally and confluent towards apex, up to 20 cm long; axes with indumentum as stems or denser and with long capitate glands intermixed towards apex; cymes up to 8-flowered (fewer upwards), sessile or basal with peduncle up to 1 mm long; bracts lanceolate, up to 1.5 cm long; pedicels ± 1 mm long; bracteoles linear-lanceolate, ± 9 mm long. Calyx puberulous with broad glossy hairs and with capitate glands, ± 11 mm long of which the fused basal part ± 1 mm; lobes linear-lanceolate, dorsal slightly longer than rest. Corolla white with pale pink lower lip, without discernible "herring-bones", sparsely finely puberulous, ± 12 mm long of which the tube ± 8 mm long; upper lip ovate, with two short rounded lobes; lower lip broadly obovate, without orange spots and without long stiff hairs, lobes ± 2 mm long, rounded. Only 2 stamens developed with no trace of staminodes; filaments ± 5 mm long, with long thin hairs; anthers ± 1 mm long. Capsule 7–8 mm long, finely puberulous near apex, with 20–24 seeds per locule. Seed ± 0.5 mm long.

TANZANIA. Ulanga District: Selous Game Reserve, 25 km S of Ruaha Camp (7°58'S 37°05' E), 20 Oct. 1975, *Vollesen* MRC2903!
DISTR. **T** 6; not known elsewhere
HAB. Small swampy area in *Brachystegia* woodland on grey sandy soil; ± 300 m

SYN. [*Hygrophila cataractae* sensu Vollesen in Opera Bot. 59: 80 (1980), *non* S.Moore (1909)]

NOTE. Known only from this collection. With a superficial resemblance to *H. cataractae*, but differing in a number of important characters. It has only two fertile stamens with no staminodes, no orange spots and no long stiff hairs on the lower corolla-lip. The indumentum is also considerably denser than in *H. cataractae*.

5. **Hygrophila pilosa** *Burkill* in F.T.A. 5: 35 (1899); Martineau, Rhod. Wild Fl.: 79 (1953); Meyer in Prodr. Fl. SW Afr. 130: 31 (1968); Lebrun & Stork, Enum. Pl. Afr. Trop. 4: 482 (1997). Types: Zambia, Batoka Highlands, *Kirk* s.n. (K!, syn.); Malawi, Fort Hill, *Whyte* s.n. (K!, syn.)

Perennial herb with creeping rhizome; stems quadrangular, sparsely finely puberulous, with few to many broad (flattened) pilose hairs and with stalked capitate glands, in pyrophytic forms erect and up to 25 cm long, in non-pyrophytic forms becoming ± procumbent with ascending flowering branches and up to 75 cm long; whole plant usually sticky, aromatic when crushed. Leaves ovate to elliptic or narrowly so or obovate, largest 0.7–7.5 × 0.3–1.5 cm, apex acute to broadly rounded, base attenuate to cuneate or subcordate towards apex, on both sides puberulous or sparsely so and with few to many capitate glands, ± scattered to dense pilose hairs. Inflorescence paniculate to racemiform or spiciform and then interrupted to ± confluent apically, ± side branches, up to 35 cm long (usually 5–10 cm in pyrophytic plants); axes with indumentum as stems; flowers solitary or in 2–3(–5)-flowered clusters or in open 2–3(–5)-flowered dichasial cymes; peduncles long basally, shorter upwards with upper cymes subsessile, up to 3(–4) cm long; bracts up to 1(–1.3) cm long; pedicels 0.5–2(–3) mm long. Calyx finely puberulous, with long pilose hairs and with few to many capitate glands, 5–12 mm long of which fused basal part ± 1 mm; ventral and lateral lobes linear-lancolate, dorsal elliptic to obovate or narrowly so (rarely lanceolate or ovate) and 1–3 mm longer than the rest. Corolla white or with pale mauve to purple or blue lower lip, "herring-bones" indistinct, finely puberulous and with minute capitate glands, usually with scattered long hairs near apex of lobes on both lips, 9–17 mm long of which the linear tube 3–6 mm and expanded tube plus limb 6–11 mm; upper lip ovate or narrowly so, lobes rounded; lower lip obovate, with long tapering hairs, lobes 2–3 mm long, rounded. Filaments glabrous or with thin tapering hairs, upper 1–4 mm long, lower 3–6 mm long; anthers 1–1.5 mm long. Style glabrous or with a few hairs. Capsule finely puberulous, 5–9 mm long, with 10–14 seeds per locule. Seed ± 1 mm long.

TANZANIA. Chunya District: Mbogo, 17 Oct. 1932, *Geilinger* 3116!
DISTR. **T** 7; Congo-Kinshasa, Angola, Zambia, Malawi, Zimbabwe, Botswana
HAB. Seasonally flooded grassland on clay, but subject to burning in the dry season, riverbanks, riverbeds, waterholes; 900–1500 m

SYN. *Hygrophila bequaertii* De Wild. in F.R. 13: 194 (1914); Lebrun & Stork, Enum. Pl. Afr. Trop. 4: 482 (1997). Types: Congo-Kinshasa, Katanga, Sankisia, *Bequaert* 199 (BR!, syn.); Manika Plateau, *Hock* s.n. (BR!, syn.)
  H. bequaertii De Wild. var. *elliptica* De Wild. in F.R. 13: 194 (1914). Type: Congo-Kinshasa, Katanga, Ruwe, *Hock* s.n. (BR!, holo.)
  H. bequaertii De Wild. var. *reducta* De Wild. in F.R. 13: 194 (1914). Type: Congo-Kinshasa, Katanga, Nieuwendorp, *Ringoet* 4 (BR!, holo.)

6. **Hygrophila pobeguinii** *Benoist* in Not. Syst. 2: 339 (1913) & in Bull. Soc. Bot. France 60: 331 (1913); Chevalier in Rev. Bot. Appliq. 30: 266 (1950); Heine in F.W.T.A. (ed. 2) 2: 396 (1963); U.K.W.F.: 583 (1974); Blundell, Wild Fl. E. Afr.: 392 (1987); U.K.W.F. ed. 2: 268 (1994); Lebrun & Stork, Enum. Pl. Afr. Trop. 4: 482 (1997); Ensermu in F.E.E. 5: 380 (2006). Types: Guinea, Kouroussa, *Pobeguin* 1108 (P!, syn.); Congo, N'gapou, *Dybowski* 613 (P!, syn.)

Annual or short-lived perennial herb; stems erect, ascending or procumbent, basal part usually rooting, often inflated and spongy if plant growing in water, quadrangular, up to 75 cm long, glabrous apart from long flattened hairs at upper nodes. Leaves lanceolate to narrowly elliptic, largest 4.5–8 × 0.5–0.8 cm, apex subacute to rounded, base cuneate on lower leaves, subamplexicaul towards apex, glabrous to strigose-puberulous or -pubescent with flattened shiny hairs. Inflorescence racemiform, interrupted throughout, or paniculate with numerous branches, up to 35 cm long long; axes glabrous (rarely sparsely finely puberulous towards apex); floral leaves with widened base and ± clasping flowers; flowers all solitary or in 2–3(–4)-flowered sessile cymes basally; bracteoles up to 2 cm long; pedicels up to 1 mm long. Calyx strigose-ciliate or sparsely so, otherwise glabrous, 7–14 mm long of which fused basal part 4–6 mm; ventral and lateral lobes linear, dorsal lanceolate and ± 1 mm longer than the rest. Corolla white (Tanzania) or pale mauve to dull mauve with purple markings, no "herring-bones", puberulous, no capitate glands, 12–14 mm long of which the linear tube 5–6 mm and expanded tube plus limb 7–8 mm; upper lip narrowly ovate, lobes rounded, lower narrowly obovate, with stiff tapering hairs, lobes 1–2 mm long, rounded. Filaments glabrous, upper 2–3 mm long, lower 4–5 mm long; anthers ± 1 mm long. Capsule glabrous, 12–14 mm long, with 14–16 seeds per locule. Seed ± 1 mm long. Fig. 26, p. 170.

UGANDA. Teso District: Serere, Dec. 1931, *Chandler* 234! & Jan. 1933, *Chandler* 1049!; Busoga District: Hamatumba, July 1926, *Maitland* 1115!
KENYA. North Kavirondo District: Bungoma, Sep. 1968, *Tweedie* 3583!; Kisumu-Londiani District: Kisumu, 12 Dec. 1958, *McMahon* 117!
TANZANIA. Tabora District: Urambo, 17 June 1980, *Hooper & Townsend* 2015!; Songea District: E of Songea, Likonde River, 26 June 1956, *Milne-Redhead & Taylor* 10904!
DISTR. **U** 3; **K** 5; **T** 4, 7, 8; widespread in tropical Africa
HAB. Wet grassland, swamps, waterholes, roadside ditches, old ricefields; 500–1700 m

SYN. *H. vanderystii* S.Moore in J.B. 58: 46 (1920). Type: Congo-Kinshasa, Wombali, *Vanderyst* 4255 (BM!, holo.; BR!, iso.)
  H. nyassica Gilli in Ann. Naturhist. Mus. Wien 77: 51 (1973). Type: Tanzania, Rungwe District: NW shore of Lake Nyassa, Mwaya, *Gilli* 523 (K!, iso.)

NOTE. All collections from Uganda and Kenya have mauve flowers while those from Tanzania have pure white flowers. Otherwise no differences have been found, and in the Flora Zambesiaca area populations with white and pink to mauve flowers seem to occur in the same areas.

FIG. 26. *HYGROPHILA POBEGUINII* — **1**, flowering stem, × ²/₃; **2**, basal part of stem and roots,
× ¹/₃; **3**, young inflorescence, × 2; **4**, bract, × 2; **5**, flower bud, × 3; **6**, calyx and corolla, × 3;
7, calyx opened up, × 3; **8**, corolla opened up, × 3; **9**, Ovary, style and stigma, × 3; **10**, capsule,
× 3; **11**, single capsule valve, × 3; **12**, seed, × 9; **13**, seed wetted, × 9. All from *Keay* FHI 21442.
Drawn by D. R. Thomson.

7. Hygrophila sp. B

Perennial herb with short creeping stems and ascending to erect inflorescences; stems to 15 cm long, uniformly pubescent with broad glossy hairs. Leaves lanceolate to narrowly ovate-elliptic, largest ± 2.5 × 0.5 cm, apex rounded, base cuneate, densely strigose-pubescent with broad glossy hairs. Inflorescence racemoid, interrupted at base, denser upwards; axes densely pubescent with broad glossy hairs; floral leaves like vegetative; flowers solitary or paired at lower nodes; bracteoles to 1 cm long; pedicels to 1 mm long. Calyx uniformly strigose-pubescent, 8–10 mm long of which the basal fused part ± 3 mm; ventral and lateral lobes linear, dorsal lanceolate and ± 2 mm longer than the rest. Corolla mauve, without (?) markings, no "herring bones", puberulous, no capitate glands, ± 15 mm long of which the linear tube ± 5 mm and expanded tube plus limb ± 10 mm; upper lip narrowly ovate, lobes rounded; lower lip obovate, with stiff tapering hairs, lobes 2–3 mm long, rounded. Filaments minutely puberulous, 2–3 (upper) and 4–5 (lower) mm long; anthers ± 1.5 mm long. Capsule and seed not seen.

TANZANIA. Ufipa District: Sumbawanga, Mbizi Mts, Aug. 1938, *MacInnes* 499!
DISTR. **T** 4; not known elsewhere
HAB. Montane grassland; ± 1800 m

NOTE. Known only from this collection. Most closely related to *H. pobeguinii* from which it differs in the densely hairy stem, inflorescence and calyx, smaller leaves and larger corolla.

8. **Hygrophila linearis** *Burkill* in F.T.A. 5: 35 (1899); Lebrun & Stork, Enum. Pl. Afr. Trop. 4: 482 (1997). Type: Angola, *Welwitsch* 5772 (BM!, holo.; K!, iso.)

Annual or short-lived perennial herb; stems up to 50 cm long, basal part creeping and rooting, apical part ascending, glabrous. Leaves linear to lanceolate (or lowermost narrowly elliptic), largest 1.5–3.5 × 0.1–0.3(–0.5) cm, apex subacute to rounded, base cuneate to truncate, glabrous. Inflorescence an open panicle with numerous branches, up to 30 cm long; axes glabrous or finely puberulous towards apex; flowers in open dichasial 1–5-flowered cymes; peduncles up to 2.5 cm long, almost filiform; bracts up to 1 cm long; pedicels 0.5–4 mm long. Calyx finely ciliate, otherwise glabrous or very sparsely finely puberulous, 5–11 mm long of which fused basal part 2–3 mm; lobes all linear-lanceolate, dorsal 1–2 mm longer than the rest. Corolla pale mauve with white tube, no "herring-bones", no orange spots, 7–9 mm long of which the linear tube 4–5 mm and the expanded tube plus limb 3–4 mm, finely puberulous, no glands; upper lip narrowly ovate, lobes subacute; lower lip with a few long tapering hairs, obovate, lobes 1–2 mm long, rounded; only the 2 lower stamens fertile, glabrous, 1–2 mm long, upper reduced to ± 0.5 mm long staminodes; anthers 0.5–1 mm long. Capsule glabrous, 6–9 mm long, with 16–24 seeds per locule; retinaculae straight or very slightly curved. Seed ± 0.5 mm long.

TANZANIA. Songea District: 12 km W of Songea, Kimarampaka stream, 19 March 1956, *Milne-Redhead & Taylor* 9247! & N of Songea, Luhira River, 29 Apr. 1956, *Milne-Redhead & Taylor* 9954!
DISTR. **T** 8; Cameroon, Congo-Kinshasa, Angola, Zambia, Zimbabwe, Botswana
HAB. Waterholes, swamps, roadside ditches, old fields; 950–1050 m

NOTE. Cramer's arguments in Nord. J. Bot. 9: 262 (1989) for keeping *Hemiadelphis* separate from *Hygrophila* are completely negated by this and a few other species which have only 2 fertile stamens and at the same time have hairs on the lower lip. See also Heine in Adansonia 11: 674 (1971).

9. **Hygrophila spiciformis** *Lindau* in E.J. 20: 5 (1894) & in P.O.A. C: 366 (1895); Burkill in F.T.A. 5: 33 (1800); Jex-Blake, Some Wild Fl. Kenya: 104 (1948); Milne-Redhead in Mem. N.Y. Bot. Gard. 9, 1: 20 (1954); E.P.A.: 931 (1964); Binns, Checklist Herb. Fl. Malawi: 13 (1968); U.K.W.F.: 583 (1974); Champluvier in Fl. Rwanda 3: 457, fig. 143, 2 (1985); Blundell, Wild Fl. E. Afr.: 392 (1987); U.K.W.F. ed. 2: 268 (1994); Lebrun & Stork, Enum. Pl. Afr. Trop. 4: 482 (1997); Ensermu in F.E.E. 5: 380 (2006). Types: Ethiopia, Gondar, *Steudner* 1523 (not seen); "Ostafrika", no locality, *Fischer* 479 (K!, isosyn.); Tanzania, Dodoma/Mpwapwa District: Ugogo, Masswejo, *Stuhlmann* 333 (not seen) & Dodoma District: Ugogo, Pungusi, *Stuhlmann* 424 (not seen) & Unyamwezi, Mahale stream, *Stuhlmann* 461 (not seen)

Perennial herb, usually sticky glandular and aromatic if crushed; stems erect, ascending or procumbent with erect flowering branches, up to 1 m long, puberulous or densely so with long stalked capitate glands and with flattened glossy pilose hairs. Leaves ovate-elliptic or narrowly so, largest 4–15 × 0.8–3.8 cm, apex acute or subacute, base attenuate and decurrent to rounded or truncate near apex (rarely subamplexicaul), below with sparse to dense pilose hairs (densest along veins), without or with few (rarely many) capitate glands, above with uniformly scattered hairs. Inflorescence spiciform, interrupted near base and ± continuous apically, up to 25 cm long; axes puberulous or densely so with dense capitate glands and with scattered to dense pilose hairs; flowers solitary or in 2–6-flowered cymes which are sessile or rarely with peduncle up to 5 mm long near base; bracts up to 2.5 cm long; pedicels 0–1(–2) mm long. Calyx puberulous, with dense stalked capitate glands and with few to many shiny flattened pilose hairs, (10–)12–16(–19) mm long of which fused basal part 1–2(–3) mm, ventral lobes linear to narrowly oblanceolate, lateral linear, dorsal oblanceolate 2–5 mm longer than the rest (rarely same length as lateral but then still longer than ventral). Corolla pale mauve to mauve, blue or purple (rarely cream), "herring-bones" conspicuous, white with dark veins, sparsely puberulous and with scattered glands, 20–33 mm long of which linear part of tube 7–15 mm and expanded tube plus limb 11–18 mm; upper lip ovate, lobes rounded; lower lip elliptic, with numerous long clavate hairs, lobes 2–3 mm long, rounded. Filaments puberulous in basal part, upper 5–9 mm long, lower 7–12 mm long; anthers 2–3 mm long. Capsule finely puberulous and with capitate glands near apex, 10–12 mm long, with 10–14 seeds per locule. Seed ± 1 mm long.

UGANDA. Karamoja District: Lozut, 5 Nov. 1939, *A.S. Thomas* 3197!; Kigezi District: Virunga Mts, Lake Bunyoni, 27 Nov. 1934, *G. Taylor* 2155!
KENYA. Turkana District: Murua Nysigar, 21 Sep. 1963, *Paulo* 975!; Trans Nzoia District: Kitale, 15 Aug. 1956, *Bogdan* 4208!; Machakos District: Kibwezi, Chai Springs, 1963, *J. Brown* s.n.!
TANZANIA. North Mara District: Mara River, 10 Nov. 1953, *Tanner* 1767!; Ufipa District: Muse to Sumbawanga, 10 Nov. 1963, *Richards* 18390!; Mpwapwa District: W of Logi, 25 Aug. 1970, *Thulin & Mhoro* 790!
DISTR. **U** 1, 2; **K** 2–6; **T** 1, 2, 4–7; Ethiopia, Congo-Kinshasa, Rwanda, Zambia, Malawi, Mozambique, Zimbabwe
HAB. Wet grassland, swamps, riverbanks, riverbeds, waterholes, riverine forest, roadside ditches; (700–)1150–2300 m

NOTE. The isosyntype of *Fischer* 479 has a larger corolla (33 mm long) and longer upper calyx-lobe (19 mm long) than any other collection seen.
   Faden et al. 82/34 from Kenya, Thika is extremely glandular and has very narrow leaves with subamplexicaul base. It seems to be typical in all essential respects, but more collections from this area would be useful for ascertaining whether it is a form worthy of recognition.

10. **Hygrophila asteracanthoides** *Lindau* in E.J. 20: 6 (1894); Burkill in F.T.A. 5: 34 (1899); Pichi Sermolli, Miss. Stud. Lago Tana: 137 (1951); E.P.A.: 931 (1964); U.K.W.F. ed. 2: 268 (1994); Lebrun & Stork, Enum. Pl. Afr. Trop. 4: 482 (1997); Ensermu in F.E.E. 5: 380 (2006). Type: Ethiopia, between Keren and Gondar, *Steudner* 1498 (B†, holo.)

Perennial herb; stems erect, ascending or procumbent, up to 1 m long, puberulous or densely so with flattened curly glandular hairs, occasionally also with longer pilose non-glandular hairs. Leaves narrowly ovate-elliptic, largest 3.5–11 × 0.8–2 cm, apex subacute, base subamplexicaul, glabrous to uniformly puberulous (non-glandular curly hairs) on both sides. Inflorescence spiciform, up to 15 cm long, interrupted basally or ± continuous; axes densely glandular-puberulous to tomentellous, no pilose hairs; flowers in 2–5-flowered subsessile cymes or solitary towards apex; bracts up to 1.7 cm long, often purplish tinged, densely glandular-puberulous and with long glossy hairs; pedicels 1–3(–4) mm long. Calyx glandular-puberulous with glossy flattened curly hairs and with few to many long pilose hairs, 10–14 mm long of which fused basal part 1–2 mm; lobes all linear and ± same length or dorsal slightly longer. Corolla mauve to dark or bluish purple, without orange spots, "herring-bones" distinct, with yellow to orange veins, glandular-puberulous and with longer curly hairs especially towards tip of lobes, 17–20(–23) mm long of which the linear part of tube 7–10 mm and expanded tube plus limb 10–12(–15) mm; upper lip ovate, lobes rounded; lower lip elliptic, with stiff tapering hairs, lobes 2–4 mm long, broadly rounded. Filaments finely puberulous in basal part, upper 3–5 mm long, lower 7–9 mm long; anthers 1–1.5 (upper) and 2–2.5 (lower) mm long. Stigmatic lobe ± 1 mm long. Capsule finely puberulous in apical part, 8–11 mm long, with 24–28 seeds per locule. Seed ± 0.5 mm long.

Kenya. Uasin Gishu District: Kipkarren, Jan. 1932, *Brodhurst-Hill* 705!; Trans Nzoia District: Endebess, 5 Jan. 1955, *Irwin* 148! & 10 km W of Kitale, Maboonde, Dec. 1971, *Tweedie* 4203!
Distr. **K** 3; Ethiopia
Hab. Seasonally inundated grassy swamps; 1650–2100(–2300) m

Note. Because of the similar inflorescences *H. asteracanthoides* superficially looks very much like *H. spiciformis*, but the two are really quite distinct. *H. asteracanthoides* has a different calyx, a smaller corolla without orange spots and differently coloured palate and also different corolla indumentum. There are many more ovules per locule and the seeds consequently are much smaller.

11. **Hygrophila albobracteata** *Vollesen* **sp. nov.** a *H. spiciforme* floribus in capitulis pedunculatis (nec cymis racemiformibus) aggregatis, parte basali bractearum alba (nec bracteis uniformiter viridibus), corolla alba vel cremea (nec malvina neque caerulea neque purpurea) maiore 3.3–3.7 cm (nec 2–3.3 cm) longa et capsula maiore 1.1–1.3 cm (nec 1–1.2 cm) longa seminibus paucioribus (in loculi 6–8 nec 12–14) maioribus 1.5–2 mm (nec 0.5–1 mm) longis differt. Type: Tanzania, Mpanda District: Kapapa, Msaginya River, *Richards* 25862 (K!, holo.; EA!, WAG!, iso.)

Viscid glandular perennial herb, forming dense clumps, aromatic when crushed; stems erect or scrambling, up to 1.3 m long, quadrangular, densely glandular and with scattered non-glandular pilose hairs when young. Leaves narrowly ovate-elliptic, largest 7–14.5 × 0.8–2.3 cm, apex acute, base attenuate, finely puberulous or sparsely so on both sides and with usually dense capitate glands and scattered pilose hairs. Flowers in globose pedunculate axillary heads; peduncle up to 1.5(–2) cm long, indumentum as stems; bracts ovate to elliptic, up to 3.5 cm long, basal part white (rarely all green), indumentum as leaves. Calyx finely puberulous and also with capitate glands towards apex, densely pilose-ciliate on edges and also midrib towards base, 13–19 mm long of which fused basal part ± 1 mm; ventral and lateral lobes linear-lanceolate, dorsal lanceolate to oblanceolate and 1–3 mm longer than the rest. Corolla white or cream, "herring-bones" conspicuous, with pale red veins, sparsely finely puberulous and with scattered glands, 33–37 mm long of which the linear tube 14–16 mm and expanded tube plus limb 19–21 mm; upper lip oblong, ± hooded, lobes 3–4 mm long, broadly rounded; lower lip with dense long tapering hairs, broadly obovate-spathulate, lobes 5–6 mm long, central retuse, two lateral rounded. Filaments with short stubby hairs in basal part, upper 9–10 mm long, lower 13–15 mm long; anthers 3–4 mm long. Capsule sparsely finely puberulous and with a few glands near apex, 11–13 mm long, with 6–8 seeds per locule. Seed 1.5–2 mm long.

Tanzania. Mpanda District: Kapapa, Msaginya River, 12 Sep. 1970, *Richards* 25862!; Chunya
District: 15 km from Chiningo to Ngwala Mine, 14 Aug. 1968, *Sanane* 250! & Simipala, 15 Oct.
1932, *Geilinger* 3072!
Distr. **T** 4, 7; Zambia
Hab. Riverbanks and riverbeds; 850–1400 m

Note. Differs from the closely related *H. spiciformis* in the different inflorescence and larger
corolla which is white or cream, a very rare colour in *H. spiciformis*.

12. **Hygrophila abyssinica** (*Nees*) *T.Anderson* in J.L.S. 7: 22 (1863); Heine in K.B. 16:
174 (1962) & in F.W.T.A. (ed. 2) 2: 395 (1963); E.P.A.: 930 (1964); Vollesen in Opera
Bot. 59: 80 (1980); Lebrun & Stork, Enum. Pl. Afr. Trop. 4: 481 (1997); Ensermu in
F.E.E. 5: 381 (2006). Types: Sudan, Sennar, *Kotschy* 293 (BM!, K!, isosyn.); Ethiopia,
Schimper s.n. (not seen)

Annual or short-lived perennial herb, sometimes viscid, said to smell like catmint
when crushed; stems suberect, ascending or procumbent, up to 50(–75) cm long,
rounded to quadrangular, glabrous to puberulous. Leaves narrowly ovate to ovate
or narrowly elliptic to elliptic, largest (0.8–)1.2–5.5(–8) × 0.4–1.6(–2) cm, apex
rounded to acute (rarely acuminate), base attenuate to cuneate, margin entire to
crenate-dentate with large triangular teeth, glabrous to puberulous. Inflorescence
spiciform to paniculate, interrupted basally, confluent apically, up to 30 cm long;
axes finely puberulous, also with capitate glands and longer hairs; flowers in 1–12-
flowered cymes, these sessile or with peduncle up to 5 mm long; bracts up to 1 cm
long, glabrous to puberulous, ± capitate glands and with longer white hairs on
edges; pedicels up to 1 mm long. Calyx finely puberulous, also with capitate glands
and with longer hairs on edges, 3–7 mm long of which fused basal part ± 1 mm;
lobes all linear-lanceolate or dorsal slightly wider and oblanceolate, all same length
or dorsal up to 1 mm longer. Corolla subactinomorphic, pale mauve or pale blue
to deep blue or purplish with white or yellow throat (rarely pure white), no orange
spots, no long stiff hairs and no "herring-bones", finely puberulous and
occasionally with capitate glands, 4–8(–10) mm long of which the elliptic lobes
1–3 mm; tube linear with a constriction about ¹/₃ above base and slightly widened
upwards; stamens 4, didynamous, inserted just above constriction, included in
tube; filaments 0.5–1 mm long, glabrous; anthers 0.5–1 mm long. Style ± 1 mm
long, hairy, both stigmatic lobes equally developed, flattened. Capsule finely
puberulous, ± a few glands near apex, 4–7 mm long, with 8–10 seeds per locule.
Seed ± 0.5 mm long.

Tanzania. Ulanga District: Selous Game Reserve, Mlahi, Kilombero River, 13 Nov. 1976,
Vollesen in MRC 4089! & 6 km N of Mlahi (8°14'S, 37°05'E), 16 Nov. 1976, *Vollesen*
MRC4106!
Distr. **T** 6; Ghana, Nigeria, Cameroon, Central African Republic, Chad, Congo-Kinshasa,
Sudan, Ethiopia, Angola, Zambia, Malawi, Mozambique, Zimbabwe, Botswana, Namibia
Hab. Seepage area in *Brachystegia* woodland, riverbanks, sandy soil; ± 275 m

Syn. *Polyechma abyssinicum* Nees in DC., Prodr. 11: 83 (1847); Richard, Tent. Fl. Abyss. 2: 141 (1850)
Cardanthera justicioides S.Moore in J.B. 18: 6 (1880). Type: Sudan, *Schweinfurth* 972 (BM!,
holo.; K!, P!, iso.)
C. africana (T.Anderson) Benth. var. *schweinfurthii* S.Moore in Journ. Bot.18: 7 (1880) pro
parte quoad *Schweinfurth* 2799
Hemigraphis abyssinica (Nees) C.B.Clarke in F.T.A. 5: 58 (1899); Benoist in Bol. Soc. Biol.,
Ser. 2, 24: 34 (1950); F.P.S. 3: 174 (1955); Morton in Journ. W. Afr. Sci. Ass. 3: 175
(1957), quoad *Morton* 8830.
H. schweinfurthii (S. Moore) C.B.Clarke in F.T.A. 5: 58 (1899) pro parte quoad *Schweinfurth*
2799; Schnell in Mem. I.F.A.N. 18: 13 (1952)
Synnema hygrophiloides Lindau in E.J. 43: 349 (1909). Type: Zambia/ Zimbabwe, Victoria
Falls, *Engler* 2982 (B†, holo.)
[*S. acinos* sensu Lindau in Schwed. Rhod.-Kongo-Exp. 1: 304 (1916), *non* S.Moore (1908)]

S. *abyssinicum* (Nees) Bremek. in Verh. Nederl. Akad. Wetensch., Afd. Natuurk., ser. 2, 41: 136 (1944)

S. *schweinfurthii* (S.Moore) Bremek. in Verh. Nederl. Akad. Wetensch., Afd. Natuurk., ser. 2, 41: 141 (1944) pro parte quoad *Schweinfurth* 2799.

Hygrophila hygrophiloides (Lindau) Heine in Adansonia 11: 659 (1971); Lebrun & Stork, Enum. Pl. Afr. Trop. 4: 482 (1997)

NOTE. An extremely variable species as can be guessed from the lengthy synonymy. But on closer examination the variation is only in growth form and vigour. Calyx, corolla, fruit and seeds are constant throughout, and I have been unable to separate out any infraspecific taxa despite its large distribution and large disjunction from northern to southern Africa.

H. abyssinica is very close to the Indian *H. polysperma* (Roxb.) T.Anders. which differs mainly in having two fertile stamens plus two staminodes. It is quite possible that a thorough study will reveal that the two are at most subspecifically distinct. See also note after *H. linearis*.

Of the three collections cited as types for *Hemigraphis schweinfurthii* the other two (*Schweinfurth* 2708 and 2764) are *Hygrophila africana* (T.Anderson) Heine. Heine in Adansonia 11: 657 (1971) erroneously cites *Schweinfurth* 2708 as *H. abyssinica*.

H. abyssinica differs from all other species in the usually toothed leaves, the subactinomorphic corolla without long stiff hairs but with included subsessile stamens and in the short stigma with both stigmatic lobes equally developed and flattened.

13. **Hygrophila hippuroides** *Lindau* in Wiss. Ergebn. Schwed. Rhod.-Kongo- Exp. 1: 304 (1916); Lebrun & Stork, Enum. Pl. Afr. Trop. 4: 482 (1997). Type: Zambia, Lake Bangweolo, Kamindas, *R.E. Fries* 857 (UPS!, holo.)

Aquatic herb with floating spongy reddish stems with long trailing roots, creeping and rooting if in dried out places; flowering stems erect, up to 50 cm long, quadrangular, glabrous. Leaves opposite (rarely in whorls of 3); lamina lanceolate to ovate or elliptic, largest 1.5–3.3 × 0.5–1 cm, apex subacute to rounded, base truncate to auriculate, glabrous. Inflorescence spicate, up to 9(–18) cm long; axes glabrous or puberulous on edges; floral leaves ± imbricate, cordate-reniform with auriculate base, glabrous or with ciliate margin; flowers all solitary; pedicels 0–1 mm long. Calyx pilose-ciliate, otherwise glabrous, 6–9 mm long of which united basal part 0.5–1 mm; lobes all about same length, ventral and lateral lanceolate, dorsal slightly wider and oblanceolate. Corolla pure white, "herring-bones" with orange veins, finely puberulous (puberulous on lobes), 11–14 mm long, widened from base (no linear tube), tube with band of hairs near base; upper lip lanceolate to narrowly ovate, lobes acute; lower lip with long clavate hairs, broadly obovate-spathulate, lobes 3–5 mm long, rounded to truncate. Filaments with thick stubby hairs in apical half, upper ± 2 mm long, lower ± 3 mm long; anthers ± 1.5 mm long, purple edged. Stigmatic lobe flattened. Capsule 5–9 mm long, sparsely puberulent and with sparse glands near apex, with 24–42 seeds per locule. Seed ± 0.5 mm long.

TANZANIA. Tabora District: N of Usinge, 28 Nov. 1933, *Michelmore* 776!; Kigoma District: Malagarasi Swamps, south of Nguruka Station, Katare, 26 Aug. 1952, *Lowe* 553!; Ufipa District: Sumbawanga, Chapota, Empeta Swamp, 8 March 1957, *Richards* 8587!
DISTR. **T** 4; Zambia
HAB. Floating in lakes and swamps, on drying mud; 1200–1650 m

NOTE. The Tanzanian material differs from the Zambian in the white corolla, and in having more ovules per locule. But there is an overlap in ovule numbers, and the total amount of material available is too small to judge whether two separate subspecies should be recognised.

Very isolated species in *Hygrophila*. It lacks the basal linear corolla tube and has thick clavate hairs on the lower lip. It also has a band of hairs in the corolla tube.

Some of the shoots on *Bullock* 3117 have leaves in whorls of three, but seems otherwise to be quite typical.

14. **Hygrophila schulli** (*Buch.-Ham.*) *M.R.Almeida* & *S.M.Almeida* in J. Bombay Nat. Hist. Soc. 83, Suppl.: 221 (1987); Ensermu in F.E.E. 5: 379 (2006). Type: "India", Hortus Malabaricus 2: Pl. 87, fig. 45

Erect unbranched or branched annual herb, occasionally with basal part of stem creeping and rooting; stems up to 1.3(–1.8) m long, quadrangular to rounded, glabrous to densely hirsute-strigose upwards with broad glossy hairs, and usually with finely puberulous indumentum underneath. Leaves linear-lanceolate to narrowly ovate or narrowly elliptic (rarely elliptic), largest 7–32 × 0.3–3(–4.2) cm, apex subacute to acuminate, base attenuate to hastate and auriculate, glabrous to hirsute-strigose with glossy white hairs. Flowers in dense sessile many-flowered fascicles in the upper part of the stem, each with up to 6 hard yellowish to brownish finely puberulous spines (modified branches) up to 5 cm long (rarely absent); bracts numerous, becoming hard and ± bony in basal part, lanceolate to ovate or elliptic, up to 5(–7) cm long, finely puberulous and ciliate or hirsute-strigose in basal part. Calyx finely puberulous or hirsute-strigose apically and along midribs, 12–25(–30) mm long; lobes all strongly ribbed, ventral pair fused almost to apex, the others for 2–3 mm at base, ventral and lateral linear-lanceolate, dorsal 1–6 mm longer and lanceolate to oblanceolate. Corolla white or pale mauve, pale lilac, blue or purple, "herring-bones" conspicuous, white with reddish to purplish veins, finely puberulous or sparsely so and usually with capitate glands, 22–40 mm long of which linear tube 8–14 mm and expanded tube plus limb 13–27 mm; upper lip ovate to elliptic, lobes 2–5 mm long, rounded; lower lip with few to many long tapering hairs, broadly obovate, lobes 4–8 mm long, rounded. Filaments glabrous or with a few short stubby hairs near base, upper 4–10 mm long, lower 7–14 mm; anthers 2–3 mm long. Stigmatic lobe 1–2 mm long. Capsule enclosed in the hardened basal part of the bracts, retained on plant, glabrous, 8–14 mm long, with 2–6 seeds per locule. Seeds all 1.5–2 mm long or with 2 large seeds 3–4 mm long and 2 small seeds ± 2 mm long. Fig. 27, p. 177.

UGANDA. Karamoja District: Moroto, 12 Sep. 1956, *Bally* 10790!; Mbale District: near Busano, 21 Jan. 1969, *Lye* 1700!; Mengo District: Wabusana to Kalungi, 29 July 1956, *Langdale-Brown* 2258!
KENYA: Northern Frontier District: 26 km on Garissa–Hagadera road, 29 May 1977, *Gillett* 21188!; Machakos District: 35 km NE of Machakos, 21 Oct. 1967, *Mwangangi* 283!; Kwale District: near Shimoni, Kivumoni, 20 Aug. 1953, *Drummond* & *Hemsley* 3927!
TANZANIA. Musoma District: Ushashi, 24 Apr. 1959, *Tanner* 4203!; Dodoma District: 10 km N of Manyoni, 15 Apr. 1988, *Bidgood et al.* 1128!; Mbeya District: Igawa, 13 Apr. 1962, *Polhill* & *Paulo* 2002!; Pemba, Chanjana, 14 Feb. 1929, *Greenway* 1434!
DISTR. **U** 1–4; **K** 1, 3–5, 7; **T** 1–8; **Z**; **P**; widespread in tropical Africa and in India
HAB. Seasonally flooded grassland, usually on clay soils, but occasionally on wet sand, alluvial *Acacia* woodland and bushland, riverbanks, riverbeds, old ricefields; near sea level to 1650 m

SYN. *Barleria longifolia* L., Amoen. Acad. 4: 320 (1759) & Sp. Pl. (ed. 2) 2: 887 (1763). Type: India (LINN, holo.; K!, microfiche)
     Bahel schulli Buch.-Ham. in Trans. Linn. Soc. London 14: 288 (1824)
     Barleria auriculata Schumach., Beskr. Guinea Pl.: 285 (1827). Type: Guinea, *Thonning* (C!, holo.)
     Ruellia longifolia (L.) Roxb., Fl. Ind. (ed. 2) 3: 50 (1832)
     Asteracantha longifolia (L.) Nees in Wall., Pl. As. Rar. 3: 90 (1832) & in DC., Prodr. 11: 247 (1847); Lindau in P.O.A. C: 367 (1895) & in E. & P. Pf. IV, 3b: 297 (1895); F.P.N.A. 2: 274 (1947); F.P.S. 3: 166 (1956)
     A. auriculata (Schumach.) Nees in DC., Prodr. 11: 248 (1847); A.Richard, Tent. Fl. Abyss. 2: 146 (1850)
     Hygrophila spinosa T.Anderson in Thwaites, Enum. Pl. Zeyl.: 225 (1860) & in J.L.S. 7: 22 (1863); C.B. Clarke in Fl. Brit. Ind. 4: 408 (1885); Burkill in F.T.A. 5: 31 (1899); S. Moore in J.B. 45: 89 (1907); Th. & H.Durand, Syll. Fl. Cong.: 416 (1909); Benoist in Bull. Soc. Bot. France 60: 330 (1913); De Wild., Pl. Beq. 4: 20 (1926); Hutchinson & Dalziell, F.W.T.A. 2: 247 (1931); Binns, Checklist Herb. Fl. Malawi: 13 (1968). Type as for *Barleria longifolia*

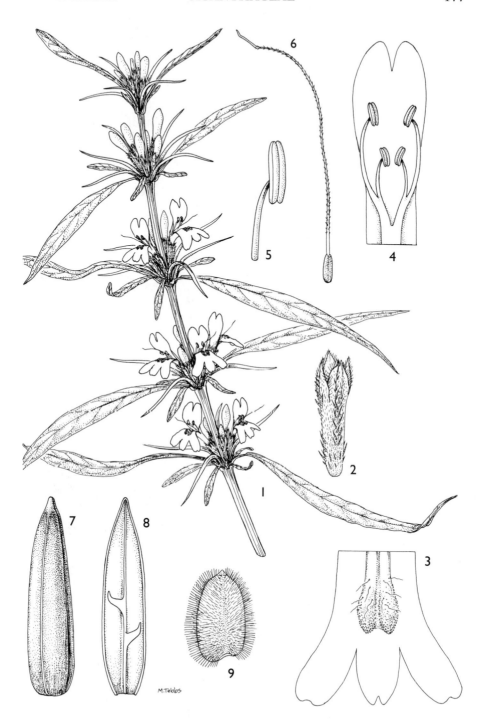

FIG. 27. *HYGROPHILA SCHULLI* — **1**, flowering stem, × ²/₃; **2**, calyx, × 3; **3**, lower corolla lip with swellings and hairs, × 3; **4**, upper lip of corolla with stamens, × 3; **5**, anther, × 4; **6**, style and stigma, × 3; **7**, whole capsule, × 6; **8**, single capsule valve, × 6; **9**, seed, × 10. 1 from *Carter et al.* 2635, 2–6 from *Polhill & Paulo* 2002, 7–9 from *Ruffo* 2499. Drawn by Margaret Tebbs.

H. longifolia (L.) Kurz in Journ. As. Soc. Beng. 39(2): 78 (1870); Engler, Hochgebirgsfl.
   Trop. Afr.: 387 (1892); Lugard in K.B. 1933: 94 (1934); Benoist in Bol. Soc. Brot., Ser. 2,
   24: 34 (1950); Berhaut, Fl. Senegal: 75 (1954); Binns, Checklist Herb. Fl. Malawi: 13
   (1968), *non* Nees (1847)
Asteracantha lindaviana De Wild. & T.Durand in Comp. Rend. Soc. Bot. Belg. 8: 100 (1899).
   Type: Congo-Kinshasa, near Businga, Evamkayo, *Thonner* s.n. (BR!, holo.; BR!, iso.)
Hygrophila lindaviana (De Wild. & T.Durand) Burkill in F.T.A. 5: 509 (1900); Lebrun &
   Stork, Enum. Pl. Afr. Trop. 4: 482 (1997)
H. auriculata (Schumach.) Heine in K.B. 16: 172 (1962) & in F.W.T.A. (ed. 2) 2: 395
   (1963); Binns, Checklist Herb. Fl. Malawi: 13 (1968); U.K.W.F.: 583 (1974); Vollesen in
   Opera Bot. 59: 80 (1980); Champluvier in Fl. Rwanda 3: 456, fig. 143,1 (1985); Blundell,
   Wild Fl. E. Afr.: 391, pl. 607 (1987); Lebrun & Stork, Enum. Pl. Afr. Trop. 4: 482 (1997);
   Friis & Vollesen in Biol. Skr. 51(2): 443 (2005)

NOTE. Material from Uganda, W Kenya and W Tanzania (*H. auriculata*, sensu stricto) has
hastate auriculate leaf bases at least on the upper part of the stem, while material from the
rest of the Flora area (*H. spinosa*, sensu stricto) has attenuate to cuneate leaf bases. The
western plants also tend to be taller and stouter.

It is quite possible that a future detailed study also including the Indian material will allow
these two forms to be separated, most probably as subspecies. But at the moment I have
decided to keep all the material under one broad species as there is a number of what seems
to be intermediates, and – in India and West Africa – a fair number of specimens tending
towards the taxon endemic to the other part of the world.

In central and east Tanzania other patterns of variation are seen. Several different forms
(e.g. white flowered and mauve flowered or subglabrous or densely hairy) can often be found
together in the same mbuga with few or no intermediates. But in the next mbuga a different
pattern is usually found.

H. schulli and *richardsiae* differ from the rest of the species in the hardened spines which
surround the inflorescences. These are modified branches as can be seen from the scars or
even minute leaves which often appear. It also differs in having the two lower sepals fused,
and in the hardened basal part of the bracts which retain the capsules on the plant. The
number of ovules (2–6) is also unusually low in the genus as are the relatively large seeds. It
may be that further work will show that *Asteracantha* can be resurrected.

15. **Hygrophila richardsiae** *Vollesen* **sp. nov.** a *H. schulli* herba perenni (nec annua),
corolla maiore 4–4.4 cm (nec 2.2–4 cm) longa, labio inferiore corollae lobis longioribus
9–12 mm (nec 4–8 mm) longis ornato atque lobo medio retuso (nec rotundato), et
filamentis pilosis differt. Type: Tanzania, Ufipa District: Tatanda Mission, *Bidgood et al.*
2418 (K!, holo.; BR!, C!, CAS!, DSM!, EA!, K!, NHT!, P!, UPS!, WAG!, iso.)

Erect unbranched short lived perennial herb, drying blackish; stems up to 50 cm
long, quadrangular to rounded, hirsute or densely so and with finely puberulous
indumentum underneath. Leaves elliptic or narrowly so, largest 6.5–14 × 1–4 cm,
apex acute, base cuneate to subhastate, whitish hirsute-strigose on both sides.
Flowers in dense sessile many-flowered fascicles in the upper part of the stem,
without or with short slender finely puberulous spines (usually less than 6) up to
2 cm long; bracts numerous, becoming hard and ± bony in basal part, lanceolate to
ovate or elliptic, up to 3 cm long, hirsute-strigose. Calyx hirsute-strigose apically and
long-ciliate, ± finely puberulous hairs apically, 13–22 mm long; lobes all strongly
ribbed, ventral pair fused almost to apex, the others for 2–3 mm at base, ventral and
lateral linear-lanceolate, dorsal 2–5 mm longer and lanceolate to oblanceolate.
Corolla white, "herring-bones" conspicuous, white with reddish to purplish veins,
sparsely crisped-puberulous and with scattered capitate glands, 40–44 mm long of
which linear tube 12–13 mm and expanded tube plus limb 27–31 mm; upper lip
ovate to elliptic, lobes 5–7 mm long, rounded to retuse; lower lip with scattered long
tapering hairs, broadly obovate, lobes 9–12 mm long, rounded or retuse (central).
Filaments with long downwardly directed hairs along their whole length, upper 6–9
mm long, lower 12–16 mm; anthers 2–3 mm long. Stigmatic lobe 1–2 mm long.
Capsule and seed not seen.

TANZANIA. Ufipa District: Sumbawanga, 4 Feb. 1957, *Richards* 8435! & Old Sumbawanga road, M'wimbe Dambo, 21 Apr. 1962, *Richards* 16356! & 50 km on Mbala–Mwasye road, 16 March 1950, *Bullock* 2644!

DISTR. **T** 4; not known elsewhere

HAB. Swampy grassland; 1500–2100 m

NOTE. This species, which is only known from a small area around Sumbawanga, is best interpreted as a high altitude replacement for *H. schulli*. But with the present material it is sufficiently distinct to be treated as a separate species. It differs in being perennial, in the larger corolla with longer lobes of which the central is retuse, and in the hairy filaments. It also invariably dries blackish, a character not usually seen in *H. schulli*. The spines surrounding the inflorescences are either absent or much reduced.

16. **Hygrophila sp. C.**

Perennial herb with short erect or ascending stems from branched rhizome; stems up to 10 cm long, hirsute to tomentose. Leaves sharply reflexed; lamina narrowly elliptic, largest 5–8 × 1–1.5 cm, apex acute, base subhastate, whitish hirsute-strigose above, below on veins and margins and subglabrous on lamina. Flowers in dense sessile fascicles merging into heads in the upper part of the stem, without spines; floral leaves and bracts foliaceous, gradually smaller upwards, hard and ± bony in basal part, inner lanceolate, to 1.5 cm long, hirsute-strigose. Calyx hirsute-strigose apically and long-ciliate, 11–13 mm long; lobes all strongly ribbed, all fused for ± 8 mm at base, ventral and lateral narrowly triangular, dorsal 2–3 mm longer and oblanceolate-spathulate. Corolla bright blue, "herring-bones" inconspicuous, crisped-puberulous, no capitate glands, ± 27 mm long of which linear tube ± 12 mm and expanded tube plus limb ± 15 mm; upper lip obovate, lobes 8–9 mm long, rounded; lower lip with scattered long curly hairs, broadly obovate, lobes 7–8 mm long, rounded or retuse (central); some flowers with 4 subequal stamens, others with 2 fertile stamens (dorsal) and two staminodes without anthers; filaments glabrous, ± 4 mm long; anthers ± 3.5 mm long. Capsule and seed not seen.

TANZANIA. Ufipa District: 22 km on Sumbawanga-Mpanda road, 16 June 1996, *Faden et al.* 96/236!

DISTR. **T** 4; not known elsewhere

HAB. Short valley bottom grassland on sandy-loamy soil; ± 1700 m

NOTE. Clearly related to *H. richardsiae* with which it shares the perennial habit and absence of spines in the inflorescences. It differs in the much smaller flowers. From spineless forms of *H. schulli* it differs in the perennial habit. It differs from both in not having the two dorsal sepals fused almost to the apex and in the relatively long linear corolla tube.

17. DYSCHORISTE

Nees in Wall., Pl. As. Rar. 3: 75, 81 (1832) & in DC., Prodr. 11: 106 (1847); Lindau in E. & P. Pf. IV, 3b: 302 (1895); C.B. Clarke in F.T.A. 5: 71 (1899)

Calophanes D. Don in Sweet, Brit. Fl. Gard., ser. 2, 2: t. 181 (1833); Nees in DC., Prodr. 11: 107 (1847); G.P. 2: 1077 (1876); C.B. Clarke in Fl. Brit. Ind. 4: 410 (1884)

Linostylis Sond. in Linnaea 23: 94 (1850)

Phillipsia Rolfe in K.B. 1895: 223 (1895)

Perennial or shrubby herbs or shrubs, with usually conspicuous linear cystoliths (especially on calyx); stems terete to distinctly angular. Leaves opposite, base decurrent into the petiole which has a distinct articulation at or up to 2 mm above base. Flowers in condensed axillary cluster-like cymes which sometimes merges into spiciform terminal panicles, more rarely solitary or in lax axillary cymes, shortly pedicellate; bracts various, outer foliaceous; bracteoles 2, filiform to linear-lanceolate. Calyx 5-lobed, the lobes subequal, linear to narrowly triangular, acuminate to cuspidate, with strong often rib-like central vein. Corolla contorted in bud; tube with linear basal part, widened into a ± distinst inflated upper part (throat); limb from ± subequally 5-lobed to distinctly 2-lipped with 2-lobed upper and 3-lobed lower lip; lower lip with or without differently coloured pleated base, without (very rarely with) long stiff retrorse hairs; lobes oblong or obovate, rounded to truncate or irregularly toothed. Stamens 4, slightly didynamous, all fertile or two sterile or reduced to staminodes, held in throat; filaments linear, fused with tube at base and free part of each short and long stamen fused at base, part fused with tube hairy; anthers 2-thecous, ellipsoid, usually glabrous, with spurs at base (rarely without). Ovary with 2 ovules per locule; style filiform, hairy throughout, upper stigma-lobe usually minute, lower 1–2 mm long, linear-oblong, flattened, glandular. Capsule 4-seeded (or 2–3-seeded by abortion), glabrous or sparsely glandular at apex, oblong to slightly clavate, cylindric, valves sulcate dorsally, often persisting after release of seeds; retinaculae strong, curved. Seeds discoid, ovoid-oblong in outline, with dense hygroscopic hairs.

About 75 species, widespread in all tropical regions and subtropical in N America.

1. Corolla-limb usually over twice as long as
   inflated part of tube (throat), at least
   3 mm longer; lobes obovate 2
   Corolla-limb about same length as throat
   (within 1–2 mm) or up to 6 mm shorter;
   lobes usually oblong 5
2. Corolla 21–41 mm long of which the linear
   tube 10–24 mm long 3
   Corolla 12–17 mm long of which the linear
   tube 5–8 mm long 4
3. Young stems with indumentum solely of
   stalked capitate glands or with a few non-
   glandular hairs intermixed; leaves
   acuminate; flowers in open monochasial
   cymes with branches 5–17 mm long;
   corolla white . . . .   18. *D. kitongaensis* (p. 199)
   Young stems puberulent to puberulous or
   sericeous, sometimes with intermixed
   stalked capitate glands; leaves acute to
   retuse; flowers usually in condensed
   cymes with branches up to 4 mm long,
   more rarely in open dichasial cymes;
   corolla not white . . . .   17. *D. hildebrandtii* (p. 198)
4. Young stems, bracts and calyx with stalked
   capitate glands; calyx-lobes 5–8(–9) mm
   long, longer than calyx-tube . . . .   16b. *D. keniensis* subsp. *glandulifera* (p. 197)
   Young stems, bracts and calyx without
   stalked capitate glands; calyx-lobes
   4–6 mm long, shorter than calyx-tube   16a. *D. keniensis* subsp. *keniense* (p. 197)
5. Perennial herbs with trailing stems (often
   ascending or erect apically), often
   rooting at nodes (very rarely fully erect,
   then corolla less than 10 mm long);
   corolla 6–19 mm long 6
   Erect (rarely decumbent) perennial
   herbs, shrubby herbs or shrubs (if
   trailing then corolla over 40 mm long);
   corolla 11–65 mm long 8

6. Ripe capsule longer than calyx; leaf-margin glabrous or minutely puberulent, lamina glabrous on both sides or with hairs on midrib; calyx 4–7(–9 in fruit) mm long; corolla 6–9(–12) mm long . . . . 1. *D. nagchana* (p. 183)
Ripe capsule shorter than calyx (rarely same length or slightly longer); leaf-margin distinctly ciliate or lamina hairy on one or both sides or calyx and/or corolla longer; calyx 6–12(–18 in fruit) mm long; corolla 7–19 mm long 7

7. Calyx and/or bracts and sometimes young leaves with stalked capitate glands; lamina ovate to elliptic, uniformly hairy on upper surface; corolla 7–11 mm long . . . . 2. *D. multicaulis* (p. 184)
Calyx, bracts and young leaves without capitate glands; leaves obovate (more rarely some or all elliptic), glabrous (more rarely hairy) on upper surface; corolla 10–19 mm long . . . . 3. *D. radicans* (p. 186)

8. Lower corolla lip with long stiff retrorse hairs; dorsal stigma lobe well-developed, about half as long as upper lobe; branches, leaves and inflorescence axes glabrous or with scattered subsessile stalked capitate glands . . . . 13. *D.* sp. *B* (p. 195)
Lower corolla lip without long stiff hairs; dorsal stigma lobe a minute tooth; either branches, leaves or inflorescence axes with non-glandular hairs 9

9. Corolla 28–65 mm long of which the throat (7–)11–26 mm and the limb 9–20 mm, lobes obovate-spathulate; capsule (where known) 15–18 mm long; apical part of filaments and style with stalked capitate glands 10
Corolla 11–30(–32) mm long of which the throat 3–10 mm and the limb 2–10 mm, lobes oblong (rarely slighty obovate); capsule 8–13 mm long; apical part of filaments and style glabrous or hairy, never with capitate glands 11

10. Much branched erect or scrambling subshrub or shrub; dry bushland in NE Uganda and Kenya, usually on stony soil or on rocky hills . . . . 14. *D. thunbergiiflora* (p. 195)
Perennial herb with several trailing unbranched or sparsely branched stems from woody rootstock; plant from *Brachystegia* woodland in W Tanzania, on sandy–loamy soil . . . . 15. *D. sallyae* (p. 196)

11. Corolla pure white or with pink to
    purple spots or lines at base of lower
    lip and in throat 12
    Corolla pale mauve to mauve, purple or
    blue, usually darker in throat 17
12. Inflorescences ± covered by upper leaves
    and bracts; leaves with strongly raised
    rib-like veins; corolla 11–18 mm long;
    anthers without spurs ....   4. *D. mutica* (p. 187)
    Inflorescences not covered by upper
    leaves and bracts; leaves not with strong
    rib -like veins; corolla 14–30(–32) mm
    long; anthers with distinct spurs 13
13. Basal part of lower corolla-lip (palate)
    strongly bullate, usually with distinct
    purple spots and lines 14
    Basal part of lower corolla-lip not or
    weakly bullate, with very pale markings
    or a pale tinge 15
14. Pyrophytic herb with 1–25 cm tall stems;
    indumentum on all parts fine,
    puberulent with hairs and glands under
    0.5 mm long ....   12. *D. tanganyikensis* (p. 194)
    Plant 0.2–1.5 m tall, not pyrophytic;
    young stems and/or calyx with some
    pilose hairs 0.5–1.5 mm long ....   11b. *D. trichocalyx* subsp. *verticillaris* (p. 192)
15. Young stems without stalked capitate
    glands; calyx without (rarely with)
    capitate glands, subglabrous to
    puberulous ....   5. *D. albiflora* (p. 188)
    Young stems with stalked capitate
    glands; calyx glandular-puberulent
    to -puberulous or densely so 16
16. Young stems with long pilose hairs some
    of which over 2 mm long; calyx
    11–18(–22 in fruit) mm long; corolla
    20–26 mm long ....   6. *D. pilifera* (p. 189)
    Young stems with hairs up to 1 mm long
    (very rarely with long pilose hairs to
    2 mm long but then calyx less than
    10 mm long and corolla less than 20 mm
    long); calyx 6–12(–14 in fruit) mm
    long; corolla 14–17 mm long ....   7. *D. tanzaniensis* (p. 189)
17. Pyrophytic herb with 1–25 cm tall stems 18
    Plant 0.2–1.5 m tall, not pyrophytic 19
18. No part of plant with viscid capitate
    glands; corolla 10–19 mm long .....   3. *D. radicans* (p. 186)
    Indumentum on all parts with viscid
    capitate glands and non-glandular
    hairs; corolla (19–)23–30 mm long ...   12. *D. tanganyikensis* (p. 194)
19. Calyx viscid glandular-puberulous or
    densely so and with long non-
    glandular pilose hairs (rarely not viscid
    and without long pilose hairs, then
    calyx-lobes distinctly longer than tube
    and corolla over 20 mm long); young
    stem not densely puberulous ....   11b. *D. trichocalyx* subsp. *verticillaris* (p. 192)

Calyx eglandular or with scattered not
viscid glands, usually without long
pilose hairs; calyx-lobes ± same length
as or shorter than tube or corolla less
than 18 mm long or young stems
densely puberulous 20
20. Calyx-lobes distinctly longer (at least 2 mm)
than tube; young stems uniformly
puberulous or densely so; corolla
14–18 mm long . . . . 11a. *D. trichocalyx* subsp. *trichocalyx* (p. 192)
Calyx-lobes shorter than or same ± length
(at most 1 mm longer) as tube; young
stems puberulent to finely sericeous 21
21. Calyx with scattered long pilose hairs;
corolla 25–30 mm long of which the
linear tube 12–15 mm; young stems
and leaves with (rarely without) long
pilose hairs . . . . 9. *D. nyassica* (p. 191)
Calyx and all other parts without long
pilose hairs; corolla 14–27 mm long of
which the linear tube 5–11 mm 22
22. Flowers in lax dichasial cymes with
peduncle and branches 1–1.5 cm long   10. *D.* sp. *A* (p. 191)
Flowers in condensed axillary cymes or
solitary towards tip of stems with
peduncles and branches 1–2(–8) mm
long 23
23. Corolla 19–27 mm long, throat 6–9 mm
across apically; limb shorter than
throat (rarely same length); calyx
6–10(–14 in fruit) mm long; leaves
elliptic to obovate . . . . 8. *D. subquadrangularis* (p. 190)
Corolla 14–18(–22) mm long, throat
4–5(–6) mm across apically; limb and
throat ± same length; calyx 8–13(–17 in
fruit) mm long; leaves ovate to elliptic   7. *D. tanzaniensis* (p. 189)

1. **Dyschoriste nagchana** (*Nees*) *Bennet* in Indian Forester 109: 220 (1983);
U.K.W.F. ed. 2: 269 (1994); Friis & Vollesen in Biol. Skr. 51(2): 442 (2005); Ensermu
in F.E.E. 5: 377 (2006). Type: India, Patna, *Herb. Wallich* (ex Herb. Hamilton) 2396
(K!, holo.)

Perennial herb with weakly developed rootstock, basal part of stems creeping and
rooting, apical part decumbent to ascending or erect; stems up to 65 cm long,
glabrous to sparsely puberulous (densest on edges). Leaves ovate to elliptic (rarely
narrowly or broadly so), largest 2–8(–10) × 0.7–2.6 cm, apex subacute to broadly
rounded (rarely retuse), glabrous (rarely with a few puberulent hairs along midrib)
beneath, with puberulous midrib above (very rarely with scattered capitate glands),
margin minutely puberulent-ciliate or glabrous. Flowers in condensed axillary cymes
with up to 12 flowers or some solitary, in lush specimens sometimes 2 clusters per
axil; peduncle and branches up to 1 mm long; outer bracts up to 1.5(–2.5) cm long;
inner bracts and bracteoles filiform or linear, up to 5 mm long. Calyx 4–8(–9 in
fruit) mm long, glabrous (rarely with a few hairs along veins); teeth usually distinctly
ciliate, 3–5 mm long, shorter than, same length or longer than tube. Corolla with
almost symmetrical limb, white or pale mauve, mauve, purple or lilac, tube usually
paler, puberulent, 6–9(–12) mm long of which the linear tube 2–3(–4) mm, the

throat 2–3(–5) mm and the limb 2–3 mm; lobes oblong, recurved. One pair of stamens sometimes reduced to staminodes; free part of filaments 1–2(–3) mm long, hairy at base, glabrous upwards; anthers ± 1 mm long, spurs less tham $^1/_4$ mm long. Capsule 7–10 mm long. Seed 2–3 mm long. Fig. 28, 11, p. 185.

UGANDA. Karamoja District: Kailekong, Sep. 1958, *Wilson* 517!; Busoga District: 40 km NNW of Jinja, 24 Sep. 1952, *Wood* 385!; Teso District: Serere, Dec. 1931, *Chandler* 168!
KENYA. Nairobi District: Eastleigh Section One, 4 Sep. 1971, *Mwangangi* 1765!; North Nyeri District: 7 km on Nyeri-Kiganjo road, Zawadi Estate, 2 June 1974, *Faden* 74/697!; North Kavirondo District: Busia, Amukura Shopping Centre, 12 Dec. 1968, *Kimani* 130!
TANZANIA. Arusha District: Sakila Usa Forestry Plantation, 13 Sep. 1971, *Richards* 27071!; Lushoto District: W Usambara Mts, Bumbuli, 11 May 1953, *Drummond & Hemsley* 2491!; Kondoa District: Kondoa-Irangi–Lake Iseria, Bubu River, 20 Aug. 1926, *Peter* 44515!; Zanzibar, Ndagaa, 27 Sep. 1963, *Faulkner* 3280!
DISTR. U 1–4; K 3–7; T 1–5; Z; widespread in tropical Africa and India
HAB. Grasslands and swamps, riverbanks, riverine forest, also secondary grassland and bushland, roadsides, old cultivations, plantations; (0–)800–1800 m

SYN. *Dipteracanthus nagchana* Nees in Wall., Pl. As. Rar. 3: 82 (1832)
　　　Dyschoriste depressa Nees in Wall., Pl. As. Rar. 3: 81 (1832) & in DC., Prodr. 11: 106 (1847); C.B.Clarke in F.T.A. 5: 72 (1899), p.p. & in Fl. Cap. 5: 16 (1901); Eyles in Trans. Roy. Soc. S. Afr. 5: 482 (1916); Ross, Fl. Natal: 323 (1973), *non Ruellia depressa* L. (1781). Types: India, *Herb. Wight* 22 (not seen); Ceylon, *Herb. Wallich* (ex Herb. Madras) 2379 (K!, syn.)
　　　Calophanes nagchana (Nees) Nees in DC., Prodr. 11: 109 (1847); T. Anderson in J.L.S. 9: 459 (1867); C.B. Clarke in Fl. Brit. Ind. 4: 410 (1884); Trimen, Handb. Fl. Ceylon 3: 294 (1895)
　　　C. perrottetii Nees in DC., Prodr. 11: 111 (1847); T. Anderson in J.L.S. 7: 23 (1863); Engler, Hochgebirgsfl. Trop. Afr.: 388 (1892). Types: Senegal, *Perrottet* in Herb. DC. 515 (not seen); Sudan, Fazogl, *Kotschy* 558 (K!, syn.; BM!, MO!, iso.)
　　　Linostylis fasciculiflora Sond. in Linnaea 23: 95 (1850); Solms-Laubert in Schweinf., Beitr. Fl. Aeth.: 112, 244 (1867). Type: Sudan, Fazogl, *Kotschy* 558 (BM!, K!, MO!, iso.)
　　　Calophanes depressa (Nees) T. Anderson in Thw., Enum. Pl. Zeyl.: 225 (1860) & in J.L.S. 9: 459 (1867)
　　　C. fasciculiflora (Sond.) Martelli in Nuov. Giorn. Bot. Ital. 20: 390 (1888)
　　　Dyschoriste perrottetii (Nees) Kuntze, Rev. Gen. 2: 486 (1891); C.B. Clarke in F.T.A. 5: 72 (1899); De Wildeman in Etud. Fl. Katanga 1: 144 (1903); Eyles in Trans. Roy. Soc. S. Afr. 5: 482 (1916); De Wildeman in Contrib. Fl. Katanga: 197 (1921); F.P.N.A. 2: 283 (1947); F.P.S. 3: 173 (1956); Heine in F.W.T.A., ed. 2, 2: 404 (1963); E.P.A.: 934 (1964); U.K.W.F.: 585 (1974); Champluvier in Fl. Rwanda 3: 454, fig. 141, 2 (1985); Iversen in Symb. Bot. Ups. 29, 3: 161 (1991); Lebrun & Stork, Enum. Pl. Afr. Trop. 4: 479 (1997)
　　　[*D. erecta* sensu Cuf., Enum. Pl. Aeth.: 933 (1964), *non* (Burm. f.) Kuntze (1891)]

NOTE. *D. nagchana* grows naturally in wet grassland, but is now often also found in secondary bushland and grassland, including lawns. In these habitats it often superficially resembles *D. radicans* but is normally easily distinguished by being almost glabrous, and by the short calyx (much shorter than fruits).
　　　Specimens intermediate with *D. radicans* have been found from these secondary habitats, mostly from the Kenyan highlands, and a limited amount of hybridisation cannot be ruled out. Examples are *Napier* 1520 from Nairobi, Kenya and *Forbes* 220 from Mbale, Uganda.
　　　Intermediates between *D. nagchana* and *D. multicaulis* are equally rare. They have glandular calyx or bracts, but the capsule is longer than the calyx. An example is *Graham* 2274 from Embu, Kenya.
　　　Hornby 1072 from Maboko Island in Lake Victoria has a very large corolla (12 mm long), but seems in all other respects to be typical *D. nagchana*.

2. **Dyschoriste multicaulis** (*A.Rich.*) *Kuntze*, Rev. Gen. 2: 486 (1891); Lindau in P.O.A. C: 367 (1895); C.B. Clarke in F.T.A. 5: 74 (1899); E.P.A.: 934 (1964); Lebrun & Stork, Enum. Pl. Afr. Trop. 4: 479 (1997); Ensermu in F.E.E. 5: 376 (2006). Type: Ethiopia, Scholoda, *Schimper* I.43 (K!, iso.)

Perennial herb with weakly developed rootstock, basal part of stems procumbent or trailing (rarely erect), rooting or not, up to 75 cm long, apical flowering part erect, up to 25 cm long, sparsely to densely puberulous to pubescent, often densest

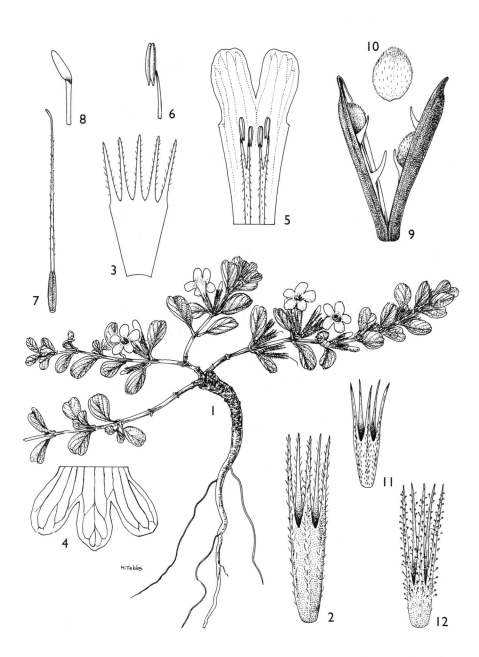

FIG. 28. *DYSCHORISTE RADICANS* — **1**, habit, × ²/₃; **2**, calyx, × 6; **3**, calyx, opened up, × 4; **4**, lower corolla lip, × 4; **5**, upper corolla lip and stamens, × 4; **6**, stamen, × 9; **7**, ovary and style, × 4; **8**, stigma, × 9; **9**, capsule, × 4; **10**, seed, × 6. *DYSCHORISTE NAGCHANA* — **11**, calyx, × 6. *DYSCHORISTE MULTICAULIS* — **12**, calyx, × 6. 1–10 from *Drummond & Hemsley* 4436, 11 from *Kimani* 130, 12 from *Faulkner* 4667. Drawn by Margaret Tebbs.

(or only) in a broad band. Leaves ovate to elliptic or broadly so, largest 2–6 ×
0.8–2.5 cm, apex subacute to rounded, beneath puberulous to pubescent or
sparsely so (sometimes only along midrib and veins) with many-celled glossy hairs,
above with midrib sericeous-puberulous and lamina with scattered to dense long
many-celled glossy hairs, towards tip of stem often with stalked capitate glands,
margin distinctly crisped-ciliate. Flowers in condensed axillary cymes with up to 7
flowers or some solitary, sometimes 2 cymes per axil, sometimes condensed towards
tip of stems; peduncle and branches up to 1 mm long, glabrous; outer bracts up to
1 cm long, usually with capitate glands; inner bracts and bracteoles linear, up to
7 mm long, often with glands. Calyx 6–10(–13 in fruit) mm long, glabrous or hairy
on veins and lobes, often with stalked capitate glands; teeth distinctly ciliate, 4–7
mm long, longer than tube. Corolla slightly 2-lipped, white to mauve or blue with
(? always) darker spots in throat, tube usually paler, puberulent and usually with
scattered capitate glands, 7–11 mm long of which the linear tube 3–4 mm, the
throat 2–4 mm and the limb 2–3 mm; lobes subequal, oblong, erect or recurved in
lower lip. Free part of filaments 2–3 mm long, with scattered hairs throughout;
anthers ± 1 mm long, spurs less than $^1\!/_4$ mm long, sometimes almost absent.
Capsule 7–9(–11) mm long, shorter than calyx (rarely same length). Seed 2–3 mm
long. Fig. 28, 12, p. 185.

UGANDA. Kigezi District: Kachwekano Farm, June 1949, *Purseglove* 2906!; Mbale District:
Busigu, near Busano, 21 Jan. 1969, *Lye* 1683!; Mengo District: Mawokota, Feb. 1905, *E.
Brown* 161!
KENYA. Trans-Nzoia District: S Cherengani, 27 Jan. 1958, *Symes* 262!; Kiambu District: Muguga,
6 Feb. 1956, *Verdcourt* 1442!; North Kavirondo District: Kakamega Forest Reserve, 10 July
1960, *Paulo* 523!
TANZANIA. Arusha District: Arusha National Park, Ziwa la Mbogo, 5 Nov. 1969, *Richards* 24599!;
Lushoto District: 6 km NE of Lushoto, Mkuzi, 21 Apr. 1953, *Drummond & Hemsley* 2163!;
Rungwe District: Mt Rungwe, 25 Oct. 1910, *Stolz* 374!
DISTR. **U** 1–4; **K** 3–6; **T** 1–3, 7; Ethiopia, Sudan, Congo-Kinshasa, Rwanda, Burundi
HAB. Montane grassland and bushland, swamps, forest clearings, lakeshores, riverbanks, also
secondary grassland, roadsides, old cultivations, lawns; (900–)1050–2500 m

SYN. *Ruellia multicaulis* A.Rich., Tent. Fl. Abyss. 2: 142 (1851)
  Calophanes multicaulis (A.Rich.) T. Anderson in J.L.S. 7: 23 (1863)
  [*Dyschoriste radicans* sensu C.B.Clarke in F.T.A. 5: 73 (1899) pro parte, *non sensu stricto*]
  D. clinopodioides Mildbr. in N.B.G.B. 9: 495 (1926); F.P.N.A. 2: 283 (1947); E.P.A.: 933
    (1964); Champluvier in Fl. Rwanda 3: 454, fig. 141, 3 (1985); Iversen in Symb. Bot. Ups.
    29, 3: 160 (1991); U.K.W.F. ed. 2: 269 (1994); Lebrun & Stork, Enum. Pl. Afr. Trop. 4:
    479 (1997); Friis & Vollesen in Biol. Skr. 51(2): 442 (2005). Type: Kenya, Nyeri, *R.E. &
    Th.C.E. Fries* 53 (B†, holo.; K!, iso.)
  D. broiloi Pic.Serm., Miss. Stud. Lago Tana 7, Ric. Bot. 1: 256 (1951); E.P.A.: 933 (1964).
    Type: Ethiopia, Gondar, *Pichi Sermolli* 2126 (FT, holo.)

NOTE. *D. multicaulis* mostly occurs at higher altitudes than *D. nagchana* and *D. radicans*, but has
  been spread secondarily into many disturbed man-made habitats and occasionally
  intermediate specimens are encountered. *Hancock* 2402 from Budadiri, Uganda is an example
  of an intermediate between *D. multicaulis* and *D. radicans*. See also note after *D. nagchana*.

3. **Dyschoriste radicans** *Nees* in DC., Prodr. 11: 106 (1847); Richard, Tent. Fl. Abyss.
2: 141 (1850); Lindau in P.O.A. C: 367 (1895) & in E. & P. Pf. IV, 3b: 302 (1895) &
in Ann. Ist. Bot. Roma 6: 68 (1896); C.B. Clarke in F.T.A. 5: 73 (1899) pro parte;
Mildbraed in N.B.G.B. 9: 495 (1926); F.P.N.A. 2: 283 (1947); E.P.A.: 935 (1964);
U.K.W.F.: 585 (1974); Wickens, Fl. Jebel Mara: 149 (1976); Champluvier in Fl.
Rwanda 3: 454, fig. 141, 1 (1985); Blundell, Wild Fl. E. Afr.: 390, pl. 605 (1987);
Iversen in Symb. Bot. Ups. 29(3): 161 (1991); U.K.W.F. ed. 2: 269 (1994); Lebrun &
Stork, Enum. Pl. Afr. Trop. 4: 480 (1997); Friis & Vollesen in Biol. Skr. 51(2): 442
(2005); Ensermu in F.E.E. 5: 376 (2006); Thulin in Fl. Somalia 3: 388 (2006). Type
not indicated

Perennial (rarely shrubby) herb with well-developed rootstock, often mat-forming; stems procumbent or trailing, rooting or not (rarely erect in pyrophytic forms), up to 80 cm long, subglabrous to densely whitish pubescent, often densest (or only) in a broad band. Leaves obovate (more rarely elliptic or broadly so), largest 1.2–5(–6) × 0.6–2(–2.8) cm, apex broadly rounded to retuse with a small triangular tip, on both sides glabrous to sparsely pubescent with ± curled hairs, often only along midrib and veins with glabrous lamina (more rarely uniformly pubescent or densely so), margin distinctly crisped-ciliate (rarely not). Flowers in condensed axillary cymes with 1–3 flowers, not condensed towards tip of stems; peduncle 3(–12) mm long, branches up to 1 mm long, puberulous; outer bracts up to 1.2 cm long; inner bracts and bracteoles linear, up to 7 mm long, often with glands. Calyx 8–12(–18 in fruit) mm long, glabrous to puberulous, often with glabrous tube or only hairy on veins; teeth distinctly ciliate (rarely not), 5–10 mm long, longer than tube or same length. Corolla with almost symmetrical limb (tube sometimes slightly curved), white, pale blue to blue or mauve to violet, tube usually paler, puberulent, 10–19 mm long of which the linear tube 4–8 mm, the throat 3–7 mm and the limb (2–)3–5 mm; lobes oblong, erect or recurved in lower lip. One pair of stamens sometimes reduced to staminodes (see fig. 141, 1 in Fl. Rwanda 3: 454 (1985)); free part of filaments 2–4 mm long, with scattered hairs throughout or glabrous upwards; anthers 1–2 mm long, muticous or with minute spurs. Capsule (8–)10–14 mm long, shorter than calyx (rarely same length). Seed 2–3 mm long. Fig. 28, 1–10, p. 185.

UGANDA. Karamoja District: near Moroto, Sep. 1955, *Wilson* 194!; Ankole District: Ruizi River, 18 Nov. 1950, *Jarrett* 138!; Mengo District: 2 km N of Wakyato, 2 Aug. 1956, *Langdale-Brown* 2293!
KENYA. Northern Frontier District: Mt Marsabit, Gof Redo, 2 June 1960, *Oteke* 9!; Ravine District: Eldama Ravine to Maji Mazuri, 26 Sep. 1953, *Drummond & Hemsley* 4436!; Kiambu District: Nairobi–Thika road, 19 Feb. 1953, *Drummond & Hemsley* 1235!
TANZANIA. Musoma/Maswa District: Serengeti Plains, Kisawira, 27 Apr. 1965, *Richards* 20304!; Masai District: 11 km on Loliondo-Narok road, 8 Nov. 1964, *Gillett* 16309!; Arusha District: Arusha National Park, Mt Meru, Tulusia Gorge, 6 Jan. 1970, *Richards* 25040!
DISTR. U 1–4; K 1–7; T 1–3, 5, 7; Congo-Kinshasa, Rwanda, Burundi, Sudan, Somalia, Ethiopia; Yemen
HAB. Montane bushland and grassland, *Acacia* and *Combretum* wooded grassland and bushland, grassland on black cotton soil (sometimes fireswept), secondary grassland and bushland, eroded areas, roadsides, old cultivations; (900–)1100–2500 m

SYN. *Ruellia radicans* Hochst. in Flora 24, Intell.-Bl. 1: 24 (1841), *nom. nud.*
    Dyschoriste radicans Nees var. *minor* Nees in DC., Prodr. 11: 106 (1847); Fiori, Miss. Biol. Boran. 4: 215 (1939). Type: Ethiopia, near Adua, *Schimper* I.17 (K!, P!, iso.)
    D. radicans Nees var. *diffusa* Nees in DC., Prodr. 11: 106 (1847). Type: Cult. in Hort. Bratislava from seeds from Ethiopia (not seen)
    Calophanes radicans (Nees) T.Anderson in J.L.S. 7: 23 (1863); Oliver in Trans. Linn. Soc. 29: 126 (1875); Engler, Hochgebirgsfl. Trop. Afr.: 388 (1892)
    [*Dyschoriste depressa* sensu C.B.Clarke in F.T.A. 5: 72 (1899) pro parte, *non* Nees (1832)]

NOTE. Usually easily distinguished from *D. multicaulis* and *D. nagchana* by its obovate leaves and larger flowers. But in areas where the three meet in man-made secondary habitats intermediate forms occur. See also notes after the preceeding species.

4. **Dyschoriste mutica** (*S.Moore*) *C.B.Clarke* in F.T.A. 5: 73 (1899); Gilli in Ann. Naturhist. Mus. Wien 77: 50 (1973); Brummitt in K.B. 38: 134 (1983); Lebrun & Stork, Enum. Pl. Afr. Trop. 4: 479 (1997). Types: Angola, Pungo Andongo, *Welwitsch* 5089 (BM!, syn.) & *Welwitsch* 5094 (BM!, syn.; K!, P!, iso.)

Perennial herb with solitary or a few erect unbranched or sparsely branched stems from a slender creeping rootstock; stems up to 45 cm long, antrorsely sericeous-puberulent to -puberulous when young. Leaves with strongly raised rib-like veins; lamina elliptic or narrowly so, largest 2.8–7 × 0.8–1.6 cm, apex subacute to rounded, glabrous to sparsely puberulous (densest along veins) when mature and finely ciliate. Flowers in condensed 3–7-flowered axillary cymes which are congested towards end of branches and ± concealed under leaves and bracts; peduncle and branches up to 2 mm long; outer bracts elliptic to obovate, up to 2 cm long; inner bracts and bracteoles linear to oblanceolate, up to 1.5 cm long. Calyx 8–15 mm long, glabrous to sparsely puberulous; teeth 4–8 mm long, longer than or same length as tube, finely ciliate. Corolla indistinctly 2-lipped, pure white or with pale yellow throat, sparsely puberulent, 11–18 mm long of which the linear tube 4–8 mm, the throat 3–6 mm and the limb 2–4 mm; lobes subequal (fused higher up in upper lip and middle in lower lip slightly wider), recurved in mature flowers. Free part of filaments 2–4 mm long, hairy in basal part, glabrous upwards; anthers ± 1 mm long, muticous at base but not spurred. Capsule 8–12 mm long. Seed ± 2.5 mm long.

TANZANIA. Ufipa District: 10 km on Tatanda–Mbala road, 24 Apr. 2006, *Bidgood et al.* 5649!; Songea District: Kwamponjore valley, 14 March 1956, *Milne-Redhead & Taylor* 9161! & 40 km on Songea-Njombe road, 22 March 1991, *Bidgood et al.* 2092!
DISTR. **T** 4, 8; Congo-Kinshasa, Angola, Zambia, Malawi, Mozambique
HAB. *Brachystegia* and *Brachystegia-Uapaca* woodland on red loam; 850–1700 m

SYN. *Calophanes radicans* (Nees) T.Anderson var. *mutica* S.Moore in J.B. 18: 198 (1880)
    Dyschoriste petalidioides S.Moore in J.B. 49: 294 (1911). Type: Angola, between Cului River and Kubanque, Fort Maria Pia, *Gossweiler* 2917 (BM!, holo.)

NOTE. A very distinctive species easily recognised by its small white flowers with muticous anthers, by the condensed inflorescence more or less covered by leaves and bracts, and by the strongly raised rib-like leaf-veins. *D. mutica* sensu C.B.Clarke in Fl. Cap. 5: 16 (1901) is a quite different species.

5. **Dyschoriste albiflora** *Lindau* in E.J. 57: 21 (1920); Lebrun & Stork, Enum. Pl. Afr. Trop. 4: 479 (1997). Type: Tanzania, Rungwe District: Kyimbila, Bulambya, *Stolz* 1938 (B†, holo.; BM!, C!, K!, P!, iso.)

Shrubby perennial herb, subshrub or shrub; stems erect to decumbent, sparsely to freely branched, up to 1(–1.5) m tall, subglabrous to puberulous or sericeous-puberulous when young. Leaves ovate to elliptic, largest 4.3–10 × 1.7–4.5 cm, apex acute to subacuminate (rarely rounded), subglabrous to sparsely puberulous, densest along veins with lamina often glabrous. Flowers in condensed axillary 1–12-flowered cymes, sometimes 2 cymes per axil (very rarely in elongated dichasial cymes with up to 15 flowers and with peduncle and branches up to 1.5(–2) cm long); peduncle and branches up to 4 mm long, indumentum as branches; outer bracts lanceolate to elliptic, up to 2 cm long; inner bracts and bracteoles linear to lanceolate, up to 1.2 cm long. Calyx subglabrous to puberulous, densest on lobes and veins, tube often glabrous, hairs up to 1 mm long, (very rarely also with capitate glands), 10–16(–19 in fruit) mm long; teeth 4–8(–10 in fruit) mm long, shorter than, same length as or longer than tube. Corolla slightly 2-lipped, pure white (very rarely with faint mauve markings on lower lip), puberulent or sparsely so (rarely with intermixed capitate glands), 14–28 mm long of which the linear tube (4–)7–11 mm, the throat 4–10 mm and the limb 4–8 mm; throat from slightly longer to up to 2 mm shorter than limb; lobes oblong, erect or reflexed in lower lip. Free part of filaments 3–7 mm long, with short stubby hairs near base, glabrous upwards; anthers 1–2 mm long, spurs ± ¹/₂ mm long. Capsule 11–13 mm long. Seed 2–3 mm long.

TANZANIA. Ufipa District: Kanda Hills, 10 March 1959, *Richards* 11114!; Njombe District: N slopes of Poroto Mts above Chimala, 24 March 1991, *Bidgood, et al.* 2108!; Mbeya District: Songwe Valley 2 km N of Mbeya–Tunduma road, 25 March 1988, *Bidgood et al.* 689!
DISTR. **T** 4, 6–8; Zambia, Malawi, Mozambique
HAB. *Brachystegia* woodland, often on rocky slopes or rocky outcrops but also on sandy to loamy soil, *Combretum* bushland on rocky limestone slopes, riverine thicket and woodland; (150–)1200–2100 m

SYN. [*Dyschoriste perrottetii* sensu Gilli in Ann. Naturhist. Mus. Wien 77: 50 (1973), *non* (Nees) Kuntze (1891)]

NOTE. A single collection (*Richards* 8753) has a few capitate glands on the calyx, but otherwise this species is consistently separated from *D. pilifera* and *D.tanzaniensis* by being eglandular.

6. **Dyschoriste pilifera** *Hutch.*, Botanist in S. Afr.: 484 (1946); Lebrun & Stork, Enum. Pl. Afr. Trop. 4: 479 (1997). Type: Zambia, 78 km NE of Livingstone, *Hutchinson & Gillett* 3501 (K!, holo.; BM!, K!, iso.)

Perennial or shrubby herb to 0.5 m tall with several erect to decumbent sparsely branched stems from a woody rootstock (rarely a shrub); stems glandular-puberulent to -puberulous or densely so, also with scattered to dense pilose hairs up to 3 mm long. Leaves ovate to elliptic (rarely narrowly so), largest 3–7 × 1–3.3 cm, apex acute to subacuminate, subglabrous to puberulent or puberulous, with stalked capitate glands when young, often also with scattered pilose hairs. Flowers in condensed axillary 1–3(–12)-flowered cymes, more rarely in elongated dichasial cymes; peduncles up to 2(–13) mm long, branches up to 7(–20 in open dichasia) mm long; outer bracts lanceolate to elliptic, up to 1.8 cm long or as inner bracts and bracteoles: linear to lanceolate, up to 8 mm long, with scattered to dense capitate glands. Calyx glandular-puberulent to -puberulous or densely so and with scattered to dense long glossy pilose hairs some or all of which over 1 mm long, 12–18(–22 in fruit) mm long; teeth 5–9(–10 in fruit) mm long, shorter than tube (rarely same length). Corolla slightly 2-lipped, pure white with pale yellow throat, puberulent and with scattered capitate glands, 20–26 mm long of which the linear tube 7–11 mm, the throat 6–10 mm and the limb 4–7 mm, throat longer than or same length as limb; lobes oblong to ovate, erect or spreading in lower lip. Free part of filaments 3–6 mm long, hairy at base, glabrous at apex; anthers 1.5–2 mm long, spurs ± ½ mm long. Capsule 11–13 mm long. Seed ± 2.5 mm long.

TANZANIA. Dodoma District: 6 km on Manyoni–Kilimatinde road, 14 Apr. 1988, *Bidgood et al.* 1089!; Iringa District: Ruaha National Park, 10 km on Msembe–Nyamakuru Rapids track, 6 Apr. 1970, *Greenway & Kanuri* 14275 & 55 km on Mafinga–Madibira road, 27 Jan. 1991, *Bidgood et al.* 1300!
DISTR. **T** 5, 7; Zambia, Zimbabwe
HAB. A wide range of woodland, bushland and thicket on an equally wide range of soils from clayey flood-plains to rocky slopes; 850–1400 m

SYN. [*Dyschoriste trichocalyx* sensu Björnstad in Serengeti Res. Inst. Publ. 215: 25 (1976), *non* (Oliv.) Lindau (1895)]

NOTE. Closely related to *D. albiflora* with which it shares the white flowers. It differs most conspicuously in the usually dense sticky glandular indumentum with long pilose hairs intermixed. The sparse material from Zambia and Zimbabwe has a considerable shorter calyx (8–13 mm long). But more material is needed to decide whether two infraspecific taxa can be recognised.

7. **Dyschoriste tanzaniensis** *Vollesen* **sp. nov.** a *D. pilifera* pilis caulium usque 1 mm tantum longis (nec longioribus aliquibus plus quam 2 mm longis), calyce minore tempora florendi 6–13 mm tantum (nec 12–18 mm) longo et corolla minore 14–18(–22) mm (nec 20–26 mm) longa differt. Type: Tanzania, Tabora District: 22 km on Tabora–Sikonge road, *Bidgood et al.* 5880 (K!, holo.; BR!, CAS!, DSM!, K!, NHT!, iso.)

Perennial or shrubby perennial herb to 0.5(–1) m tall with several erect to decumbent branched stems from a woody rootstock; stems sparsely to densely puberulous or sericeous-puberulous with a mixture of glands and non-glandular hairs up to 0.5 mm long, no long pilose hairs (very rarely with pilose hairs to 2 mm long).

Leaves ovate to elliptic, largest 2–6.5 × 0.7–3 cm, apex subacute to subacuminate, subglabrous to puberulous, sometimes with stalked capitate glands when young, no long pilose hairs. Flowers in condensed axillary 1–7-flowered cymes or solitary towards tip of stems; peduncles and branches 1–2(–8) mm long, puberulous; outer bracts ovate, up to 1.2 cm long or as inner bracts and bracteoles: filiform to linear, up to 5 mm long, puberulent and with capitate glands. Calyx glandular-puberulent or densely so, sometimes with non-glandular hairs intermixed or mostly non-glandular, without (rarely with a few) pilose hairs over 1 mm long, 6–13(–17 in fruit) mm long; teeth 3–6 mm long, shorter than tube or same length. Corolla slightly 2-lipped, white with pale yellow throat or pale blue to blue or mauve sometimes with darker markings in throat, sparsely puberulent and with scattered capitate glands, 14–18(–22) mm long of which the linear tube 5–9 mm, the throat 4–5(–8) mm long and 4–5(–6) mm across apically, the limb 4–5(–8) mm long, throat longer than or same length as limb; lobes oblong or slightly obovate, erect or reflexed in lower lip. Free part of filaments 3–5 mm long, hairy at base, glabrous at apex; anthers 1–1.5 mm long, spurs ± ¼ mm long. Capsule 10–13 mm long. Seed ± 2 × 1.5 mm.

TANZANIA. Mwanza District: Mbarika, 15 Apr. 1953, *Tanner* 1380!; Tabora District: 22 km on Tabora–Sikonge road, 11 May 2006, *Bidgood et al.* 5880!; Dodoma District: 6 km on Manyoni–Kilimatinde road, 14 Apr. 1988, *Bidgood et al.* 1083!
DISTR. T 1, 2, 4, 5, 7; not known elsewhere
HAB. *Brachystegia* woodland on sandy to loamy soil and rocky slopes, *Brachystegia microphylla-Bussea-Commiphora* bushland, *Acacia-Commiphora* thicket on rocky hills, thicket clumps in *Brachystegia* woodland, riverine forest and thicket, old cultivations; 1000–1500(–1700) m

SYN. *Dyschoriste sp.* of T.T.C.L.: 8 (1949)

8. **Dyschoriste subquadrangularis** (*Lindau*) *C.B.Clarke* in F.T.A. 5: 79 (1899); Iversen in Symb. Bot. Ups. 29, 3: 161 (1991). Type: Tanzania, Lushoto District: Usambara Mts, Kongeni, *Buchwald* 94 (B†, holo.; BM!, BR!, COI!, K!, iso.)

Much branched shrubby herb or subshrub to 1 m tall, with several stems from a woody rootstock; stems puberulent to sericeous-puberulent or finely sericeous or sparsely so (rarely with scattered capitate glands). Leaves elliptic to obovate, largest 1.5–5 × 0.6–1.9 cm, apex rounded to truncate with a small triangular tooth, subglabrous to densely and uniformly puberulent (or denser along veins). Flowers in condensed to somewhat lax 1–7-flowered axillary cymes or solitary towards tip of branches; peduncle and branches 1–4(–6) mm long; outer bracts elliptic, up to 1.5 cm long; inner bracts and bracteoles filiform to linear, up to 5 mm long, indumentum as leaves. Calyx subglabrous to puberulent or sericeous-puberulent, often with short (shorter than hairs) capitate glands intermixed, 6–10(–14 in fruit) mm long; teeth 2–4(–6) mm long, shorter than tube (rarely same length). Corolla slightly 2-lipped, bright blue to deep purple, with darker spots and lines at base of lower lip, tube yellowish, puberulent or sparsely so and with scattered capitate glands, 19–27 mm long of which the linear tube 7–11 mm, the throat 7–9 mm long and 6–9 mm across apically, the limb 4–7 mm long, limb shorter than throat (rarely same length); lobes oblong-obovate, erect or deflexed in lower lip. Free part of filaments 3–5 mm long, hairy near base, glabrous upwards; anthers 1–2 mm long, spurs less than ¼ mm long. Capsule 11–13 mm long. Seed ± 2.5 mm long.

TANZANIA. Lushoto District: Korogwe to Mombo, Makauni, 9 March 1956, *Greenway* 8977!; Handeni District: 15 km on Korogwe-Handeni road, 19 Nov. 1955, *Milne-Redhead & Taylor* 7334! & 32 km on Korogwe-Handeni road, Sindeni, 12 Sep. 1933, *Burtt* 4855!
DISTR. T 3; not known elsewhere
HAB. *Combretum* wooded grassland, *Acacia-Combretum-Dobera* thicket clumps in tall grassland, semi-deciduous lowland forest, dry riverine forest, on orange loamy to dark grey loamy to clayey soil; 300–600(?–1000) m

SYN. *Hygrophila subquadrangularis* Lindau in E.J. 24: 314 (1897)

9. **Dyschoriste nyassica** *Gilli* in Ann. Naturhist. Mus. Wien 77: 50 (1973). Type: Tanzania, Njombe District: NE shore of Lake Malawi, Lumbila, *Gilli* 522 (W, holo.; K!, iso.)

Shrubby herb or shrub; stems erect or decumbent (then also rooting), up to 75 cm long, puberulent or sparsely so, with (rarely without) long pilose hairs. Leaves ovate to elliptic, largest 5.5–8 × 2.3–4 cm, apex acute or drawn out into a short obtuse tip, glabrous or sericeous-puberulous along veins when mature or with scattered long pilose hairs. Flowers in condensed to lax 1–5-flowered axillary cymes; peduncle and branches 1–4 mm long or up to 25 mm in plants with lax cymes, puberulent; outer bracts elliptic, up to 8 mm long, or as inner bracts and bracteoles: linear, up to 5 mm long, with long pilose hairs. Calyx subglabrous to sericeous-puberulent, with scattered long pilose hairs and with capitate glands, 10–14 mm long; teeth 4–6 mm long, shorter than tube. Corolla slightly 2-lipped, pale lilac or purple, sparsely puberulent and with scattered capitate glands, 25–30 mm long of which the linear tube 12–15 mm, the throat 7–8 mm and the limb 7–8 mm; lobes oblong, erect or deflexed in lower lip. Free part of filaments 5–7 mm long, hairy near base, glabrous upwards; anthers ± 2 mm long, spurs ± 0.5 mm long. Capsule (immature) 10–12 mm long. Seed not seen.

TANZANIA. Njombe District: NE shore of Lake Malawi, Lumbila, 8 Aug. 1958, *Gilli* 522!
DISTR. **T** 7; Malawi
HAB. Riverine forest; ± 525 m

NOTE. A very distinct species known from only three collections around the northern end of Lake Malawi. Distinguished by the large purple flowers combined with a sparse eglandular (apart from calyx and corolla) indumentum with scattered long pilose hairs.

10. **Dyschoriste sp. A**

Brittle-stemmed shrub; young branches finely sericeous and with scattered stalked capitate glands. Leaves ovate, largest ± 4 × 2 cm, apex acute to subacute, sparsely sericeous along veins otherwise glabrous, towards tip of stem with scattered capitate glands. Flowers in open dichasial 3–7-flowered cymes (one branch often more developed after first branching); peduncle and branches 1–1.5 cm long, sparsely and finely sericeous and with scattered capitate glands; bracts and bracteoles linear-lanceolate, up to 1 cm long, with similar indumentum; pedicels 0.5–1 mm long. Calyx finely sericeous and with scattered stalked capitate glands, 9–11 mm long; teeth narrowly triangular, dorsal slightly longer, cuspidate, same length as tube. Corolla slightly 2-lipped, reddish purple, puberulent, ± 22 mm long of which the linear tube + 9 mm, the throat ± 6 mm and the limb ± 7 mm; lobes oblong; anthers ± 2 mm long, spurs ± ¹⁄₃ mm long. Capsule and seed not seen.

TANZANIA. Iringa District: 65 km on Iringa–Morogoro road, Kitonga Gorge, Apr. 1966, *Procter* 3327!
DISTR. **T** 7; not known elsewhere
HAB. Dry *Acacia* bushland; no altitude given

NOTE. Known only from this collection. I have personally looked for this species several times without success and it is obviously rare. It is immediately recognised among all other species of *Dyschoriste* by its open dichasial inflorescences.

11. **Dyschoriste trichocalyx** (*Oliv.*) *Lindau* in P.O.A. C: 367 (1895); C.B. Clarke in F.T.A. 5: 75 (1899); De Wildeman, Notes Fl. Katanga 2: 71 (1913) & Contrib. Fl. Katanga: 197 (1921); T.T.C.L.: 8 (1949); U.K.W.F.: 585 (1974); Champluvier in Fl. Rwanda 3: 454, fig. 141, 4 (1985); U.K.W.F. ed. 2: 269 (1994); Lebrun & Stork, Enum. Pl. Afr. Trop. 4: 480 (1997). Type: Tanzania, Bukoba District: Karagwe, *Speke & Grant* 404 (K!, holo.)

Perennial or shrubby herb with several unbranched or sparsely branched stems from woody rootstock; stems erect (rarely scrambling or decumbent), to 1.5 m tall, sparsely to densely uniformly puberulous (rarely sericeous), in subsp. *verticillaris* usually also with stalked capitate glands and long pilose hairs. Mature leaves ovate to elliptic or narrowly so (rarely obovate), largest 3–10.5 × 0.9–3.2(–4) cm, apex subacuminate to rounded, beneath uniformly puberulous or denser along veins or along veins only, above uniformly puberulous or sparsely so. Flowers in condensed axillary 3–15-flowered cymes, usually 2 cymes per axil or solitary towards apex, cymes often congested towards tip of stems; peduncle and branches up to 3(–5) mm long, indumentum as stems; outer bracts elliptic or narrowly so, up to 2 cm long, sometimes with capitate glands; inner bracts and bracteoles linear to lanceolate, up to 1 cm long. Calyx sparsely to densely puberulous with non-glandular hairs (subsp. *trichocalyx*) or viscid glandular-puberulous (subsp. *verticillaris*), also with or without long pilose non-glandular hairs, (8–)10–19(–21 in fruit) mm long; teeth (4–)5–12(–14 in fruit) mm long, same length or longer than tube. Corolla distinctly 2-lipped, pale mauve to mauve, pale purple to purple, blue to bluish purple (rarely white), usually with darker lines or spots on lower lip and in throat, palate distinctly bullate between veins, puberulent and with scattered capitate glands, 14–29(–37) mm long of which the linear tube 5–12(–14) mm, the throat 3–9(–12) mm and the limb 4–10(–13) mm, throat same length or 1–2 mm shorter than limb; lobes oblong, erect in upper lip, deflexed in lower. Free part of filaments 2–8 mm long, hairy near base, glabrous upwards or hairy to apex; anthers pale violet or mauve, 1–3 mm long, spurs ± $\frac{1}{4}$–$\frac{1}{2}$ mm long often with 1–2 small subsidiary spurs or main spurs branched. Capsule 8–12 mm long. Seed 2–3 mm long. Fig. 29, p. 193.

11a. subsp. **trichocalyx**

Calyx sparsely to densely puberulous with non-glandular hairs or with scattered non-viscid glands, usually without long pilose hairs; corolla 14–18 mm long.

UGANDA. Kigezi District: Rukungiri, 24 Apr. 1941, *Thomas* 3827!; Ankole District: Igara, March 1939, *Purseglove* 634!; Mengo District: Singo, 1915, *Snowden* 176!
KENYA. South Kavirondo District: Gori Kisii, Sep. 1933, *Napier* 2905! & 9 km on Migori–Kisii road, 5 Nov. 1980, *Gilbert* 5988!
TANZANIA. Bukoba District: Bugene, Oct. 1931, *Haarer* 2289!; Ngara District: Rusumo Falls, March 1960, *Tanner* 4766! & Bugufi, Jan. 1961, *Tanner* 5569!
DISTR. **U** 2, 4; **K** 5; **T** 1, 4; Congo-Kinshasa, Rwanda, Burundi
HAB. Grassland, bushland and wooded grassland; 1350–1800 m

SYN. *Calophanes trichocalyx* Oliv. in Trans. Linn. Soc. 29: 126, t. 126 (1875)
    Dyschoriste decora S.Moore in J.B. 51: 212 (1913); De Wildeman, Etud. Fl. Katanga 2: 145 (1913); Lebrun & Stork, Enum. Pl. Afr. Trop. 4: 479 (1997). Type: Congo-Kinshasa, Kasonema, *Kässner* 2563 (BM!, holo.; K!, iso.)
    Hygrophila quadrangularis De Wild. in F.R. 13: 194 (1914) & Notes Fl. Katanga 4: 80 (1914) & Contrib. Fl. Katanga: 196 (1921) *non* (Klotzsch) Lindau (1895). Type: Congo-Kinshasa, Katanga, Lubumbashi [Elisabethville], *Homblé* 116 (BR!, holo.; BR!, iso.)
    [*Dyschoriste verticillaris* sensu Lindau in Z.A.E.: 296 (1914), *non* C.B.Clarke (1899)]

11b. subsp. **verticillaris** (C.B.Clarke) *Vollesen* **comb. nov.** Type: Malawi, Shire Highlands, *Buchanan* 74 (K!, lecto., selected here)

Calyx usually viscid glandular-puberulous or densely so and usually with long non-glandular pilose hairs, if non-viscid then corolla over 20 mm long; corolla 17–29(–37) mm long.

TANZANIA. Ufipa District: Mbisi Forest, 11 Aug. 1960, *Richards* 13082!; Morogoro District: road to Morningside, 25 Oct. 1934, *Bruce* 33!; Songea District: 3 km NE of Kigonsera, 13 Apr. 1956, *Milne-Redhead & Taylor* 9599!
DISTR. **T** 4–8; Congo-Kinshasa, Zambia, Malawi, Mozambique, Zimbabwe
HAB. *Brachystegia* woodland, *Combretum* and *Acacia* bushland, grassland, dry riverine thicket; 175–2100 m

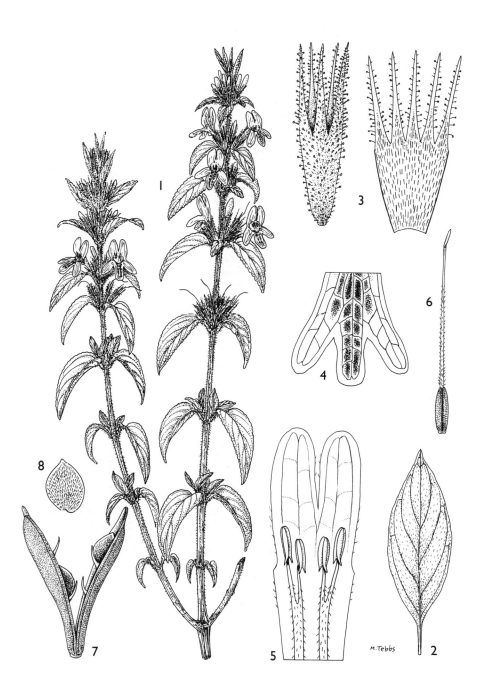

FIG. 29. *DYSCHORISTE TRICHOCALYX* subsp. *VERTICILLARIS* — **1** habit, × ²/₃; **2** vegetative leaf, × ²/₃; **3** calyx, whole and opened up, × 4; **4** lower corolla lip, × 4; **5** upper corolla lip and stamens, × 6; **6** style and stigma, × 4; **7** capsule, × 4; **8** seed, × 6. 1 & 3 from *Bidgood et al.* 3132, 2 from *Milne-Redhead & Taylor* 11089, 4–6 from *Richards* 13082, 7 & 8 from *Hornby* 913. Drawn by Margaret Tebbs

SYN. *Calophanes verticillaris* Oliv. in Trans. Linn. Soc. 29: 126 (1875), *nom. nud.*

Nomaphila glandulosa Klotzsch in Peters, Reise Mossamb. 2: 198 (1861), *non Dyschoriste glandulosa* (Nees) Kuntze (1891). Type: Mozambique, Rios de Sena, *Peters* s.n. (B†, holo.)

Hygrophila glandulosa (Klotzsch) Lindau in P.O.A. C: 367 (1895)

Mellera parvifolia Lindau in E.J. 24: 314 (1897); C.B. Clarke in F.T.A. 5: 52 (1899); S. Moore in J.B. 67: 229 (1929). Type: Congo-Kinshasa, Lufira River, *Descamps* s.n. (BR!, holo.)

Dyschoriste verticillaris C.B.Clarke in F.T.A. 5: 75 (1899); De Wildeman, Etud. Fl. Katanga 1: 145 (1903) & Contrib. Fl. Katanga: 197 (1921); T.T.C.L.: 8 (1949); Milne-Redhead in Mem. N.Y. Bot. Gard. 9: 22 (1954); Binns, Checklist Herb. Fl. Malawi: 13 (1968); Vollesen in Opera Bot. 59: 80 (1980); Lebrun & Stork, Enum. Pl. Afr. Trop. 4: 480 (1997)

D. verdickii De Wild., Etud. Fl. Katanga 1: 144 (1903). Type: Congo-Kinshasa, Shaba, Lukafu, *Verdick* 375 (BR!, holo.)

Hygrophila homblei De Wild. in F.R. 13: 193 (1914) & Notes Fl. Katanga 4: 79 (1914) & Contrib. Fl. Katanga: 195 (1921). Type: Congo-Kinshasa, Katanga, Kapiri Valley, *Homblé* 1193 (BR!, holo.; BR!, iso.)

H. ringoetii De Wild. in F.R. 13:193 (1914) & Notes Fl. Katanga 4: 81 (1914) & Contrib. Fl. Katanga: 196 (1921).Type: Congo-Kinshasa, Katanga, Nieuwdorp, *Ringoet* 5 (BR!, holo.; BR!, iso.)

H. kyimbilensis Lindau in E.J. 57: 20 (1920); T.T.C.L.: 8 (1949). Type: Tanzania, Rungwe District: Kyimbila, Lumkali River, *Stolz* 2099 (B†, holo.; BM!, C!, P!, iso.)

Dyschoriste kyimbilensis (Lindau) S.Moore in J. Bot. 67: 229 (1929); Lebrun & Stork, Enum. Pl. Afr. Trop. 4: 479 (1997)

Lepidagathis glaucifolia Gilli in Ann. Naturhist. Mus. Wien 77: 52 (1973). Type: Tanzania, NE shore of Lake Malawi, Lupingu, *Gilli* 533 (W, holo.; K!, iso.)

[*Dyschoriste fischeri* sensu Vollesen in Opera Bot. 59: 80 (1980), *non* Lindau (1894)]

[*D. tanganyikensis* sensu Vollesen in Opera Bot. 59: 80 (1980), *non* C.B.Clarke (1899)]

D. sp. aff. verticillaris sensu Vollesen in Opera Bot. 59: 80 (1980)

12. **Dyschoriste tanganyikensis** *C.B.Clarke* in F.T.A. 5: 77 (1899); Binns, Checklist Herb. Fl. Malawi: 13 (1968); Richards & Morony, Checklist Mbala Distr.: 229 (1969); Björnstad in Serengeti Res. Inst., Publ. 215: 25 (1976); Lebrun & Stork, Enum. Pl. Afr. Trop. 4: 480 (1997). Type: Zambia, Urungu, Fwambo, *Carson* s.n. (K!, holo.)

Viscid pyrophytic herb with several usually unbranched stems from large woody rootstock; stems erect (becoming decumbent in unburnt plants), 1–25 cm tall (up to 70 cm in Southern Africa), sparsely to densely glandular-puberulent and with non-glandular hairs (less than 0.5 mm long) intermixed. Leaves elliptic to obovate or narrowly so, 1–3 × 0.3–1 cm in flowering pyrophytic plants, up to 6 × 2 cm in mature plants, apex subacuminate to rounded with central triangular tooth, young leaves puberulent or sparsely so with capitate glands and non-glandular hairs or puberulent along veins only and without glands, mature leaves glabrous or with scattered hairs along midrib and on margins. Flowers solitary or in 3–7-flowered condensed axillary cymes (rarely two per axil); peduncle and branches up to 4 mm long, indumentum as stems or of glands only; outer bracts narrowly elliptic, up to 1(–2.5) cm long, or as inner bracts and bracteoles: linear-lanceolate, up to 8 mm long, glandular-puberulent. Calyx sparsely to densely viscid glandular-puberulent, often with puberulent (less than 0.5 mm long) non-glandular hairs intermixed, 9–13(–15 in fruit) mm long; teeth 4–7(–8 in fruit) mm long, shorter than, same length or longer than tube. Corolla distinctly 2-lipped, white with pale mauve to purple spots and lines on basal part of lower lip to mauve with purple spots, palate not or slightly bullate between veins, sparsely puberulent and with capitate glands, (19–)23–30 mm long of which the linear tube (7–)9–13 mm, the throat (5–)7–10 mm and the limb 7–9 mm, throat from 1–2 mm shorter to slightly longer than limb; lobes oblong, erect in upper lip, spreading or deflexed in lower. Free part of filaments 4–7 mm long, hairy throughout or glabrous upwards; anthers 1.5–2.5 mm long, spur ¼–½ mm long. Capsule 11–13 mm long. Seed ± 3 mm long.

TANZANIA. Mwanza District: Mbarika, Ihelele, 11 Sep. 1952, *Tanner* 988!; Ufipa District: Mbisi Forest, 13 Aug. 1960, *Richards* 13095!; Iringa District: Ruaha National Park, Naganjwa, 26 Aug. 1969, *Greenway & Kanuri* 13792!

DISTR. **T** 1, 4, 7; Congo-Kinshasa, Zambia, Malawi

HAB. Burnt *Brachystegia-Julbernardia* woodland, burnt montane grassland, on sandy to loamy, clayey or gritty soil, appearing shortly after burning; 1200–2400 m

SYN. *Hygrophila microthamnia* Lindau in E.J. 30: 408 (1901); De Wildeman, Etud. Fl. Katanga: 144 (1913) & Notes Fl. Katanga 2: 71 (1913) & Contrib. Fl. Katanga: 196 (1921). Type: Tanzania, Mbeya District: Unyika, Fingano, *Goetze* 1375 (B†, holo.; BR!, P!, iso.)

NOTE. *D. tanganyikensis* is closely related to *D. trichocalyx* subsp. *verticillaris.* The main differences are the pyrophytic habit, and the very fine puberulent indumentum without long non-glandular hairs on any part.

13. **Dyschoriste sp. B**

Shrub; young branches conspicuously ridged, with scattered subsessile capitate glands, otherwise glabrous. Leaves drying greyish green; lamina broadly obovate, largest ± 5 × 3.5 cm, apex truncate to retuse with a recurved apical tooth, glabrous. Flowers in 3–7-flowered condensed axillary cymes (often with several of the flowers aborting); peduncle and branches 1–3 mm long, with scattered subsessile capitate glands; outer bracts foliaceous, to 1.5 cm long, inner bracts and bracteoles lanceolate, up to 8 mm long, with scattered subsessile capitate glands. Calyx with scattered subsessile capitate glands and sparsely ciliate teeth, ± 13 mm long; teeth ± 7 mm long. Corolla nearly symmetrical (trumpet-shaped with bell-shaped throat expanded ventrally), no information on colour but probably mauve or purple, palate distinctly bullate between veins, with sparse short-stalked capitate glands and palate with long stiff retrorse hairs, ± 30 mm long of which the linear tube ± 10 mm, the throat ± 7 mm and the limb ± 13 mm long and ± 11 mm across apically; lobes oblong, erect in upper lip, spreading or deflexed in lower. Free part of filaments 6–8 mm long, with a few hairs near base and glabrous upwards; anthers ± 3.5 mm long, spur ± ½ mm long. Upper stigma lobe well developed, about ½ the length of lower. Capsule and seed not seen.

TANZANIA. Masai District: Eastern Masailand, Shambarai, 5 June 1964, *G. B. M.* 1/5/6/64!
DISTR. **T** 2; not known elsewhere
HAB. Probably *Acacia* bushland; no altitude given

NOTE. A very unusual species with characters unique in *Dyschoriste*, i.e. stiff retrorse hairs on the palate and a well-developed dorsal stigma lobe. These two characters would seem to point towards a relationship with the genera around *Mellera*, particularly *Duosperma* with which genus it also shares the greyish green colour.
   The equally 5-toothed calyx, corolla shape and spurred anthers are typically *Dyschoriste* and the ovary clearly has more than 2 ovules.
   For the moment this species is best kept in *Dyschoriste* but further material is urgently required.

14. **Dyschoriste thunbergiiflora** (*S.Moore*) *Lindau* in E. & P. Pf. IV, 3b: 302 (1895) & in P.O.A. C: 367 (1895); C.B. Clarke in F.T.A. 5: 78 (1899); K.T.S.: 17 (1961); U.K.W.F.: 585 (1974); Blundell, Wild Fl. E. Afr.: 391, pl. 606 (1987); K.T.S.L.: 602 (1994); U.K.W.F. ed. 2: 269 (1994); Lebrun & Stork, Enum. Pl. Afr. Trop. 4: 480 (1997). Type: Kenya, Ukamba, Kitui, *Hildebrandt* 2719 (BM!, holo.; K!, P!, iso.)

Much branched erect or scrambling sticky subshrub or shrub to 2 m tall; branches brittle, sparsely to densely glandular-puberulent when young, often with non-glandular hairs intermixed (very rarely without glands). Leaves ovate to elliptic, largest 1.5–5.5(–7.5) × 0.6–2.7(–3.2) cm, apex acute to truncate, puberulent to puberulous or sparsely so, sometimes densest along veins, often with intermixed capitate glands. Flowers in open axillary dichasial 3–18-flowered cymes or solitary towards tip of branches; peduncle and branches 1–15(–20) mm long, becoming woody when old, sparsely to densely glandular-puberulent (rarely also with longer pilose hairs); outer bracts elliptic, up to 1.5 cm long or like inner bracts and bracteoles: linear and up to 7 mm long, indumentum as leaves. Calyx sparsely to densely glandular-puberulent, usually also with non-glandular hairs along veins and on lobes, (8–)10–22 mm long; teeth 4–11 mm long, from shorter to longer than tube. Corolla with limb nearly symmetric, trumpet-shaped with throat expanded downwards, white to mauve, purple or lilac or pale to dark blue, with darker patches at base of lower lip and in throat, tube yellowish, puberulent or sparsely so and with capitate glands, 28–65 mm long of which the linear tube 11–23 mm, the throat (7–)11–26 mm and the limb 9–20 mm, limb shorter than throat (rarely longer); lobes obovate-spathulate, recurved in mature flowers. Free part of filaments 7–15 mm long, hairy towards base, glandular or with mixture of glands and hairs towards apex; anthers (1.5–)2–2.5 mm long, spurs $\frac{1}{4}$–$\frac{1}{2}$ mm long. Style apically with mixture of hairs and stalked capitate glands. Capsule 15–18 mm long, glandular near apex. Seed ± 3 mm long.

UGANDA. Karamoja District: Nabilatuk, Feb. 1927, *Eggeling* 2797! & Rupa–Moroto road, Sep. 1955, *Wilson* 197! & N of Moroto, Rupa, 12 June 1970, *Katende & Lye* 427!
KENYA. Northern Frontier District: Ndoto Mts, E of Nguronit Mission, 9 Aug. 1979, *Gilbert et al.* 5576!; West Suk District: Kapenguria to Sigor, Wakor, 29 Aug. 1969, *Mabberley & McCall* 265!; Kitui District: Mutomo, 24 Apr., 1969, *Napper* 2073!
DISTR. U 1; K 1–4, 6, 7; not known elsewhere
HAB. Dry *Acacia*, *Acacia-Commiphora* and *Commiphora-Euphorbia* bushland, often on stony soil or on rocky hills; (500–)800–1700(–1850)

SYN. *Calophanes thunbergiiflora* S.Moore in J.B. 18: 8 (1880)

NOTE. This large-flowered attractive shrub is widely cultivated both within and outside its natural distribution in East Africa. Cultivated specimens have also been seen from W Uganda and NE Tanzania. Outside East Africa cultivated specimens have been seen from Zambia, Zimbabwe, Namibia and South Africa.
   Iversen in Symb. Bot. Ups. 29, 3: 161 (1991) records this species from the E Usambara Mts (T 3), but the specimen cited by him (*Ngoundai* 400) is clearly of cultivated origin as are other collections from this area. It is possible that the species occurs as wild in NE Tanzania, but no definite records exist.

15. **Dyschoriste sallyae** *Vollesen* **sp. nov.** a *D. thunbergiiflora* habitu herbaceo perenni caulibus prostratis e caudice lignoso orientibus (nec habitu erecto fruticoso nec suffruticoso) differt, atque in silvis humidis *Brachystegiae* nec in dumosis aridis *Acaciae* crescit. Type: Tanzania, Mpanda District: Chala–Mpanda road, *Bidgood et al.* 3877 (K!, holo.; BR!, C!, CAS!, DSM!, EA!, K!, NHT!, P!, iso.)

Perennial herb with several trailing stems from woody rootstock; stems to 75 cm long, puberulous or retrorsely sericeous-puberulous when young, with or without stalked capitate glands intermixed. Leaves ovate to elliptic, largest 4.5–8 × 2–3.2 cm, apex acute to subacuminate, sparsely sericeous-puberulous along veins. Flowers in open axillary dichasial 1–7-flowered cymes, often 1 flower and two aborted buds towards tip of stems, cymes sometimes aggregated towards tip of stems; peduncle and branches 2–15 mm long, sparsely to densely glandular-puberulous; outer bracts elliptic, up to 2 cm long or like inner bracts and bracteoles: linear-lanceolate and up to 6 mm long, indumentum as leaves; pedicels to 2 cm long. Calyx puberulous with mixture of stalked glands and non-glandular hairs, ciliate on teeth, 13–20 mm long; teeth 7–10 mm long, same length or slightly longer than tube, cuspidate. Corolla with limb nearly symmetric (trumpet-shaped with throat expanded downwards), white (rarely very pale mauve), with orange tube and orange patches at base of lower lip and in throat, sparsely puberulous and with scattered capitate glands, 35–50 mm long of which the linear tube 12–20 mm, the throat 12–15 mm and the limb 11–15 mm, limb and throat same length; lobes obovate-spathulate, eventually recurved. Free part of filaments 6–8 and 9–10 mm long, with scattered capitate glands towards apex; anthers 2–2.5 mm long, spurs ± $\frac{1}{2}$ mm long. Style apically with mixture of hairs and stalked capitate glands. Capsule and seed not seen.

Tanzania. Mpanda District: Urema [Karema], Homo Hills, June 1928, *C.H.B. Grant* s.n.! & 78 km on Chala–Mpanda road, 11 May 1997, *Bidgood et al.* 3877! & 24 May 2000, *Bidgood et al.* 4428!
Distr. **T** 4; not known elsewhere
Hab. Tall mature *Brachystegia* woodland on grey sandy-loamy soil; 1300 m

Note. This species differs from *D. thunbergiiflora* in little more than its herbaceous growth form with trailing branches. But it also has a different ecology and very disjunct distribution. Despite its large spectacular flowers it is known only from these three collections.

16. **Dyschoriste keniensis** *Malombe, Mwachala & Vollesen* in K.B. 62: 433, fig. 1 (2006). Type: Kenya, Kitui District: 45 km on Thika–Kitui road, *Gillett* 19809 (EA!, holo.; BR!, K!, NHT!, iso.)

Erect (rarely scrambling) much branched shrub or subshrub to 1 m tall; stems brittle, subglabrous to puberulous or finely sericeous or glandular-puberulous with non-glandular hairs intermixed. Leaves elliptic to obovate, largest 1–5.5 × 0.5–2.3 cm, apex subacute to retuse, with a small triangular tooth, glabrous to puberulous or sericeous, densest along veins. Flowers in condensed 3–7(–12)-flowered axillary cymes or solitary towards tip of branches, often condensed towards tip of stems; peduncle and branches 0–2 mm long, subglabrous to finely sericeous, often also with stalked capitate glands; outer bracts elliptic to obovate, up to 1 cm long, or as inner bracts and bracteoles: linear to lanceolate, up to 5 mm long, glabrous to sericeous or with sparse to dense stalked capitate glands. Calyx subglabrous to sparsely sericeous or sericeous-puberulent or uniformly glandular-puberulous, 8–14(–16 in fruit) mm long; teeth 4–8(–9 in fruit) mm long, shorter or longer than tube. Corolla slightly 2-lipped, mauve to purple, bluish purple or blue to dark blue, with darker patches and lines at base of lower lip and in throat, sericeous-puberulent and often with scattered capitate glands, 12–17 mm long of which the linear tube 5–8 mm, the throat 2–3 mm and the limb 5–7 mm, throat indistinctly delimited, much shorter than limb; lobes narrowly oblong or slightly obovate, erect or deflexed in lower lip. Free part of filaments 2–4 mm long, hairy near base, glabrous or with a few scattered hairs upwards; anthers 1–1.5 mm long, spurs less than $^1/_4$ mm long. Capsule 10–13 mm long. Seed ± 2.5 mm long.

16a. subsp. **keniensis**

Flowering stems, bracts and bracteoles subglabrous to puberulous or finely sericeous, without stalked capitate glands; calyx subglabrous to sparsely sericeous, without stalked capitate glands.

Kenya. Kiambu District: 1 km NE of Ruiru, Lions Inn, 7 Dec. 1966, *Perdue & Kibuwa* 8182!; Kitui District: 45 km on Thika-Kitui road, 24 June 1972, *Gillett* 19809!; Machakos District: 3 km on Yatta–Kangonde road, 18 May 1969, *Kimani* 171!
Distr. **K** 4; not known elsewhere
Hab. *Combretum*, *Acacia-Combretum* and *Acacia-Commiphora* wooded grassland, bushland and thicket on sandy to clayey soil, grassland, old fields and generally disturbed areas; 1150–1850 m

Syn. [*D. depressa* sensu C.B.Clarke in F.T.A. 5: 72 (1899) quoad *Scott Elliot* 7706, *non* Nees (1832)]
D. sp. A of U.K.W.F.: 585 (1974); Blundell, Wild Fl. E. Afr.: 390, pl. 603 (1987); U.K.W.F., ed. 2: 269 (1994)

16b. subsp. **glandulifera** *Malombe, Mwachala & Vollesen* in K.B. 62: 436 (2006). Type: Kenya, Teita District: Wusi–Mwatate road, *Drummond & Hemsley* 4409 (EA!, holo.; BR!, K!, iso.)

Flowering stems, bracts and bracteoles glandular-puberulous with stalked capitate glands; calyx puberulous with dense stalked capitate glands.

Kenya. Teita District: Wusi, 15 May 1931, *Napier* 1089! & Wusi–Mwatate road, 18 Sep. 1953, *Drummond & Hemsley* 4409! & Mwatate, 17 Apr., 1960, *Verdcourt & Polhill* 2738!

TANZANIA. Moshi District: Moshi, 19 Feb. 1924, *Durham* s.n.!; Lushoto District: Mkomazi Game Reserve, Viteweni Hill, 28 Apr. 1995, *Abdallah & Vollesen* 95/32! & Mkomazi Game Reserve, Zange to Ibaya Camp, 29 Apr. 1995, *Abdallah & Vollesen* 95/54!
DISTR. **K** 7; **T** 2, 3; not known elsewhere
HAB. In Kenya in *Combretum, Terminalia* and *Acacia* woodland, wooded grassland and bushland, persisting in old cultivation. In Tanzania in *Acacia-Commiphora* bushland on red loam, dry evergreen hilltop scrub and in dry forest partly reduced by burning to wooded grassland; (600–)800–1350(–1650) m

SYN. *Dyschoriste* sp. nov. (= *L.D. Harris* 129) of Vollesen *et al.*, Checklist Vasc. Pl. and Pter. Mkomazi: 82 (1999)

17. **Dyschoriste hildebrandtii** (*S.Moore*) *S.Moore* in J.B. 32: 132 (1894); Lindau in P.O.A. C: 367 (1895); C.B. Clarke in F.T.A. 5: 76 (1899); T.T.C.L.: 8 (1949); U.K.W.F.: 585 (1974); Moriarty, Wild Fl. Malawi: 85, pl. 43, fig. 2 (1975); Blundell, Wild Fl. E. Afr.: 390, pl. 604 (1987); Iversen in Symb. Bot. Ups. 29, 3: 160 (1991); U.K.W.F. ed. 2: 269 (1994); Lebrun & Stork, Enum. Pl. Afr. Trop. 4: 479 (1997); Ensermu in F.E.E. 5: 375 (2006); Thulin in Fl. Somalia 3: 388 (2006) & in Nord. J. Bot. 24: 387 (2007). Type: Kenya, Ukamba, Kitui, *Hildebrandt* 2718 (BM!, holo.; K!, iso.)

Shrubby herb or shrub usually much branched or erect perennial herb with several unbranched to sparsely branched stems from woody rootstock; stems erect to decumbent (then sometimes rooting), up to 1 m long, finely sericeous to puberulous or densely so, without or with with scattered to dense stalked capitate glands (plant then sticky). Leaves ovate to elliptic or obovate, largest 1–6 × 0.5–2.6 cm, apex acute to retuse with a central triangular tooth, subglabrous to finely sericeous or puberulent to densely puberulous, without or with scattered to dense capitate glands. Flowers in 3–12(–15)-flowered condensed to open axillary cymes (sometimes two cymes per axil) or solitary towards tip of branches, sometimes cymes congested towards tip of branches; peduncle and branches up to 1.2 cm long, indumentum as branches; outer bracts elliptic to obovate, up to 1.5 cm long, or like inner bracts and bracteoles: linear to oblanceolate, up to 1 cm long, indumentum as young leaves. Calyx subglabrous to puberulous or sericeous-puberulous, with or without scattered to dense stalked viscid capitate glands, (6–)10–19(–22 in fruit) mm long; teeth 2–9 mm long, shorter than tube or same length (rarely longer). Corolla 2-lipped, cream to livid yellow or dark yellow, yellow tinged with pale purple, cream turning purple, pinkish white to pale or deep mauve, pale blue to bright blue or vivid purplish blue, with purple spots and longitudinal markings in throat, puberulent and with scattered to dense capitate glands, 21–41(–45) mm long of which the linear tube 10–24(–30) mm, the throat 3–7 mm and the limb 7–13(–16) mm, limb much longer than throat; lobes obovate, erect in upper lip, spreading or recurved in lower lip. One or two of the shorter stamens (rarely all four) sometimes reduced to staminodes; free part of filaments 3–6 mm long, hairy at base, glabrous upwards; anthers 1–2 mm long, spurs minute (under ¼ mm long) or absent on some or all anthers. Capsule 10–16 mm long. Seed ± 3 mm long.

KENYA. Northern Frontier District: Moyale, 10 July 1952, *Gillett* 13555!; Masai District: Magadi road, S of Ngong Hills, 11 Aug. 1951, *Verdcourt* 577!; Kwale District: between Samburu and Mackinnon road, near Taru, 3 Sep. 1953, *Drummond & Hemsley* 4124!
TANZANIA. Masai District: Olongogo, 14 Dec. 1962, *Newbould* 6382!; Handeni District: Handeni, 18 July 1966, *Archbold* 860!; Morogoro District: Lusunguru Forest Reserve, 22 Oct. 1959, *Mgaza* 343!
DISTR. **K** 1, 3, 4, 6, 7; **T** 1–3, 5, 6; Ethiopia, Somalia, Zambia, Malawi, Mozambique, Zimbabwe, Botswana, South Africa
HAB. In Kenya and N Tanzania in *Acacia, Acacia-Commiphora* and *Combretum* woodland, wooded grassland and bushland, further south usually in *Brachystegia* woodland, on a wide range of soils and on rocky hills; 100–1700(–2000) m

SYN. *Calophanes hildebrandtii* S.Moore in J.B. 18: 8 (1880)

> *Hygrophila volkensii* Lindau in E.J. 19, Beibl. 47: 46 (1894) & in P.O.A. C: 366 (1895). Type: Tanzania, Kilimanjaro, Machame, *Volkens* 1627 (B†, holo.; BM!, K!, iso.)
>
> *Dyschoriste fischeri* Lindau in E.J. 20: 11 (1894) & in P.O.A. C: 367 (1895); C.B. Clarke in F.T.A. 5: 77 (1899); Eyles in Trans. Roy. Soc. S. Afr. 5: 482 (1916); Fiori, Miss. Biol. Boran. 4: 214 (1939); T.T.C.L.: 8 (1949); & F.F.N.R.: 382 (1962); E.P.A.: 933 (1964); Binns, Checklist Herb. Fl. Malawi: 13 (1968); Iversen in Symb. Bot. Ups. 29, 3: 160 (1991); Lebrun & Stork, Enum. Pl. Afr. Trop. 4: 479 (1997). Types: Tanzania, Masai Steppe, *Fischer* 485 (B†, syn.); Mozambique, Rios de Sena, *Peters* s.n. (B†, syn.; K!, iso.)
>
> *D. hildebrandtii* (S.Moore) S.Moore var. *mollis* S.Moore in J.B. 32: 132 (1894). Type: Kenya, Thika Hills, Tana, *Gregory* s.n. (BM!, holo.)
>
> *Phillipsia fruticulosa* Rolfe in K.B. 1895: 223 (1895). Type: Somalia, Golis Range, *Lort Phillips* s.n. (K!, holo.)
>
> *Dyschoriste somalensis* Rendle in J.B. 34: 413 (1896); C.B. Clarke in F.T.A. 5: 78 (1899). Type: Ethiopia, Sheik Hussein, *Donaldson Smith* s.n. (BM!, holo.)
>
> *Satanocrater fruticulosa* (Rolfe) Lindau in E. & P. Pf. IV, 3b, Nachtr. I: 305 (1897)
>
> *Dyschoriste nobilior* C.B.Clarke in F.T.A. 5: 74 (1899) pro parte, quoad *Holst* 3204a (K!, syn.). The other syntypes from Angola represent a quite different species.
>
> *D. mollis* (S.Moore) C.B.Clarke in F.T.A. 5: 77 (1899); T.T.C.L.: 8 (1949); Binns, Checklist Herb. Fl. Malawi: 13 (1968); Iversen in Symb. Bot. Ups. 29, 3: 160 (1991); Lebrun & Stork, Enum. Pl. Afr. Trop. 4: 479 (1997)
>
> *D. volkensii* (Lindau) C.B.Clarke in Fl. Trop. Afr. 5: 77 (1899); T.T.C.L.: 8 (1949); Lebrun & Stork, Enum. Pl. Afr. Trop. 4: 479 (1997)
>
> *D. fruticulosa* (Rolfe) Chiov., Fl. Somala: 249 (1929); E.P.A.: 934 (1964); Kuchar, Pl. Somalia (CRDP Techn. Rep. No. 16): 246 (1986); Lebrun & Stork, Enum. Pl. Afr. Trop. 4: 479 (1997)
>
> *D. decumbens* E.A.Bruce in K.B. 1932: 99 (1932), *non* (A. Gray) Kuntze (1891). Type: Kenya, Masai District: Rift Valley, Narok road, *Napier* 413 (K!, holo.)
>
> *D. procumbens* E.A.Bruce in K.B. 1934: 304 (1934); E.P.A.: 935 (1964). Type as for *D. decumbens*
>
> *Dyschoriste* sp. 1 of White, F.F.N.R.: 382 (1962)

NOTE. There is a bewildering range of variation in this species, most notably in indumentum and corolla-colour, but I have been unable to find a satisfactory way of dividing it into any infraspecific taxa. There are two major groups.

In **K** 6 and **T** 1, 2, 5 plants with usually appressed eglandular (or almost so) indumentum and with yellowish to greenish flowers. In **K** 4, 7 and **T** 3, 6 plants with spreading often glandular indumentum and usually bluish or purplish flowers. But there are far too many intermediate specimens (including most of the material from **K** 1, Ethiopia and Somalia) to allow any reasonably distinct taxa to be defined. In south tropical Africa the variation is even more bewildering.

18. **Dyschoriste kitongaensis** *Vollesen* **sp. nov.** a *D. hildebrandtii* indumento caulium atque axium inflorescentarium glanduloso (nec e glandulis pilisque intermixtis sistenti), foliis acuminatis (nec acutis neque retusis), cymis elongatis monochasialibus et corolla alba differt. Type: Tanzania, Iringa District: Kitonga Gorge, *Bidgood, Congdon & Vollesen* 2211 (K!, holo.; C!, CAS!, DSM!, EA!, K!, MO!, NHT!, UPS!, iso.)

Slender spindly branched shrubby herb or shrub; stems brittle, erect or straggling, up to 1 m long, glandular-puberulent or sparsely so (slightly sticky), without or with a few scattered non-glandular hairs intermixed. Leaves ovate, largest 3–7 × 1.2–2.7 cm, apex acuminate or subacuminate, subglabrous to glandular-puberulent beneath, above subglabrous to puberulent with non-glandular hairs. Flowers in open elongated monochasial (occasionally some dichasial) cymes with up to 7 flowers or solitary towards tip of branches; peduncle and branches 0.5–1.7 cm long, indumentum as stems; bracts up to 1.5 cm long at basal cymes, smaller upwards, indumentum as leaves; bracteoles not differing. Calyx subglabrous to sparsely glandular-puberulent and with scattered non-glandular hairs on veins, 13–17(–19 in fruit) mm long; teeth 6–8(–9 in fruit) mm long, from slightly shorter to slightly longer than tube. Corolla 2-

lipped, pure white or with faint pink lines into throat, puberulent and with scattered capitate glands, 26–32 mm long of which the linear tube 13–18 mm, the throat 4–5 mm and the limb 7–8 mm, limb much longer than throat; lobes obovate, erect in upper lip, spreading or recurved in lower lip. Free part of filaments 3–5 mm long, glabrous throughout; anthers pale pink, 1–2 mm long, spurs ± $^1/_4$ mm long or absent on short anthers. Capsule 13–17 mm long. Seed ± 3 mm long.

TANZANIA. Iringa District: 70 km on Iringa–Dodoma road, between Kilindimo and Izazi, 18 Apr. 1962, *Polhill & Paulo* 2055! & Kitonga Gorge, 21 Apr. 1991, *Bidgood et al.* 2211! & 7 July 1991, *Congdon* 313!
DISTR. **T** 5, 7; not known elsewhere
HAB. Dry riverine forest and scrub on rocky slopes, *Euphorbia-Adansonia* succulent woodland on sand; 600–900 m

NOTE. This species, which is locally common in the drier parts of the Ruaha-system, is obviously related to the widespread and variable *D. hildebrandtii*. It differs in the purely glandular indumentum of stems and leaves, the elongated monochasial cymes and the pure white corolla.

18. **EREMOMASTAX**

Lindau in E.J. 20: 8 (1894)

Paulowilhelmia Hochst. in Flora 27, Bes. Beibl.: 4 (late 1844); Nees in DC., Prodr. 11: 208 (1847); G.P. 2: 1079 (1876); Lindau in E.J. 17: 104 (1893); C.B. Clarke in F.T.A. 5: 52 (1899); F.W.T.A. 2: 246 (1931); Benoist in Fl. Madag. 182, 1: 83 (1967), *non Paulowilhelmia* Hochst. in Flora 27: 17 (January 1844)

Perennial or shrubby herb; cystoliths conspicuous; stems indistinctly quadrangular. Leaves opposite. Flowers in open to dense dichasial cymes aggregated into small to large open to condensed terminal panicles, pedicellate; bracteoles usually conspicuous, often caducous. Calyx 5-lobed, divided almost to base; lobes subequal or dorsal slightly wider, linear, narrowing slightly upwards. Corolla contorted in bud; tube long, cylindric, slightly curved, without clearly delimited throat, split dorsally to give a 5-lobed lower lip with horizontally spreading lobes, limb without stiff hairs, not pleated. Stamens 4, subequal or slightly didynamous, exserted, inserted just inside tube; filaments forming indistinct glabrous ridges in tube, free part of each pair fused at base; anthers 2-thecous, thecae linear-oblong, slightly curved, longitudinally twisted when old, truncate at apex, apiculate at base. Ovary with 4–8 ovules per locule; style filiform, hairy almost to apex. Dorsal stigma lobe a minute tooth, ventral large, linear, flattened. Capsule (4–)6–16-seeded, ellipsoid, thin-walled, glossy, slightly laterally compressed, valves sulcate dorsally; retinaculae strong, hooked. Seed discoid, ovoid-suborbicular in outline, covered in hygroscopic hairs.

Monotypic genus from the Guineo-Congolian forest regions of Africa and on Madagascar.

Eremomastax speciosa (*Hochst.*) *Cuf.*, E.P.A.: 931 (1964); Heine in Fl. Gabon: 30 (1966); Lebrun & Stork, Enum. Pl. Afr. Trop. 4: 481 (1997); Friis & Vollesen in Biol. Skr. 51(2): 443 (2005); Ensermu in F.E.E. 5: 378 (2006). Type: Ethiopia, no locality, *Schimper* III.1954 (P!, iso.)

Erect or somewhat scrambling perennial herb to 1.3(–2) m tall; stems glabrous to sericeous-puberulent when young. Leaves with petiole 3–9.5 cm long; lamina ovate or broadly so, largest 7–11.5 × 4–7 cm, margin subentire to grossly crenate-dentate, usually becoming less toothed upwards, apex acuminate to cuspidate, base truncate to shallowly cordate, crisped-puberulous to pubescent along major veins with glossy broad hairs, sometimes also on lamina beneath. Panicles up to 20 cm long, with indumentum as stems or also with long capitate glands; bracts leafy near base,

Fig. 30. *EREMOMASTAX SPECIOSA* — **1**, habit, × 1; **2**, basal leaf, × ²/₃; **3**, calyx, × 3; **4**, calyx, opened up, × 3; **5**, corolla, tube opened up, × 3; **6**, stamens, × 6; **7**, ovary, style and stigma, × 3; **8**, capsule, × 4.5; **9**, seed, × 9. 1 from *Jacques-Felix* 9074, 2 from *Breteler* 434, 3–7 from *Hepper* 1522, 8–9 from *Dawe* 269. Drawn by Margaret Tebbs.

gradually smaller upwards; dichasia many-flowered at base, with fewer flowers towards tip of branches; peduncle and branches up to 15 mm long; pedicels 0.5–1 mm long; bracteoles up to 7 mm long. Calyx puberulent and with scattered to dense long broad glossy hairs, sometimes also with short to long capitate glands, 10–19 mm long of which the basal fused part 1–3 mm; lobes linear, acute. Corolla blue to mauve or purple, throat white with maroon shading, sericeous-puberulent outside, 25–40 mm long of which the tube 15–28 mm long and split for up to 5 mm dorsally; lobes oblong, parallel-sided, truncate to broadly emarginate. Free part of filaments 6–8 mm long, glabrous; anthers 3–4 mm long. Capsule 15–18 mm long, with scattered short capitate glands near apex. Seed ovoid in outline, ± 2 × 1.5 mm. Fig. 30, p. 201.

UGANDA. Toro District: Unyoro, Bugoma Forest, 24 Dec. 1906, *Bagshawe* 1397!; Masaka District: Sese Islands, Lusoge, June 1925, *Maitland* 807!; Mengo District: Entebbe, 1905, *E. Brown* 318!
KENYA. Trans-Nzoia District: Kitale, Nov. 1964, *Tweedie* 2927!; North Kavirondo District: Kakamega Forest, Yale River, 25 Nov. 1969, *Bally* 13679!
TANZANIA. Bukoba District: Minziro Forest, Nyanya Etagera, 24 May 1999, *Festo & Simon* 215! & Minziro Forest, 5 July 2000, *Bidgood et al.* 4862!
DISTR. **U** 2, 4; **K** 3, 5; **T** 1; widespread from West Africa through Central African Republic and N Congo-Kinshasa to S Sudan and SW Ethiopia, Madagascar
HAB. Evergreen and riverine forest, swamp forest, forest edges and clearings; 900–1900 m

SYN. *Paulowilhelmia speciosa* Hochst. in Flora 27, Bes. Beibl.: 5 (1844); Nees in DC., Prodr. 11: 208 (1847); N. E. Brown in Gard. Chron. 6: 749, fig. 106 (1889); Lindau in E. & P. Pf. IV, 3b: 300, fig. 100 H (1895); C.B. Clarke in F.T.A. 5: 53 (1899)
  P. polysperma Benth. in Hooker, Niger Fl.: 479 (1849); Lindau in E. & P. Pf. IV, 3b: 301 (1895); C.B. Clarke in F.T.A. 5: 52 (1899); S. Moore in J.B. 45: 91 (1907); F.W.T.A. 2: 246 (1931). Type: Sierra Leone, Sugarloaf Mountain, *Don* 118 (BM!, holo.)
  Ruellia sclerochiton S.Moore in J.B. 18: 7 (1880). Type: Sudan, Niamniam, Assika River, *Schweinfurth* 3257 (BM!, holo.; K!, P!, iso.)
  Eremomastax crossandriflora Lindau in E.J. 20: 8 (1894) & in E. & P. Pf. IV, 3b: 297 (1895) & in P.O.A. C: 367, t. 42 (1895). Types: Cameroon, Jaunde, *Zenker* 384 (B†, syn.); Tanzania, Bukoba, *Stuhlmann* 4013 (B†, syn.)
  Paulowilhelmia sclerochiton (S.Moore) Lindau in E.J. 17: 105 (1895); C.B. Clarke in F.T.A. 5: 53 (1899); F.W.T.A. 2: 246 (1931)
  Eremomastax polysperma (Benth.) Dandy in F.P.S. 3: 174 (1956); Heine in F.W.T.A. (ed. 2) 2: 397 (1963); U.K.W.F.: 587 (1974)

19. **RUELLIA**

L., Sp. Pl.: 634 (1753); Nees in DC., Prodr. 11: 143 (1847); G.P. 2: 1077 (1876); C.B. Clarke in F.T.A. 5: 44 (1899)

Dipteracanthus Nees in Wallich, Pl. As. Rar. 3: 75 (1832) & in DC., Prodr. 11: 115 (1847); Bremek. in Ned. Akad. Weten., Verh. 45: 14 (1948) & in Acta Bot. Neerl. 11: 195 (1962)

Perennial or shrubby herbs or shrubs; cystoliths conspicuous. Leaves opposite, entire to crenate. Flowers solitary or paired in leaf axils or in variously branched cymes or panicles; bracts persistent, leaf-like; bracteoles persistent (rarely caducous), usually large and leafy. Calyx 5-lobed, deeply divided into 5 subequal lobes or 2–3 lobes fused higher up; lobes usually with distinct dorsal keel, dorsal usually slightly longer than the rest. Corolla hairy on the outside, contorted in bud, subactinomorphic; basal cylindric part of tube short to long, throat short to long, usually much widened upwards; lobes 5, subequal, spreading. Stamens 4, subequal to didynamous, inserted near base of throat, exserted or included in throat; free part of filaments fused at base, below insertion forming two ridges running to base of tube; anthers 2-thecous, thecae oblong, rounded at base, without spurs, connective with small curved appendage at apex. Ovary with 3 to many ovules per locule; style filiform, hairy throughout or glabrous towards apex; stigma with dorsal lobe reduced to a small tooth (rarely well developed), ventral lobe large, flattened to spoon-shaped with wavy margin. Capsule cylindric to clavate, valves eventually splitting lengthwise from bottom up along dorsal suture, with 4 to over 20 seeds; retinaculae strongly hooked. Seed discoid, faces glabrous or with discoid hairs, rim strongly raised and with dense hygroscopic hairs.

About 250 species widely distributed in all tropical regions, most numerous in South America.

The native East African species all belong to Subgen. *Dipteracanthus* (Nees) C.B.Clarke which is characterised by having a clavate capsule with sterile basal part.

1. Plant with indumentum of stellate hairs . . . . 11. *R. discifolia* (p. 213)
   Plant with indumentum of simple hairs or glabrous 2
2. Cylindric part of corolla tube 4.5–13.5 cm long 3
   Cylindric part of corolla tube 0.3–3.5(–4.2) cm long 5
3. Bracteoles linear to slightly spathulate, caducous, ± 1 mm wide . . . . 10. *R. amabilis* (p. 213)
   Bracteoles ovate to orbicular or cordiform, persistent, 8–23 mm wide 4
4. Calyx 13–25 mm long; cylindric part of corolla tube 4.5–11 cm long; stamens included in throat; anthers 5–7 mm long . . . . 8. *R. bignoniiflora* (p. 210)
   Calyx 32–38 mm long; cylindric part of corolla tube 12.5–13.5 cm long; stamens exserted; anthers ± 8 mm long . . . . 9. *R. richardsiae* (p. 212)
5. Flowers in 3-flowered dichasia with peduncle 1.5–3.5 cm long; capsule linear, without sterile basal part, with more than 20 seeds . . . . 1. *R. tuberosa* (p. 204)
   Flowers solitary or in 2–3-flowered cymes with peduncles up to 0.6(–1.2) cm long; capsule oblanceolate to clavate, with sterile basal part, with up to 16 seeds 6
6. Capsule ± 27 mm long; cylindric part of corolla tube ± 3.5 cm long; anthers ± 5.5 mm long; seed ± 6 × 5.5 mm . . . . 7. *R.* sp. *A* (p. 210)
   Capsule (8–)11–24 mm long; cylindric part of corolla tube 0.3–2(–4.2) cm long; anthers 1.5–3(–4) mm long; seed 3–5 × 2.5–4.5 mm 7
7. Ovary and capsule hairy 8
   Ovary and capsule glabrous 9
8. Calyx (6–)9–23(–27) mm long, divided to 1–3 mm from base, lobes filiform; capsule clavate, densely puberulous to sericeous; petiole not with broad glossy curly hairs . . . . 2. *R. prostrata* (p. 204)
   Calyx 6–11 mm long, divided to 3–5 mm from base, lobes triangular; capsule narrowly oblanceolate, with scattered appressed hairs; petiole with broad glossy curly hairs . . . . 3. *R. linearibracteolata* (p. 205)
9. Calyx ± 10 mm long, lobes filiform (rare form) . . . 2. *R. prostrata* (p. 204)
   Calyx 3–10(–12) mm long, lobes narrowly triangular 10
10. Corolla tube 37–64 mm long of which the basal cylindric part 20–42 mm, cylindric part always longer than throat; seed 4.5–5 × 4–4.5 mm . . . 6. *R. burttii* (p. 209)
    Corolla tube (10–)12–34 mm long of which the basal cylindric part 3–12(–20) mm, cylindric part normally shorter than throat; seed 3–4.5 × 2.5–4.5 mm 11

11. Young stems, leaf margins and bracts with long
broad curly glandular hairs to 2 mm long;
longer pair of filaments 3–5 mm longer than
shorter pair; corolla open during the day . . . .   5. *R. praetermissa* (p. 208)
Young stems, leaf margins and bracts without or
with much shorter glandular hairs; longer pair
of filaments 1–2 mm longer than shorter pair;
corolla falling early in the morning . . . .   4. *R. patula* (p. 207)

1. **Ruellia tuberosa** *L.*, Sp. Pl.: 635 (1753); Leonard in J. Washington Acad. Sci. 17: 510 (1927); U.O.P.Z.: 436 (1949); Iversen in Symb. Bot. Ups. 29(3): 162 (1991); Lebrun & Stork, Enum. Pl. Afr. Trop. 4: 501 (1997). Type: Jamaica, *Herb. Linn.* 804.8 (K, microfiche!)

Perennial herb with several erect stems to 30 cm tall from woody rootstock with tuberous roots; young stems with sparse to dense long and short broad glossy hairs. Leaves ovate to elliptic, largest (including petiole) 5–10.5 × 2–4 cm, apex subacute to broadly rounded, base attenuate, long-decurrent, glabrous or with scattered broad glossy hairs along veins. Flowers in 3-flowered axillary dichasia (or 1-flowered with two undeveloping buds); primary peduncle 1.5–3.5 cm long, secondary 0.7–1.5 cm, subglabrous to sparsely puberulous; pedicels 10–20(–26) mm (central flower) or 3–10 mm (lateral flowers) long, puberulous; bracts and bracteoles linear, 3–7 mm long. Calyx 17–32 mm long, divided to 2–4 mm from base, lobes filiform, ciliate with broad curly glossy hairs, otherwise glabrous. Corolla open during the day, mauve to bluish purple; tube 32–40 mm long of which the basal cylindric part 8–12 mm and the throat 24–28 mm long; lobes 18–20 × 20–22 mm, orbicular to transversely elliptic with entire to crenate-dentate edge. Stamens included in throat, strongly didynamous, anthers not overlapping; filaments fused for 1–2 mm at base, free parts 2–3 and 6–7 mm long; anthers 3–4 mm long. Ovary glabrous; ventral stigma lobe 1.5–2 × 0.5–1 mm. Capsule linear, glabrous, 20–32 mm long, with more than 20 seeds. Seed orbicular, brown, ± 2.5 × 2 mm.

KENYA. Mombasa District: Frere Town, 28 May 1934, *Napier* 6195!; Kwale District: Mazeras, 26 Feb. 1992, *Luke* 3076!
TANZANIA. Tanga District: Tanga, 13 Apr. 1979, *Grabner* 326!; Uzaramo District: Dar es Salaam, Oyster Bay, 30 Nov. 1968, *Batty* 316A! & Pugu Hills, 8 km S of Pugu, 13 Nov. 1994, *Goyder et al.* 3733!; Pemba, Banani Mission, 23 Dec. 1930, *Greenway* 2778!
DISTR. **K** 7; **T** 3, 6, 8; **Z**; **P**; native from southern USA through Central America and the Caribbean to N South America, introduced and naturalized sporadically in Africa and Asia
HAB. Roadside ditches, old abandoned gardens, coconut plantations, secondary grassland; near sea level to 150 m

2. **Ruellia prostrata** *Poir.*, Encycl. Meth. 6: 349 (1804); T. Anderson in J.L.S. 7: 24 (1863); C.B. Clarke in F.T.A. 5: 46 (1899); E.P.A.: 939 (1964); Binns, Checklist Herb. Fl. Malawi: 15 (1968); U.K.W.F.: 589 (1974); Vollesen in Opera Bot. 59: 81 (1980); Blundell, Wild Fl. E. Afr.: 395 (1987); Iversen in Symb. Bot. Ups. 29(3): 162 (1991); U.K.W.F., ed. 2: 270 (1994); Lebrun & Stork, Enum. Pl. Afr. Trop. 4: 501 (1997); Friis & Vollesen in Biol. Skr. 51(2): 453 (2005); Ensermu in F.E.E. 5: 395 (2006); Hedrén in Fl. Somalia 3: 395 (2006). Type: India, *Dupuis* s.n. (P, holo.)

Erect to decumbent perennial herb to 50 cm tall with several stems from woody rootstock or erect to scrambling subshrub to 1.5 m tall; young stems subglabrous to densely puberulous, pubescent or pilose with broad glossy straight or curly spreading or upwardly directed non-glandular hairs. Leaves with petiole (0–)3–20(–30) mm long; lamina lanceolate to broadly ovate or broadly elliptic, largest (3–)4–10(–14) × (0.5–)1–6(–7) cm, apex acute to rounded, base attenuate to truncate, subglabrous to pubescent, densest along veins. Flowers solitary or in 2–3-flowered axillary cymes; pedicels 1–2.5 mm long, glabrous to puberulous; bracteoles (bracts in cymes) lanceolate to broadly ovate, 7–23 × 2–16 mm. Calyx (6–)9–23(–27) mm long, divided to 1–3 mm from base, with filiform pubescent- to pilose-ciliate lobes with hairs to 2 mm long, otherwise glabrous to puberulous. Corolla falling early in the morning, white to mauve or purple; tube 14–28 mm long of which the basal cylindric part 4–10 mm and the throat 11–21 mm; lobes 6–16 × 5–14 mm, elliptic or broadly so with entire to crenate-dentate edge. Stamens included in throat, didynamous, anthers not or slightly overlapping; filaments fused for 1–2 mm at base, free parts 2–4 and 5–8 mm long; anthers 1.5–3 mm long. Ovary densely puberulous; ventral stigma lobe 1–2 × ± 0.5 mm. Capsule narrowly oblanceolate to clavate, puberulous to sericeous (very rarely glabrous), (13–)15–24 mm long, (6–)8–16-seeded. Seed ovoid to orbicular, grey to brown, 3–4 × 3–4 mm. Fig. 31, 10–11, p. 206.

UGANDA. Acholi District: Kitgum, Matidi, Apr. 1943, *Purseglove* 1507!; Karamoja District: Mt Moroto, 10 Oct. 1952, *Verdcourt* 810!; Mbale District: Sebei, Buligenyi, Muyembe, 19 May 1955, *Norman* 266!
KENYA. Northern Frontier District: Dandu, 2 May 1952, *Gillett* 13005!; Baringo District: Chebloch, Jan. 1962, *Tweedie* 2281!; Tana River District: Tana River National Primate Reserve, 22 March 1990, *Luke et al.* TPR775!
TANZANIA. Mbulu District: Lake Manyara National Park, Msasa, 6 Dec. 1963, *Greenway & Kirrika* 11143!; Dodoma District: 75 km N of Dodoma, Chenene, 24 Jan. 1962, *Polhill & Paulo* 1254!; Masasi District: 8 km NE of Masasi, Mkwera Hill, 16 March 1991, *Bidgood et al.* 2026!
DISTR. U 1–4; K 1–7; T 1–8; Sudan, Ethiopia, Somalia, Angola, Zambia, Malawi, Mozambique, Zimbabwe, Botswana, Namibia, South Africa; Yemen, Oman, India, Sri Lanka, Christmas Island, New Guinea, New Caledonia, Vanuatu, Samoa; possibly introduced in SE Asia
HAB. In a wide variety of grassland, bushland and woodland on an equally wide variety of soil types; near sea level to 1500 m

SYN. *Ruellia nubica* Delile, Cent. Pl. Meroe: 74 (1826). Type: Sudan, Sennar, Meroe, *Cailliaud* s.n. (P, holo.)
    Dipteracanthus prostratus (Poir.) Nees in Wall., Pl. As. Rar. 3: 81 (1832) & in DC., Prodr. 11: 124 (1847)
    Dipteracanthus genduanus Schweinf. in Verh. Zool-Bot. Ges. Wien 18: 680 (1868). Type: Sudan, Gallabat, Gendua River, *Schweinfurth* 132 (BM!, P!, iso.)
    D. sudanicus Schweinf. in Verh. Zool.-Bot. Ges. Wien 18: 679 (1868) pro parte, quoad *Schweinfurth* 131
    R. genduana (Schweinf.) C.B.Clarke in F.T.A. 5: 47 (1899); F.P.S. 3: 187 (1956); Lebrun & Stork, Enum. Pl. Afr. Trop. 4: 501 (1997)
    R. patula Jacq. var. *prostrata* (Poir.) Chiov. in Atti. Ist. Bot. Pavia, Ser. 4, 7: 146 (1936)
    R. sp. aff. prostrata sensu Vollesen in Opera Bot. 59: 81 (1980)

NOTE. A many-formed species which may eventually prove to contain several infra-specific taxa, but extensive field studies will be needed to confirm this. All plants under this species are linked by the filiform calyx lobes and the hairy capsule. The leaves are generally larger and more acute than in *R. patula*.
    C.B.Clarke in F.T.A. expresses some doubt as to whether the type specimen of *R. prostrata* is distinct from *R. patula*. It is however apparent that the Indian material of *R. prostrata* generally has a shorter calyx than the East African material. But the shape of the calyx lobes and calyx indumentum is still distinct.
    Burtt 1134, Tanzania, Kirangi, Mburi has a glabrous capsule but seems from other characters to be *R. prostrata*. For the moment it is best kept in this species.

3. **Ruellia linearibracteolata** *Lindau* in Ann. R. Ist. Bot. Roma 6: 70 (1897); C.B. Clarke in F.T.A. 5: 47 (1899); E.P.A.: 936 (1964); Ensermu in F.E.E. 5: 390 (2006); Hedrén in Fl. Somalia 3: 394 (2006). Type: Ethiopia, Ganale, *Ruspoli & Riva* 1152 (FT, holo.)

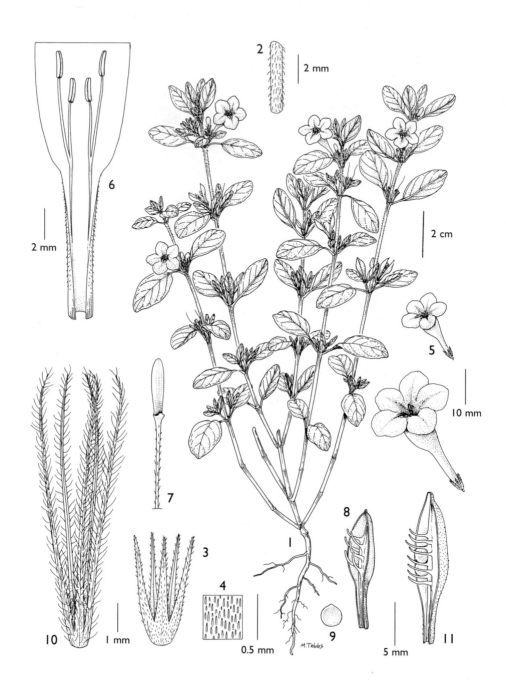

FIG. 31. *RUELLIA PATULA* — **1**, habit; **2**, stem indumentum; **3**, calyx; **4**, calyx indumentum; **5**, variation in corolla size; **6**, corolla opened up showing stamens; **7**, style and stigma; **8**, capsule; **9**, seed. *R. PROSTRATA* — **10**, calyx; **11**, capsule. 1–5 (small corolla) from *Faulkner* 1159, 5 (large corolla) from *Faulkner* 3507, 6 from *Mathew* 6259, 7 from *Faulkner* 543, 8–9 from *Newbould* 6766, 10 from *Mungai* 79/73, 11 from *Verdcourt* 5253. Drawn by Margaret Tebbs.

Shrubby perennial herb or shrub to 0.5(–1) m tall; young stems glabrous except often with long broad curly glossy hairs to 3 mm long near nodes. Leaves with petiole 0–10 mm long, with long broad curly glossy hairs to 3 mm long; lamina narrowly ovate to ovate or elliptic, largest 2.5–8.5 × 0.6–2.2 cm, apex acute, base attenuate, decurrent on petiole, pubescent or sparsely so, densest along veins, margin distinctly ciliate. Flowers solitary or in 3-flowered axillary dichasia; pedicels 0.5–1.5 mm long, glabrous to puberulous; bracteoles (bracts in cymes) narrowly ovate to ovate, 10–20 × 3–8 mm. Calyx 6–11 mm long, divided to 3–5 mm from base (i.e. normally about half way), with ciliate lobes, otherwise glabrous or with a few hairs along ribs; lobes triangular or narrowly so. Corolla falling early in the morning, pale mauve to purple, with darker lines in throat; tube 15–17 mm long of which the basal cylindric part 4–5 mm and the throat 11–12 mm long; lobes 8–11 × 7–10 mm, ovate-elliptic or broadly so, with entire to crenate margin. Stamens included in throat, didynamous, anthers not overlapping; filaments fused for ± 0.5 mm at base, free parts 1–3 and 4–6 mm long; anthers ± 2 mm long. Ovary with sparse appressed hairs; ventral stigma lobe 1–1.5 × ± 0.5 mm. Capsule narrowly oblanceolate, with appressed hairs, eventually glabrous, 17–22 mm long, 8–12-seeded. Mature seed not seen.

UGANDA. Karamoja District: Kanamugit, no date, *Eggeling* 2996! & 32 km N of Kacheliba, 8 May 1953, *Padwa* 92!
KENYA. Northern Frontier District: Dandu, 13 May 1952, *Gillett* 13162!; Turkana District: W of Turkwell, 19 Apr. 1954, *Popov* 1497!; Masai District: 21 km on Ol Tukai–Namanga road, 15 Dec. 1959, *Verdcourt* 2574!
DISTR. **U** 1; **K** 1, 2, 6, 7; Ethiopia, Somalia; Yemen, India
HAB. *Acacia, Acacia-Commiphora* and *Acacia-Euphorbia* bushland on sandy or alluvial soil; (50–)200–950 m

NOTE. This species has characters from both *R. prostrata* (hairy capsule) and *R. patula* (short triangular calyx lobes). It differs from both in the shrubby habit and the narrowly oblanceolate capsule with appressed hairs.

4. **Ruellia patula** *Jacq.*, Misc. Bot. 2: 358 (1781); C.B. Clarke in F.T.A. 5: 45 (1899); Fiori, Boschi e Piante Legn. Eritrea: 345 (1912); Chiovenda, Fl. Somala 2: 348 (1932); Lugard in K.B. 1933: 94 (1934); F.P.S. 3: 187, fig. 50 (1956); E.P.A.: 938 (1964); Napper in Hooker, Ic. Pl., ser. 5, 7: t. 3690 (1971); U.K.W.F.: 583 (1974); Vollesen in Opera Bot. 59: 81 (1980); Champluvier in Fl. Rwanda 3: 483, fig. 147 (1985); Blundell, Wild Fl. E. Afr.: 395, pl. 612 & 802 (1987); Hepper & Friis, Pl. Forssk. Fl. Aeg.-Arab.: 69 (1994); U.K.W.F., ed. 2: 270 (1994); Lebrun & Stork, Enum. Pl. Afr. Trop. 4: 501 (1997); Friis & Vollesen in Biol. Skr. 51(2): 453 (2005); Ensermu in F.E.E. 5: 389 (2006); Hedrén in Fl. Somalia 3: 395 (2006). Type: Cult. in Hort. Vienna from seeds from India, probably not preserved

Perennial or shrubby herb or subshrub; stems to 0.6(–1) m long, erect to trailing, when young subglabrous to densely puberulous, pubescent or pilose, sometimes with stalked capitate glands. Leaves with petiole 2–32 mm long; lamina ovate to elliptic or narrowly so (rarely lanceolate), largest 1–6 × 0.5–3.7 cm, apex acute to rounded (rarely retuse), base cuneate to truncate, sparsely to densely puberulous to pubescent (rarely subglabrous), densest along veins, sometimes with scattered capitate glands. Flowers solitary or in 3(–5)-flowered axillary cymes; peduncles in cymes 1–6(–12) mm long; pedicels 0–2(–5) mm long, glabrous to puberulous; bracteoles (bracts in cymes) ovate-elliptic or narrowly so, 7–18 × 3–8 mm. Calyx 3–8(–9) mm long, divided to 1–2 mm from base, uniformly puberulous or with ciliate lobes and midribs or glabrous; lobes narrowly triangular. Corolla falling early in the morning, white to mauve or purple; tube (10–)12–34 mm long of which the basal cylindric part 3–12(–20) mm and the throat (7–)9–19 mm long; lobes 4–15 × 3–15 mm, ovate-elliptic or broadly so with entire to crenate-dentate margin. Stamens included in throat (rarely slightly exserted), didynamous, anthers not or slightly overlapping; filaments fused for 1–3 mm at base, free parts 2–5 and 3–7 mm long;

anthers 1.5–2.5 mm long. Ovary glabrous; ventral stigma lobe 1–1.5 × ± 0.5 mm. Capsule clavate, glabrous, 11–20(–22) mm long, (4–)6–14(–16)-seeded. Seed broadly ellipsoid to circular, dark brown, 3–4.5 × 2.5–4.5 mm. Fig. 31, 1–9, p. 206.

UGANDA. Karamoja District: Kangole, Dec. 1957, *Wilson* 410!; Ankole District: Sanga Camp, Oct. 1932, *Eggeling* 604!; Teso District: Agu Swamp, Sep. 1932, *Chandler* 981!
KENYA. Northern Frontier District: Furroli, 13 Sep. 1952, *Gillett* 13836!; Trans-Nzoia District: Mt Elgon, above Endebess, 30 Apr. 1961, *Polhill* 402!; Kwale District: Shimba Hills, Longomwagandi, 13 Nov. 1992, *Harvey & Vollesen* 52!
TANZANIA. Mwanza District: Massanza, Bujingwa, 20 Nov. 1951, *Tanner* 483!; Arusha District: Engari Nairobi, no date, *Greenway & Kanuri* 14824!; Dodoma District: Salango Forest, 20 March 1974, *Richards & Arasululu* 29010!; Zanzibar, Mloni, Apr. 1953, *R.O. Williams* 177!
DISTR. **U** 1–4; **K** 1–7; **T** 1–3, 5–8; **Z**; **P**; Niger, Central African Republic, Congo-Kinshasa, Rwanda, Burundi, Sudan, Ethiopia, Somalia, Zambia, Malawi, Mozambique, Zimbabwe, Botswana, South Africa; Egypt, Saudi Arabia, Yemen, Oman, India
HAB. In a wide range of grassland, bushland, woodland, forest margins and riverine forest, persisting in disturbed areas, on an equally wide range of soil types; near sea level to 2100 m

SYN. *Dipteracanthus patulus* (Jacq.) Nees in Wall., Pl. As. Rar. 3: 82 (1832) & in DC., Prodr. 11: 126 (1847); Wight, Ic. Pl. 4: t.1505 (1850)
    Dipteracanthus patulus (Jacq.) Nees var. *obtusior* Nees in DC., Prodr. 11: 126 (1847); Solms-Laubert in Schweinf., Fl. Aeth.: 109 & 243 (1867). Type: Sudan, Cordofan, Arasch Cool, *Kotschy* 119 (K!, lecto.; selected here; BM!, iso.)
    Dipteracanthus sudanicus Schweinf. in Verh. Zool.-Bot. Ges. Wien 18: 679 (1868). Type: Eritrea, Keren, Ainsaba, *Steudner* 1806 (B†, holo.)
    Ruellia patula Jacq. var. *erythraea* Terrac. in Ann. R. Ist. Bot. Roma 5: 102 (1893); Fiori, Boschi e Piante Legn. Eritrea: 345 (1912). Types: Eritrea, Anfilah Bay, Mt. Ferehan, *Terraciano* (FT, syn.) & Midir Island, *Terraciano* (FT, syn.)
    R. sudanica (Schweinf.) Lindau in E.J. 20: 15 (1894); C.B. Clarke in F.T.A. 5: 46 (1899); F.P.S. 3: 187 (1956); E.P.A.: 939 (1964); Iversen in Symb. Bot. Ups. 29(3): 162 (1991); Lebrun & Stork, Enum. Pl. Afr. Trop. 4: 501 (1997)
    R. ibbensis Lindau in E.J. 20: 15 (1894). Type: Sudan, Niamniam, Ibba, *Schweinfurth* 3978 (B†, holo.)
    R. gongodes Lindau in E.J. 33: 187 (1902); Lebrun & Stork, Enum. Pl. Afr. Trop. 4: 501 (1997). Type: Tanzania, Kilwa District: Dondeland, Kwa Likemba, *Busse* 567 (B†, holo.)
    R. patula Jacq. var. *dumicola* Chiov., Fl. Somala 2: 348 (1932). Type: Somalia, Mogadishu, *Senni* 390 (FT, holo.)
    [*R. ovata* sensu Mildbraed in N.B.G.B. 9: 494 (1926), *non* Thunb. (1800)]

NOTE. *R. patula* is an excessively variable species, but without extensive field-work not much progress can be made in separating out infraspecific (? or specific) taxa.
    Plants from the dry northern, eastern and central parts are usually erect, have a short often dense indumentum usually with capitate glands intermixed. Plants from the central and western highland areas are erect to decumbent and have a sparse to dense indumentum of longer pilose hairs. Similar trailing plants also occur along the coast where they have sometimes been named as *R. cordata*. But this is a species with – as the name implies – cordate or subcordate leaves from southern Africa only extending north into Zimbabwe and Mozambique
    The lower cylindrical part of the corolla tube is normally shorter than the throat. But in scattered collections (e.g. *Polhill & Paulo* 2267, Singida District: Singida–Sekenke road) throughout Kenya and Tanzania the lower part of the tube is longer than the throat. In all other characters these collections are typical and I see no reason to treat them as a separate taxon. Further south in tropical Africa this latter condition becomes the norm; here scattered specimens with short lower tube occur throughout the area.

5. **Ruellia praetermissa** *Lindau* in E.J. 20: 15 (1894) & in E. & P. Pf. IV,3b: 310 (1895); C.B. Clarke in F.T.A. 5: 45 (1899); F.P.S. 3: 187 (1956); Heine in F.W.T.A. (ed. 2) 2: 396 (1963); E.P.A.: 939 (1964); Heine in Fl. Gabon 13: 11 (1966); Burkill, Useful Pl. W. Trop. Afr. 1: 26 (1985); Lebrun & Stork, Enum. Pl. Afr. Trop. 4: 501 (1997). Types: Sudan, Bongo, Gir, *Schweinfurth* 2155 (K!, isosyn.); Niamniam Nobambisso, *Schweinfurth* 3754 (not seen); Tukamis, Saiba Indimma, *Schweinfurth* 3789 (not seen)

Erect perennial with single stems to 60 cm tall from creeping rhizome; young stems pilose or sparsely so with broad curly glossy glandular hairs to 2 mm long. Leaves with petiole 5–25 mm long; lamina ovate or broadly so, largest 4–7.5 × 2.2–4.2 cm, apex acute, base cuneate to truncate (rarely subcordate), pubescent to pilose or sparsely so. Flowers solitary or in 2-flowered axillary cymes; peduncle in cymes 3–7 mm long; pedicels 1–3 mm long, glabrous to pilose; bracteoles (bracts in cymes) ovate-elliptic or narrowly so, 11–27 × 3–12 mm. Calyx 5–7 mm long, divided to (1–)2–3 mm from base, glabrous or lobes with scattered pilose hairs; lobes narrowly triangular. Corolla open during the day, mauve; tube 16–23 mm long of which the basal cylindric part 6–8 mm and the throat 8–13 mm long; lobes 8–12 × 6–11 mm, ovate-elliptic with entire to slightly crenate margin. Stamens included in throat, didynamous, anthers not overlapping; filaments fused for 1–2 mm at base, free parts 2–3 and 5–8 mm long; anthers 1.5–2 mm long. Ovary glabrous; ventral stigma lobe 1–1.5 × 0.5–1 mm. Capsule clavate, glabrous, 13–18 mm long, 8–14-seeded. Seed ellipsoid to circular, brown, 3–3.5 × 2.5–3.5 mm.

UGANDA. West Nile District: near Congo border W of Oleiba, 2 Aug. 1953, *Chancellor* 90a!
TANZANIA. Kilwa District: Selous Game Reserve, Madaba, 7 March 1976, *Vollesen* MRC 3346!
DISTR. **U** 1; **T** 8; Senegal, Mali, Sierra Leone, Ivory Coast, Ghana, Togo, Benin, Cameroon, Gabon, Central African Republic, Congo-Kinshasa, Rwanda, Burundi, Sudan, Zambia, Malawi
HAB. Wetter types of grassland, bushland and woodland; 300–1450 m

SYN. [*Ruellia patula* sensu C.B.Clarke in F.T.A. 5: 45 (1899), quoad spec. ex W. Afr.; F.W.T.A. 2: 246 (1931), *non* Jacq. (1781)]
    [*Ruellia prostrata* sensu Vollesen in Opera Bot. 59: 81(1980) pro parte, *non* Poir. (1804)]

NOTE. Very closely related to *R. patula* from which it differs in the habit, in the long broad curly glossy glandular hairs and in the longer pair of filaments being 3–5 mm longer than the shorter pair. *R. praetermissa* also has the flower open during the day while in *R. patula* it drops early in the morning.
    Almost certainly also in W and SW Tanzania. It has been collected close to the Zambia/Tanzania border.

6. **Ruellia burttii** *Vollesen* **sp. nov.** a *R. patula* corollae seminaque majora distinguenda. Tubus corollae 3.7–6.4 cm longus (parte cylindrica basali 2–4.2 cm). In *R. patula* tubus corollae (1–)1.2–3.4 cm longus (parte cylindrica basali 0.3–1.2(–2) cm). Semina majora 4.5–5 × 4–4.5 mm (in *R. patula* 3–4.5 × 2.5–4.5 mm). Type: Tanzania, Dodoma District: Kazikazi, *Burtt* 4596 (K!, holo.)

Perennial herb with several erect stems to 40 cm tall from woody rootstock; stems glabrous (with a fringe of hairs at nodes) to sparsely puberulous or with a few scattered curly hairs. Leaves with petiole 0–10 mm long; lamina elliptic-obovate or narrowly so, largest 3.5–7 × 1–2.8 cm, apex acute to rounded, base attenuate to cuneate, decurrent on petiole, glabrous or with scattered hairs on midrib (rarely sparsely puberulous). Flowers in 2-flowered axillary cymes (flower away from stem developing first) or solitary towards apex; peduncle 0–2 mm long; pedicels 0–1 mm long, glabrous (rarely puberulous); bracts and bracteoles lanceolate to elliptic, 11–30 × 1–9 mm. Calyx 8–10(–12) mm long, divided to 2–3 mm from base, glabrous with ciliate lobes or glabrous except for a terminal hair on each lobe. Corolla opening in the evening, fallen by morning, white to pale mauve; tube 37–64 mm long of which the basal cylindric part 20–42 mm and the throat 13–22 mm long; lobes 12–16 × 11–16 mm, broadly ovate with entire to dentate crenate margin. Stamens included in throat, didynamous, anthers slightly overlapping; filaments fused for 1–2 mm at base, free parts 2–4 and 4–6 mm long; anthers 2.5–4 mm long. Ovary glabrous; ventral stigma lobe 1–1.5 × ± 0.5 mm. Capsule clavate, glabrous, 17–22 mm long, (4–)6–14-seeded. Seed broadly ellipsoid to circular, grey to brown, 4.5–5 × 4–4.5 mm.

TANZANIA. Shinyanga District: Shinyanga, Feb. 1935, *Burtt* 5086!; Mbulu/Singida District: Yaida Valley, 23 Jan. 1970, *Richards* 25298!; Dodoma District: 10 km on Manyoni–Singida road, 15 Apr. 1988, *Bidgood et al.* 1120!

DISTR. **T** 1, 2, 5; not known elsewhere

HAB. Seasonally waterlogged short grassland on black cotton soil, *Acacia drepanolobium* grassland on clayey loam; 1100–1300 m

NOTE. Closely related to the widespread and variable *R. patula*. It differs in the larger corolla, larger seeds and different inflorescence.

According to *Burtt* 4596 this species is pollinated by hawkmoths.

7. **Ruellia** sp. A.

Erect shrubby perennial herb, 1 m tall; young stems sparsely puberulous and with scattered glands in two bands below nodes. Leaves with petiole 1–3.5 cm long, densely puberulous; lamina ovate, largest 9 × 4.2 cm, apex gradually narrowed into an elongated acute tip, base cuneate, unequal-sided, sparsely puberulous and with scattered glands. Flowers solitary in leaf axils towards apex of stem; pedicels 0–1 mm long; bracteoles narrowly ovate, up to 2.2 × 0.5 cm, decurrent into a petiole which is longer than the lamina. Calyx 16–20 mm long, divided to 1–2 mm from base; puberulous with glandular and non-glandular hairs; lobes linear. Corolla open during the day, blue (? mauve); tube ± 5 cm long of which the basal cylindric part ± 3.5 cm and the throat ± 1.5 cm long; lobes ± 1 × 1 cm, broadly ovate with rounded apex. Stamens included in throat, slightly didynamous; filaments fused for ± 3 mm at base, free parts ± 6 and 8 mm long, longer with scattered spreading hairs near base; anthers ± 5.5 mm long. Ovary not seen; ventral stigma lobe ± 3 mm long. Capsule strongly clavate, with scattered capitate glands in upper ¹/₄, ± 27 mm long, 8-seeded. Seed orbicular, dark brown, ± 6 × 5.5 mm.

TANZANIA. Uzaramo District: Pande Forest Reserve, 15 May 1979, *Mwasumbi* 11737!

DISTR. **T** 6; not known elsewhere

HAB. Lowland semi-evergreen forest; ± 175 m

NOTE. Known only from this collection. Superficially close to *R. amabilis* from which it differs in the cuneate leaf-base, broader bracteoles, shorter corolla tube, sessile flowers and blue corolla. These two last characters indicate that its real affinity is possibly more with the *R. patula*-group. From the species in this group it differs in the large anthers, large capsule and large seeds.

8. **Ruellia bignoniiflora** *S.Moore* in J.B. 18: 198 (1880); C.B. Clarke in F.T.A. 5: 48 (1899); Lebrun & Stork, Enum. Pl. Afr. Trop. 4: 501 (1997); Friis & Vollesen in Biol. Skr. 51(2): 453 (2005); Ensermu in F.E.E. 5: 392 (2006). Type: Angola, Loanda, *Welwitsch* 5063 (BM!, lecto., selected here; C!, K!, P!, iso.)

Erect to decumbent or scrambling shrubby herb or shrub to 1(–1.5) m tall, often with several stems from woody rootstock; young stems puberulous or densely so with glandular and non-glandular hairs (or without glands), often also with scattered to dense long thin pilose hairs. Leaves with petiole 0.5–3(–4.2) cm long; lamina ovate to cordiform or broadly so, largest 2–7.5(–10) × 1.7–4.5(–7) cm, apex subacute to rounded (rarely acute), base subcordate to truncate, puberulous to pubescent, densest along veins, often with scattered glands. Flowers solitary (rarely 2 per axil on separate pedicels), axillary; pedicels (2–)4–25(–40) mm long; bracteoles foliaceous, persistent, ovate to orbicular or cordiform, 10–17(–20) × 8–18(–21) mm; petiole 2–5 mm long. Calyx 13–25 mm long, divided to 1–2 mm from base, glandular-puberulous or with non-glandular hairs intermixed, often also with long thin pilose hairs; lobes linear. Corolla opening in the evening, fallen by morning, white to cream; tube 7–13 cm long of which the basal cylindric part 4.5–11 cm and the throat 2–3.5 cm long, the upper 5 mm widely expanded, almost saucer-like; lobes (10–)13–20 ×

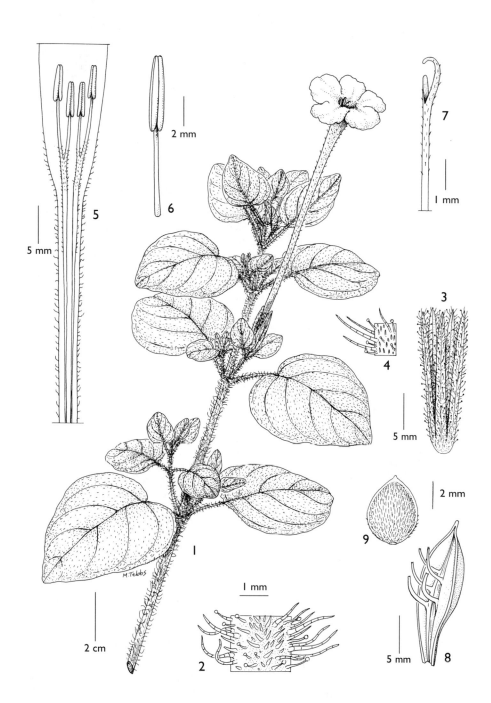

FIG. 32. *RUELLIA BIGNONIIFLORA* — **1**, habit; **2**, stem indumentum; **3**, calyx; **4**, calyx indumentum; **5**, corolla opened up showing stamens; **6**, stamen; **7**, style and stigma; **8**, capsule; **9**, seed. 1–2 from *Tweedie* 4033, 3–4 & 6 from *Verdcourt* 2319, 5 from *Hooper & Townsend* 1278, 7–8 from *Beesley* 264, 9 from *Carter & Stannard* 442. Drawn by Margaret Tebbs.

(10–)14–21 mm, broadly ovate-reniform, subacute to rounded. Stamens included in throat (rarely slightly exserted), subequal or slightly didynamous; filaments fused for 2–4(–5) mm at base, free parts (4–)6–8 mm long, glabrous; anthers 5–7 mm long. Ovary glabrous; stigma with two subequal lobes 2.5–4 mm long or dorsal shorter and down to 1 mm long. Capsule strongly clavate, glabrous or with a few hairs at apex, 20–23 mm long, 10-seeded. Seed broadly ovoid to circular, dark grey, 5–6 × 4.5–6 mm. Fig. 32, p. 211.

UGANDA. Karamoja District: S of Kapendongor, 10 June 1970, *Lye* 5589! & Moroto, Kasuweri Estate, June 1972, *Wilson* 2145!; Teso District: Serere, Tira, July 1926, *Maitland* 1294!
KENYA. Northern Frontier District: 65 km on Maralal–Baragoi road, Lopet Plateau, 16 Nov. 1977, *Carter & Stannard* 442!; Machakos District: Kibwezi, 2 Jan. 1959, *Verdcourt* 2319!; Teita District: 12 km on Voi–Taveta road, 11 Feb. 1966, *Gillett & Burtt* 17178!
TANZANIA. Musoma District: Banagi, Ikoma, 28 Nov. 1953, *Tanner* 1857!; Mbulu District: Lake Manyara National Park, 7 March 1969, *Richards* 24286!; Mpwapwa District: 11 km S of Gulwe, 9 Apr. 1988, *Bidgood et al.* 975!
DISTR. U 1, 3; K 1, 3, 4, 6, 7; T 1–3, 5; SE Sudan, S Ethiopia, Angola, Zambia, Malawi, Zimbabwe
HAB. *Acacia-* and *Acacia-Commiphora* woodland, bushland and scrub, often on rocky hills but also on black cotton soil; 600–1700 m

SYN. *Ruellia megachlamys* S.Moore in J.B. 32: 134 (1894); C.B. Clarke in F.T.A. 5: 48 (1899); U.K.W.F.: 589 (1974); Iversen in Symb. Bot. Ups. 29(3): 162 (1991); K.T.S.L.: 607 (1994); U.K.W.F., ed. 2: 270, pl. 117 (1994); Lebrun & Stork, Enum. Pl. Afr. Trop. 4: 501 (1997). Type: Kenya, Taita Hills, Ndara Ndi, Ngurunga Kifaniko, *Gregory* s.n. (BM!, holo.)
Dischistocayx bignoniiflorus (S.Moore) Lindau in E. & P. Pf. IV, 3b: 307 (1895)
Ruellia cygniflora Lindau in E.J. 33: 186 (1902); E.P.A.: 937 (1964); Lebrun & Stork, Enum. Pl. Afr. Trop. 4: 501 (1997). Type: Ethiopia, Hararghe, Gara Haqim, *Ellenbeck* 963 (B†, holo.)
Dipteracanthus bignoniiflorus (S.Moore) Bremek. in E.J. 73: 149 (1943)

NOTE. *Bidgood et al.* 975 from Mpwapwa District: S of Gulwe, is a long way outside the rest of the East African distribution. It has very large leaves (values in brackets) and thus superficially resembles the following species. But it has the small flowers of *R. bignoniiflora* with included stamens.

9. **Ruellia richardsiae** *Vollesen* **sp. nov.** a *R. bignoniiflora* praecipue floribus majoribus differt: calyx 3.2–3.8 cm longus; tubus corollae 14–15 cm long (parte cylindrica basali 12.5–13.5 cm longa fauce 1.5–2.5 cm longa, lobis 2–2.5 cm longis et 2–2.8 cm latis). In *R. bignoniiflora* calyx 1.3–2.5 cm longus; tubus corollae 7–13 cm longus (parte cylindrica basali 4.5–11 cm longa fauce 2–3.5 cm longa, lobis (1–)1.3–2 cm longis et (1–)1.4–2.1 cm latis). Stamina exserta (in *R. bignoniiflora* stamina in tubo inclusa); antherae 8 mm longa (5–7 mm in *R. bignoniiflora*). Type: Tanzania, Iringa District: Ruaha National Park, Mbage Camp, *Richards* 20950 (K!, holo.; EA!, K!, iso.)

Erect shrubby perennial herb to 60 cm tall with several stems from woody rootstock; young stems with sparse to dense spreading thin pilose hairs to 4 mm long and sparse to dense stalked capitate glands to 2 mm long. Leaves with petiole 2.5–3.8 cm long; lamina ovate to cordiform, largest 6–9 × 4.5–6 cm, apex acute or narrowing into a weakly defined tip, base truncate to subcordate, sparsely pubescent, densest along veins. Flowers solitary, axillary; pedicels 7–17 mm long; bracteoles foliaceous, 2–3 × 1–2.3 cm; petiole 5–10 mm long. Calyx 32–38 mm long, divided to 2–3 mm from base, pilose with white glossy hairs and usually also with dense stalked capitate glands; lobes linear, 2–3.5 mm wide. Corolla opening in the evening, fallen by early morning, pure white; tube 14–15 cm long of which the basal cylindric part 12.5–13.5 cm and the throat 1.5–2.5 cm long; lobes 2–2.5 × 2–2.8 cm, broadly ovate to reniform with broadly rounded apex. Stamens exserted, slightly didynamous; filaments fused for ± 4 mm at base, free parts ± 10 and 11 mm long, all four hairy throughout; anthers ± 8 mm long. Ovary not seen; ventral stigma lobe ± 2 mm long, dorsal ± 0.5 mm long. Capsule and seed not seen.

TANZANIA. Iringa District: Ruaha National Park, Mbage Camp, 10 Jan. 1966, *Richards* 20950! & Msembe–Kimiramatonge track, 4 March 1970, *Greenway & Kanuri* 14012! & park boundary on Idodi road, 28 Jan. 1972, *Björnstad* 1295!

DISTR. **T** 7; not known elsewhere

HAB. *Acacia-Commiphora-Combretum* woodland and wooded grassland; 800–900 m

SYN. [*Ruellia bignoniiflora* sensu Björnstad in Serengeti Res. Inst. Publ. 215: 26 (1976), *non* S.Moore (1880)]

NOTE. I have only seen these three collections but Björnstad (l.c.) cites additional collections. Differs from *R. bignoniiflora* – which occurs both to the north and south – in the much larger flower parts. It might eventually be treated as a subspecies.

10. **Ruellia amabilis** *S.Moore* in J.B. 18: 7 (1880); Lindau in P.O.A. C: 368 (1895) & in E. & P. Pf. IV, 3b: 309 (1895); C.B. Clarke in F.T.A. 5: 47 (1899); Lebrun & Stork, Enum. Pl. Afr. Trop. 4: 501 (1997); Ensermu in F.E.E. 5: 393 (2006). Type: Kenya, Teita District: Teita, Voi River, *Hildebrandt* 2480 (BM!, holo.; K!, P!, iso.)

Scandent or trailing shrubby herb to 2 m tall; young stems pubescent to pilose or sparsely so with shiny white hairs and with shorter glandular and non-glandular hairs. Leaves with petiole 0.5–5.5 cm long; lamina ovate to cordiform, largest 4–8.5 × 2.5–6.5 cm, apex gradually or abruptly narrowed into an acute to acuminate tip, base truncate to cordate, puberulous to pubescent or sparsely so, densest along veins, and with scattered glands. Flowers solitary or in 3-flowered axillary cymes; pedicels (peduncles in cymes) 8–20(–25) mm long; bracteoles (bracts in cymes) linear to slightly spathulate, caducous, 7–14 × ± 1 mm. Calyx 22–26 mm long, divided to 2–3 mm from base, puberulous with glandular and non-glandular hairs and with scattered to dense longer pilose hairs; lobes filiform. Corolla opening in the evening, fallen by early morning, pure white; tube 6.5–8.5 cm long of which the basal cylindric part 5.5–7.5 cm and the throat ± 1 cm long; lobes 13–17 × 12–18 mm, broadly ovate-reniform with rounded apex. Stamens exserted, didynamous; filaments fused for 1–2 mm at base, free parts 6–9 and 9–12 mm long, glabrous; anthers 4–5 mm long. Ovary with dense short-stalked capitate glands; ventral stigma lobe 1.5–2 × ± 0.5 mm, dorsal 0.5–1 mm long. Capsule clavate, glandular-puberulous or sparsely so all over, 24–30 mm long, 6–8-seeded. Seed oblong to orbicular, brown, 3.5–4 × 3–3.5 mm.

KENYA. Turkana District: no locality, Aug. 1982, *Ohta* 258!; Tana River District: 7 km NE of Garsen, 16–17 July 1972, *Gillett & Kibuwa* 19949!; Teita District: Voi River, 5 Feb. 1953, *Bally* 8735!

TANZANIA. Lushoto District: Daluni, Makifati, 3 Feb. 1954, *Faulkner* 1342!

DISTR. **K** 1/2, 7; **T** 3; Ethiopia

HAB. Undergrowth of evergreen lowland forest and riverine forest; 25–450 m (but clearly higher in NW Kenya).

NOTE. The Ethiopian material has much larger flowers than the East African material and should possibly be considered a distinct subspecies.

11. **Ruellia discifolia** *Oliv.* in Hooker, Ic. Pl., ser. 3, 6: t. 1511 (1886); James, Unknown Horn Afr.: 321, t.2 (1888); C.B. Clarke in F.T.A. 5: 47 (1899); Chiovenda, Fl. Somala: 248 (1929); E.P.A.: 937 (1964); Kuchar, Pl. Somalia (CRDP Techn. Rep. Ser., no. 16): 250 (1986); Lebrun & Stork, Enum. Pl. Afr. Trop. 4: 501 (1997); Ensermu in F.E.E. 5: 388 (2006); Hedrén in Fl. Somalia 3: 393 (2006). Type: Ethiopia, Adda Galla, *James & Thrupp* s.n. (K!, holo.)

Erect or straggling shrubby herb or shrub to 1 m tall; young stems densely whitish pubescent to tomentellous with many-rayed stellate hairs, often also with dense glandular hairs and rarely with long thin pilose hairs. Leaves with petiole 0.5–2(–3) cm long; lamina ovate to broadly cordiform or reniform, largest 1.7–6 × 1.5–5 cm, apex subacute to broadly rounded, base truncate to cordate, densely whitish stellate-pubescent to tomentellous beneath, sparser above. Flowers solitary (or with two flowers on distinct pedicels), axillary; pedicels 2–25(–35) mm long; bracteoles foliaceous, 6–15 × 4–12 mm. Calyx 13–25 mm long, divided to 2–3 mm from base, stellate-tomentellous and often also with stalked glands; lobes, linear, acute. Corolla opening in the evening, fallen by early morning, white (rarely with a faint purple tinge); tube 4–8 cm long of which the basal cylindric part 3–6 cm and the throat 1–3 cm long; lobes 8–13 × 7–14 mm, broadly triangular-reniform with wavy margin and rounded apex. Stamens didynamous; filaments fused for 2–3 mm at base, free parts 4–6 and 7–9 mm long, glabrous or longer pair with a few hairs; anthers 3–5 mm long. Overy densely pubescent; ventral stigma lobe 1.5–2.5 × ± 0.5 mm, dorsal ± 0.5 mm long. Capsule strongly clavate, pubescent all over with non-glandular hairs, 18–22 mm long, 6-seeded. Seed orbicular, dark grey, ± 6 × 6 mm.

KENYA. Northern Frontier District: Emurri, 28 June 1951, *Kirrika* 108! & 30 km on Ramu–Malka Mari road, 8 May 1978, *Gilbert & Thulin* 1586!
DISTR. **K** 1; Ethiopia, Somalia; Saudi Arabia, Yemen
HAB. *Acacia-Commiphora* woodland and bushland with *Sterculia* and *Terminalia* on rocky limestone hills; 400–600 m

20. ACANTHOPALE

C.B.Clarke in F.T.A. 5: 62 (1899), excl. distrib. India; Bremekamp in E.J. 73: 142 (1943)

Shrubby herbs or shrubs; cystoliths conspicuous. Leaves opposite, slightly to distinctly anisophyllous, entire to crenate. Flowers in axillary elongated or condensed racemoid cymes, often also with extra solitary flowers; bracts persistent, foliaceous; bracteoles persistent. Calyx deeply divided into 5 subequal (dorsal often slightly longer and wider) lobes. Corolla outside hairy on lobes and in a dorsal band, more rarely all over, often also with scattered stalked capitate glands, inside with a broad band of hairs ventrally and a transverse band below lobes, contorted in bud, zygomorphic; basal tube widening upwards, distinctly curved; throat long, funnel-shaped; lobes 5, erect or reflexed, broadly triangular-oblong, widest in upper lip. Stamens 4, didynamous, inserted at base of throat, included in throat; free part of filaments fused at base and running like two ridges to base of tube; anthers 2-thecous, thecae linear-oblong, rounded or apiculate at base and apex, glabrous. Ovary with 2 ovules per locule; style filiform, hairy throughout; dorsal stigma lobe reduced to a small tooth, ventral lobe large linear flattened. Capsule 2-seeded, ellipsoid, glabrous, thin-walled and often breaking irregularly and splitting lengthwise from base; retinaculae strong, curved. Seed discoid, densely covered with hygroscopic hairs.

6–8 species in tropical Africa and about 5 in Madagascar. A complete revision of the genus may eventually result in the merging of some Madagascan and African species.

1. Bracts, bracteoles and calyces with long spreading glossy white hairs on margins 2
   Bracts, bracteoles and calyces glabrous or with indistinct appressed or spreading hairs on margins 3
2. Calyx 5–9 mm long, whitish sericeous-pubescent or densely so apically; stems with appressed straight hairs; all four filaments densely hairy in lower half . . . .   1. *A. pubescens*
   Calyx 10–14 mm long, densely pubescent in apical half with long curly glossy hairs; stems with spreading curly hairs; two short filaments only hairy near base, two longer thinly hairy in lower half . . . .   2. *A. confertiflora*

3. Bracteoles and calyces glabrous; filaments densely hairy
   along their whole length; corolla tube below filament
   attachment 3–5 mm long; capsule 15–18 mm long . . . . . .   4. *A. macrocarpa*
   Bracteoles and calyces hairy at least on margins; filaments
   only hairy at base or at most halfway up; corolla tube
   below filament attachment 5–15 mm long; capsule
   11–14 mm long 4
4. Calyx with spreading hairs on margins; flowers in elongated
   racemes up to 16 cm long; corolla tube below filament
   attachment 5–8 mm long . . . .   3. *A. laxiflora*
   Calyx with appressed hairs on margins; flowers in dense
   heads (rarely in racemes up to 3 cm long); corolla tube
   below filament attachment 10–15 mm long . . . .   1. *A. pubescens*

1. **Acanthopale pubescens** (*Engl.*) *C.B.Clarke* in F.T.A. 5: 64 (1899); F.P.N.A. 2: 285
(1947); T.T.C.L.: 1 (1949); Binns, Checklist Herb. Fl. Malawi: 12 (1968); Champluvier in
Fl. Rwanda 3: 430 (1985); U.K.W.F., ed. 2: 270 (1994); K.T.S.L.: 596 (1994); Lebrun &
Stork, Enum. Pl. Afr. Trop. 4: 466 (1997); White *et al.*, For. Fl. Malawi: 112 (2001); Friis
& Vollesen in Biol. Skr. 51(2): 435 (2005); Ensermu in F.E.E. 5: 369 (2006). Type:
Congo-Kinshasa, N of Lake Kivu, Mt Ninagongo, *von Götzen* 58 (B†, holo.)

Shrubby herb or soft-wooded shrub to 2.5(–3) m tall, sometimes with stilt-roots;
young stems quadrangular and longitudinally furrowed, subglabrous to densely
sericeous with closely appressed often (if sparse) almost invisible hairs. Leaves with
petiole (1–)2–6.5 cm long on larger leaves with sparse to dense long spreading hairs
on edges; lamina ovate to elliptic, largest 8–24 × 3.5–9.5 cm, apex acuminate to
cuspidate, base attenuate, beneath subglabrous to pubescent with broad glossy hairs
along midrib and veins, sparser (rarely dense) on lamina, above with sparse to dense
long broad appressed hairs all over, denser on midrib (rarely glabrous). Flowers in
dense capitate racemoid heads or elongated racemes to 3 cm long; peduncle 0.5–23 mm
long, winged, glabrous to sericeous; bracts often purplish; pedicels to 1 mm long;
bracteoles linear to spathulate, 5–14(–18) mm long, sometimes with apical part
recurved, conspicuously whitish sericeous-pubescent (more rarely sparsely purplish-
sericeous-pubescent on midrib and margins) on edges (rarely also with stalked
capitate glands). Calyx 5–10 mm long, usually conspicuously whitish sericeous-
pubescent or densely so apically, more rarely sparsely sericeous in apical half with
purple appressed hairs, glabrous towards base (rarely also with stalked capitate
glands), lobes linear-lanceolate or slightly spathulate, acute to obtuse. Corolla pure
white (rarely) or white with pink to purple lines in throat, pubescent with broad
glossy hairs, densest dorsally and sometimes only in a dorsal band; tube below
attachment of stamens 8–15 mm long, throat 10–25 mm long and 8–20 mm in
diameter apically, lobes 4–8 mm long, acute or bifid at apex. Free part of filaments
4–8 and 6–12 mm long, densely hairy in lower half; anthers 2–3 mm long; ventral
stigma lobe 2–3 mm long. Capsule 11–14 mm long. Seed ellipsoid to circular, 4–5 ×
3–4 mm. Fig. 33, p. 216.

UGANDA. Ankole District: Igara, Kalinzu Forest, 6 Jan. 1953, *Osmaston* 2675!; Toro District:
   Ruwenzori, Bunyangabo, 9 Sep. 1951, *Osmaston* 1195!; Mbale District: NW Elgon, above Sipi,
   1939, *Dale* 50!
KENYA. Meru District: Nyambeni Range, Itiani, 15 Nov. 1970, *Mabberley* 427!; Kericho District:
   SW Mau Forest, Sambret, 16 Aug. 1962, *Kerfoot* 4073!; North Kavirondo District: Kakamega
   Forest, 15 Oct. 1953, *Drummond & Hemsley* 4773!
TANZANIA. Bukoba District: Minziro Forest Reserve, Nyakashenye, 5 Feb. 2001, *Festo* 876!;
   Morogoro District: Nguru Mts, Maskati, 16 March 1988, *Bidgood et al.* 455!; Njombe District:
   Livingstone Mts, Bumbigi, 28 Feb. 1991, *Gereau & Kayombo* 4137!
DISTR. **U** 2–4; **K** 1, 3–5; **T** 1, 4–7; Congo-Kinshasa, Rwanda, Burundi, Ethiopia, Sudan, Malawi,
   Mozambique, Zimbabwe

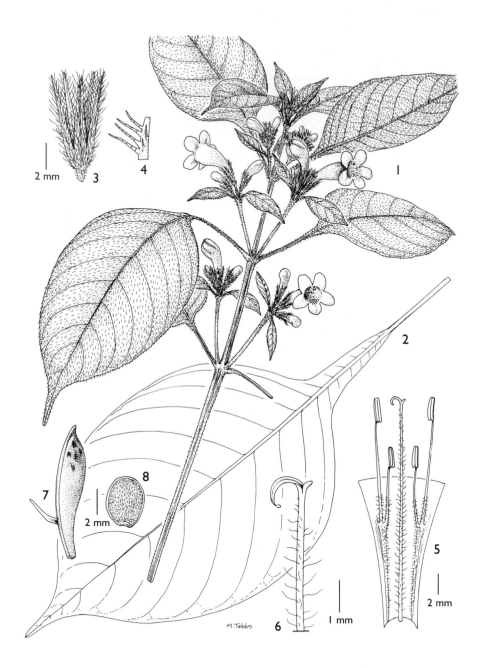

FIG. 33. *ACANTHOPALE PUBESCENS* — **1**, habit, × ²⁄₃; **2**, basal leaf, × ²⁄₃; **3**, calyx; **4**, detail of calyx indumentum; **5**, corolla tube opened with stamens and style; **6**, apical part of style and stigma; **7**, capsule; **8**, seed. 1 & 3–4 from *Tweedie* 2434, 2 from *Fries* 1424, 5–6 from *Congdon* 336, 7–8 from *Battiscombe* 1164. Drawn by Margaret Tebbs.

HAB. Medium altitude and montane wet evergreen forest, swamp forest; 1150–2750 m

SYN. *Dischistocalyx pubescens* Engl. in Götzen, Durch Afrika, Sonderabdr.: 9 (1895)
    Acanthopale albosetulosa C.B.Clarke in F.T.A. 5: 64 (1899); Binns, Checklist Herb. Fl. Malawi:
    12 (1968); Lebrun & Stork, Enum. Pl. Afr. Trop. 4: 466 (1997). Type: Malawi, Masuku
    Plateau, *Whyte* s.n. (K!, holo.)
    Dischistocalyx pubescens Engl. var. *longipilosa* Mildbr. in N.B.G.B. 11: 1082 (1934). Type:
    Tanzania, Morogoro District: Uluguru Mts, Lukwangule Plateau, *Schlieben* 3568 (B†,
    holo.; BR!, P!, iso.)
    Acanthopale longipilosa (Mildbr.) Bremek. in E.J. 73: 146 (1943); Lebrun & Stork, Enum. Pl.
    Afr. Trop. 4: 466 (1997)

NOTE. The anthers of the shorter pair of stamens are occasionally sterile.

2. **Acanthopale confertiflora** (*Lindau*) *C.B.Clarke* in F.T.A. 5: 64 (1899);
Bremekamp in E.J. 73: 144 (1943); Binns, Checklist Herb. Fl. Malawi: 12 (1968);
Champluvier in Fl. Rwanda 3: 430 (1985); Lebrun & Stork, Enum. Pl. Afr. Trop. 4:
466 (1997); White *et al.*, For. Fl. Malawi: 112 (2001). Types: Malawi, *Buchanan* 22 (B†,
syn.; K!, iso.); Congo-Kinshasa, Ituri, *Stuhlmann* 2687 (B†, syn.)

Shrubby herb or soft-wooded shrub to 4 m tall, often with stilt-roots; young stems
with sparse spreading broad glossy hairs, often denser at nodes, without appressed
hairs. Leaves with petiole 1–6 cm long on larger leaves with sparse to dense long
spreading hairs on edges; lamina ovate to elliptic, largest 9–18 × 3.5–8.5 cm, apex
acuminate to cuspidate, base attenuate, beneath subglabrous to pubescent along
midrib and veins, sparser or glabrous on lamina, above with sparse long broad
appressed hairs all over, denser on midrib. Flowers in condensed head-like racemes;
peduncle 5–15(–30) mm long, winged, sparsely pubescent with curly glossy hairs;
bracts sometimes purplish; pedicels to 1 mm long; bracteoles oblanceolate to
slightly spathulate, 12–18 mm long, sometimes with apical part recurved, pubescent
or densely so with broad glossy hairs. Calyx 10–14 mm long, pubescent or densely
so in apical half with long curly glossy hairs, glabrous towards base, lobes linear to
oblanceolate, subacute to obtuse. Corolla white with pink to purple lines in throat,
pubescent with broad glossy hairs in a dorsal band and sometimes along veins; tube
below attachment of stamens 8–12 mm long, throat 18–25 mm long and 18–25 mm
in diameter apically, lobes 8–12 mm long, emarginate or bifid at apex. Free part of
filaments 8–10 and 11–15 mm long, two short filaments only hairy near base, two
long ones thinly hairy halfway up; anthers ± 3.5 mm long; ventral stigma lobe 1.5–2 mm
long. Capsule 12–13 mm long. Seed ellipsoid to circular, ± 6 × 5 mm.

UGANDA. Kigezi District: 3 km W of Buhoma, Impenetrable Forest, 18 Feb. 1997, *Katende* 5155!
TANZANIA. Njombe District: Livingstone Mts, Milo, Soroto Forest, 1 Feb. 1961, *Richards* 14071!
    & Msalaba, 21 March 1991, *Gereau & Kayombo* 4423!; Mbeya District: Poroto Mts, above
    Irambo, Chumvi Forest, 8 Feb. 1979, *Cribb et al.* 11349!
DISTR. **U** 2; **T** 7; Congo-Kinshasa, Rwanda, Burundi, Zambia, Malawi
HAB. Wet evergreen montane forest; 2100–2250 m

SYN. *Dischistocalyx confertiflora* Lindau in E.J. 20: 13 (1894) & in P.O.A. C: 368 (1895)

NOTE. The anthers of the shorter pair of stamens (rarely of all) are occasionally sterile.

3. **Acanthopale laxiflora** (*Lindau*) *C.B.Clarke* in F.T.A. 5: 63 (1899), excl. specim.
ex Cameroon; T.T.C.L.: 1 (1949); Iversen in Symb. Bot. Ups. 29(3): 160 (1991); Ruffo
et al., Cat. Lushoto Herb. Tanzania: 1 (1996), as *taxiflora*. Type: Tanzania, Lushoto
District: Usambara Mts, Magamba Forest, *Holst* 3840 (B†, holo.)

Shrubby herb or soft-wooded shrub to 3 m tall; young stems winged, with appressed hairs at nodes (rarely along whole internode or glabrous). Leaves with petiole 1–4 cm long on larger leaves without or with a few long spreading hairs on edges; lamina ovate to elliptic, largest 6.5–15(–18.5) × 2.5–5(–7) cm, apex acuminate to cuspidate, base attenuate, with scattered hairs along midrib and veins beneath and with a few hairs on lamina above. Flowers in elongated racemes up to 16 cm long; peduncle up to 1.5 cm long, winged, glabrous or with sparse appressed hairs; internodes to 3.5 cm long, winged; bracts sometimes purplish; pedicels to 1 mm long; bracteoles linear to oblanceolate, 10–16 mm long, sometimes with apical part recurved, glabrous or with appressed hairs on midrib, with a few spreading hairs on edges near base, sometimes also with stalked capitate glands. Calyx 7–12 mm long, glabrous or sparsely sericeous along midrib, with short spreading hairs on margin, sometimes also with stalked capitate glands, lobes linear, subacute or obtuse. Corolla white with purple throat or with purple lines in throat, sparsely hairy in a narrow a dorsal band; tube below attachment of stamens 5–8 mm long, throat 13–20 mm long and 10–20 mm in diameter apically, lobes 4–7 mm long, acute or bifid at apex. Free part of filaments 7–10 and 10–13 mm long, only hairy at the very base; anthers 3–4 mm long; ventral stigma lobe ± 3 mm long. Capsule 13–14 mm long. Seed not seen.

TANZANIA. Arusha District: Mt Meru, 31 Oct. 1965, *Beesley* 169!; Moshi District: Mt Kilimanjaro, 16 Oct. 1993, *Grimshaw* 931008!; Lushoto District: W Usambara Mts, Shagayu Forest Reserve, 14 March 1984, *Borhidi et al.* 84837!
DISTR. **T** 2, 3; not known elsewhere
HAB. Wet evergreen montane forest; (1250–)1800–2300 m

SYN. *Dischistocalyx laxiflorus* Lindau in E.J. 20: 13 (1894) & in P.O.A. C: 368 (1895)
    Acanthopale azaleoides C.B.Clarke in F.T.A. 5: 63 (1899); T.T.C.L.: 1 (1949); V.C. Gilbert, Pl. Mt Kilimanjaro: 79 (1970). Type: Tanzania, Kilimanjaro, *Johnston* 12 (K!, holo.)

NOTE. Mass flowering and subsequent dying back occurred on Mt Kilimanjaro in 1992 and 1993 (Grimshaw, pers. comm.).
    Superficially close to the West African *A. decempedalis* but with a different bract- and calyx indumentum.

4. **Acanthopale macrocarpa** *Vollesen* **sp. nov.** ab *A. laxiflora* bracteolis calycibusque omnino glabris, filamentis per longitudinem totam (non tantum in dimidio basali) dense indumentosis, tubo basali corollae breviore (3–5 mm longo non 5–15 mm), capsula majore (15–18 mm longo non 11–14 mm) distincta. Type: Kenya, N Kavirondo District: Kakamega Forest, *Drummond & Hemsley* 4772 (K!, holo.; EA!, K!, iso.)

Shrubby herb or soft-wooded shrub to 5 m tall; young stems quadrangular and longitudinally furrowed but not winged, glabrous or with a few hairs at nodes. Leaves with petiole 1–3 cm long on larger leaves without or with a few long spreading hairs on edges; lamina ovate to elliptic, largest 11–24 × 4–8.5 cm, apex acuminate to cuspidate, base attenuate, glabrous or with a few hairs along midrib and veins beneath, without (rarely a few) hairs on edges. Inflorescences at first capitate but soon elongating into racemes to 5(–15 in fruit) cm long; peduncle up to 2.5 cm long, winged, glabrous; internodes to 2 cm long in elongated racemes, winged; pedicels to 1 mm long; bracteoles lanceolate to oblanceolate, 10–15 mm long, never with apical part recurved, glabrous. Calyx 10–14 mm long, glabrous, lobes linear-lanceolate, subacute, apical part often recurved. Corolla white, hairy on lobes and in a thin dorsal band; tube below attachment of stamens 3–5 mm long, throat 20–25 mm long and 15–20 mm in diameter apically, lobes 4–7 mm long, acute or bifid at apex. Free part of filaments 8–15 and 10–17 mm long, densely hairy along their whole length; anthers 3–4 mm long; ventral stigma lobe ± 2 mm long. Capsule 15–18 mm long. Seed (immature) 4–5 × 3–4 mm.

UGANDA. Toro District: Kibale Forest, Kanyawara, Dec. 1974, *Struhsaker* 176A! & 11 Oct. 1982, *Struhsaker* 377!
KENYA. North Kavirondo District: Kakamega Forest, 15 Oct. 1953, *Drummond & Hemsley* 4772! & 26 Nov. 1969, *Bally* 13667!
DISTR. **U** 2; **K** 5; not known elsewhere
HAB. Medium altitude wet evergreen forest; 1500–1600 m

NOTE. Differs from *A. laxiflora* in the glabrous bracteoles and calyces, the very short basal corolla tube, the densely hairy filaments and the larger capsules.
Mass flowering and subsequent dying back occurred in Uganda in 1974 and 1982.

21. **MIMULOPSIS**

Schweinf. in Verh. Zool.-Bot. Ges. Wien 18: 677 (1868); G.P. 2: 1080 (1876); C.B. Clarke in F.T.A. 5: 54 (1899)

Perennial or shrubby herbs, shrubs or small trees; stems quadrangular; cystoliths conspicuous. Leaves opposiste, often with conspicuous "blisters" along veins beneath. Flowers in condensed racemiform to open panicles; bracts caducous or persistent; bracteoles 2, linear to narrowly ligulate, caducous or persistent. Calyx 5-lobed, divided almost to the base into linear-lanceolate to narrowly spathulate-ligulate lobes, dorsal lobe slightly to distinctly longer and wider then the rest. Corolla glabrous on the outside, contorted in bud, subactinomorphic, basal cylindric part of tube usually very short, throat long and much expanded at the mouth, with ill-defined herring-bones ventrally; lobes 5, two lateral in lower lip held horizontally to more or less vertically, similar to the two in upper lip, central in lower lip similar or longer and narrower, deflexed, with a papillate area above, throat with dense band of long stiff retrorse hairs on lower lip, this band becoming thinner on lateral and upper lip but often more or less continuous, also with longitudinal band from lower lip to base of tube. Stamens 4, didynamous, inserted near base of throat, free part of filaments fused at base, below insertion forming two hairy ridges to base of tube; anthers 2-thecous, thecae oblong, glandular and often also hairy dorsally along connective, rounded at apex, outer theca of longer pair with long spur at base, other six thecae mucronate. Ovary with 6–8 ovules per locule; style filiform, hairy but for apical few mm; stigma with dorsal lobe, absent or much reduced (rarely well-developed), ventral lobe linear. Capsule 6–8-seeded, ellipsoid-obovoid, valves strongly sulcate; retinaculae strong, hooked. Seed discoid, circular in outline, glabrous on faces, with dense hygroscopic hairs on rim.

About 15 species in tropical Africa and Madagascar.

1. Basal cylindric part of corolla tube 8–13 mm long; papillate area on lower lip 10–13 mm long, almost as long as the lip; capsule 35–50 mm long; bracts white, caducous . . . . 1. *M. arborescens*
   Basal cylindric part of corolla tube 0.5–4 mm long; papillate area on lower lip much smaller; capsule 14–26 mm long 2
2. Bracts white, caducous; dorsal calyx-lobe ± 2 mm longer than the others . . . . 2. *M. runssorica*
   Bracts green, persistent (rarely caducous towards apex of inflorescence); dorsal calyx-lobe 2–8 mm longer than the others (except *M. solmsii*) 3
3. Inflorescence axes and pedicels without stalked capitate glands . . . . 5. *M. elliotii*
   Inflorescence axes and pedicels with stalked capitate glands 4

4. Lower bracts persistent, upper caducous; inflorescence
   an open panicle or contracted in apical part   . . . . .   3. *M. excellens*
   All bracts persistent   5
5. Flowers in open panicles with long spreading branches;
   dorsal calyx-lobe same length or up to 2 mm longer
   than the others   . . . .   4. *M. solmsii*
   Flowers in condensed racemiform panicles with more
   or less erect short branches, but often with two long
   branches from basal nodes; dorsal calyx-lobe
   2–5(–7) mm longer than the others   6
6. Apical 2–7 mm of calyx lobes without stalked capitate
   glands; inflorescence axes and pedicels with
   capitate glands up to 1 mm long; upper part of
   stems glabrous or with short straight hairs   . . . .   8. *M. kilimandscharica*
   Calyx lobes with stalked capitate glands to within 1 mm
   of apex; inflorescence axes and pedicels with at least
   some glands 2 mm or longer; upper part of stems
   with long curly many-celled hairs   7
7. Calyx lobes 1–3 mm wide; corolla lobes about as wide
   as long or wider; E Uganda, Kenya, N Tanzania   . . .   6. *M. alpina*
   Calyx lobes 0.2–0.5 mm wide; corolla lobes distinctly
   longer than wide; W Uganda   . . . .   7. *M. sp. A*

1. **Mimulopsis arborescens** *C.B.Clarke* in F.T.A. 5: 57 (1899); Lindau in Wiss. Ergebn. Schwed. Rhodesia-Kongo-Exp. 1: 305 (1916); Chiovenda in Racc. Bot. Miss. Cons. Kenya: 94 (1935); T.S.K.: 161 (1936); F.P.N.A. 2: 277 (1947); K.T.S.: 17 (1961); U.K.W.F.: 587 (1974); K.T.S.L.: 605 (1994); U.K.W.F., ed. 2: 268 (1995); Lebrun & Stork, Enum. Pl. Afr. Trop. 4: 494 (1997). Type: Uganda, Ruwenzori Mts, Kivata, *Scott Elliot* 7666 (K!, holo.; BM!, iso.)

Strongly aromatic shrub or tree to 8 m tall, often thicket-forming, sometimes with stilt roots, basal "trunk" to 10 cm in diameter; young stems glabrous (rarely with scattered long curly many-celled broad hairs). Leaves with petiole of stem-leaves 5–20(–25) cm long; lamina of stem-leaves broadly cordiform-reniform, largest 11–26(–33) × 10–27(–32) cm, upwards smaller and narrower, apex triangular, base cordate or deeply so (upper subcordate to truncate) with sinus 1–3.5 cm deep, margin grossly doubly dentate with large triangular teeth 2–5 cm long, glabrous or with scattered long curly many-celled broad hairs, usually only along veins. Inflorescence narrowly racemiform, (5–)10–30(–40) cm long; branches erect, 0.5–2.5 cm long near base, shorter upwards (rarely with two longer lateral branches from basal nodes); flowers in 3-flowered cymes; bracts caducous, greenish white, boat-like and clasping partial inflorescences, up to 4.5 × 3.5 cm; branches and bracts glabrous to densely glandular-pubescent; pedicels to 5(–10) mm long. Calyx glandular-pubescent with broad glossy hairs, 19–38(–55 in fruit) mm long along dorsal lobe which is 1–2 mm longer than the others; lobes linear to oblanceolate, 1–5(–7) mm wide, acute or subacute. Corolla white to cream with dark brown throat, lower lip yellow with white hairs, 30–41 mm long, cylindric tube 8–13 mm, throat 8–12 mm, dorsal and lateral lobes broadly elliptic, 8–11 × 8–12 mm, lower lip oblong 14–17 × 8–10 mm, dorsal lobes without ring of stiff hairs and lateral only with a few; papillate area on lower lip ovate, 10–13 mm long, covering almost the whole of the lobe. Free part of filaments 3–5 mm long and free fused part similar length, hairy; anthers 5–7 mm long, spur ± 1 mm long. Stigma 2–3 mm long, dorsal lobe absent; ovary hairy near apex. Capsule 35–50 mm long, glabrous but for a few hairs near apex. Seed 7–9 × 6–8 mm.

UGANDA. Kigezi District: Kinkizi, Amahenga, Nov. 1946, *Purseglove* 2275!; Toro District: Ruwenzori Mts, Namwamba Valley, 5 Jan. 1935, *G. Taylor* 2857!; Mbale District: Mt Elgon, Bugugarga Forest, Sep. 1937, *Hancock* 216/37!

KENYA. Trans-Nzoia District: NE Elgon, Dec. 1953, *Tweedie* 1150!; South Nyeri District: S slopes of Mt Kenya, Thiba Fishing Camp, 21 Oct. 1979, *M.G. Gilbert* 5784!; Kericho District: 35 km NE of Kericho, Kimugu Tea Estate, 4 Dec. 1967, *Perdue & Kibuwa* 9259!

TANZANIA. Morogoro District: Uluguru Mts, Bondwa, 20 July 1972, *Mabberley* 1215!; Iringa District: Udzungwa Mts, Udekwa Village, Dec. 1982, *Rodgers & Hall* 2291! & above Sanje Falls, 24 July 1983, *Polhill & Lovett* 5149!

DISTR. **U** 2, 3; **K** 3–5; **T** 6, 7; Congo-Kinshasa, Rwanda, Burundi

HAB. Wet montane (rarely lowland) forest, including bamboo forest, usually in gaps or clearings or along streams; (600–)1250–2500 m

SYN. *Mimulopsis longisepala* Mildbr. in N.B.G.B. 11: 1081 (1934); T.T.C.L.: 13 (1949); Lebrun & Stork, Enum. Pl. Afr. Trop. 4: 494 (1997). Type: Tanzania, Morogoro District: Uluguru Mts, Kinole, *Schlieben* 2870 (B†, holo.; BR!, iso.)

NOTE. Mass-germination in this species was seen in the Mufindi area in March 2006 (pers. obs.) following flowering in late 2005 (C. Congdon, pers. comm.). There are other reports of the species flowering gregariously, but not of it actually dying back after flowering.

The populations in S Tanzania are very disjunct from the main area of the species but no significant morphological differences can be found.

2. **Mimulopsis runssorica** *Lindau* in E.J. 20: 10 (1894) & in P.O.A. C: 367 (1895) & in E. & P. Pf. IV, 3b: 301 (1895); C.B. Clarke in F.T.A. 5: 56 (1899); Chiovenda, Ruwenzori 1: 449 (1909); Mildbraed in N.B.G.B. 9: 497 (1926); Lebrun & Stork, Enum. Pl. Afr. Trop. 4: 494 (1997). Type: Congo-Kinshasa, Ruwenzori [Runssoro], *Stuhlmann* 2424 (B†, holo.)

Erect or scrambling shrubby herb or shrub to 3(?–5) m tall, often thicket-forming; young branches glabrous or with scattered long glandular many-celled hairs (rarely with fine downwardly directed sericeous hairs). Leaves with petiole of stem-leaves (4–)6.5–12 cm long; lamina of stem-leaves ovate-cordiform or broadly so, 10–18 × 7.5–15 cm, upwards smaller and narrower, apex acuminate, base of stem-leaves cordate (rarely subcordate) with sinus (0.2–)0.5–1 cm deep, of upper leaves cuneate to subcordate, margin grossly double-dentate with large triangular teeth 0.5–1 cm long, glabrous or with scattered hairs along veins and near margins, with distinct domatia in angles of main lateral veins beneath. Inflorescence narrow and racemiform when young, an open panicle later, 7–20 cm long; flowers in 3(–5)-flowered cymes or solitary towards tip of branches; bracts white, soon caducous, clasping young partial inflorescences; branches, bracts and pedicels glandular-tomentose with long many-celled hairs and with a fine "undercoat" of non-glandular hairs; pedicels to 6 mm long. Calyx glandular-pubescent or densely so with broad glossy hairs and with a fine non-glandular "undercoat", 9–18(–22 in fruit) mm long along dorsal lobe which is ± 2 mm longer than the others; lobes linear, 1–1.5(–2) mm wide. Corolla white to pale mauve with light to dark brown throat and yellow patch on lower lip, throat with dense long white hairs on lower lip and two lateral lobes, sparser or absent on dorsal lobes, 20–24 mm long, cylindric tube 1–2(–4) mm long, throat 9–10 mm long; lobes all transversely elliptic, 6–8 × 7–8 mm; papillate area on lower lip semi-circular or divided into two patches. Free part of filaments 2–5 mm long and free fused part ± 1 mm, glabrous or longer with a few hairs; anthers 3–4 mm long, spur 1–1.5 mm long. Stigma 1–2 mm long, dorsal lobe present, minute or absent; ovary densely hairy all over. Capsule ± 15 mm long, glandular and hairy all over. Seed ± 4 × 3.5 mm.

UGANDA. Ankole District: Kasyoha-Kitomi Forest Reserve, 24 Nov. 1994, *Poulsen* 713!; Toro District: Ruwenzori Mts, Mobuku Valley, near Nyinabitabu, 8 July 1952, *Ross* 438 & Lume Valley, 23 Aug. 1952, *Ross* 964!

DISTR. **U** 2; Congo-Kinshasa (Ruwenzori Mts)

HAB. *Podocarpus* forest on ridgetops, bamboo forest; (1200–)2500–3000 m

SYN. [*Mimulopsis arborescens* sensu Lindau in Z.A.E.: 295 (1911), *non* C.B.Clarke (1899)]
    [*Mimulopsis excellens* sensu Staner in Bull. Agr. Congo Belge 25: 427 (1934), *non* Lindau (1911)]

NOTE. I have not seen any authentic material of this species but C.B.Clarke (l.c.), who saw the type, annotated *Scott Elliot* s.n. (BM, K) as the same species. There are, however, some discrepancies between the material seen for this account and Lindau's original description. He describes the stems as tomentose and the leaves as being ovate and truncate at base. It is most likely that the specimen collected by Stuhlmann consisted only of the apical part of a stem, and the "tomentose" stem thus refers to the indumentum of the inflorescence axes. The description of the "leaf" as ovate with truncate base would also seem to refer to one of the narrow leaves immediately below the inflorescence.

According to Axel Poulsen (pers. comm.) this species flowered gregariously in 1994 on the Uganda side of the Ruwenzori.

3. **Mimulopsis excellens** *Lindau* in Z.A.E.: 295, t. 33 (1911); F.P.N.A. 2: 278 (1947); Champluvier in Fl. Rwanda 3: 474 (1985); Lebrun & Stork, Enum. Pl. Afr. Trop. 4: 494 (1997). Types: Rwanda, Rugege, *Mildbraed* 910 (B†, syn.; BR!, isosyn.) & Mt Karisimbi, *Mildbraed* 1581 (B†, syn.).

Strongly aromatic erect or scrambling shrubby herb to 3(?–5) m tall, sometimes forming large thickets; branches glabrous or laxly tomentose on upper nodes with long curly many-celled hairs to 5 mm long. Leaves with petiole of stem-leaves 4.5–9(–11) cm long; lamina of stem-leaves cordiform or broadly so, 9–20 × 5–18 cm, upwards smaller and narrower, apex acuminate to cuspidate, base cordate with sinus 0.5–2 cm deep, margin of stem-leaves sharply dentate (rarely crenate) with teeth to 0.5(–1) cm deep, subglabrous to pubescent on both sides. Inflorescence an open panicle when well-developed, but often racemiform when young or racemiform towards apex, often with longer branches from basal nodes, 10–40 cm long; flowers in 3–7-flowered cymes towards base of inflorescence, fewer or solitary upwards; bracts leafy in basal part of inflorescence, gradually more bract-like upwards and upper ones caducous in fruit; lower bracts with indumentum as leaves, upper branches and pedicels pubescent to tomentose with long curly glandular and non-glandular hairs (longest 3–5 mm); pedicels up to 8 mm long. Calyx glandular-pubescent or densely so; dorsal lobe 13–23(–28 in fruit) mm long, ventral and lateral lobes 11–15(–20 in fruit) mm long; lobes lanceolate to narrowly oblanceolate, 1.5–3 (dorsal rarely to 5) mm wide, subacute. Corolla white to pale mauve with brown throat and yellow spot on lower lip, throat with dense band of long hairs on lower and lateral lobes, without hairs on dorsal lobes, 22–29 mm long, cylindric tube 1–3.5 mm long, throat 12–18 mm long; upper and lateral lobes transversely elliptic, 8–10 × 9–12 mm, lower 10–14 × 8–11 mm; papillate area on lower lip crescent-shaped or interrupted along median line. Free part of filaments 3–6 mm long and free fused part 1–2 mm, glabrous or longer with a few hairs; anthers 4–6 mm long, spur 1.5–2 mm long. Stigma 1–2 mm long, dorsal lobe absent; ovary with dense short glands all over. Capsule 22–26 mm long, uniformly glandular hairy all over. Seed (immature) ± 6 × 5 mm.

UGANDA. Toro District: Ruwenzori Mts, Mobuku Valley, Bikoni, 30 Dec. 1934, *G. Taylor* 2736! & Nyabitaba, June 1953, *Osmaston* 3848! & Busongora, 5 km W of Kilembe, 3 June 1970, *Lye* 5527!
DISTR. U 2; Congo-Kinshasa (Ruwenzori Mts), Rwanda, Burundi
HAB. Wet montane forest; 2100–2600 m (1900–3000 m in Rwanda and Burundi)

NOTE. Lindau's (l.c.) illustration shows a plants which has only just started flowering thus showing a rather condensed inflorescence.

Specimens from Uganda and Congo have subglabrous leaves while specimens from Rwanda and Burundi have pubescent leaves and also generally more densely hairy inflorescences.

Osmaston records on his 3848: "no flowers seen in the last 3 years".

4. **Mimulopsis solmsii** *Schweinf.* in Verh. Zool.-Bot. Ges. Wien 18: 677 (1868); Engler, Hochgebirgsfl. Trop. Afr.: 388 (1892); Lindau in E.J. 18: 63 (1893); C.B. Clarke in F.T.A. 5: 55 (1899); F.P.S. 3: 183 (1956); Heine in F.W.T.A., ed. 2, 2: 403 (1963); E.P.A.: 932 (1964); Binns, Checklist Herb. Fl. Malawi: 15 (1968); U.K.W.F.:

587 (1974); Champluvier in Fl. Rwanda 3: 474 (1985); Blundell, Wild Fl. E. Afr.: 394, pl. 126 (1987); Iversen in Symb. Bot. Ups. 29(3): 162 (1991); U.K.W.F. ed. 2: 269, pl. 117 (1994); K.T.S.L.: 605 (1994); Lebrun & Stork, Enum. Pl. Afr. Trop. 4: 495 (1997); White *et al.*, For. Fl. Malawi: 118 (2001); Friis & Vollesen in Biol. Skr. 51(2): 450 (2005); Ensermu in F.E.E. 5: 383 (2006). Type: Ethiopia, Semien Mts, Ghaba Valley, *Steudner* 1497 (K!, iso.)

Erect or scrambling/scandent shrubby herb to 3(–5) m tall, often forming large almost impenetrable thickets and covering large areas as a "monoculture"; whole plant usually with an overpowering aromatic or foetid smell; young stems subglabrous to sericeous-puberulent with downwardly directed hairs (very rarely tomentose). Leaves with petiole of stem-leaves 2.5–10(–11.5) cm long; lamina of stem-leaves ovate to broadly cordiform, 7–21 × 3–15 cm, apex acuminate to cuspidate, base subcordate to cordate with sinus to 1 cm deep, margin grossly crenate-dentate with teeth to 1(–2) cm long, sparsely sericeous-puberulent to -pubescent, usually only along midrib and main veins (rarely subglabrous). Inflorescence an open branched panicle with long spreading branches, usually also with long branches developing lateral panicles from basal nodes or upper leaf axils, (10–)15–35 cm long; flowers in lax dichasia with up to 9 flowers or solitary towards apex of panicle; bracts green, leafy, persistent, indumentum as leaves; branches and pedicels sericeous-puberulent to puberulous or densely so and with sparse to dense long capitate glands; pediels to 5(–10) mm long. Calyx puberulent to puberulous and with scattered to dense long purplish glands (rarely subsessile or absent); dorsal lobe (5–)8–21(–30 in fruit) mm long, ventral and lateral (5–)7–19(–28 in fruit) mm long; lobes linear to lanceolate, narrowly elliptic or narrowly spathulate, 0.5–2 mm wide, subacute. Corolla white to pale mauve with brown throat and one or two yellow patches on lower lip, throat with dense band of long hairs on lower lip, sparser on lateral lobes and only a few or none on dorsal lobes, 17–30(–33) mm long, cylindric part of tube 1–2(–3) mm long, throat 7–19 mm long; papillate area on lower lip of two separate elipsoid patches or fused to one; lobes all similar, broadly elliptic, 8–12(–15) × 6–10(–14) mm. Free part of filaments 2–5 mm long and fused free part ± 1 mm, glabrous or longer with scattered hairs; anthers 3–5 mm long, spur 1.5–2 mm long. Stigma ± 2 mm long, dorsal lobe absent; ovary densely hairy and glandular in apical $^1/_2$–$^2/_3$ or all over. Capsule 15–24(–26) mm long, hairy and glandular all over or glabrous in basal half, more rarely (in Tanzania) only hairy at the very tip. Seed 4–6 × 3.5–5 mm. Fig. 34, p. 224.

UGANDA. Toro District: Isunga, 26 Nov. 1918, *Snowden* 658!; Kigezi District: Mafuga, June 1950, *Purseglove* 3459!; Mbale District: Mt Elgon, Nkokonjeru, 20 Dec. 1924, *Snowden* 947!
KENYA. Trans-Nzoia District: Mt Elgon, Mberri bridge, Oct. 1961, *Tweedie* 2234!; Kiambu District: between Kerita Forest Station and Gatamayu, 25 Jan. 1969, *Napper* 1817!; Kericho District: SW Mau Forest, Saosa Catchment, 14 Jan. 1959, *Kerfoot* 736!
TANZANIA. Morogoro District: Mt Kanga, 2 Dec. 1987, *Lovett & D. Thomas* 2630!; Kilosa District: Ukaguru Mts, Mt Mnyera, 1 June 1978, *Thulin & Mhoro* 2818!: Rungwe District: Ngozi Crater, 4 July 1991, *Kayombo* 1103!
DISTR. U 2–4; K 3–7; T 6, 7; Guinea, Sierra Leone, Liberia, Nigeria, Cameroon, Bioko, Congo-Kinshasa, Rwanda, Burundi, Sudan, Ethiopia, Zambia, Malawi, Zimbabwe
HAB. Evergreen montane forest, including bamboo forest, often on edges or in clearings, also in secondary forest and upland grassland and bushland derived from forest; (1200–)1500–2700 m

SYN. *Mimulopsis violacea* Lindau in E.J. 17: 105 (1893) & in E. & P. Pf. IV, 3b: 301, fig. 121 D-H (1895); C.B. Clarke in F.T.A. 5: 55 (1899); Lindau in Z.A.E.: 296 (1911); F.W.T.A. 2: 252 (1931); F.P.N.A. 2: 281 (1947); T.T.C.L.: 13 (1949). Type: Cameroon, Buea, *Preuss* 814 (B†, holo.; BM!, K!, iso.)
   M. sesamoides S.Moore in Trans. Linn. Soc., ser. 2, 4: 30 (1894); Lindau in P.O.A. C: 367 (1895); C.B. Clarke in F.T.A. 5: 56 (1899); Binns, Checklist Herb. Fl. Malawi: 15 (1968); Lebrun & Stork, Enum. Pl. Afr. Trop. 4: 494 (1997). Type: Malawi, Mt Mlanje, *Whyte* 89 (BM!, holo.)

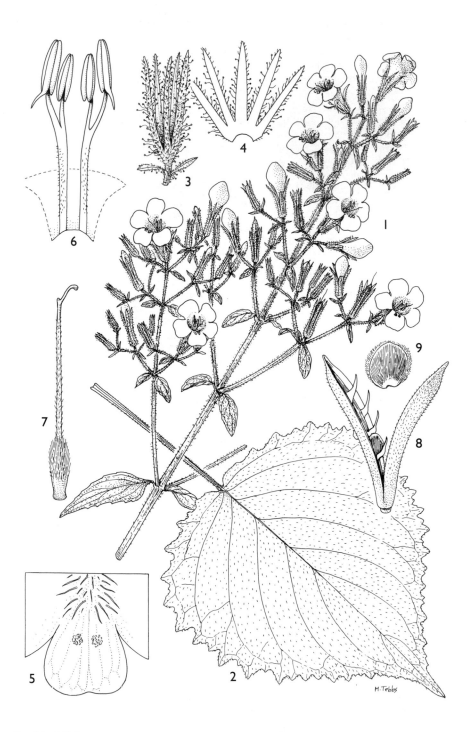

FIG. 34. *MIMULOPSIS SOLMSII* — **1**, inflorescence, × ²⁄₃; **2**, basal leaf, × ²⁄₃; **3**, calyx, × 2; **4**, calyx opened up, × 2; **5**, lower corolla lip, × 4; **6**, stamens, × 4; **7**, ovary, style and stigma, × 4; **8**, capsule, × 2; **9**, seed, × 2. 1 & 5–7 from *Tweedie* 2241, 2–4 & 8 from *Napper* 2171, 9 from *Boughey* 87. Drawn by Margaret Tebbs.

M. spathulata C.B.Clarke in F.T.A. 5: 55 (1899); Lugard in K.B. 1933: 94 (1934); Lebrun & Stork, Enum. Pl. Afr. Trop. 4: 495 (1997). Type: Kenya, Nakuru/Masai District: Mau, *Scott Elliot* 6960 (B†, holo.; BM!, K!, iso.) See note.

M. thomsonii C.B.Clarke in F.T.A. 5: 55 (1899), excl. syn.; Lebrun & Stork, Enum. Pl. Afr. Trop. 4: 495 (1997). Types: Tanzania/Malawi, N of Lake Nyasa, *Thomson* s.n. (K!, syn.) & between Lake Nyasa and Lake Tanganayika, *Thomson* s.n. (K!, syn.)

M. bagshawei S.Moore in J.B. 45: 90 (1907); Lebrun & Stork, Enum. Pl. Afr. Trop. 4: 494 (1997). Type: Uganda, Mengo District: Entebbe to Hoima, *Bagshawe* 801 (BM!, holo.)

M. usumburensis Lindau in E.J. 43: 350 (1909); Lebrun & Stork, Enum. Pl. Afr. Trop. 4: 495 (1997), as *usambarensis*. Type: Burundi, Bujumbura [Usumbura], Mt Ludyona, *Keil* 270 (B†, holo.)

M. velutinella Mildbr. in B.J.B.B. 17: 86 (1943); Lebrun & Stork, Enum. Pl. Afr. Trop. 4: 494 (1997). Type: Congo-Kinshasa, Kivu, Kamatembe, *de Witte* 1506 (BR!, holo.; BR!, iso.)

M. violacea Lindau var. *mikenica* Mildbr. in B.J.B.B. 17: 87 (1943); F.P.N.A. 2: 282 (1947). Type: Congo-Kinshasa, Kivu, near Kabara, Mikeno, *de Witte* 1688 pro parte (BR!, holo.)

M. violacea Lindau var. *kivuensis* Mildbr. in B.J.B.B. 17: 87 (1943); F.P.N.A. 2: 282 (1947). Type: Congo-Kinshasa, Kivu, Kirorirwe, *de Witte* 1449 (BR!, holo.)

M. solmsii Schweinf. var. *mikenica* (Mildbr.) Troupin in B.J.B.B. 52: 463 (1982)

M. solmsii Schweinf. var. *kivuensis* (Mildbr.) Troupin in B.J.B.B. 52: 463 (1982)

M. solmsii Schweinf. var. *orophila* Troupin in B.J.B.B. 52: 464 (1982); Champluvier in Fl. Rwanda 3: 475 (1985). Type: Burundi, Muramuya Prov., Mt Teza, *Reekmans* 6991 (BR!, holo.; C!, EA!, K!, P!, iso.)

NOTE. C.B.Clarke (l.c.) cites the type collection of *Mimulopsis spathulata* as *Scott Elliot* 6060, but there seems little doubt that this was a transcription error on the duplicate which he saw in Berlin.

The most common and widespread species in the genus and consequently also the most variable, but there seems to be no satisfactory way of dividing the multitude of local forms into distinct taxonomic entities. Reported several times from different parts of its area as flowering at intervals of seven years.

5. **Mimulopsis elliotii** *C.B.Clarke* in F.T.A. 5: 56 (1899); F.P.N.A. 2: 278 (1947); Lebrun & Stork, Enum. Pl. Afr. Trop. 4: 494 (1997). Types: Uganda, Ruwenzori Mts, Yeria Valley, *Scott Elliot* 7867 (K!, syn.) & Butaga Valley, *Scott Elliot* 8013 (K!, BM!, syn.)

Erect or scrambling shrubby herb or shrub to 3(?–5) m tall, often covering large areas as a "monoculture" and forming dense tangles; branches sparsely to densely sericeous or sericeous-pubescent with downwardly directed non-glandular many-celled hairs. Leaves with petiole of stem-leaves 2–5.5 cm long, sericeous with downwardly directed many-celled purplish hairs; lamina of stem-leaves ovate or broadly so, 7–11 × 4–7 cm, upwards smaller and narrower, apex acuminate, base truncate to cordate with sinus up to 5 mm deep, margin of stem-leaves dentate or grossly dentate with teeth to 5 mm long, beneath sericeous or sparsely so, densest along veins with lamina often almost glabrous, often with domatia or with distinctly longer hairs along midrib, above puberulous to pubescent or sparsely so on lamina and sericeous on veins. Inflorescence narrow, racemiform or also with long racemiform branches from lower nodes, 5–30 cm long; flowers in 3–7-flowered cymes towards base and solitary towards apex of inflorescence; bracts green, leafy, persistent or upper caducous in fruit; lower bracts with indumentum as leaves, upper bracts, branches and pedicels sericeous-tomentellous with downwardly directed non-glandular many-celled purplish hairs; pedicels to 5 mm long. Calyx reddish-tinged, glandular-puberulous to pubescent or densely so or with non-glandular hairs intermixed to almost non-glandular; dorsal lobe 14–29 mm long, ventral and lateral lobes 12–22 mm long; lobes linear-lanceolate or dorsal slightly oblanceolate, 1.5–3 mm wide, acute or subacute. Corolla white with yellowish brown throat and yellow patches on lower lip, without or with scattered hairs on lateral lobes, no hairs on dorsal lobes, 26–36 mm long, cylindric tube 0.5–1 mm long, throat 16–23 mm long; lobes all transversely elliptic, 9–14 × 10–14 mm; papillate area on lower lip ellipsoid. Free part of filaments 5–8 mm long and free fused part

1–2 mm, longer pair hairy, short pair glabrous; anthers 4–6 mm long, spur 1.5–2 mm long. Stigma 2–3 mm long, dorsal lobe absent or present (rarely up to 1 mm long); ovary with dense long hairs all over. Capsule 14–19 mm long, hairy and glandular all over. Seed (immature) ± 4.5 × 4 mm.

UGANDA. Toro District: Ruwenzori Mts National Park, Namwamba Valley, 5 Jan. 1935, *G. Taylor* 2885! & Mubuku Valley, 4 Feb. 1997, *Lye* 22405! & 19 Feb. 1997, *Lye* 22579!
DISTR. **U** 2; Congo-Kinshasa (Ruwenzori Mts)
HAB. *Erica* forest and bushland, bamboo forest; (2500–)2700–3500 m

SYN. [*Mimulopsis kilimandscharica* sensu Lindau in Z.A.E.: 295 (1911); Robyns in F.P.N.A. 2: 278 (1947), *non* Lindau (1894)]
　　[*Mimulopsis runssorica* sensu F.P.N.A. 2: 280 (1947), *non* Lindau (1894)]

NOTE. Known only from high altitude *Erica* forest and bamboo forest on the Ruwenzori Mts. This seems to be the highest recorded member of the Acanthaceae in the FTEA-area. It is easily recognised be the dense sericeous-pubescent indumentum on the inflorescence and by the long corolla-throat relative to the lobes.

6. **Mimulopsis alpina** *Chiov.* in Racc. Bot. Miss. Consol. Kenya: 94 (1935); U.K.W.F.: 587 (1974) & ed. 2: 269, pl. 116 (1994); K.T.S.L.: 605 (1994); Lebrun & Stork, Enum. Pl. Afr. Trop. 4: 494 (1997). Type: Kenya, Aberdare Mts, Kinangop, *Balbo* 37 (FT, holo.).

Erect or scrambling/scandent foetid shrubby herb to 3(?–4) m tall, often forming large thickets; branches sparsely to densely sericeous-pubescent (hairs downwardly directed) to pubescent with long curly white to purplish many-celled hairs. Leaves with petiole of stem-leaves 0.5–5 cm long; lamina of stem-leaves ovate to elliptic or cordiform, 1.3–14 × 0.7–9.5 cm, apex broadly rounded to acuminate, base truncate to cordate with sinus to 0.5(–1) cm deep, margin crenate to dentate with teeth to 5 mm long, beneath sparsely pubescent, densest along veins and lamina often glabrous, above sparsely pubescent all over. Inflorescence narrow, racemiform, sometimes with longer branches from basal nodes or with lateral inflorescences from upper leaf-axils, in depauperate forms sometimes reduced to a few flowers only, (1–)2–16(–21) cm long; flowers all solitary or in 3-flowered cymes basally; bracts green and leafy basally in inflorescence, gradually more bract-like and often purplish upwards, all persistent; basal bracts with indumentum as leaves, upper bracts, branches and pedicels pubescent to tomentose with long (some 2 mm or more) curly glandular and non-glandular white to purple many-celled hairs; pedicels to 2 mm long. Calyx glandular-pubescent or densely so with glands to tip of lobes; dorsal lobe (10–)14–24 mm long, ventral and lateral lobes (7–)12–20 mm long, dorsal 2–5 mm longer; lobes lanceolate to narrowly oblanceolate, 1–3 mm wide, subacute. Corolla mauve to purplish blue (rarely white with a mauve tinge) with brown throat and yellow patch on lower lip, throat with dense band of long hairs on lower and lateral lips, thinner to almost absent on dorsal lobes, 17–30(–32) mm long, cylindric tube 1–2 mm long, throat 11–18(–20) mm long; upper and lateral lobes transversely elliptic, 7–13 × 8–13 mm, lower lip ovate-elliptic, 11–15 × 10–13 mm; papillate area on lower lip crescent-shaped. Free part of filaments 2–6 mm and free fused part 1–2 mm, longer with scattered hairs, short glabrous; anthers 3–5.5 mm long, spur 1–1.5 mm long. Stigma 1–2 mm long, dorsal lobe absent (rarely present); ovary densely glandular apically, thinner downwards and base glabrous. Capsule 17–20 mm long, glandular apically, glabrous towards base (rarely glabrous but for a few glands at apex). Mature seed not seen.

UGANDA. Mbale District: Mt Elgon, Jan. 1918, *Dümmer* 3464! & Bulambuli, 11 Nov. 1933, *Tothill* 2278! & 16 Dec. 1938, *Thomas* 2744!
KENYA. Trans Nzoia District: NE Elgon, Dec. 1955, *Tweedie* 1379!; Naivasha District: Aberdare Mts, Kinangop, Dec. 1926, *Dale* 2154!; Narok District: E side of Nasampolai Valley, 10 March 1973, *Greenway & Kanuri* 15089!

TANZANIA. Pare District: S Pare Mts, Chome, 14 Aug. 1971, *Ngonyani* 38!; Morogoro District: Uluguru Mts, Bondwa, 26 Sep. 1970, *B.J. Harris et al.* 5081!; Iringa District: Udzungwa Mts, Mt Luhomero, Msisimwana Ridge, 19 Aug. 1985, *Rodgers & Hall* 4553!

DISTR. **U** 3; **K** 1, 3–6; **T** 2, 3, 6, 7; Malawi

HAB. A variety of evergreen montane forest and bush and, including bamboo forest, extending into the ericaceous zone; 2000–3300 m

NOTE. Closely related to *Mimulopsis excellens* which has a longer uniformly hairy capsule, a corolla without a ring of hairs on the upper lobes and generally larger and wider leaves with longer petiole.

M. alpina shows a very large morphological variation from large robust large-leaved plants (e.g. *Tweedie* 1379) to small only 20 cm tall slender herbs with a few flowers only (e.g. *Tweedie* 895). It seems that these depauperate forms, which occur throughout the distribution area, are either from exposed habitats or are abnormal specimens flowering out of season. They are connected to the robust forms by all sorts of intermediates.

In its core area in Kenya the flowers are always pale mauve to purple, but around the "edges" in Tanzania and Uganda, the species is occasionally reported as having white or almost white flowers.

B.J. Harris et al. 5081, the only collection seen from the Uluguru Mts has an almost completely glabrous capsule but is otherwise entirely typical. Material from the Udzungwa Mts has the usual capsule indumentum.

Tweedie records mass-flowering of *M. alpina* on Mt Elgon in 1947, 1955 and 1963 with scattered individuals flowering in the intervening years. Similar mass-flowering has been recorded from the Aberdares and Rift Valley forests but seems to occur in different cycles on different mountains.

7. **Mimulopsis sp. A**

Erect shrubby herb, 2 m tall; stems densely glandular-pubescent on upper nodes. Leaves with petiole of stem-leaves 4–5 cm long; lamina of stem-leaves cordiform, ± 11 × 9 cm, apex cuspidate, base cordate with sinus ± 1.5 cm deep, margin grossly doubly dentate with teeth to 1 cm long, sparsely pubescent on both sides, densest along veins, with longer spreading hairs along midrib beneath. Inflorescence narrow, racemiform, with long branches from basal nodes, ± 12 cm long; flowers in 3-flowered cymes towards base of inflorescence, solitary upwards; bracts green and leafy towards base of inflorescence, bract like upwards, all persistent, with indumentum as leaves; branches and pedicels densely glandular-pubescent to tomentose (hairs to 2 mm long); pedicels to 3 mm long. Calyx glandular-pubescent all over (to within 1 mm of apex of lobes); dorsal lobe ± 17 mm long, ventral and lateral lobes ± 13 mm long; lobes filiform to linear, 0.2–0.5 mm wide, subacute. Corolla white (no notes on throat and patch on lower lip), with dense band of long hairs on lower lip, sparser on lateral lobes and absent on dorsal lobes, ± 17 mm long, cylindric tube ± 1 mm long, throat ± 8 mm long; lobes all similar, elliptic, ± 8 × 6 mm; papillate area on lower lip crescent shaped. Free part of filaments 2–3 mm and fused free part ± 1 mm, glabrous; anthers 3–4 mm long, spur ± 1 mm. Stigma ± 1.5 mm long, dorsal lobe absent; ovary with dense short stubby non-glandular hairs in apical $^2/_3$, glabrous towards base. Capsule and seed not seen.

UGANDA. Kigezi District: Rubuguli, 24 May 1992, *Cunningham* 4020!

DISTR. **U** 2; not known elsewhere

HAB. Margins of wet montane forest; no altitude recorded

NOTE. A very distinctive species known only from this one collection. It is clearly related to the *M. alpina–kilimandscharica* complex. It differs most conspicuously in its very narrow sepals and small corolla with elliptic lobes.

8. **Mimulopsis kilimandscharica** *Lindau* in E.J. 20: 10 (1894) & in P.O.A. C: 367 (1895); C.B. Clarke in F.T.A. 5: 56 (1899); V.C. Gilbert, Pl. Mt Kilimanjaro: 82 (1970); Iversen in Symb. Bot. Ups. 29, 3: 162 (1991); Lebrun & Stork, Enum. Pl. Afr. Trop. 4: 494 (1997). Type: Tanzania, Mt Kilimanjaro, Mawenzi, Orna Ndogo River, *Volkens* 963 (B†, holo.)

Erect or scrambling/scandent shrubby herb or shrub to 2(–3) m tall, often forming large stands; stems glabrous to retrorsely sericeous in two bands with short straight hairs. Leaves with petiole of stem-leaves 2.5–9 cm long; lamina of stem-leaves ovate to cordiform, 4–17 × 2–11 cm, apex acuminate to cuspidate, base truncate to cordate with sinus to 0.5(–1) cm deep, margin crenate to dentate with teeth to 0.5(–1) cm long, sparsely to densely sericeous along veins and lamina with scattered hairs, often also with long many-celled thin hairs along midrib. Inflorescence narrow, racemiform, usually with longer branches from basal nodes or with lateral inflorescences from upper leaf-axils, in depauperate forms sometimes reduced to a few flowers, 2–29 cm long; flowers all solitary or in 3-flowered cymes basally in the inflorescence; bracts green and leafy basally, gradually more bract-like upwards, all persistent, with indumentum as leaves but denser upwards; branches and pedicels puberulous or densely so to sericeous (hairs to 1 mm long) with glandular and non-glandular hairs; pedicels to 3(–5) mm long. Calyx puberulous to pubescent (hairs to 1 mm long), densest in basal part and here with glandular hairs intermixed, often more or less subglabrous at apex and apical 2–7 mm of lobes without capitate glands; dorsal lobe 13–25 mm long, ventral and lateral lobes 11–23 mm long, dorsal 2–5(–7) mm longer; lobes linear-lanceolate to oblong, 1–2(–3) mm wide, sometimes with a few teeth, subacute. Corolla white to pale mauve with brownish throat and yellow patch on lower lip, with dense band of long hairs on lower lip, sparser on lateral lobes, without or with a few scattered hairs on dorsal lobes, 20–29 mm long, cylindric tube ± 1 mm long, throat 11–19 mm long; upper and lateral lobes transversely elliptic, 6–11 × 7–12 mm, lower ovate-oblong, 8–15 × 8–11 mm; papillate area on lower lip crescent-shaped. Free part of filaments 3–6 mm long and free fused part ± 1 mm, glabrous or longer with a few hairs; anthers 3–5 mm long, spur ± 1 mm long. Stigma 1–2 mm long, dosal lobe absent or minute; ovary densely glandular apically, thinner downwards and base glabrous. Capsule 18–21 mm long, glandular apically, glabrous towards base. Mature seed not seen.

TANZANIA. Moshi District: Mt Kilimanjaro, Kifinika, Nov. 1893, *Volkens* 1337! & 8 Dec. 1932, *Geilinger* s.n.!; Lushoto District: W Usambara Mts, Mto wa Nguruwe, 20 Sep. 1961, *Semsei* 3310! DISTR. **T** 2, 3; not known elsewhere HAB. *Podocarpus* forest extending into *Erica* scrub; 1700–2900 m

SYN. *Mimulopsis sp.* A of Iversen in Symb. Bot. Ups. 29, 3: 162 (1991)

NOTE. *Mimulopsis kilimandscharica* differs most conspicuously from the closely related *M. alpina* in the different indumentum on stems, inflorescences and calyx.
      Known only from W Usambara Mts, Mt Kilimanjaro and Mt Loliondo.

22. **EPICLASTOPELMA**

Lindau in E.J. 22: 114 (1895) & in E. & P. Pf. IV, 3b: 353 (1895); Mildbraed in N.B.G.B. 11: 1081 (1934); Pócs in Acta Bot. Acad. Sci. Hung. 19: 463 (1973)

Sooia Pócs in Acta Bot. Acad. Sci. Hung. 19: 461 (1973)

Shrubby herbs; stems quadrangular; cystoliths conspicuous. Leaves opposite. Flowers in open much-branched panicles with long spreading branches; bracts leafy, persistent; bracteoles 2, linear to narrowly ligulate, caducous or persistent; pedicels straight or sharply bent or geniculate where attached to branch. Calyx 5-lobed, divided almost to the base into linear-lanceolate lobes, dorsal lobe longer than the rest but same width. Corolla straight or resupinate through backwards genuflexion of the pedicel, bright red to crimson or maroon, glabrous or puberulous on the outside, contorted in bud, subactinomorphic to distinctly 2-lipped, basal cylindric part of tube short or long, throat long, expanded at the mouth, with ill-defined herring-bones ventrally; lobes 5, the two in the upper lip and the two lateral in lower lip subequal, lateral in lower lip held vertically and together with two dorsal forming a "hood", middle in lower lip spreading to deflexed, without a papillate area above, throat without ring of long stiff retrorse hairs. Stamens 4, didynamous, inserted near base of throat, held in the "hood" formed by the dorsal and lateral lobes; free part of filaments fused at base, below insertion forming two glabrous ridges to base of tube; anthers 2-thecous, thecae oblong, glabrous, rounded at apex, outer theca of longer pair with or without spur at base, other six thecae mucronate. Ovary with 6–8 ovules per locule; style filiform, hairy but for apical few mm; stigma with dorsal lobe absent, ventral lobe linear. Capsule 4-seeded, ellipsoid-obovoid, valves strongly sulcate; retinaculae strong, hooked. Seed discoid, circular in outline, glabrous on faces, with dense hygroscopic hairs on rim.

3 species in Tanzania.

Epiclastopelma is closely related to *Mimulopsis* and might with some justification be considered congeneric. But the peculiar red flowers without long stiff hairs (obviously adapted to bird pollination) are strikingly different from anything seen in *Mimulopsis*. Also the fruit is constantly 4-seeded as opposed to 6–8-seeded in *Mimulopsis*.

1. Corolla ± 4 cm long of which cylindric tube ± 2 mm and
    throat ± 2.8 cm; throat ± 1.2 cm across in dried flowers;
    outer theca of longer stamens with ± 1.5 mm long spur . . .   1. *E. glandulosum*
  Corolla 4.5–6 cm long of which cylindric tube 0.8–1.1 cm
    and throat 2.7–3.7 cm; throat 7–9 mm across in dried
    flowers; all thecae without spur 2
2. Pedicels to 3 mm long; calyx lobes 0.7–1.5 cm long in
    flower; corolla crimson red, resupinate, 4.5–5.6 cm long,
    dorsal and lateral lobes 6–9 mm long, ventral lobe ±
    2.2 cm long . . . .   2. *E. macranthum*
  Pedicels 0.5–1.8 cm long; calyx lobes 1.8–2.5 cm long in
    flower; corolla dark maroon, not resupinate, ± 6 cm long,
    dorsal lobes ± 2.5 cm long, lateral lobes ± 2.2 cm long,
    ventral lobe ± 1.7 cm long . . . .   3. *E. marroninus*

1. **Epiclastopelma glandulosum** *Lindau* in E.J. 22: 114 (1895); Mildbraed in N.B.G.B. 11: 1081 (1934). Type: Tanzania, Morogoro District: Uluguru Mts, Bukubuku, *Stuhlmann* 8781 (B†, holo.)

Straggly shrubby herb to 2 m tall with sticky inflorescences; stems towards apex sparsely and finely puberulous in two bands. Leaves with petiole of stem-leaves 1.5–3 cm long; lamina of stem-leaves narrowly ovate to ovate, largest ± 9 × 3.5 cm, apex acuminate, base unequal-sided, cordate with a sinus to 3 mm deep, margin shallowly crenate-dentate, with finely puberulous midrib above and with domatia in axils of main lateral veins beneath, otherwise glabrous. Inflorescence a 10–30 cm long open much-branched panicle with long spreading branches and flowers in lax dichasial cymes with 3–7 flowers or solitary towards apex; bracts leafy, persistent, green, upwards gradually smaller and narrower; branches and pedicels at base of inflorescence with indumentum as branches, upwards gradually more densely puberulous and with scattered to dense long capitate glands; pedicels to 4 mm long, sharply bent where attached to branch. Calyx puberulous and with long broad glossy many-celled capitate glands (apical 2–3 mm of lobes without glands); lobes linear-lanceolate, 0.5–1 mm wide, ventral and lateral 10–12(–16 in fruit) mm long, dorsal 13–15(–20 in fruit) mm long, 2–3(–5 in fruit) mm longer than the others. Corolla resupinate, bright red (or when young whitish outside with reddish throat), ± 40 mm long, glabrous on the outside, cylindric tube ± 2 mm long, throat ± 28 mm long and ± 12 mm across at mouth in dried state, slightly curved along dorsal edge (ventrally in open flower); dorsal and lateral lobes broadly ovate, ± 10 × 8 mm, ventral lobe ± 13 × 10 mm, erect. Free part of filaments 3–4 mm long and free fused part ± 4 mm, glabrous; anthers 3.5–4.5 mm long, spur on outer theca of longer anthers ± 1.5 mm long. Stigma ± 2 mm long; ovary hairy in apical half. Capsule ± 21 mm long, sparsely hairy towards apex. Seed (immature) ± 4 × 3 mm.

TANZANIA. Morogoro District: Uluguru Mts, Bondwa Ridge, 17 July 1972, *Mabberley* 1144! & Morningside to Bondwa, 12 Sep. 1972, *Mwasumbi* DSM2724!
DISTR. **T** 6; not known elsewhere
HAB. Ground-floor in primary *Allanblackia* forest; 1600 m

SYN. *Mimulopsis glandulosa* (Lindau) Milne-Redh. in K.B. 1931: 275 (1931), *non M. glandulosa* Baker (1890)

NOTE. Known from the the type (now destroyed) and these two collections made within two months of each other. Mabberley notes: "very common and dominant over large areas of the forest floor". Although it can not be proved on such a slender basis, this species almost certainly has periodic mass-flowering.
 Despite the absence of original material, the extensive original description expanded by Mildbraed (l.c.) leaves little doubt as to the identity of this species.

2. **Epiclastopelma macranthum** *Mildbr.* in N.B.G.B. 11: 1079 (1934); T.T.C.L.: 9 (1949); Lebrun & Stork, Enum. Pl. Afr. Trop. 4: 481 (1997), as *micranthum*. Type: Tanzania, Morogoro District: Uluguru Mts, NE side, *Schlieben* 2839 (B†, holo.; BM!, BR!, HBG!, P!, iso.)

Scrambling shrubby herb forming dense bushes to 2 m tall; stems on upper nodes with two thin lines of puberulous hairs. Leaves with petiole of stem-leaves 3–5 cm long; lamina of stem-leaves ovate, largest ± 14 × 6.5 cm, apex acuminate, base subcordate or cordate with sinus to 3 mm deep, margin crenate-dentate, puberulous along midrib on both sides and with distinct domatia in axils of main veins beneath, otherwise glabrous. Inflorescence a 20–25 cm long open much branched panicle with long spreading branches and flowers in 3-flowered dichasial cymes or solitary towards apex of panicle; bracts green, at base of panicle leafy and persistent, upwards narrower and caducous; branches and pedicels in basal part of inflorescence with indumentum as branches, upwards gradually more densely puberulous and with scattered to dense long capitate glands; pedicels to 3 mm long, at point where joined to branch sharply geniculate. Calyx finely puberulous and with long many-celled capitate glands; lobes linear-lanceolate, 0.2–1 mm wide, ventral and lateral 7–10(–13 in fruit) mm long, dorsal ± 15(–20 in fruit) mm long. Corolla resupinate, held erect, crimson red with brownish tube, 45–56 mm long, glabrous on the outside, cylindric tube 8–11 mm long, throat 30–37 mm long and 7–9 mm across at mouth in dried state, distinctly curved along dorsal edge (ventrally in open flower); dorsal and lateral lobes oblong, 6–9 × 4–7 mm, ventral lobe elliptic, ± 22 × 6 mm. Free part of filaments 3–6 mm long and free fused part ± 4 mm, glabrous; anthers 5–6 mm long, outer theca of longer pair without spur, rounded at base. Stigma ± 5 mm long; ovary hairy at apex. Capsule ± 20 mm long, sparsely hairy at apex. Seed (immature) ± 6 × 4 mm. Fig. 35, p. 231.

TANZANIA. Morogoro District: Uluguru Mts, NE side, *Schlieben* 2839! & Uluguru Mts, above Bunduki, Kikododo, 21 Sep. 1971, *Pócs & Mwanjabe* 6463/C! & *Pócs & Mwanjabe* 6464/C!
DISTR. **T** 6; not known elsewhere
HAB. Montane *Ocotea* forest; 1600–1950 m

SYN. *Sooia ulugurica* Pócs in Acta Bot. Acad. Sci. Hung. 19: 461, fig. 1 (1973); Lebrun & Stork, Enum. Pl. Afr. Trop. 4: 481 (1997). Type: Tanzania, Morogoro District: Uluguru Mts, above Bunduki, Kikododo, *Pócs & Mwanjabe* 6463/C (BP, holo.; BR!, DSM!, EA!, K!, P!, iso.)

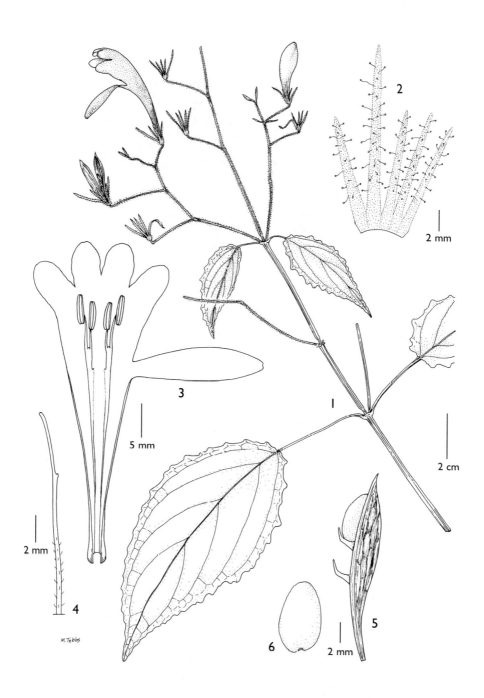

FIG. 35. *EPICLASTOPELMA MACRANTHUM* — **1**, habit; **2**, calyx; **3**, corolla opened up with stamens; **4**, style and stigma; **5**, single capsule valve; **6**, seed. 1 & 3–6 from *Pócs & Mwanjabe* 6463/C, 2 from *Pócs & Mwanjabe* 6464/C. Drawn by Margaret Tebbs.

3. **Epiclastopelma marroninus** *Vollesen* **sp. nov.** ab *E. macrantho* pedicellis longioribus, lobis calycis longioribis, corolla majore atro-marronina non-resupinata lobis dorsalibus longioribus et lobis ventralibus brevioribus differt. Type: Tanzania, Iringa District: Udzungwa Mts National Park, *Luke et al.* 7883 (EA!, holo.; K!, iso.)

Scrambling shrubby herb; young stems glabrous. Leaves with petiole of stem-leaves 3–5.5 cm long; lamina of stem-leaves elliptic, largest ± 14.5 × 5.5 cm, apex acuminate, base subcordate, margin crenate, glabrous apart from distinct domatia in axils of main veins beneath. Inflorescence a ± 8 cm long open much branched densely glandular-puberulous panicle with spreading branches and flowers in 3-flowered dichasial cymes or solitary towards apex of panicle; peduncle ± 4.5 cm long; secondary branches 1–1.5 cm long; bracts green, at base of panicle leafy and persistent, upwards narrower and caducous; pedicels 0.5–1.8 mm long, not geniculate; bracteoles filiform, to 2 cm long. Calyx densely glandular-pubescent; lobes filiform, ± 0.5 mm wide, 1.8–2.5 cm long. Corolla not resupinate, maroon, ± 6 cm long, puberulous on the outside, cylindric tube ± 1 cm long, throat ± 2.7 cm long and ± 7 mm across at mouth in dried state, straight along dorsal edge; lobes oblong, dorsals ± 2.5 × 1 cm, laterals ± 2.2 × 0.4 cm, ventral ± 1.7 × 0.4 cm. Free part of filaments ± 1.7 cm long, puberulous; anthers ± 7 mm long, outer theca of longer pair without spur, rounded at base. Stigma ± 3 mm long; ovary hairy at apex. Capsule and seed not seen.

TANZANIA. Iringa District: Udzungwa Mts National Park [7°46' S 36°49' E], 27 Sep. 2001, *Luke et al.* 7883!
DISTR. **T** 7; not known elsewhere
HAB. Evergreen submontane forest, along stream; 1100 m

23. MELLERA

S.Moore in J.B. 17: 225 (1879) & in J.B. 32: 133 (1894); Lindau in E. & P. Pf. IV, 3b: 297 (1895) & in P.O.A. C: 367 (1895); C.B. Clarke in F.T.A. 5: 50 (1899)

Onus Gilli in Österr. Bot. Zeit. 118: 560 (1970)

Strongly aromatic (unpleasantly or not depending on taste) perennial or shrubby herbs or shrubs; cystoliths usually conspicuous; stems quadrangular. Leaves opposite. Flowers pedicellate, solitary or in few- to many-flowered dichasia aggregated into terminal spiciform racemes or in contracted to open panicles; bracts leafy, gradually smaller towards apex of inflorescence; bracteoles conspicuous. Calyx 5-lobed, divided almost to base or basal part fused; lobes linear to narrowly oblong or spathulate, subequal or dorsal wider and longer, two ventral sometimes fused slightly higher up. Corolla contorted in bud; sparsely puberulent outside, mainly on veins; basal part of tube cylindric widening into a clearly defined strongly pleated throat; limb distinctly 2-lipped, middle lobe in lower lip horizontal or deflexed, with usually numerous long stiff reflexed hairs, lobes in upper lip flat. Stamens 4, inserted dorsally in throat and held under upper lip, slightly didynamous; filaments fused with tube and running like ridges to base of tube, free part of filaments of each pair fused at base; anthers 2-thecous, thecae oblong or ellipsoid, straight, of different length, rounded at apex, all thecae with long spurs or apiculate at base, or only one of the locules (longer) spurred and the other rounded. Ovary with (2–)4–6 ovules per locule; style hairy throughout or glabrous near apex, with one (ventral) linear stigma-lobe, the dorsal reduced (rarely with subequal lobes). Capsule with more than 4 seeds (rarely 2-seeded), ellipsoid (rarely ovoid-ellipsoid), thin-walled (rarely woody), glossy, valves sulcate dorsally; retinaculae strong, hooked. Seed discoid, faces glabrous or with hairs, edges with dense long hygroscopic hairs.

7 species in Central, Eastern and Southern Tropical Africa.

1. Capsule 2-seeded, obovoid, hard and glossy, retained
unopened until the following rainy season; stigma lobe
± 0.5 mm long ....   6. *M. congdonii*
Capsule with more than 4 seeds, thin-walled, not retained
on plant; stigma lobe 1–5 mm long 2
2. Flowers in narrow racemiform cymes, solitary or in 2–3-
flowered dichasia with peduncles up to 3(–6) mm long;
lateral lobes in lower corolla-lip horizontal; anther-
thecae with long spurs ....   1. *M. lobulata*
Flowers in condensed to open paniculate cymes; some or
all peduncles > 1 cm long; some or all dichasia with 5
or more flowers 3
3. Panicle ± condensed with peduncles up to 17 mm long;
corolla 35–43 mm long (measured along upper lip),
lateral lobes in lower lip vertical; anther-thecae apiculate
or with minute spurs; stigma 1–2 mm long ....   5. *M. submutica*
Panicle very open with peduncles 1–4 cm long; corolla
30–55 mm long, lateral lobes in lower lip horizontal; one
or both anther-thecae with spurs 0.5–1.5 mm long;
stigma 2–5 mm 4
4. Only one of the thecae spurred, the other rounded;
corolla limb subactinomorphic, expanded part of tube
(throat) ± 22 mm long and 22–28 mm across in dried
state; largest leaf ± 5.5 cm wide ....   4. *M. insignis*
All eight thecae spurred; corolla limb distinctly 2-lipped,
expanded part of tube (throat) 5–16 mm long and
5–12 mm across in dried state; largest leaf 8–12 cm
wide 5
5. Corolla 40–55 mm long, pale blue with dark blue central
part of lower lip; cylindric tube 12–17 mm long, throat
12–15 mm long and 10–12 mm across in dried state;
calyx 16–26 mm long, the lobes over 1 mm wide near
apex; young branches quadrangular ....   2. *M. menthiodora*
Corolla ± 30 mm long, white with mauve central part of
lower lip, cylindric tube ± 8 mm long, throat ± 5 mm
long and 5 mm across in dried state; calyx 13–18 mm
long, the lobes less than 0.5 mm wide near apex; young
branches rounded ....   3. *M.* sp. *A*

1. **Mellera lobulata** *S.Moore* in J.B. 17: 225, t. 203 (1879); Lindau in E.J. 18: 63, t. 1,
fig. 12 (1893) & in P.O.A. C: 367 (1895); C.B. Clarke in F.T.A. 5: 50 (1899); De
Wildeman, Etud. Fl. Katanga 1: 143 (1903) & Etud. Fl. Congo 1: 315 (1906); S.
Moore in J.B. 45: 91 (1907); Th. & H. Durand, Syll. Fl. Congo: 418 (1909); Lindau in
Z.A.E.: 293 (1911); Eyles in Trans. Roy. Soc. South Africa 5: 481 (1916); De
Wildeman, Contrib. Etud. Fl. Katanga: 196 (1921); T.T.C.L.: 13 (1949); E.P.A.: 931
(1964); Binns, Checklist Herb. Fl. Malawi: 15 (1968); Fanshawe, Checklist Woody Pl.
Zambia: 35 (1970); Champluvier in Fl. Rwanda 3: 472 (1985); Iversen in Symb. Bot.
Ups. 29(3): 161 (1991); Lebrun & Stork, Enum. Pl. Afr. Trop. 4: 494 (1997); White
et al., For. Fl. Malawi: 116 (2001); Friis & Vollesen in Biol. Skr. 51(2): 449 (2005);
Ensermu in F.E.E. 5: 382 (2006). Types: Mozambique, Moramballa, *Waller* s.n. (K!,
syn.); Malawi, Manganja Hills, *Meller* s.n. (K!, syn.)

Erect or scrambling (sometimes forming tangles) perennial or shrubby herb to
2(–3) m tall; stems puberulous and sometimes with long pilose hairs and/or stalked
glands on upper nodes below inflorescence. Leaves with petiole distinct, up to 13.5 cm
long; lamina ovate to broadly so, elliptic or cordiform, largest 6–20.5 × 3–15 cm,

margin subentire to grossly dentate, apex acuminate to cuspidate, base cuneate to cordate, the two sides often unequal, puberulous, beneath densest along veins with lamina often ± glabrous, above only along veins. Raceme up to 14 cm long, sometimes with 2 long side branches at lowermost node; axes, peduncles and pedicels sticky glandular-puberulous to pubescent or densely so, sometimes also with long pilose hairs; upper bracts sometimes with white basal part; flowers solitary or in 2–3-flowered dichasia; peduncle and pedicels 1–3 mm long (peduncle rarely to 6 mm); bracteoles up to 2.2 cm long. Calyx sticky, often whitish with green tips, 10–23 mm long of which the fused basal part 2–5 mm, sparsely to densely glandular-puberulous and with (rarely without) long glossy pilose hairs; lobes linear-oblong or slightly spathulate, acute to rounded. Corolla white to pale mauve, lower lip with darker front edge to fully dark purple, pleated area yellow to orange, setae white with dark tips, 25–42 mm long, cylindric tube 6–15 mm, throat 5–9 mm, middle lobe in lower lip triangular, truncate, 10–21 × 4–9 mm, lateral lobes oblong, rounded, 10–20 × 3–6 mm, lobes in upper lip ovate-oblong, rounded, 11–25 × 3–6 mm. Free part of filaments 5–16 mm long, fused for 1–2 mm at base, hairy at the very base, glabrous upwards; anthers 3.5–6 mm long, oblong, hairy and also with capitate glands on connective; spurs ± 1 mm long. Stigma 1–2 mm long. Capsule 11–19 mm long, apical part with scattered capitate glands. Seed oblong, ± 2.5 × 2 mm. Fig. 36, p. 235.

UGANDA. Acholi District: Mt Rom, 17 Nov. 1945, *Thomas* 4402!; Bunyoro District: Budongo Forest, Sonso River, 15. Feb. 1999, *Lye et al.* 23476!; Toro District: Bwamba, Sempaya Hot Springs, 26 Oct. 1953, *Dawkins* 812!
TANZANIA. Lushoto District: Amani, Sigi, 6 March 1953, *Drummond & Hemsley* 1429!; Kigoma District: 15 km N of Kigoma, Kakombe, 8 July 1959, *Newbould & Harley* 4310!; Morogoro District: Uluguru Mts, S of Mtebwa, 4 Nov. 1947, *Greenway & Brenan* 8294!
DISTR. U 1–4; T 1–4, 6–8; Central African Republic, Sudan, Ethiopia, Congo-Kinshasa, Rwanda, Burundi, Zambia, Malawi, Mozambique, Zimbabwe
HAB. Lowland and montane evergreen forest and riverine forest, often along tracks or in clearings; 300–1600 m

SYN. *Mellera angustata* Lindau in Z.A.E.: 294, pl. 32 (1911); F.P.N.A. 2: 274 (1947); Lebrun & Stork, Enum. Pl. Afr. Trop. 4: 494 (1997). Type: Congo-Kinshasa, W shore of Lake Albert Edward, *Mildbraed* 1979 (B†, holo.; BR!, iso.)

2. **Mellera menthiodora** *Lindau* in E.J. 57: 20 (1920); T.T.C.L.: 13 (1949); Lebrun & Stork, Enum. Pl. Afr. Trop. 4: 494 (1997). Type: Rwanda, Bihembe to Kanjanampa, *Braun* 5549 (B†, holo.)

Erect to scrambling perennial herb to 1 m tall, with a strong smell of mint; stems distinctly quadrangular, sparsely puberulous or sericeous-puberulous and with broad bands of long pilose hairs at nodes. Leaves with petiole distinct, up to 14 cm long; lamina cordiform, largest 10–16 × 8–12 cm, margin grossly and irregularly dentate, apex acute, base subcordate or cordate, often unequal, beneath glabrous but for scattered hairs on veins, above with uniformly scattered broad glossy hairs. Panicle very lax, pyramidal, 5–20 cm long and wide, sticky; axes, peduncles and pedicels glandular-pubescent; dichasia many-flowered at base of panicle, gradually fewer upwards; peduncles 1–4 cm long; pedicels 0.5–3 mm long; bracteoles up to 15 mm long. Calyx 16–26 mm long of which the fused basal part 1–2 mm, glandular-pubescent; lobes linear, over 1 mm wide near apex, parallel-sided or very slightly spathulate, rounded. Corolla pale blue or blue, central part of lower lip dark blue, orange in throat, setae few, 40–55 mm long, cylindric tube 12–17 mm long, throat 12–16 mm long and 10–12 mm across in dried state, middle lobe in lower lip ovate-triangular, lateral elliptic, rounded, all 15–20 × 7–8 mm, lobes in upper lip elliptic, rounded, 15–22 × 7–8 mm. Free part of filaments 13–15 and 15–17 mm long, fused for ± 5 mm at base, two long hairy along their whole length, two short glabrous; anthers oblong, 4–6 mm long, shorter glabrous, longer glandular and with short hairs near apex, spurs 1–1.5 mm long. Stigma 4–5 mm long. Capsule 15–22 mm long, with a few glands near apex. Seed oblong, 2–2.5 × 2 mm.

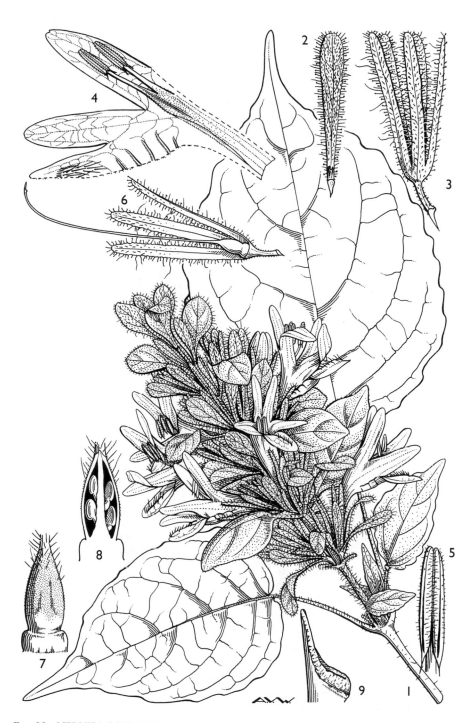

FIG. 36. *MELLERA LOBULATA* — **1**, inflorescence and leaves, × 1; **2**, bract, × 2; **3**, calyx, × 2; **4**, longitudinal section of corolla, × 2; **5**, anther, × 6; **6**, calyx, ovary, style and stigma, × 2; **7**, ovary, × 6; **8**, section of ovary, × 6; **9**, stigma, × 20. 1–9 from *Drummond & Hemsley* 1429. Drawn by Ann Webster.

TANZANIA. Biharamulo District: 50 km on Biharamulo–Muleba road, 11 July 2000, *Bidgood et al.* 4896!; Kigoma District: Malagarassi, Bubanda Escarpment, July 1926, *C.H.B. Grant* s.n.! & Mt Kungwe, Kasieha River, 20 July 1959, *Newbould & Harley* 4489!
DISTR. **T** 1, 4; Rwanda
HAB. In rock crevice in riverine forest, riverine scrub in woodland; 900–1300 m

NOTE. Lindau (l.c.) cited the type locality as being in Bukoba District in Tanzania, but it is clearly inside present day Rwanda. This no doubt led Brenan & Greenway (l.c.) to include the species without having seen any material, and probably also accounts for the species being omitted by Champluvier in Fl. Rwanda 3 (1985).
  No type material has been traced and the identity of this species is therefore by no means certain, but the material seen fits the original description in all important respects.

3. **Mellera sp. A**

Perennial herb, 1 m tall; stems rounded, sparsely puberulous and with long pilose hairs above nodes. Leaves with petiole distinct, up to 5.5 cm long, with long pilose hairs and capitate glands; lamina cordiform, largest ± 13 × 10 cm, margin grossly and irregularly dentate, apex subacuminate, base cordate, on both sides with scattered broad glossy hairs, beneath also with scattered stalked capitate glands. Panicle very lax, pyramidal, 8–10 × 10–12 cm; axes, peduncles and pedicels densely glandular-puberulous; dichasia many-flowered at base of panicle, fewer upwards; peduncles 2.5–3.5 cm long; pedicels ± 0.5 mm long; bracteoles up to 12 mm long, linear-oblanceolate. Calyx densely glandular-pubescent, 13–18 mm long of which the fused basal part ± 1 mm; lobes filiform to oblinear, under 0.5 mm wide near apex, parallel-sided or very slightly spathulate, rounded. Corolla white, tinged pale mauve outside and central part of lower lip mauve, setae numerous, ± 3 cm long, cylindric tube ± 8 mm long, throat ± 5 mm long and ± 5 mm across in dried state, middle lobe in lower lip ovate-triangular, lateral elliptic, rounded, all ± 13 × 5 mm, lobes in upper lip elliptic, rounded, ± 10 × 5 mm. Free part of filaments ± 11 and 12 mm long, fused for ± 7 mm at base, all with scattered hairs and glands along their whole length; anthers narrowly ellipsoid, 4–5 mm long, hairy and with capitate glands on connective, spurs ± 0.5 mm long. Stigma ± 2 mm long. Capsule and seed not seen.

TANZANIA. Lindi District: Rondo Plateau, Rondo Forest Reserve, 2 Nov. 2005, *Kayombo et al.* 5086!
DISTR. **T** 8; not known elsewhere
HAB. Semi-evergreen lowland forest; 700 m

NOTE. Known only from one incomplete collection. Seems to be closest to *M. menthiodora* from NW Tanzania. The open inflorescence is also reminiscent of *M. nyassana*, a woodland species from Zambia and Malawi not yet recorded from the Flora area. This has an eglandular inflorescence with much larger flowers.

4. **Mellera insignis** *Vollesen* **sp. nov.** a *M. menthiodora* antheris solum una theca cujusque paris calcarata obsitis, parte expansa tubi corollae longissima latissimaque, stigmati brevi difert. Type: Tanzania, Buha District: Kibondo, *Bullock* 3098 (K!, holo.; K!, iso.)

Straggling shrubby herb to 1.8 m tall; stems puberulous or sericeous-puberulous, upwards also with scattered broad glossy many-celled hairs. Leaves with petiole of stem leaves 2.5–5 cm long; lamina ovate, largest ± 11.5 × 5.5 cm, margin crenate-dentate, apex subacuminate with rounded tip, base cordate, unequal-sided, beneath sparsely uniformly pubescent on lamina, denser along veins, above uniformly sparsely pubescent with broad glossy hairs. Flowers in lax dichasial cymes to 8 cm long from upper leaf-axils; axes and pedicels puberulous with mixture of non-glandular hairs and capitate glands and with scattered long many-celled hairs; dichasia 5–7-flowered; peduncle 1.8–3.2 cm long; pedicels 1–5 mm long; bracteoles narrowly elliptic to oblanceolate, up to 1.8 cm long. Calyx 14–24 mm long of which the fused basal part 1–2 mm, puberulous with mixture of non-glandular hairs and capitate glands and with long shiny many-celled hairs to 2 mm long; lobes linear or slightly widened towards apex, rounded. Corolla blue with purple patch on lower lip and purple spots on lateral lobes, tube yellow-orange, lower lip ± 55 mm long with numerous setae in a broad ± 1.5 cm long band; cylindric tube 16–17 mm long, throat strongly expanded, 22–23 mm long and 22–28 mm across at mouth (in dried state); all lobes similar, transversely elliptic, 12–16 × 13–17 mm. Free part of filaments ± 7 and ± 11 mm long, fused for ± 3 mm at base, with two very distinct flanges, free part of long filaments hairy and with capitate glands, short with scattered hairs; anthers 6–6.5 mm long, oblong, glabrous, longer locule on all four with ± 1.5 mm long spur, shorter locule on short pair without spur, on long pair with $^1/_4$ mm long mucro. Stigma ± 2.5 mm long. Capsule and seed not seen

TANZANIA. Buha District: Kibondo, 7 Aug. 1950, *Bullock* 3098!
DISTR. **T** 4; not known elsewhere
HAB. Riverine forest; ± 1375 m

NOTE. A spectacular plant which differs from the other species in the genus in having one theca (the longer) of each anther long-spurred and the other rounded or with a minute mucro.
   Known only from the type collection but it is so distinct that I have no hesitation in describing it as new.

5. **M. submutica** *C.B.Clarke* in F.T.A. 5: 51 (1899); De Wildeman, Etud. Fl. Katanga 1: 143, pl. 11, fig. 1 (1903); Th. & H. Durand, Syll. Fl. Congo: 418 (1909); Binns, Checklist Herb. Fl. Malawi: 15 (1968); Richards & Morony, Checklist Fl. Pl. Mbala Distr.: 233 (1969); Fanshawe, Checklist Woody Pl. Zambia: 35 (1970). Types: Malawi, Blantyre, *Buchanan* 103 (K!, BM!, syn.) & *Dewèvre* s.n. (BR!, syn.); Mt Chiradzulu, *Whyte* s.n. (K!, syn.); Nyika Plateau, *Whyte* s.n. (K!, syn.); Manganja Hills, *Meller* s.n. (K!, syn.)

Perennial or shrubby herb, sometimes scrambling, to 1.5 m tall; stems subglabrous to densely pubescent with broad glossy hairs. Leaves with petiole distinct, up to 10 cm long; lamina ovate to cordiform or broadly so, largest 8–14 × 5.5–10.5 cm, margin grossly and often irregularly dentate, apex acute to acuminate, base subcordate or cordate, often with unequal sides, beneath sparsely pubescent to tomentose with broad glossy hairs, above subglabrous to sparsely pubescent, densest along veins. Panicle narrowly to broadly pyramidal, up to 30 cm long, sticky; axes, peduncles and pedicels glandular-pubescent or densely so; dichasia many-flowered at base of panicle, gradually fewer upwards; peduncles up to 17 mm long; pedicels 1–2 mm long; bracteoles up to 15 mm long. Calyx 16–25 mm long of which the basal fused part 1–3 mm, glandular-puberulous to pubescent or densely so; lobes linear-oblong (rarely) to spathulate, rounded to truncate. Corolla blue to purple, pleated area yellow, 35–43 mm long, cylindric tube 10–15 mm, throat 8–11 mm, middle lobe of lower lip triangular, acute, 10–15 × 12–14 mm, lateral lobes triangular, acute, 10–14 × 6–10 mm, held vertically, lobes in upper lip elliptic-oblong or obovate, rounded, 10–15 × 4–6 mm. Free part of filaments 3–5 mm long, fused for ± 1 mm at base, sparsely glandular throughout; anthers ellipsoid, 3–4.5 mm long, glandular on connective, otherwise glabrous, apiculate or with minute spurs to $^1/_4$ mm long. Stigma 1–2 mm long. Capsule 15–19 mm long, glandular near apex. Seed oblong, ± 3 × 2.5 mm.

TANZANIA. Kigoma District: Mahali Mts, E of Utahya, 2 Aug. 1958, *Newbould & Jefford* 1302!; Mpwapwa District: Mpwapwa, without date, *Hornby* 286A!; Mbeya District: Mbozi, Kantesia Estate, 28 Sep. 1936, *Burtt* 6220!
DISTR. **T** 2, 4, 5, 7; Congo-Kinshasa, Zambia, Malawi, Zimbabwe
HAB. Ravines and rocky slopes, riverine forest and scrub; 1050–1700(–1900) m

SYN. *Pseudobarleria lindaui* Dewèvre in Bull. Soc. Roy. Bot. Belg. 33, 2: 104 (1894), *nom. nud.*
Mellera submutica C.B.Clarke var. *grandiflora* De Wild., Etud. Fl. Katanga 1: 143, pl. 11, fig.
2–13 (1903); Th. & H. Durand, Syll. Fl. Congo: 418 (1909); De Wildeman, Contrib. Fl.
Katanga: 196 (1912) & Etud. Fl. Katanga 2: 145 (1913). Type: Congo-Kinshasa, Katanga,
Lukafu, *Verdick* 566 (BR!, holo.)
M. submutica C.B.Clarke var. *grandiflora* De Wild. forma *latifolia* De Wild., Etud. Fl. Katanga
2: 145 (1913) & Contrib. Fl. Katanga: 196 (1921). Type: Congo-Kinshasa, Jamba-Jamba,
Hock s.n. (BR!, holo.)
Onus submuticus (C.B.Clarke) Gilli in Österr. Bot. Zeit.118: 561 (1970); Lebrun & Stork,
Enum. Pl. Afr. Trop. 4: 496 (1997)
O. cochlearibracteatus Gilli in Österr. Bot. Zeit. 118: 561, fig. 2 (1970). Type: Tanzania, Mbeya
District: Mbeya Mountain, *Gilli* 535 (W!, holo.)

6. **Mellera congdonii** *Vollesen* **sp. nov.** a ceteris speciebus *Mellerae* stigmati
brevissimo, capsula obovoidea biseminatis differt. In ceteris speciebus, capsula
anguste ellipsoidea plus quam 4 semina gerentibus est. Type: Tanzania, Iringa
District: Ruaha National Park, Mwagusi River, *Congdon* 526 (K!, holo.; BR!, C!, CAS!,
DSM!, EA!, K!, MO!, NHT!, P!, UPS!, iso.)

Erect shrubby herb or shrub to 2 m tall; stems quadrangular, straw-coloured, when
young glabrous to puberulous or pubescent or glandular-puberulous. Leaves with
basal part of petiole forming a short stipule-like sheath around stem; lamina ovate to
broadly ovate, largest 6–14 × 3–8.5 cm, apex acute to acuminate, base decurrent
almost to stem, glabrous or with scattered broad glossy hairs along veins and
sometimes also on lamina. Flowers in axillary dichasia which sometimes merge into
more or less distinct paniculate cymes, dichasia glandular-puberulous or densely so;
peduncles up to 18 cm long; branches up to 2.5(–3.5) cm long; pedicels up to 3 mm
long; bracts and bracteoles elliptic to obovate, up to 15(–25) × 3(–8) mm. Calyx
(15–)20–37 mm long of which the fused basal part 2–7 mm, densely glandular-
puberulous with pale yellow stalked capitate glands; lobes linear to narrowly
spathulate, dorsal (13–)20–30 × 2–4 mm, ventral and lateral (10–)12–25 × 1–1.5 mm.
Corolla white to pale mauve with orange-brown pleated throat, sparsely glandular-
puberulous outside, 25–34 mm long of which cylindric tube 7–10 mm, throat 6–8
mm and lips 8–18 mm; lobes in upper lip 5–7 mm long, ovate, in lower lip 6–11 mm
long, narrowly ovate-oblong. Free part of filaments hairy at base or almost to apex
(densest downwards), fused for 2–4 mm basally, short 8–10 mm and long 12–14 mm;
anthers 4.5–5.5 mm long, all thecae with spurs 0.5–1 mm long, apiculate at apex.
Stigma ± 0.5 mm long. Capsule 2-seeded, obovoid, 12–15 mm long, pale to dark
brown, glossy, retained on plant. Seed ± 5 × 4 mm.

TANZANIA. Mpwapwa District: S of Gulwe, 29 July 1937, *Hornby* 815!; Iringa District: Ruaha
National Park, 28 km from Msembe, Mwagusi [Wangusi] Sand River, 4 Aug. 1969, *Greenway*
& *Kanuri* 13689! & 12 km from Ruaha Bridge, Mwagusi Sand River, Mwagusi-Ikuka
confluence, 30 Sep. 1998, *Congdon* 526!
DISTR. T 5, 7; not known elsewhere
HAB. Grassland on black cotton soil, alluvial *Acacia* grassland and thicket, degraded Itigi
thicket, riverine scrub; 750–1400 m

SYN. [*Hygrophila glutinifolia* sensu Björnstad in Serengeti Res. Inst. Publ. 215: 25 (1976) *pro parte*,
non Lindau (1903)]

NOTE. *Mellera congdonii* differs conspicuously from all other species in the genus by its obovoid
2-seeded capsule which is retained unopened on the plant until the following rainy season.
All other species have narrowly ellipsoid capsules with more than 4 seeds. All other species
are basically from forest or wetter types of woodland and I consider this to be an adaptation
to a dry country environment. Similar retention of the fruits are seen in other dry country
Acanthaceae, e.g. *Blepharis, Duosperma* and *Dyschoriste*.

24. DUOSPERMA

Dayton in Rhodora 47: 262 (1945); Brummitt in K.B. 29: 411 (1974); Dyer, Gen. S. Afr. Fl. Pl. 1: 587 (1975); Vollesen in K.B. 62: 289 (2006)

Disperma C.B.Clarke in F.T.A. 5: 79 (1899); Phillips, Gen. S. Afr. Fl. Pl. (ed. 2): 703 (1951), *non Disperma* J.F. Gmel. (1791)

Perennial or shrubby herbs or shrubs, with usually conspicuous linear cystoliths (especially on calyx); stems terete to distinctly quadrangular; bark on older branches usually peeling off in flakes or strips. Leaves opposite, with strong unpleasant smell when crushed; basal part of petiole forming a stipule-like sheath around the stem. Flowers subsessile or petiolate, solitary or in condensed axillary cymes which sometimes merge into large panicles; bracteoles usually large and conspicuous, often foliaceous. Calyx leathery, 5-lobed, fused in basal part, splitting in fruit; lobes lanceolate to triangular, with strong rib-like central vein and with scarious margins, dorsal lobe usually distinctly longer and wider than the rest, 2 ventral lobes fused higher up. Corolla contorted in bud, distinctly 2-lipped; tube cylindric at base, widening into a dictinct throat which is hairy ventrally; upper lip 2-lobed, erect, sometimes slightly hooded, lower lip 3-lobed, deflexed, lobes all similar, but wider in lower lip, lower lip pleated at base, with or without long stiff hairs. Stamens 4, inserted just inside cylindric part of tube, didynamous (longer pair 1–2 mm longer than short pair), held under upper lip, the two short stamens occasionally reduced to staminodes; filaments fused with tube, fused part running like two puberulous longitudinal ridges to base of tube, free part of each long and short filament fused at base; anthers 2-thecous, thecae oblong, rounded or apiculate at both ends, never with spurs. Ovary with 1–2 ovules per locule; style filiform, with upwardly directed hairs; stigma lobes 2, linear, equal or dorsal about half the length of ventral. Capsule (1–)2-seeded, ellipsoid-obovoid, woody, glossy, laterally compressed, usually glabrous, valves sulcate dorsally; retinaculae strong, hooked. Seed discoid, ovoid to orbicular in outline, with hygroscopic hairs.

26 species in eastern and southern tropical Africa from Somalia and S Ethiopia to Botswana and Angola. One species extends to Yemen. Particularly diverse from central and SW Tanzania into N Zambia and SE Congo.

Duosperma is morphologically very similar to *Dyschoriste* and *Hygrophila* but palynological work (Vollesen l.c.) strongly implies that it is closer to *Mellera*, *Mimulopsis* and *Eremomastax*. It differs from *Dyschoriste* and *Hygrophila* in the laterally compressed 2-seeded capsule and in having a style with both stigma lobes well-developed. The large conspicuous bracteoles also provide a useful spot-character.

The capsules usually persist on the plant until next rainy season. The apical part (base of style) is thickened and slightly bulbous and only tardily dehiscent.

1. Calyx (and often also young stems, bracts or bracteoles) with stalked capitate glands 2
   No vegetative part of plant with stalked capitate glands 7
2. Leaves with stellate hairs; corolla 3–4(–5) cm long, with indumentum of capitate glands only .... 6. *D. grandiflorum* (p. 248)
   Leaf indumentum of simple hairs; corolla 1.2–2.4 cm long, with indumentum of non-glandular hairs or a mixture of hairs and glands 3

3. Flowers in condensed axillary widely spaced
   cymes; corolla white, with or without red to
   mauve markings on lower lip, palate on lower
   lip without or with a few long stiff hairs 4
   Flowers in racemoid cymes which sometimes
   merge into large panicles; corolla mauve to
   purple (more rarely white), palate with many
   long stiff (rarely curly) hairs 5
4. Calyx 6–8 mm long in flower, teeth triangular,
   acute, dorsal 2–3 mm long, ventral 2–3 mm
   long of which the free part 1–2 mm; corolla
   12–14 mm long . . . .   10. *D. trachyphyllum* (p. 252)
   Calyx 8–15 mm long in flower, teeth lanceolate to
   narrowly triangular, acuminate to cuspidate,
   dorsal 5–8 mm long, ventral 4–7 mm long of
   which the free part 2–5 mm; corolla
   12–17(–20) mm long . . . .   11. *D. tanzaniense* (p. 253)
5. Young stems and bracts with stalked capitate
   glands . . . .   1. *D. quadrangulare* (p. 242)
   Young stems and bracts without stalked
   capitate glands 6
6. Largest leaf 8–16 cm wide; calyx 11–15 mm
   long in flower, dorsal tooth 8–9 mm long
   and the two ventral teeth 5–6 mm long;
   corolla 18–24 mm long; erect or scrambling
   shrub to 4 m tall . . . .   3. *D. densiflorum* (p. 244)
   Largest leaf 2.5–4.5 cm wide; calyx 5–8 mm
   long in flower, dorsal tooth 2.5–4 mm long
   and the two ventral teeth 1–2 mm long;
   corolla 13–17 mm long; perennial or shrubby
   herbs to 75 cm tall . . . .   2. *D. subquadrangulare* (p. 244)
7. Leaves (sometimes only very youngest) with
   stellate hairs 8
   Leaves glabrous or with simple hairs 10
8. Corolla 3–4(–5) cm long, with indumentum of
   stalked capitate glands only or with
   intermixed non-glandular hairs . . . .   6. *D. grandiflorum* (p. 248)
   Corolla 0.9–2.2(–3.1) cm long, with indumentum
   of non-glandular hairs 8
9. Corolla 1.8–2.2(–3.1) cm long, mauve to
   purple or deep blue; vegetative parts almost
   totally glabrous; "stipule sheath" 1.5–3 mm
   long . . . .   5. *D. stoloniferum* (p. 247)
   Corolla 0.9–1.6(–1.8) cm long, white with
   purple or red patches on lower lip;
   vegetative parts distinctly hairy; "stipule
   sheath" 0.5–1 mm long . . . .   4. *D. longicalyx* (p. 245)
10. Corolla mauve to purple or blue; plant with
    arching or stoloniferous branches rooting at
    the nodes . . . .   5. *D. stoloniferum* (p. 247)
    Corolla white, usually with red to purple
    patches on lower lip; plant not with
    stoloniferous rooting branches 11
11. Corolla 3–4 (–5) cm long, with indumentum of
    capitate glands only or with intermixed non-
    glandular hairs . . . .   6. *D. grandiflorum* (p. 248)
    Corolla 0.6–2.4 cm long, with indumentum of
    non-glandular hairs only 12

12. Leaves broadly obovate, retuse with triangular recurved apical tooth; rare form from northern Kenya .... **4. *D. longicalyx*** (p. 245)
    Leaves if broadly obovate not retuse nor with recurved apical tooth; plant not from northern Kenya 13
13. Corolla 6–11 mm long with tube 4–7 mm long; always some cymes with more than 5 flowers 14
    Corolla 10–24 mm long with tube 6–13 mm long; if less than 12 mm then flowers all solitary or in 3(–5)-flowered cymes 15
14. Young branches finely sericeous or sparsely so (rarely puberulous); calyx without long broad glossy hairs; palate on lower corolla lip with long stiff hairs .... **7. *D. kilimandscharicum*** (p. 249)
    Young branches glabrous to strigose-pubescent with long broad glossy hairs; calyx with long broad glossy hairs; palate on lower corolla-lip without long stiff hairs .... **8. *D. crenatum*** (p. 249)
15. Leaves less than twice as long as wide, largest 2–4.5 × 1.5–2.7 cm; corolla 10–14 mm long; flowers solitary or in 3(–5)-flowered cymes . **9. *D. latifolium*** (p. 252)
    Leaves more than twice as long as wide, largest 3.5–11 × 1.3–3 cm; corolla 12–24 mm long; some cymes usually with more than 5 flowers, if not then corolla over 15 mm long 16
16. Calyx teeth with long (1.5–2 mm) broad glossy hairs from base to apex 17
    Calyx teeth with long (1–1.5 mm) broad glossy hairs at base but with shorter (less than 1 mm) hairs at apex 18
17. Dorsal calyx tooth 4–8 mm long, about twice as long as the other teeth and about same length as calyx tube, ventral teeth 2–4 mm long, fused 0.5–1 mm higher than dorsal and lateral teeth; largest leaf 4–6 cm long . **14. *D. nudantherum*** (p. 255)
    Dorsal calyx tooth (5–)6–9 mm long, at most 2 mm longer than the other teeth and ± 2 mm longer than calyx tube, ventral teeth 4–7 mm long, fused ± 2 mm higher than dorsal and lateral teeth; largest leaf 6–11 cm long ..... **15. *D. fimbriatum*** (p. 256)
18. Corolla 21–24 mm long, lips 9–11 mm long and lobes 6–7 mm wide, anthers ± 2.5 mm long; leaves with a few hairs on midrib and finely ciliate, otherwise glabrous .... **13. *D. livingstoniense*** (p. 254)
    Corolla 12–23 mm long, lips 4–9 mm long and lobes 1.5–5(–6) mm wide, anthers 1–2 mm long; leaves uniformly (sometimes sparsely) hairy above and at least along main veins beneath 19
19. Corolla (1.5–)1.8–2.3 cm long, tube 1–1.4 cm long, filaments hairy to apex; capsule ± 12 mm long; calyx (1.2–)1.4–1.6 cm long in flower, basal part of tube puberulous or densely so . **12. *D. porotoense*** (p. 254)
    Corolla 1.2–1.7(–2) cm long, tube 0.7–1.2 cm long, filaments glabrous or with scattered hairs towards base; capsule 7–9 mm long; calyx 0.8–1.5 cm long, basal part of tube glabrous (rarely puberulous) .... **11. *D. tanzaniense*** (p. 253)

1. **Duosperma quadrangulare** (*Klotzsch*) *Brummitt* in K.B. 29: 412 (1974); Vollesen in Opera Bot. 59: 80 (1980); Ruffo, Cat. Lushoto Herb. Tanzania: 4 (1996); Lebrun & Stork, Enum. Pl. Afr. Trop. 4: 479 (1997); Welman in Strelitzia 14: 98 (2003); Mapura & Timberlake, Checklist Zimb. Vasc. Pl.: 13 (2004); Champluvier in Syst. Geogr. Pl. 75: 56 (2005); Phiri, Checklist Zamb. Vasc. Pl.: 18 (2005); Setshogo, Prelim. Checklist Botswana: 18 (2005); Vollesen in K.B. 62: 300, fig. 3 (2006). Type: Mozambique, Rios de Sena, *Peters* s.n. (B†, holo.). Neotype: Mozambique, Gaza, Lago M'fucue, *Barbosa & de Lemos* 8585 (K!, neo, selected by Vollesen, l.c.; COI!, LISC!, PRE!, SRGH!, iso.)

Erect (rarely decumbent) viscid glandular aromatic (when crushed) perennial or shrubby herb (often pyrophytic) from creeping rootstock and tuberous roots, or shrub; stems to 1(–1.5) m tall, usually much branched, quadrangular, when young sparsely to densely glandular-puberulous and with scattered longer glands (to 1 mm) intermixed, often also with long pilose non-glandular hairs up to 3 mm and with puberulous non-glandular hairs. Leaves often immature at time of flowering, elliptic to slightly obovate, largest (mature) 3–15 × 1.3–5.5(–6.5) cm, apex subacute to rounded, base long-decurrent, without distinct petiole, margin subentire to dentate, glabrous or with scattered broad long glossy hairs on veins, more rarely uniformly puberulous. Flowers solitary or in 3-flowered cymules, aggregated into racemiform cymes which are again aggregated into branched panicles, indumentum as young stems but with denser glands; bracts foliaceous towards base, gradually smaller upwards; peduncles of cymules to 10(–30) mm long; pedicels 0–2 mm long; bracteoles elliptic to orbicular or obovate, 3–11 mm long, with recurved tip. Calyx 6–11(–14 in fruit) mm long, glandular-puberulous or densely so, usually also with long pilose hairs; tube 5–6.5 mm long; teeth narrowly triangular to lanceolate, acute, dorsal 1–4.5 mm long, ventral 1.5–4 mm long with the free part 1–3 mm. Corolla white to mauve or purple, with darker patches on lower lip and with orange-brown palate, 14–23 mm long, finely sericeous or sparsely so and with scattered capitate glands outside; tube 8–13 mm long of which the throat 2–4 mm; lips 5–10 mm long; lobes oblong-obovate, 1.5–3.5 mm wide, retuse to truncate; palate on lower lip with numerous long stiff hairs directed down into throat. Filaments 2–5 mm long, sparsely puberulous at base and with scattered capitate glands towards apex; anthers 1.5–2 mm long, connective with usually dense captate glands. Stigma lobes subequal. Capsule 7–10 mm long, obovoid, glabrous or with scattered glands near apex. Seed 3–4 mm long. Fig. 37, p. 243.

TANZANIA. Dodoma District: Kazikazi, 12 June 1932, *Burtt* 3670!; Rufiji District: 12 km W of Lake Utunge, 6 Aug. 1976, *Vollesen* MRC 3880!; Kilwa District: Lungonya Plain, Kingupira Forest, 11 Sep. 1977, *Vollesen* MRC 4681!
DISTR. T 5–8; Congo-Kinshasa, Angola, Zambia, Malawi, Mozambique, Zimbabwe, Botswana, Namibia.
HAB. Grassland on grey to black alluvial clay, swamps, riverbanks, open *Acacia* bushland on white sandy-loamy hardpan, riverine woodland; often reported as being heavily grazed; (50–)200–1500 m

SYN. *Nomaphila quadrangularis* Klotzsch in Peters, Reise Mossamb., Bot.: 197 (1861)
    Hygrophila quadrangularis (Klotzsch) Lindau in P.O.A. C: 367 (1895)
    Disperma quadrangulare (Klotzsch) C.B.Clarke in F.T.A. 5: 81 (1899); T.T.C.L.: 7 (1949); Binns, Checklist Herb. Fl. Malawi: 13 (1968); Fanshawe, Checklist Woody Pl. Zambia: 19 (1970)
    D. dentatum C.B.Clarke in F.T.A. 5: 81 (1899); Binns, Checklist Herb. Fl. Malawi: 13 (1968). Type: Malawi, *Whyte* s.n. (K!, lecto., selected by Vollesen, l.c.)
    [*Hygrophila sessilifolia* sensu Lindau in Schwed.Rhod.-Kongo-Exped. 1: 305 (1916), *non* Lindau (1903)]
    [*H. glutinifolia* sensu Björnstad in Serengeti Res. Inst. Publ. 215: 25 (1976) p.p. *non* Lindau (1903)]

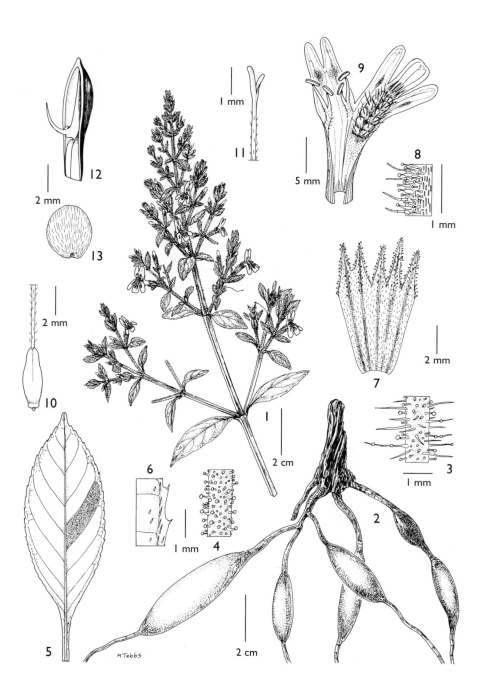

FIG. 37. *DUOSPERMA QUADRANGULARE* — **1**, habit; **2**, tuberous roots; **3–4**, stem indumentum; **5**, large leaf; **6**, leaf indumentum; **7**, calyx opened up; **8**, calyx indumentum; **9**, corolla opened up; **10**, ovary and basal part of style; **11**, apical part of style and stigma; **12**, capsule; **13**, seed. 1 & 5–6 from *Barbosa* 8585, 2 from *Richards* 26353, 3 from *Thulin* 586, 4 & 7–8 from *Greenway* 13675, 9–11 from *Greenway* 13698, 12–13 from *Batty* 729. Drawn by Margaret Tebbs.

2. **Duosperma subquadrangulare** *Vollesen* in K.B. 62: 305, fig. 4 (2006). Type: Tanzania, Masai District: Ardai Plains, *Greenway* 6814 (EA!, holo.; K!, PRE!, iso.)

Much branched perennial or shrubby herb to 75 cm tall with erect to procumbent stems, sometimes rooting at lower nodes, forming dense clumps; young stems quadrangular, subglabrous to puberulous or sericeous-puberulous and with slightly longer hairs on edges, no capitate glands, eventually with exfoliating bark. Leaves broadly elliptic-obovate to orbicular, largest 2–5 × 1.3–4.5 cm (less than twice as long as wide), apex subacute to emarginate, base attenuate, decurrent to stem or with an ill-defined petiole up to 1.5 cm long, margin subentire to dentate, subglabrous to sparsely and uniformly sericeous-puberulous; stipule-sheath 0.5–1 mm long. Flowers solitary or in 3-flowered cymules (peduncles up to 5 mm long) which are arranged in narrow racemiform axillary sericeous-puberulous panicles; bracts foliaceous, gradually smaller upwards; pedicels 0.5–2 mm long, puberulous and with subsessile glands; bracteoles linear-lanceolate to obovate, up to 7 mm long, puberulous and with slightly longer ciliae, with or without subsessile (shorter than hairs) capitate glands. Calyx 5–8(–9 in fruit) mm long, puberulous and with scattered to dense subsessile (shorter than or same length as hairs) capitate glands, tube 4–5 mm long; teeth triangular, acute or subacute, dorsal 2.5–4 mm long, ventral 2–3 mm long with the free part 1–2 mm long. Corolla white to pale pink or mauve with orange patches on lower lip, 13–17 mm long, sericeous-puberulent on the outside; tube 8–10 mm long of which the throat 2–4 mm; lips 4–8 mm long; lobes oblong-obovate, 2–3.5 (upper lip) to 5 (lower lip) mm wide, retuse to truncate; palate with numerous long stiff straight hairs. Filaments 2–6 mm long, with scattered hairs and glands throughout; anthers 1.5–2 mm long, rounded, connective with sparse capitate glands. Dorsal stigma lobe about half the length of ventral. Capsule and seed not seen.

KENYA. Machakos District: Emali to Simba, Merueshi, no date, *Wamukoya* 1090!; Kitui District: Kitui area, 11 Aug. 1978, *Chebii* 34!
TANZANIA. Masai District: Eleanata, 20 June 1941, *Hornby* 2115! & Ardai Plains, 10 July 1943, *Greenway* 6814!; Dodoma/Mpwapwa District: Dodoma–Mpwapwa road, 6 June 1937, *Hornby* 816!
DISTR. **K** 4; **T** 1–3, 5; not known elsewhere
HAB. Seasonally flooded grassland on grey to black or brown clayey soils, *Acacia-Commiphora* bushland on clayey soil; 1050–1400 m

SYN. *Duosperma sp. B* of U.K.W.F., ed. 2: 270 (1994)

NOTE. This species occurs on the northern fringes of the widespread *D. quadrangulare*, and was at first thought to be merely a form of it. But there are enough differences to recognise it as a distinct species; the main difference being the absence of capitate glands and long pilose hairs from most parts. Where capitate glands occur (mainly on the calyx) they are very short. The dimensions of the calyx and corolla are at the lower end of the variation in *D. quadrangulare*.

3. **Duosperma densiflorum** (*C.B.Clarke*) *Brummitt* in K.B. 29: 411 (1974); Lebrun & Stork, Enum. Pl. Afr. Trop. 4: 479 (1997); Vollesen in K.B. 62: 311, fig. 7 (2006). Type: Tanzania, without locality, *Scott Elliot* s.n. (K!, holo.)

Much branched erect or scrambling shrub to 4 m tall, sometimes thicket-forming; young stems sparsely to densely puberulous. Leaves broadly ovate to elliptic, largest (mature) 8–16 × 6–9 cm, less than twice as long as wide, apex drawn out into an elongated subacute tip, base cuneate with a distinct petiole 1–3 cm long, margin grossly crenate to dentate, puberulous along veins below, subglabrous to sparsely uniformly pubescent above; stipule sheath ± 0.5 mm long. Flowers in many-flowered condensed axillary cymes merging into narrow panicles; floral leaves gradually smaller and bract-like upwards; outer bracts sometimes leafy, up to 3.5 cm long, otherwise bracts and bracteoles somewhat scarious, lanceolate, 8–15 mm long, acuminate to cuspidate, sparsely puberulous along midrib and crisped-ciliate, bracteoles also with scattered subsessile capitate glands towards apex. Calyx 11–15 mm long, puberulous or sparsely so, often only on veins, crisped-ciliate, with scattered subsessile capitate glands, often only on lobes; tube 4–5 mm long; teeth narrowly triangular, cuspidate, dorsal 8–9 mm long, ventral 7–8 mm long of which the free part 5–6 mm long. Corolla purplish pink or bluish pink, 18–24 mm long, sericeous to finely puberulous, no capitate glands; tube 10–13 mm long of which the throat 3–4 mm; lips 7–11 mm long; lobes narrowly obovate, 2–3 mm wide, retuse; palate with long stiff hairs. Filaments 4–8 mm long, finely puberulous towards base, with scattered capitate glands towards apex; anthers 2–2.5 mm long, apex apiculate, base rounded, connective with subsessile glands. Stigma lobes subequal. Capsule 9–11 mm long, ellipsoid-obovoid. Seed ± 4 mm long.

TANZANIA. Kigoma District: Bulimba, 25 May 1975, *Kahurananga et al.* 2794!; Mpanda District: Mahali Mts, mouth of Lumbye River, 30 July 1958, *Newbould & Jefford* 1151! & 37 km on Mwese road from Mpanda-Uvinza road, 6 June 2000, *Bidgood et al.* 4610!
DISTR. **T** 4; Congo-Kinshasa
HAB. Riverine forest, edges and clearings of montane forest, grassland on black cotton soil, riverbanks, lakeshores, termite mounds in *Brachystegia* woodland; 700–1550 m

SYN. *Disperma densiflorum* C.B.Clarke in F.T.A. 5: 82 (1899); Ruffo *et al.*, Cat. Lushoto Herb. Tanzania: 4 (1996)

NOTE. Sometimes acts as a firebreak between grassland and forest and then survives for some time in pockets within the occasionally expanding forest.

4. **Duosperma longicalyx** (*Deflers*) *Vollesen* in K.B. 62: 315 (2006); Ensermu in F.E.E. 5: 373 (2006); Hedren in Fl. Somalia 3: 391 (2006). Type: Yemen, Bilad Fodhli, Serrya, *Deflers* 1032 (P!, lecto., selected by Vollesen, l.c.; P!, iso.)

Intricately branched often cushion-shaped shrubby herb or shrub to 1 m tall, often with arching branches which sometimes root on touching ground; young branches whitish puberulous or sparsely so (rarely subglabrous) with simple and/or stellate hairs, sometimes also with scattered long broad glossy hairs to 2(–3) mm long. Leaves often greyish, rough and leathery, elliptic or obovate to suborbicular, largest 1–3(–4.5) × 0.7–2.3(–2.8) cm, apex truncate to retuse (rarely acute), with a triangular usually recurved tip, base cuneate (rarely truncate), decurrent into a distinct petiole to 5(–8) mm long, margin crenate to dentate (rarely subentire), the teeth often bent downwards, sparsely to densely (when young) stellate puberulous (but see note), usually also with scattered long broad glossy hairs and sometimes with fine puberulent simple hairs; stipule sheath 0.5–1 mm long. Flowers subsessile, in 1–5-flowered condensed axillary cymes (rarely with cymes on lateral branches to 5(–35) mm long, bracts foliaceous, to 1.5 cm long, basal part scarious, middle part narrowed below an expanded leafy apical part, basal part with dense (rarely sparse) long broad glossy hairs, apical part as leaves; bracteoles oblanceolate, to 1 cm long, with similar indumentum or with long glossy hairs all over. Calyx 6–14 mm long, finely puberulous with simple and/or stellate hairs and with sparse to (usually) dense long broad glossy hairs; tube 4–10 mm long; teeth triangular or narrowly so, acute to acuminate, dorsal 2–4 mm long, ventral 1–4 mm long of which the free part 0.5–1.5 mm. Corolla white to cream with red or purple patches on lower lip, 9–19 mm long, finely sericeous-puberulous; tube 6–13 mm long of which the throat 2–4 mm; lips 2–6 mm long; lobes oblong, 1–3 mm wide, retuse (rarely truncate or rounded); palate indistinctly pleated, without long stiff hairs. Filaments 1–5 mm long, glabrous or with a few hairs near base; anthers 1.5–2.5 mm long, rounded, connective glabrous. Dorsal stigma lobe about half the length of ventral (rarely subequal). Capsule 5–10 mm long, obovoid. Seed 3–5 × 2–4 mm.

SYN. *Ruellia longicalyx* Defl. in Bull. Soc. Bot. France 43: 219 (1896); Blatter in Rec. Bot. Surv.
    India 8: 354 (1921)
  Disperma eremophilum Milne-Redh. in K.B. 1935: 282 (1935); E.P.A.: 935 (1964). Type:
    Kenya, S Turkana, *Buxton* 1026 (K!, holo.)
  Dyschoriste longicalyx (Defl.) Schwartz in Mitt. Inst. Allg. Bot. Hamburg 10: 249 (1939)
  Duosperma eremophilum (Milne-Redh.) Brummitt in K.B. 29: 411 (1974); U.K.W.F.: 587
    (1974); Kuchar, Pl. Somalia (CRDP Techn. Rep. Ser., no. 16): 229 (1986); K.T.S.L.: 601
    (1994); U.K.W.F., ed. 2: 270 (1994); Lebrun & Stork, Enum. Pl. Afr. Trop. 4: 479 (1997)

1. Capsule 5–8 mm long; bracts and bracteoles with long
    glossy hairs 2–3 mm long. Calyx 6–11(–13) mm
    long; corolla 9–16(–18) mm long; stellate hairs on
    young branches with the central hair in each cluster
    same length as the rest (rarely branches glabrous or
    subglabrous) . . . .   4a. subsp. *longicalyx*
    Capsule 8–10 mm long; bracts and bracteoles without
    or with glossy hairs to 1.5 mm long 2
2. Calyx 6–8(–10) mm long; corolla 10–12 mm long;
    capsule 8–9 mm long; young branches sericeous-
    puberulous, no stellate hairs, no long hairs; bracts
    and bracteoles without or with glossy hairs to 1 mm
    long . . . .   4b. subsp. *mkomaziense*
    Calyx 9–14 mm long; corolla 11–19 mm long; capsule
    9–10 mm long; young branches stellate-pubescent
    with the central hair in each cluster much
    elongated into a broad glossy hair; bracts and
    bracteoles with long broad glossy hairs to 1.5 mm
    long . . . .   4c. subsp. *magadiense*

4a. subsp. **longicalyx**; Vollesen in K.B. 62: 316, fig. 8, A–H (2006)

Young branches puberulous or sericeous-puberulous or sparsely so (rarely glabrous or
subglabrous) with simple and/or stellate hairs, sometimes also with scattered long broad glossy
hairs to 2(–3) mm long, stellate hairs with the central hair in each cluster same length as the
rest. Bracts and bracteoles with glossy hairs 2–3 mm long. Calyx 6–11(–13) mm long,
puberulous and with sparse to dense long glossy hair to 2 mm long, dorsal tooth 2–3 mm long.
Corolla 9–16(–18) mm long. Capsule 5–8 mm long.

UGANDA. Karamoja District: Kanamugit, no date, *Eggeling* 2925!
KENYA. Northern Frontier District: South Horr, Korungwe River, 14 Nov. 1978, *Hepper & Jaeger*
    6791! & north bank of Garissa, Korokora, 30 June 1960, *Paulo* 479!; Tana River District:
    Thika–Garissa road, Namorumat Drift, 9 June 1974, *Faden* 74/771!
DISTR. U 1; K 1, 2, 4, 6, 7; Sudan, Ethiopia, Somalia; Yemen
HAB. Dry often semi-desert areas with open *Acacia* or *Acacia-Commiphora* bushland, often on
    alluvial soils, dry riverine scrub; 200–1050(–1400) m

NOTE. Often reported as the dominant shrub over large areas of Turkana and Northern
    Frontier Districts.
      U.K.W.F. (l.c.) mentions that *D. longicalyx* seems to be poikilohydric. It is therefore an
    important source of browse as well as of water for night feeding animals in Northern Kenya.
      An almost completely glabrous form occurs around Marsabit (**K** 1), it does however, have
    stellate hairs on the very youngest leaves. It is easily recognised by the obovate leaves with
    truncate-retuse apex. Examples are *Faden* 68/554 and *J. G. Williams* EAH 11038.
      Two collections *Hucks* 305 and 1174, from the area in eastern Kenya shared with *D.
    stoloniferum*, have larger corollas (18 mm) with longer (6 mm) rounded lobes than normal *D.
    longicalyx* subsp. *longicalyx* from this area. They might be the result of hybridisation between
    the two species, but seem to have normally developed anthers.

4b. subsp. **mkomaziense** *Vollesen* in K.B. 62: 317, fig. 8, L–M (2006). Type: Tanzania, Same
District: Mkomazi Game Reserve, Ndea Hill, *Abdallah, Mboya & Vollesen* 96/150 (K!, holo.; BR!,
C!, CAS!, K!, NHT!, P!, iso.)

Young branches sericeous-puberulous, no stellate hairs, no long hairs. Bracts and bracteoles without or with glossy hairs to 1 mm long. Calyx 6–8(–10) mm long, puberulous with simple or stellate hairs, without or with slightly longer hairs on veins, dorsal tooth 1–2 mm long. Corolla 10–12 mm long. Capsule 8–9 mm long.

TANZANIA. Same District: Nyumba ya Mungu, 6 June 1970, *Mwasumbi et al.* 10841! & Mkomazi Game Reserve, Ndea area, 4 May 1995, *Abdallah & Vollesen* 95/133!
DISTR. **T** 2, 3; not known elsewhere
HAB. *Acacia-Commiphora-Cordia quercifolia* bushland and grassland on grey clayey loam; 750–850 m

SYN. *Duosperma sp. nov. aff. D. eremophilum* (Milne-Redh.) Brummitt (= *Mwasumbi* 10841) sensu Vollesen *et al.*, Checklist Vasc. Pl. and Pter. Mkomazi: 82 (1999)

NOTE. The distribution of subsp. *mkomaziense* is disjunct from that of subsp. *longicalyx*. It differs from it in the absence of stellate hairs on stems and in the absence of long glossy hairs on bracts and bracteoles as well as in the larger capsules. In its general appearance, however, it is so similar to forms of subsp. *longicalyx* that it does not warrant recognition at species level.

4c. subsp. **magadiense** *Vollesen* in K.B. 62: 318, fig. 8, J–K (2006). Type: Kenya, Kajiado District: Nairobi–Magadi road, *Vollesen* 95/215 (K!, holo.; BR!, C!, CAS!, EA!, ETH!, K!, MO!, P!, iso.)

Young branches stellate-pubescent with the central hair in each cluster much elongated. Bracts and bracteoles with long broad glossy hairs to 1.5 mm long. Calyx 9–14 mm long, steellate puberulous, basal part also with dense long broad glossy hairs to 1.5 mm long on veins, dorsal tooth 3–4 mm long. Corolla 11–19 mm long. Capsule 9–10 mm long.

KENYA. Masai District: 50 km on Nairobi–Magadi road, 4 May 1988, *Bidgood & Vollesen* 1278! & Endoinyo Siruai, 25 May 1996, *Pearce & Vollesen* 936!
DISTR. **K** 6; not known elsewhere
HAB. Open *Acacia* bushland on pale grey to dark brown volcanic clay or on black cotton soil; 1200–1750 m

NOTE. Subsp. *magadiense* has a distribution quite disjunct from the areas of both the other subspecies. It was originally thought to be a distinct species, differing from *D. longicalyx* in its larger flowers and capsules. But there is a considerable overlap in flower size, and capsule size alone hardly merits treating it as a distinct species, especially considering the intermediate capsule size of subsp. *mkomaziense.*

5. **Duosperma stoloniferum** *Vollesen* in K.B. 62: 318, fig. 9 (2006). Type: Kenya, Northern Frontier District: Garissa–Liboi road, 24 km E of junction with Mado Gashi road, *Bally & A. R. Smith* 14979 (K!, holo.; BR!, C!, EA!, P!, WAG!, iso.)

Shrubby herb to 30 cm tall with long arching or stoloniferous branches rooting at nodes; outer bark translucent, peeling in papery strips; stems up to 50 cm long, glabrous or stipule-sheaths with a few broad glossy hairs. Leaves fleshy, obovate or broadly so, largest 1.5–3.5 × 0.6–2.2 cm, apex truncate to retuse (rarely rounded), with a recurved triangular apical tooth, base attenuate, decurrent to base or with an indistinct petiole to 2 mm long, margin entire or crenate-dentate towards apex (rarely along whole margin), glabrous or immature (rarely also mature) stellate puberulous or with scattered long broad glossy hairs along midrib (rarely also along lateral veins), stipule-sheath 1.5–3 mm long. Flowers in 1–3(–7)-flowered condensed axillary cymes with peduncle to 5 mm long; bracts to 1.2 cm long, obovate, glabrous; bracteoles similar or lanceolate, to 1(–1.5) cm long, glabrous or with scattered broad glossy hairs, sometimes finely ciliate. Calyx 7–11(–13 in fruit) mm long, glabrous (rarely with a few broad glossy hairs), finely ciliate; tube 4–6 mm long; teeth narrowly triangular, acuminate, dorsal 3–5 mm long, ventral 2–3 mm long of which the free part 1–2 mm. Corolla mauve to purple or deep blue with darker patches on lower lip (almost white in bud), tube yellowish-brown, 18–22(–31) mm long, finely puberulous; tube 12–14(–26) mm long of which the throat 4–5 mm; lips 5–8 mm long; lobes obovate, 3–6 mm wide, broadly rounded; palate indistinctly pleated, without long stiff hairs. Filaments included in throat, 3–5 mm long, pale mauve, with scattered hairs throughout; anthers 1.5–2 mm long, rounded, connective glabrous. Style pale mauve; stigma lobes equal. Capsule ± 8 mm long, obovoid. Seed not seen.

KENYA. Northern Frontier District: Garissa–Liboi road, 24 km E of junction with Mado Gashi road, 28 Jan. 1972, *Bally & A.R.Smith* 14979! & 33 km on Garissa–Dadaab road, 11 May 1974, *Gillett & Gachati* 20605!; Meru District: Meru National Park, Tana–Rojeweru confluence, 24 Dec. 1969, *Gillett* 18901!

DISTR. **K** 1, 4, 7; Somalia

HAB. Open *Acacia-Commiphora* bushland on fawn sandy to fine clayey soils in areas liable to seasonal flooding, riverine forest on sandy alluvium; 200–350 m

SYN. *Duosperma sp.* (= *Gillett* 18901) of Luke & Robertson, Checklist Vasc. Pl. Coastal Kenya: 81 (1993)

NOTE. The presence of stellate hairs on the youngest leaves indicate that this species should – despite the almost complete lack of indumentum on mature parts – be classified with the species with stellate indumentum. It also differs from the other species in this group in its growth form, large purple corolla and very conspicuous stipule sheath.
    Mungai et al. 83/83 (Tana River District: Kampi ya Simba) has much larger flowers than the rest of the material (maximum dimensions in description) and is also slightly more hairy. But an even more hairy form occurs in S Somalia, and I have considered this to be an acceptable degree of variation in a rarely collected species.

6. **Duosperma grandiflorum** *Vollesen* in K.B. 62: 320, fig. 10 (2006). Type: Kenya, Meru District: Tana-Rojeweru confluence, *Gillett* 18899 (K!, holo.; B!, BR!, EA!, M!, NHT!, P!, WAG!, iso.)

Much branched erect (rarely scrambling or decumbent) shrubby herb or shrub to 2 m tall; outer bark papery and flaky; young branches subglabrous to puberulous with stellate and simple hairs. Leaves rough, ovate to elliptic or broadly so, largest 1.7–6.5 × 1.2–3.2 cm, apex acute to truncate, with a triangular central tooth, base attenuate to cuneate (rarely rounded), decurrent to a distinct petiole to 5(–8) mm long, margin crenate to dentate; stellate puberulous or sparsely so (rarely subglabrous), below also with long broad glossy hairs (middle hair in each cluster much elongated), stellate hairs sometimes only present on young leaves (rarely not at all), stipule-sheath 1–2.5 mm long. Flowers in 1–3(–7)-flowered axillary cymes with peduncle to 4(–6) mm long; bracts and bracteoles obovate-spathulate, to 2.5(–3) cm long, indumentum as leaves. Calyx 10–19 mm long, sparsely puberulent to puberulous with simple and sometimes also stellate hairs (rarely with scattered capitate glands), sometimes also with long hairs along veins or glabrous with long hairs along veins; tube 6–8 mm long; teeth lanceolate to narrowly triangular, dorsal 3–6 mm long, ventral 4–6 mm long of which the free part 1–4 mm. Corolla white to mauve, with or without purple patches on lower lip, 30–40(–50) mm long, puberulous or sparsely so with stalked capitate glands only or with intermixed non-glandular hairs; tube 13–25 mm long of which the throat 3–8(–12) mm; lips 12–25 mm long, usually deeply split between upper and lower; lobes obovate, 5–12 mm wide, broadly rounded; palate indistinctly pleated, without long stiff hairs. Filaments 7–10 and 15–18 mm long (longer pair 8–10 mm longer than shorter pair), with a few scattered hairs near base; anthers 1.5–2.5 mm long, finely apiculate, connective glabrous. Stigma lobes subequal. Capsule 7–11 mm long, ellipsoid. Seed ± 4 mm long.

KENYA. Northern Frontier District: Lowaweregoi, 15 Dec. 1958, *Newbould* 3220!; Meru District: Meru National Park, near Maua, 10 Sep. 1963, *Verdcourt* 3742!; Tana River District: Kora Game Reserve, 30 km from Kampi ya Simba towards Kamaguru, 17 May 1983, *Mungai et al.* 239/83!

DISTR. **K** 1, 4, 7; not known elsewhere

HAB. *Acacia-Commiphora* bushland, *Acacia* and *Combretum* woodland and wooded grassland, on red sandy or stony granitic soils or on rocky slopes; 300–1200(–1400) m

SYN. *Duosperma sp. A* of U.K.W.F.: 586 (1974); Luke & Robertson, Checklist Vasc. Pl. Coastal Kenya: 81 (1993); U.K.W.F., ed. 2: 269 (1994)

NOTE. It is surprising that this relatively common species, which has by far the largest flowers of any species of *Duosperma*, has until recently been without a name. Apart from the corolla size it differs from the other species with stellate indumentum in having a corolla indumentum consisting mostly of stalked capitate glands.

7. **Duosperma kilimandscharicum** (*Lindau*) *Dayton* in Rhodora 47: 262 (1945); Brummitt in K.B. 29: 411 (1974); U.K.W.F.: 587 (1974); Lebrun & Stork, Enum. Pl. Afr. Trop. 4: 479 (1997); U.K.W.F., ed. 2: 270 (1994); Vollesen *et al.*, Checklist Vasc. Pl. and Pter. Mkomazi: 82 (1999) & in K.B. 62: 323, fig. 11 M (2006). Type: Tanzania, Kilimanjaro, below Marangu, Himo, *Volkens* 1721 (B†, holo.; BM!, K!, iso.)

Much branched erect or scrambling shrub to 2(?–3) m tall; bark peeling in long papery strips; young branches finely sericeous or sparsely so (rarely finely puberulous), the hairs short and stubby and often slightly curly, no long spreading hairs. Leaves ovate, elliptic or obovate to suborbicular, largest (1.5–)2–6.2(–8.5) × (0.8–)1–4(–6) cm, apex acute to truncate (rarely retuse), with a triangular apical tooth (sometimes recurved), base attenuate to truncate, decurrent to a distinct petiole to 1.3(–2.5) cm long, subglabrous to curly puberulous on both sides (below densest along veins), margin dentate (rarely crenate); stipule sheath to 1 mm long. Flowers in few- to many-flowered (some with 5 or more flowers) condensed dichasial axillary cymes with peduncle and branches 1–3 mm long, indumentum as branches; outer bracts, lanceolate to oblong, to 1.2 cm long; inner bracts to 8 mm long, sparsely curly sericeous-puberulous; bracteoles similar, to 6 mm long. Calyx 4–9(–11 in fruit) mm long, sericeous-puberulous or sparsely so (rarely subglabrous), tube 4–5 mm long; teeth lanceolate to narrowly triangular, acuminate, dorsal 4–5 mm long, ventral 3–5 mm long of which the free part 1–3 mm long. Corolla white to cream or pale yellow with red or purple patches on lower lip, 6–10 mm long, finely sericeous-puberulous; tube 4–7 mm long of which the throat 1–3 mm, with band of long (0.5–1 mm) hairs at insertion of stamens; lips 2–4 mm long; lobes oblong, 1–2 mm wide, rounded to retuse; palate indistinctly pleated, with long stiff or curly hairs. Filaments 2–4 mm long, with scattered hairs towards base, glabrous or with scattered glands upwards; anthers ± 1 mm long, rounded, connective with scattered glands, sometimes also hairy at one end. Dorsal stigma lobe slightly shorter than ventral. Capsule 6–8 mm long, obovoid. Seed ± 3 mm long. Fig. 38, 12, p. 250.

KENYA. Machakos District: Bushwackers Camp, Masaleni, 23 Apr. 1969, *Napper & Kanuri* 2062! & Tsavo National Park West, Ngulia Hills, 4 Aug. 1963, *Verdcourt* 3696!; Teita District: west side of Voi–Tsavo National Park East road, 19 March 1974, *Faden* 74/261!
TANZANIA. Moshi District: Himo River, 24 Jan. 1936, *Greenway* 4494 & 15 km E of Moshi, 4 Nov. 1955, *Milne-Redhead & Taylor* 7224!; Pare District: Mkomazi Game Reserve, Kamakota Hill, 11 June 1996, *Abdallah, Mboya & Vollesen* 96/162!
DISTR. **K** 4, 7; **T** 2, 3; not known elsewhere
HAB. *Acacia-Commiphora* and *Combretum* woodland and bushland, on grey or brown sandy to black loamy or clayey soil or on rocky hills, rocky thickets, dry riverine scrub; 350–1000 m

SYN. *Dyschoriste kilimandscharica* Lindau in P.O.A. C: 367 (1895) & in E.J. 24: 315 (1897)
    Disperma kilimandscharicum (Lindau) C.B. Clarke in F.T.A. 5: 80 (1899); Milne-Redhead in K.B. 1935: 283 (1935); T.T.C.L.: 7 (1949); Gilbert, Pl. Mt Kilimanjaro: 80 (1970)
    Disperma kilimandscharicum (Lindau) C.B. Clarke var. *bracteolatum* C.B.Clarke in F.T.A. 5: 80 (1899). Type: Kenya, Ukamba, *Scott Elliot* 6306A (K!, holo.; BM!, iso.)

8. **Duosperma crenatum** (*Lindau*) *P.G.Meyer* in Mitt. Bot. Staatssaml. München 3: 602 (1960) & in Prodr. Fl. SW. Afr. 130. Acanthaceae: 28 (1968); Brummitt in K.B. 29: 411 & 414 (1974); Björnstad in Serengeti Res. Inst. Publ. 215: 25 (1976); Grignon & Johnsen, Checklist Vasc. Pl. Botswana: 4 (1986); U.K.W.F., ed. 2: 270 (1994);

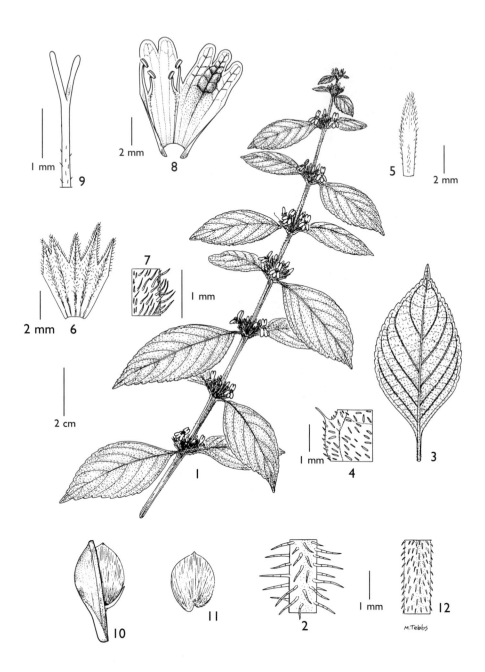

FIG. 38. *DUOSPERMA CRENATUM* — **1**, habit; **2**, stem indumentum; **3**, leaf; **4**, leaf indumentum; **5**, bract; **6**, calyx opened up; **7**, calyx indumentum; **8**, corolla opened up; **9**, style and stigma; **10**, capsule; **11**, seed. *D. KILIMANDSCHARICUM* — **12**, stem indumentum. 1–2 & 5–9 from *Bidgood et al.* 258, 3–4 from *Richards* 17597, 10–11 from *Chuwa* 3020, 12 from *Milne-Redhead & Taylor* 7224. Drawn by Margaret Tebbs.

Lebrun & Stork, Enum. Pl. Afr. Trop. 4: 479 (1997); Mapura & Timberlake, Checklist Zimb. Vasc. Pl.: 13 (2004); Phiri, Checklist Zamb. Vasc. Pl.: 18 (2005); Vollesen in K.B. 62: 326, fig. 11 A–L (2006). Type: Tanzania, Mpwapwa, *Stuhlmann* 287 (B†, holo.; K!, iso.)

Erect to decumbent or scrambling aromatic (when crushed) shrubby herb or shrub, to 2(?–3) m tall; bark peeling in papery strips; young stems glabrous to strigose-pubescent or sericeous-puberulous to -pubescent with broad shiny sometimes curly hairs. Leaves ovate to elliptic or slightly obovate, largest (2.5–)4–10.5 × 1.5–5.5 cm, normally over twice as long as wide, apex drawn out into an acute to acuminate tip or acute to acuminate (rarely rounded), base attenuate to cuneate with a more or less distinct petiole to 1(–2) cm long, subglabrous to sparsely sericeous-puberulous or -pubescent (below densest along veins, above uniformly) with broad shiny hairs, margin crenate to dentate (rarely subentire); petiole sheath to 1(–2) mm long. Flowers in few- to many-flowered condensed axillary cymes, upwards gradually with fewer flowers and near apex often with some solitary flowers; outer bracts leafy, to 2.5 cm long; inner bracts to 1.5 cm long, constricted above a hyaline base and with widened apical part or lanceolate to narrowly ovate, glabrous or with scattered to dense broad shiny hairs along midrib and glabrous to sericeous or puberulous lamina, ciliate; bracteoles lanceolate to narrowly ovate or oblanceolate, to 1 cm long, acute to acuminate. Calyx 4–8(–11 in fruit) mm long, indumentum as bracts and bracteoles; tube 2.5–6 mm long; teeth narrowly triangular, acute to acuminate, dorsal 1–3 mm long, ventral 1–3 mm long of which the free part 0.5–1 mm long. Corolla white or creamy white with two red to purple patches on lower lip, 7–11 mm long, minutely sericeous-puberulous; tube 4–6 mm long of which the throat 1–2 mm; lips 2–5 mm long; lobes oblong, 1–1.5 mm wide, truncate to retuse; palate indistinctly pleated, without long stiff hairs. Filaments 2–4 mm long, with a few hairs near base; anthers 1–1.5 mm long, rounded, connective glabrous or with scattered glands. Dorsal stigma lobe about half as long as ventral. Capsule 6–8 mm long, obovoid. Seed ± 4 mm long. Fig. 38, 1–11, p. 250.

KENYA. Meru District: Meru National Park, Muguongo, 14 June 1963, *Mathenge* 143!; Machakos District: near Kibwezi, Athi River, 23–25 May 1959, *Napper* 1253!; Masai District: Nguruman Escarpment, 5 Aug. 1962, *Glover & Samuel* 3244!
TANZANIA. Masai District: W of Lake Natron, Saleh, 24 July 1962, *Newbould* 6211!; Mpwapwa District: 5 km on Mpwapwa-Gulwe track, 9 Apr. 1988, *Bidgood et al.* 947!; Iringa District: Ruaha National Park, 10 km on Msembe–Causeway track, 23 March 1970, *Greenway & Kanuri* 14186!
DISTR. **K** 4, 6, 7; **T** 1–3, 5–7; Ethiopia, Zambia, Malawi, Mozambique, Zimbabwe, Botswana, Namibia, South Africa
HAB. *Acacia* and *Acacia-Commiphora* woodland, bushland and thicket on clayey or sandy to gravelly or stony soils, *Acacia-Euphorbia* thickets on rocky slopes, *Terminalia* bushland on alkaline alluvium, grassland; 600–1450(–1700) m

SYN. *Hygrophila crenata* Lindau in E.J. 20: 6 (1894) & in P.O.A. C: 366 (1895)
  H. parviflora Lindau in E.J. 20: 7 (1894) & in P.O.A. C: 366 (1895). Type: Malawi, no locality, *Buchanan* 556A (B†, holo.; BM!, K!, iso.)
  Disperma quadrisepalum C.B.Clarke in F.T.A. 5: 80 (1899), *nom. illeg. superf.*
  D. parviflorum (Lindau) C.B.Clarke in F.T.A. 5: 81 (1899); Binns, Checklist Herb. Fl. Malawi: 13 (1968)
  Dyschoriste quadrisepala (C.B.Clarke) Lindau in Wiss. Ergebn. Schwed. Rhod.-Kongo-Exped. 1: 306 (1914)
  Disperma crenatum (Lindau) Milne-Redh. in K.B. 1933: 477 (1933) & in K.B. 1935: 283 (1935); T.T.C.L.: 7 (1949); Binns, Checklist Herb. Fl. Malawi: 13 (1968); Fanshawe, Checklist Woody Pl. Zambia: 19 (1970); Ruffo, Cat. Lushoto Herb. Tanzania: 4 (1996)

NOTE. In the dry central parts of Tanzania this is often a common thicket-forming species in *Acacia-Commiphora* bushland. It seems hardly ever to be eaten by cattle, and is therefore favoured by slight to moderate overgrazing.

9. **Duosperma latifolium** *Vollesen* in K.B. 62: 329, fig. 12 A–J (2006). Type: Tanzania, Masai District: Ardai Plains, *Greenway* 7505 (K, holo.; EA!, NY!, PRE!, iso.)

Shrubby herb to 1 m tall with several stems from woody rootstock; stems often straggling or decumbent and rooting, when young glabrous to finely sericeous on two sides and often with broad glossy retrorse hairs on edges, later with bark peeling in papery strips. Leaves elliptic or obovate to suborbicular, largest (1.7–)2–4.5 × 1.5–2.7 cm, usually less than twice as long as wide, apex subacute to truncate, base attenuate to cuneate, decurrent to an indistinct petiole to 3 mm long which has long (to 1.5 mm) glossy hairs on edges, glabrous or with scattered broad glossy hairs along midrib, larger veins and on edges, margin crenate to dentate; stipule sheath to 1 mm long. Flowers subsessile, in 1–3(–5)-flowered contracted widely separated axillary cymes; peduncles 1–2.5 mm long, glabrous; bracts foliaceous, oblong or obovate-spathulate, to 1.8 cm long, with widened apical part, glabrous or with scattered broad glossy hairs along midrib and on edges; bracteoles lanceolate to ovate, to 8 mm long. Calyx 7–9 mm long, glabrous or with scattered broad glossy hairs on midrib, tip of lobes long-ciliate, tube 5–6 mm long; teeth triangular, acute, dorsal 2–3 mm long, ventral 2–3 mm long of which the free part 0.5–1 mm long. Corolla white with purple patches on lower lip, 10–14(–15) mm long, minutely and sparsely sericeous-puberulous; tube 6–9(–10) mm long of which the throat 1–3 mm; lips 3–5 mm long; lobes oblong, 1.5–2 mm wide, truncate to retuse; palate without long stiff hairs. Filaments 2–4 mm long, with scattered hairs near base; anthers ± 1.5 mm long, rounded, connective glabrous. Dorsal stigma lobe about half the length of ventral. Capsule ± 8 mm long, obovoid. Seed ± 3 mm long.

TANZANIA. Masai District: Eleanata, 20 June 1941, *Hornby* 2116!; Mbulu District: 14 km on Makuyni–Mto wa Mbu road, 25 May 1995, *Vollesen* 95/210!; Kondoa District: Bereko to Kikare, 2 Apr. 1974, *Richards* 29146!
DISTR. **T** 2, 5; not known elsewhere
HAB. Grassland and *Acacia* bushland on black cotton soil, lavaflows with open *Acacia* bushland on fine grey silty soil; 900–1600 m

NOTE. Differs from the closely related *D. crenatum* in the larger corolla with longer tube, in the few-flowered cymes and in the smaller and wider often suborbicular leaves.

10. **Duosperma trachyphyllum** (*Bullock*) *Dayton* in Rhodora 47: 236 (1945); Brummitt in K.B. 29: 411 (1974); Lebrun & Stork, Enum. Pl. Afr. Trop. 4: 479 (1997); Vollesen in K.B. 62: 331, fig. 12 K–M (2006). Type: Tanzania, Mpwapwa, *Greenway* 2391 (K!, holo.; EA!, iso.)

Erect shrubby herb to 1.3 m tall; young branches curly-puberulous or sparsely so, eventually pale yellow and with exfoliating bark. Leaves ovate to elliptic, largest 3–8 × 1.4–3 cm, more than twice as long as wide or about the same, apex acute, base cuneate, decurrent into a distinct petiole to 8 mm long with long broad glossy hairs on edges, uniformly curly puberulous or sparsely so on both sides, margin dentate, venation beneath distinctly raised, reticulate; stipule sheath 1.5–3 mm long. Flowers in few- to many-flowered condensed widely separated axillary cymes; outer bracts foliaceous, to 2 cm long; inner bracts oblong, to 1 cm long, uniformly puberulous or sparsely so and with long broad glossy hairs along midrib and on margins; bracteoles similar, to 8 mm long, with scattered subsessile capitate glands. Calyx 6–8(–10 in fruit) mm long, uniformly finely puberulous and with scattered to dense long broad glossy hairs along midrib and on lobes, with scattered short-stalked capitate glands, tube 4–6 mm long; teeth triangular, acute, dorsal 2–3 mm long, ventral 2–3 mm long of which the free part 1–2 mm. Corolla white, 12–14 mm long, finely sericeous-puberulous; tube 7–8 mm long of which the throat ± 2 mm; lips 4–5 mm long; lobes oblong, ± 2 mm wide, emarginate; palate without long hairs. Filaments 2–4 mm long, with a few hairs near base; anthers ± 1.5 mm long, rounded at both ends, connective with subsessile glands. Dorsal stigma lobe about half the length of ventral. Capsule and seed not seen.

TANZANIA. Mpwapwa District: Mpwapwa, 18 Aug. 1930, *Greenway* 2391!; Iringa District: David Moyer's Farm [7°48'S 35°48'E], 31 March 2006, *Bidgood et al.* 5266!
DISTR. **T** 5, 7; not known elsewhere
HAB. *Brachystegia-Combretum* woodland on brown soil or on stony hillsides; 1200–1450 m

SYN. *Disperma trachyphyllum* Bullock in K.B. 1933: 476 (1933); Milne-Redhead in K.B. 1935: 283 (1935); T.T.C.L.: 7 (1949)

NOTE. Despite the presence of glandular hairs on bracts and calyx *D. trachyphyllum* is probably closer to the group of species around *D. crenatum* than to the species around *D. quadrangulare*. It differs from the species of the *D. quadrangulare* group in the widely separated condensed axillary cymes and the white corolla without stiff hairs on the palate.
   Dayton (l.c.) possibly confuses *D. trachyphyllum* with *D. crenatum* when he writes that it is "reported to form impenetrable thickets on stony hill slopes in neighborhood (sic) of 4000 feet elevation in Dodoma Province, Tanganyika". *D. crenatum* is (pers. obs.) a very common and often thicket forming species in the Mpwapwa–Dodoma area while *D. trachyphyllum* (despite being described as "common" on the label of *Greenway* 2391) is known only from the two above mentioned collections. On the Iringa locality it was (pers. obs.) rare.

11. **Duosperma tanzaniense** *Vollesen* in K.B. 62: 333, fig. A–K (2006). Type: Tanzania, Mbeya District: Poroto Mts, Chimala, *Bidgood, Congdon & Vollesen* 2105 (K!, holo.; BR!, C!, CAS!, DSM!, K!, NHT!, UPS!, iso.)

Perennial or shrubby herb with erect (rarely decumbent) branched stems from woody rootstock; stems to 1(–1.5) m tall, glabrous to bifariously sericeous (hairs retrorse) or puberulous and with (rarely without) sparse to dense broad glossy hairs to 1.5 mm long; bark eventually peeling in large papery strips. Leaves ovate to elliptic, largest 4.5–11 × 2–4.5 cm, more than twice as long as wide, apex subacute to acuminate (rarely rounded), base attenuate to truncate, decurrent to a distinct petiole to 1(–2) cm long, with sparse long broad hairs, below densest (or only) along veins, above uniformly dispersed, rarely sericeous-pubescent or subglabrous, petiole distinctly ciliate with broad glossy hairs, margin crenate to dentate, tertiary veins distinctly raised beneath; stipule sheath to 1 mm long, ciliate and usually hairy. Flowers subsessile, in many-flowered well separated condensed axillary cymes (rarely 1- or few-flowered); peduncles 1–3(–4) mm long, glabrous to puberulous or sericeous; outer bracts foliaceous, to 2.2 cm long; inner bracts and bracteoles lanceolate to ovate, to 1.5 cm long, acuminate to cuspidate, subglabrous to puberulent, without or with sparse to dense long broad glossy hairs, ciliate. Calyx 8–15(–18 in fruit) mm long, glabrous (rarely puberulous or sparsely so), tube and basal part of teeth also with (rarely without) sparse to dense long (to 1.5 mm) broad glossy hairs, apical part of teeth with shorter hairs, sometimes with a few stalked capitate glands near base of tube, tube 4–7 mm long; teeth lanceolate to narrowly triangular, acuminate to cuspidate, dorsal 5–8 mm long (same length or longer than tube), ventral 4–7 mm long of which the free part 2–5 mm long. Corolla white with red to purple lines or patches on lower lip, 12–17(–20) mm long, finely sericeous-puberulous or sparsely so; tube 8–10(–12) mm long of which the throat 2–4(–5) mm long; lips 4–9 mm long; lobes oblong to obovate, 1.5–4(–6) mm wide, truncate to retuse; palate with or without scattered to dense stiff hairs. Filaments 2–5 mm long, with scattered hairs near base; anthers 1–2 mm long, thecae rounded or finely apiculate at base, connective with scattered to dense glands. Stigma lobes subequal or dorsal distinctly shorter. Immature capsule 7–8 mm long, obovoid. Seed not seen.

TANZANIA. Ufipa District: near Kisi, 1 Sep. 1950, *Bullock* 3310!; Iringa District: 30 km on Mafinga–Madibira road, 9 March 1986, *Bidgood et al.* 195!; Njombe District: 70 km on Makambako–Mbeya road, 20 Apr. 1991, *Bidgood & Vollesen* 2200!

DISTR. **T** 4, 5, 7, 8; not known elsewhere

HAB. *Brachystegia* woodland on brown to red gritty to stony soil on rocky slopes, *Combretum* bushland on limestone slope, *Acacia* woodland, grassland on old cultivations; (800–)1000–1650(–1800) m

12. **Duosperma porotoense** *Vollesen* in K.B. 62: 333, fig. 13 L–N (2006). Type: Tanzania, Mbeya District: Poroto Mts, Chimala, *Bidgood, Congdon & Vollesen* 2104 (K!, holo.!; BR!, C!, CAS!, EA!, K!, MO!, NHT!, P!, WAG!, iso.)

Perennial or shrubby perennial herb with erect branched stems from a creeping rootstock; stems to 50 cm tall, when young retrorsely sericeous or puberulous with hairs 0.5–1 mm long, sometimes only on two sides, eventually with bark peeling in papery strips. Leaves ovate to elliptic, largest 3.8–8 × 1.5–3.5 cm, more than twice as long as wide, apex acute to rounded, base attenuate to truncate, decurrent to an indistinctly delimited ciliate petiole to 5 mm long, uniformly puberulous or sparsely so above, below densest along veins with lamina sometimes glabrous; margin crenate to dentate, tertiary veins not or slightly raised, stipule sheath 0.5–1 mm long. Flowers in 1- to many-flowered well separated condensed axillary cymes or these sometimes congested towards tip of branches; peduncles and pedicels 1–2 mm long, sericeous or densely so; outer bracts foliaceous, to 1.5 cm long; inner bracts and bracteoles linear to oblong, to 1.5 cm long, with scattered to dense broad glossy hairs (densest along margins) and often with a sparsely puberulous "undercoat". Calyx (12–)14–16(–21 in fruit) mm long, tube 5–7 mm long, puberulous or densely so, teeth gradually with sparser indumentum upwards, tube and basal part of teeth also with scattered to dense broad glossy hairs to 1 mm long; teeth narrowly triangular, acuminate to cuspidate, dorsal 6–9 mm long (longer than tube), ventral 5–7 mm long of which the free part 2–3 mm. Corolla white with pale pink patches in throat, (15–)18–23 mm long, puberulent; tube 10–14 mm long of which the throat 4–5 mm; lips 5–9 mm long; lobes oblong, 2–3.5 mm wide, rounded to retuse; palate without long stiff hairs. Filaments 3–5 mm long, with scattered hairs throughout; anthers 1.5–2 mm long, thecae rounded, connective densely glandular or with mixtute of glands and hairs. Dorsal stigma lobe half the length of ventral. Immature capsule ± 12 mm long, obovoid. Seed not seen.

TANZANIA. Mbeya District: track above Chimala, 6 Feb. 1974, *Bally & Carter* 16457! & 24 March 1991, *Bidgood et al.* 2104!; Mbeya/Chunya District: Usafwa, 28 Feb. 1932, *Davies* 331!

DISTR. **T** 7; not known elsewhere

HAB. Open *Brachystegia* woodland on steep rocky slopes with brown gritty loamy soil; 1300–1550(?–2000) m

13. **Duosperma livingstoniense** *Vollesen* in K.B. 62: 334, fig. 14 (2006). Type: Tanzania, Njombe District: eastern flank of Livingstone Mts, 5 km N of Mbwila village on Mlangali–Ludewa track, *Gereau & Kayombo* 4350 (K!, holo.; DSM!, EA!, MO!, NHT!, PRE!, iso.)

Shrubby perennial herb with decumbent or ascending branched stems from woody rootstock; stems to 40 cm long, when young bifariously sericeous-puberulous with retrorse hairs, soon glabrescent, eventually with peeling bark. Leaves elliptic to slightly obovate, largest 3.5–4.5 × 1.4–1.8 cm, more than twice as long as wide, apex subacute to rounded, base attenuate, decurrent to an ill-defined petiole to 3 mm long, glabrous apart from a few hairs on midrib and ciliate, petiole with scattered long hairs, margin crenate, tertiary veins not raised beneath; stipule sheath 1–1.5 mm long, finely sericeous and ciliate. Flowers in 1–3-flowered axillary cymes; peduncles to 2 mm and pedicels ± 1 mm long; bracts and bracteoles leafy, obovate to 1.7 (bracts) or 1.2 (bracteoles) cm long, indumentum as leaves. Calyx 13–14 mm long, sparsely and finely puberulous or glabrous upwards, also with scattered broad glossy hairs on tube and basal part of teeth, apical part of teeth finely ciliate, tube 5–6 mm long; teeth narrowly triangular, acute, dorsal 7–8 mm long, ventral ± 6 mm long of which the free parts ± 3 mm. Corolla white with mauve patches on lower lip, 21–24 mm long, minutely sericeous-puberulous; tube 10–13 mm long of which the throat 4–5 mm; lips 9–11 mm long; lobes obovate, 6–7 mm wide, broadly rounded to truncate; palate without long stiff hairs. Filaments 6–8 mm long, with scattered hairs almost to apex; anthers ± 2.5 mm long, rounded, connective with dense glands and sparse hairs. Stigma lobes subequal. Capsule and seed not seen.

TANZANIA. Njombe District: eastern flank of Livingstone Mts, 5 km N of Mbwila village on Mlangali–Ludewa track, 19 March 1991, *Gereau & Kayombo* 4350!
DISTR. **T** 7; not known elsewhere
HAB. Valley bottom *Brachystegia* woodland on sandy to loamy soil; ± 1525 m

NOTE. The relationships of this very distinct species are not at all clear but it is probably closest to *D. tanzaniense* and *D. porotoense*, from which it differs in the larger corolla and subglabrous leaves.

14. **Duosperma nudantherum** (*C.B.Clarke*) Brummitt in K.B. 29: 412 (1974); Lebrun & Stork, Enum. Pl. Afr. Trop. 4: 479 (1997); Phiri, Checklist Zamb. Vasc. Pl.: 18 (2005); Vollesen in K.B. 62: 337, fig. 15K (2006). Type: Zambia, Lake Tanganyika, Niomkolo Island, *Carson* s.n. (K!, holo.)

Erect unbranched or sparsely branched perennial herb with several stems from woody rootstock; stems to 50 cm long, scriceous to pubescent or tomentose with long broad glossy spreading or antrorse hairs. Leaves elliptic or narrowly so to slightly obovate, largest 4–6 × 1.5–2.5 cm, apex acute to rounded, base attenuate to rounded, decurrent to stem or with a distinct petiole up to 5 mm long, pubescent to sericeous or sparsely so, margin entire to dentate; stipule sheath ± 1 mm long. Flowers in condensed axillary cymes, many-flowered towards base and with fewer flowers upwards; outer bracts leafy, to 2.5 cm long; inner bracts and bracteoles lanceolate to oblong, to 1 cm long, acuminate, with dense long broad glossy hairs. Calyx 9–14(–16 in fruit) mm long, with dense (rarely sparse) long broad glossy antrorse hairs to 3(–4) mm long all over or on teeth only, usually also with a finer puberulous-sericeous "undercoat", tube 4–7 mm long; teeth narrowly triangular, acute to cuspidate, dorsal 4–8 mm long, ventral 2–4 mm long of which the free part 1.5–3 mm. Corolla white with purple patches on lower lip, 12–16 mm long, tube sparsely and finely sericeous, lobes finely puberulous; tube 7–9 mm long of which the throat 2–3 mm; lips 5–7 mm long; lobes oblong, 1.5–2.5 mm wide, retuse; palate indistinctly pleated, without long stiff hairs. Filaments 2–4 mm long, with scattered spreading hairs; anthers 1–2 mm long, apex finely apiculate, connective with mixture of hairs and glands or with glands only. Dorsal stigma lobe about half the length of ventral. Capsule 8–9 mm long, obovoid. Seed ± 5 × 3.5 mm.

TANZANIA. Ufipa District: Sumbawanga, Chapota, 6 March 1957, *Richards* 8520! & Kawa River Gorge, 15 Feb. 1959, *Richards* 10880!; Nkansi District: 15 km on Kipili–Namanyere road, 5 March 1994, *Bidgood et al.* 2658!
DISTR. **T** 4; Congo-Kinshasa, Zambia, Malawi
HAB. *Brachystegia* and *Uapaca* woodland on sandy to loamy or stony soil; 850–1650 m

SYN. *Dyschoriste nudanthera* C.B.Clarke in F.T.A. 5: 74 (1899)
    Disperma nudanthera (C.B.Clarke) Milne-Redh. in K.B. 1934: 304 (1934); Richards & Morony, Checklist fl. Mbala: 230 (1969)

NOTE. *Bidgood et al.* 2658A from Nkansi District is somewhat intermediate between *D. nudantherum* and *D. fimbriatum* and might indicate that hybridization between these two species is possible. This specimen has a calyx which is only hairy along the edges of the lobes. But it has the small calyx of *D. nudantherum* and also densely hairy stems with antrorse hairs. It was found growing together with typical *D. nudantherum.*

Reported as occurring on soils with high copper content in Shaba region, Congo.

15. **Duosperma fimbriatum** *Brummitt* in K.B. 29: 413 (1974); Lebrun & Stork, Enum. Pl. Afr. Trop. 4: 479 (1997); Vollesen in K.B. 62: 337, fig. 15 A–J (2006). Type: Zambia, 25 km on Mbala [Abercorn]–Lunzwa Bridge road, *Richards* 22193 (K!, holo.; LISC!, iso.)

Erect perennial herb with unbranched or sparsely branched stems from a creeping woody rootstock; stems to 1 m long, when young glabrous to puberulous and with scattered to dense long broad spreading or antrorse glossy hairs. Leaves elliptic or narrowly so or slightly obovate, largest 6–11 × 1.3–2.5(–3) cm, apex acute or subacute, base attenuate, decurrent to stem or with a distinct petiole to 3(–5) mm long, subglabrous or with scattered to dense long broad glossy hairs, margin subentire to dentate; stipule sheath 0.5–2 mm long. Flowers in condensed axillary cymes, with fewer flowers upwards; outer bracts foliaceous, to 2 cm long; inner bracts lanceolate (rarely oblong), to 1.5 cm long, acute to acuminate, with dense long broad glossy hairs on midrib and margins; bracteoles similar, to 1 cm long. Calyx (9–)12–14(–16 in fruit) mm long, with long broad glossy hairs often only on lobes or tip of lobes, basal part of tube always glabrous, not with "undercoat" of fine hairs, tube (4–)5–7 mm long; teeth linear-lanceolate to narrowly triangular, acuminate to cuspidate, dorsal (5–)6–9 mm long, ventral 4–7 mm long of which the free part 2–5 mm. Corolla white to pale mauve or pinkish purple with red to mauve markings on lower lip, 14–16 mm long, tube finely sericeous, lobes finely puberulous; tube 10–12 mm long of which the throat 2–3 mm; lips 3–4 mm long; lobes oblong, 1–2 mm wide, rounded to retuse; palate indistinctly pleated, without long stiff hairs. Filaments 2–4 mm long, glabrous or with scattered spreading hairs; anthers 1–1.5 mm long, apex finely apiculate, connective with hairs or a mixture of hairs and glands. Dorsal stigma lobe about half the length of ventral. Immature capsule 8–9 mm long, obovoid. Seed not seen.

TANZANIA. Mpanda District: 8 km on Mpanda–Uvinza road, 7 March 1994, *Bidgood et al.* 2668!; Tunduru District: 55 km on Tunduru–Songea road, 21 March 1991, *Bidgood et al.* 2089!; Songea/Tunduru District: 139 km on Songea–Tunduru road, 4 June 1956, *Milne-Redhead & Taylor* 10493!
DISTR. **T** 4, 8; Congo-Kinshasa, Zambia, Mozambique
HAB. *Brachystegia* woodland on grey to brown sandy to loamy soil; 700–900 m

SYN. *Disperma* sp. near *nudanthera* (C.B.Clarke) Milne-Redh. sensu Richards & Morony, Checklist fl. Mbala: 230 (1969)

25. PHAULOPSIS

Willd., Sp. Pl. 3: 342 (1800), as *Phaylopsis*, corr. Sprengel, Anleit., ed. 2, 2: 422 (1817), *nom. et orth. cons.*; G.P. 2: 1081 (1876); Lindau in Nat. Pflanzenfam., Nachtr. Zu II-IV, 1: 305 (1897); C.B. Clarke in F.T.A. 5: 82 (1899); M. Manktelow in Symb. Bot. Ups. 31(2): 78 (1996).

Micranthus Wendl., Bot. Beobacht.: 38 (1798), *nom. rej.*, *non Micranthus* (Pers.) Eckl. (1827), *nom. cons.*; Benoist in Fl. Madagascar 182, 1: 112 (1967)

Aetheilema R. Br., Prodr. Fl. Nov. Holl. 1: 478 (1810); Nees in DC., Prodr. 11: 261 (1847)

Annual or perennial herbs; cystoliths conspicuous; stems quadrangular. Leaves opposite, often anisophyllous entire to crenate. Flowers in few- to many-flowered dichasia or monochasia condensed into into terminal and axillary densely strobilate or interrupted dorsiventral inflorescences with flowers on one side only (rarely solitary in leaf-axils); main axis bracts foliaceous, cyme bracts bracteate; bracteoles small. Calyx deeply divided into 5 sepals, 2 ventral sepals often fused slightly higher; dorsal sepal ovate-oblong, much larger than and covering the other four. Corolla contorted in bud, finely puberulous and sometimes also glandular on the outside; basal part of tube cylindric, widening into a more or less distinct throat; limb distinctly 2-lipped, lobes subequal; palate on lower lip without or with few to many retrorse hairs. Stamens 4, didynamous, inserted dorsally in throat, more or less exserted or included in throat, tube and base of filaments glabrous at point of insertion, each long and short filament fused at the base and fused part running like two glabrous ridges to base of tube; anthers 2-thecous, oblong, thecae subequal, rounded apically, at base without or with minute spurs. Ovary 2-locular, 4-ovulate; style hairy in basal part; stigma lobes unequal, linear, flattened. Capsule 4-seeded, ovoid, glandular towards apex, placentae at maturity separating elastically from fruit wall; retinaculae strong, hook-like. Seed discoid, densely covered with hygroscopic hairs.

22 species, mostly in tropical Africa, three species in Madagascar and the Mascarenes one of which also in SE Asia, one species introduced to central and S America.

1. Bracts and calyx without stalked capitate glands 2
   Bracts and calyx with stalked capitate glands 3
2. Flowers in dense dorsiventral inflorescences; corolla 7–8.5 mm
     long, hardly longer than the calyx; largest leaf 3–7.5 cm long   5. *P. angolana*
   Flowers apparently solitary or in 2-flowered cymules in axils of
     vegetative leaves or condensed towards the end of stems;
     corolla 9–12 mm long, distinctly longer than the calyx;
     largest leaf 0.7–3.3 cm long . . . .   6. *P. pulchella*
3. Corolla tube divided into a cylindric basal part and a
     conspicuously widened throat 5–7 mm in diameter at apex;
     anthers protruding from corolla tube; lower corolla lip with
     numerous retrorse hairs 4
   Apical part of corolla tube not conspicuously wider than basal
     part, 1–3 mm in diameter at apex; anthers included in
     corolla tube; lower lip without or with a few retrorse hairs 5
4. Leaves with a petiole 1–5 cm long; stems decumbent or
     scrambling, rooting from the lower nodes . . . .   2. *P. sangana*
   Leaf base decurrent to stem or leaves with a petiole to 4(–8) mm
     long; stems erect to decumbent, never rooting at the nodes   1. *P. johnstonii*
5. Corolla (8)10–15 mm long, tube clearly bent, lips 2–6 mm
     long; stems usually more than 2 mm in diameter at base,
     rooting at lower nodes . . . .   3. *P. imbricata*
   Corolla 7–11 mm long, tube bent or straight, lips 1.5–3.5 mm
     long 6
6. Plant usually erect, not rooting at lower nodes, stems coarse,
     more than 2 mm in diameter at base; largest leaf 3.3–5.5 cm
     wide; corolla tube bent . . . .   3. *P. imbricata*
   Plant prostrate, decumbent or scrambling, rooting at lower
     nodes; stems slender, to 1(–2) mm in diameter at base;
     largest leaf 0.7–2.3(–4) cm wide; corolla tube straight . . . .   4. *P. gediensis*

1. **Phaulopsis johnstonii** *C.B.Clarke* in F.T.A. 5: 86 (1899), as *Phaylopsis*; M. Manktelow in Symb. Bot. Ups. 31(2): 115 (1996); Lebrun & Stork, Enum. Pl. Afr. Trop. 4: 498 (1997). Type: Angola, Gambos, *Newton* s.n. (K!, lecto.; selected by Manktelow)

Perennial herb with one to several erect to decumbent stems from woody rootstock; stems to 1 m long, never rooting at nodes, when young puberulous to pubescent and sometimes with scattered glands. Leaves elliptic or narrowly so, largest 5–12 × 1.2–3 cm, usually more than three times as long as wide, apex acute or subacute, base cuneate to attenuate, decurrent to the stem or with petiole to 4(–8) mm long, subglabrous to pubescent-pilose with long (to 2 mm) glossy hairs. Inflorescences terminal only or terminal and lateral, ovoid to ellipsoid, 2–7 cm long; axis with long thin hairs and dense glands; axis bracts to 4 cm long, smaller upwards, with dense (rarely sparse or absent) capitate glands; cyme bracts pale green, sometimes with dark edge or flushed with purple, ovate or elliptic to orbicular, 7–14 × 5–10 mm, acute to rounded, densely (rarely sparsely) glandular. Calyx 9–13 mm long of which the basal tube 1–2 mm, ciliate on edges and veins with thin hairs to 3 mm long and with dense glands to 2 mm long; dorsal lobe elliptic, 3–5 mm wide, subacute to rounded; ventral lobes linear to oblanceolate, same length or to 2 mm shorter than dorsal; lateral lobes linear, 2–3 mm shorter than ventral. Corolla white with orange and mauve patches on callus of throat, 15–19 mm long; cylindric part of tube 5–7.5 mm long; throat conspicuously widened, 5–6 mm long and 6–7 mm in diameter at apex; lips 4–6 mm long, lobes oblong-obovate, 2–4 mm long, rounded to retuse; lower lip with many retrorse hairs. Filaments with scattered short hairs; anthers protruding, ± 1.5 mm long, with distinct spurs. Capsule 7–8.5 mm long. Seed 2–3 mm in diameter. Fig. 39, 8, p. 259.

TANZANIA. Mpanda District: 8 km on Mpanda–Uvinza road, 13 May 1997, *Bidgood et al.* 3891!; Iringa District: Mafinga–Madibira road, Ndembera River, 3 May 1986, *Congdon* 71!; Songea District: Hanga Farm, 27 June 1956, *Milne-Redhead & Taylor* 10911!
DISTR. T 4, 7, 8; Congo-Kinshasa, Angola, Zambia
HAB. *Brachystegia* and *Brachystegia-Uapaca* woodland, on sandy to gravelly soil; 1000–1700 m

SYN. *Phaulopsis betonica* S.Moore in J.B. 47: 295 (1909), as *Phaylopsis*. Type: Congo-Kinshasa, Mt Senga, *Kässner* 2977 (BM!, holo.; K!, iso.)

2. **Phaulopsis sangana** *S.Moore* in J.B. 49: 295 (1911), as *Phaylopsis*; M. Manktelow in Symb. Bot. Ups. 31(2): 109 (1996); Lebrun & Stork, Enum. Pl. Afr. Trop. 4: 499 (1997). Type: Angola, Sanga, Kuve River, *Gossweiler* 4467 (BM!, holo.; K!, iso.)

Perennial herb with one to several decumbent or scrambling (? rarely erect) stems from woody rootstock; stems to 50 cm long (to 1 m in scrambling plants), rooting at lower nodes, when young sericeous-puberulous with downwardly directed hairs and sometimes with scattered glands. Leaves ovate to elliptic, largest 3–11.5 × 1.5–5 cm, apex acuminate to subacute, base cuneate to attenuate, decurrent to the 1–5 cm long petiole, sparsely puberulous, below densest along veins, above uniformly. Inflorescences terminal and lateral, ovoid to ellipsoid, 2–4.5 cm long; axis bracts to 2.5 cm long, smaller upwards, cyme bracts pale green with dark green to purplish edge, broadly ovate to orbicular or reniform, 7.5–15 × 5–18 mm, rounded, with long thin hairs to 3(–4) mm long and with sparse to dense glands. Calyx 9–11 mm long of which the basal tube 1–1.5 mm, ciliate on edges and veins with thin glossy hairs to 2 mm long and with dense glands; dorsal lobe elliptic-obovate, 3–4 mm wide, acute to rounded; ventral lobes lanceolate or oblanceolate, 1–3 mm shorter than dorsal; lateral lobes linear, 1–3 mm shorter than ventral. Corolla white (or tip of upper lip mauve) with mauve or violet markings on callus of throat, 16–20 mm long; cylindric part of tube 4–5 mm long; throat conspicuously widened, ± 4 mm long and 5–6 mm in diameter at apex; lips 5.5–8 mm long, lobes oblong-obovate, 2–4 mm long, rounded to retuse; lower lip with many retrorse hairs. Filaments with scattered short hairs; anthers protruding, ± 1.5 mm long, with distinct spurs. Capsule and seed not seen.

TANZANIA. Njombe District: Ukinga Mts, *Goetze* 1191! & 13 km WNW of Njombe, Nyumbanyito, 11 July 1956, *Milne-Redhead & Taylor* 11050!; Mbeya District: Umalila, Ibala, *Leedal* 4507!

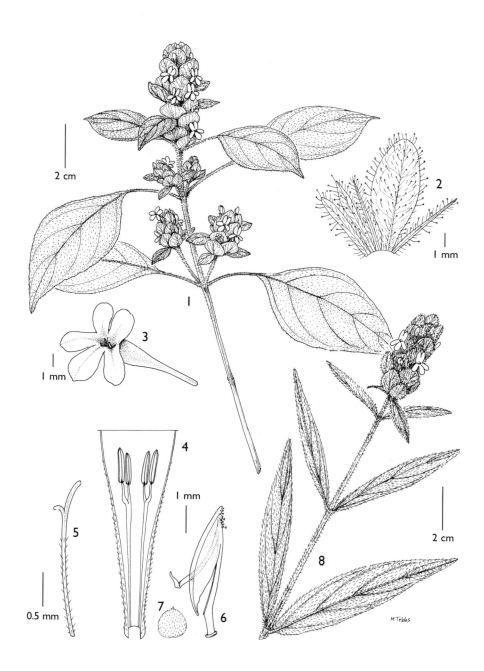

FIG. 39. *PHAULOPSIS IMBRICATA* subsp. *IMBRICATA* — **1**, habit; **2**, calyx, opened up; **3**, corolla; **4**, corolla tube opened up showing stamens; **5**, style and stigma; **6**, capsule; **7**, seed. *PHAULOPSIS JOHNSTONII* — **8**, habit. 1 from *Drummond & Hemsley* 4381, 2 from *Geilinger* 1430, 3–5 from *Tanner* 2437, 6–7 from *Borhidi* 84149, 8 from *Bidgood et al.* 3891. Drawn by Margaret Tebbs.

DISTR. **T** 7, 8; Angola, Malawi

HAB. Edges and clearings of montane forest, persisting in pine plantations; 1900–2300 m

SYN. *Phaulopsis major* Mildbr. in N.B.G.B. 15: 636 (1941). Type: Tanzania, Songea District: Matengo Highlands, Ugano, *Zerny* 735 (B†, holo.; W, iso.)

3. **Phaulopsis imbricata** (*Forssk.*) *Sweet*, Hort. Brit.: 327 (1826), as *Phaylopsis*; F.P.S. 3: 185 (1956); E.P.A.: 933 (1964); Binns, Checklist Herb. Fl. Malawi: 15 (1968), as *Phaylopsis*; Richards & Morony, Checklist Fl. Mbala Distr.: 233 (1969), as *Phaylopsis*; U.K.W.F.: 589 (1974); Vollesen in Opera Bot. 59: 81 (1980); Iversen in Symb. Bot. Ups. 29(3): 162 (1991); U.K.W.F., ed. 2: 268, pl. 116 (1994); M. Manktelow in Symb. Bot. Ups. 31(2): 126 (1996); Lebrun & Stork, Enum. Pl. Afr. Trop. 4: 498 (1997); Friis & Vollesen in Biol. Skr. 51(2): 452 (2005); Ensermu in F.E.E. 5: 385 (2006). Type: Yemen, Mokhaja, *Forsskål* 367 (C!, lecto., selected by Manktelow)

Perennial herb with erect, scrambling or procumbent stems, with or without woody rootstock; stems to 1(–2) m long, usually over 2 mm in diameter near base of plant, often rooting at nodes, when young sparsely to densely puberulous to pubescent or sericeous with hairs to 1 mm long, often densest on two sides. Leaves ovate to elliptic, largest 2.5–14 × 1.2–7 cm, apex acuminate to acute, base attenuate, decurrent to the 0.5–5(–8) cm long petiole, subglabrous to pubescent with glossy hairs to 2 mm long, below densest along veins, above uniformly. Inflorescences terminal and lateral, strobilate or interrupted in basal part, 2–7(–15) cm long; axis puberulous to pubescent or densely so, sometimes with capitate glands; axis bracts to 2.5(–3.5) cm long, smaller upwards, indumentum as leaves; cyme bracts pale green with dark edge, broadly ovate to reniform, to 11 × 16 mm, subacute to rounded, glandular or sparsely so, often only near base, and with long thin hairs to 2 mm. Calyx 6–10 mm long of which the basal tube 1–2 mm, ciliate on edges and veins with thin hairs to 2 mm long and with dense capitate glands; dorsal lobe elliptic to slightly obovate, 1.5–5 mm wide, subacute to rounded; ventral lobes linear to oblanceolate, 1–2 mm shorter than dorsal; lateral lobes linear, ± 1 mm shorter than ventral. Corolla white, more rarely with pale pink markings on lower lip or with pink tinge, 7–15 mm long; tube bent, 5.5–10 mm long of which the cylindric part about half; throat only slightly widened, 1.5–2.5 mm in diameter at apex; lips 1.5–6 mm long, lobes oblong to obovate, 1–5 mm long, rounded to retuse; lower lip without (rarely with a few) retrorse hairs. Filaments glabrous; anthers included in throat, 0.5–1.5 mm long, without or with minute appendages. Capsule 6–7 mm long. Seed 1.5–2 mm in diameter.

SYN. *Ruellia imbricata* Forssk., Descr. Aegypt.-Arab.: 113 (1775); Vahl, Symb. Bot. 2: 73 (1791)
    Aetheilema imbricata (Forssk.) Spreng., Syst. Veg., ed. 16, 2: 826 (1825); Richard, Tent. Fl. Abyss. 2: 149 (1850)
    Micranthus imbricatus (Forssk.) Kuntze, Rev. Gen. Pl. 2: 493 (1891)

a. subsp. **imbricata**

Plant usually scrambling or decumbent and rooting at the lower nodes, but occasionally erect or suberect with non-rooting stems; corolla usually pure white, but occasionally with a pale pink tinge or with pale pink markings on lower lip, (8–)10–15 mm long with tube (6–)7–10 mm long and lips 2–6 mm long. Fig. 39, 1–7, p. 259.

UGANDA. Acholi District: Mt Rom, no date, *Eggeling* 2387!; Kigezi District: Kanungu, June 1939, *Purseglove* 804!; Masaka District: Malabigambo Forest, 6 km SSW of Katera, 2 Oct. 1953, *Drummond & Hemsley* 4561!
KENYA. Northern Frontier District: Maralal, Lorok Plateau, 10 Nov. 1978, *Hepper & Jaeger* 6715!; Trans-Nzoia District: Mt Elgon, Endebess, no date, *Irwin* 53!; Nairobi District: Nairobi City Park, 26 Sep. 1971, *Mwangangi & Kasyoki* 1828!
TANZANIA. Mbulu District: Lake Manyara National Park, 28 June 1965, *Greenway & Kanuri* 11916!; Mpanda District: Mt Livandabe [Lubalisi], 4 June 1997, *Bidgood et al.* 4306!; Uzaramo District: Pugu Hills, 24 Oct. 1968, *B. J. Harris* 2482!; Pemba, no locality, no date, *Vaughan* 510!

Distr. **U** 1–4; **K** 1–7; **T** 1–8; **Z**; **P**; Congo-Kinshasa, Rwanda, Burundi, Sudan, Ethiopia, Zambia, Malawi, Mozambique, Zimbabwe, Swaziland, South Africa; Madagascar; Yemen

Hab. Lowland and montane forest, commonly on edges, in clearings, along paths, often in disturbed places, riverine forest, secondary scrub, plantations, lawns, roadsides, weed in moister areas; near sea level to 2400 m

Syn. *Aetheilema reniforme* Nees in Wall., Pl. As. Rar. 3: 94 (1832) var. *hispidosa* Nees in DC., Prodr. 11: 261 (1847). Type: Tanzania, Pemba Island, *Bojer* s.n. (K!, holo.)
    [*Micranthus longifolius* sensu Lindau in E. & P. Pf. IV, 3b: 298, fig. 120 (1895) & in P.O.A. C: 367 (1895), *non* (Sims) Kuntze (1891)]
    Phaulopsis longifolia C.B.Clarke in F.T.A. 5: 84 (1899), *non* Sims (1823)
    [*P. parviflora* sensu C.B.Clarke in F.T.A. 5: 83 (1899), *non* Willd. (1800)]
    P. inaequalis Pic.Serm. in Webbia 7: 339 (1950), *nom. nov.* for *P. longifolia* C. B. Clarke
    P. imbricata (Forssk.) Sweet var. *inaequalis* (Pic.Serm.) Cuf., E.P.A.: 932 (1964). Type: Ethiopia, Mt Scholoda, *Schimper* I.367 (K!, lecto.; BM!, BR!, K!, iso.; selected by Manktelow)

    b. subsp. **poggei** (*Lindau*) *M.Manktelow* in Symb. Bot. Ups. 31(2): 138 (1996); Lebrun & Stork, Enum. Pl. Afr. Trop. 4: 498 (1997); Friis & Vollesen in Biol. Skr. 51(2): 452 (2005); Ensermu in F.E.E. 5: 386 (2006). Type: Congo-Kinshasa, Lualaba River, Nyangwe, *Pogge* 978 (K!, lecto., selected by Manktelow).

Plant usually erect or ascending, usually not rooting at lower nodes; corolla usually with mauve markings on lower lip or pale pink, more rarely pure white, 7–11 mm long with tube 5.5–8 mm long and lips 1.5–3 mm long.

Uganda. West Nile District: Arua, Dec. 1937, *Hazel* 410!; Mbale District: Bugishu, Budadiri, Jan. 1932, *Chandler* 506!; Mengo District: Entebbe, Aug. 1905, *E. Brown* 297!
Kenya. Trans-Nzoia District: Moi's [Hoey's] Bridge, 15 Aug. 1963, *Heriz-Smith & Paulo* 856!
Tanzania. Mwanza District: Geita, Uzinza, Nyamililo, 3 July 1953, *Tanner* 1554!; Kigoma District: 15 km N of Kigoma, Kakombe, 8 July 1959, *Newbould & Harley* 4297!
Distr. **U** 1–4; **K** 3; **T** 1, 4; Guinea, Sierra Leone, Liberia, Ghana, Nigeria, Cameroon, Gabon, Central African Republic, Congo (Brazzaville), Congo-Kinshasa, Burundi, Rwanda, Sudan, Ethiopia, Angola, Zambia
Hab. Woodland, bushland, grassland, riverine forest and scrub, margins of dry forest, secondary scrub, plantations; 800–1300 m

Syn. *Micranthus poggei* Lindau in E.J. 17: 108 (1893) & in E. & P. Pf. IV, 3b: 298 (1895)
    Phaulopsis poggei (Lindau) Lindau in Nat. Pflanzenfam., Nachtr. Zu II–IV, 1: 305 (1897); C.B. Clarke in F.T.A. 5: 85 (1899); Th. & H. Durand, Syll. Fl. Congo.: 420 (1909); Heine in Fl. Gabon 13: 48 (1966)
    Phaulopsis imbricata (Forssk.) Sweet subsp. *pallidifolia* M.Manktelow in Symb. Bot. Ups. 31(2): 144 (1996); Lebrun & Stork, Enum. Pl. Afr. Trop. 4: 498 (1997). Type: Zambia, Ndola, *Fanshawe* 1160 (K!, holo.)

Note to the species. This widespread and variable species has been divided into four rather vaguely delimited subspecies by Manktelow. Two of these occur in the flora area and have been upheld here. They are reasonably easy to separate by the combination of characters used here but intermediates occur. A third subspecies described by Manktelow from Central Africa is by me considered synonymous with subsp. *poggei*. The fourth subspecies is endemic in Madagascar.

    4. **Phaulopsis gediensis** *M.Manktelow* in Symb. Bot. Ups. 31(2): 152 (1996); Lebrun & Stork, Enum. Pl. Afr. Trop. 4: 498 (1997). Type: Kenya, Kilifi District: Arabuko-Sokoke Forest, Mida, *Polhill & Paulo* 882 (K!, holo.; BR!, iso.)

Perennial herb with prostrate, decumbent or scrambling stems, without woody rootstock; stems to 1 m long, to 1(–2) mm in diameter near base of plant, rooting at lower nodes, when young sparsely to densely sericeous-puberulous (hairs downwardly curved) with hairs to 1 mm long, densest on two sides. Leaves ovate to elliptic or narrowly so, largest 1.5–5.3(–7.5) × 0.7–2.3(–4) cm, apex subacuminate to subacute, base cuneate to attenuate, often very unequal-sided, decurrent to the 0.3–3(–5) cm long petiole, subglabrous to sparsely puberulous or sparsely pubescent with glossy hairs to 1 mm long, below densest along veins, above uniformly. Inflorescences terminal only or terminal and lateral, strobilate or interrupted in basal part, 0.5–7 cm long; axis puberulous or sericeous-puberulous, sometimes with capitate glands; axis bracts to 2 cm long, smaller upwards, indumentum as leaves; cyme bracts green or pale green, suborbicular to reniform, to 10 × 14 mm, subacute to rounded, sparsely puberulous and with long thin hairs to 2 mm, also glandular or sparsely so, often only near base. Calyx 6.5–8 mm long of which the basal tube 1–1.5 mm, ciliate on edges and veins with thin hairs to 2 mm long and with scattered to dense capitate glands; dorsal lobe ovate to elliptic or broadly so, 2.5–5 mm wide, subacute to rounded; ventral lobes linear to lanceolate or oblanceolate, 0.5–1.5 mm shorter than dorsal; lateral lobes linear, 1–1.5 mm shorter than ventral. Corolla white, 8–10 mm long; tube straight, 5–6.5 mm long of which the cylindric part about half; throat only slightly widened, 1–2 mm in diameter at apex; lips 2–3.5 mm long, lobes obovate, 1–2.5 mm long, truncate to retuse; lower lip without retrorse hairs. Filaments glabrous; anthers included in throat, ± 1 mm long, without or with minute appendages. Capsule 5–6 mm long. Seed ± 1.5 mm in diameter.

KENYA. Kwale District: Cha Simba Forest, 2 Feb. 1953, *Drummond & Hemsley* 1091! & Dzombo Hill, 12 Nov. 1992, *Luke et al.* 3372!; Kilifi District: Gedi, 19 Jan. 1992, *Robertson* 6575!
TANZANIA. Lushoto District: E Usambara Mts, Maramba–Lugongo, 16 Dec. 1917, *Peter* 60631!; Tanga District: Kange Estate, 9 Jan. 1952, *Faulkner* 850!; Kilwa District: Selous Game Reserve, Nahomba, 7 May 1970, *Rodgers* MRC1060!
DISTR. **K** 7; **T** 3, 6, 8; not known elsewhere
HAB. Evergreen and semi-evergreen lowland forest and thicket; near sea level to 500 m

SYN. [*Phaulopsis imbricata* sensu Vollesen in Opera Bot. 59: 81 (1980) pro parte, *non* (Forssk.) Sweet (1826)]
    [*Phaulopsis silvestris* sensu Iversen in Symb. Bot. Ups. 29(3): 162 (1991), *non* (Lindau) Lindau (1897)]

NOTE. Closely related to *Phaulopsis imbricata* subsp. *imbricata* from which it differs in the slender habit, small leaves and corolla with a straight tube. A few collections of *P. imbricata* subsp. *imbricata* (e.g. *Abdallah & Vollesen* 95/94 from **T** 3, Same District: Mkomazi Game Reserve) are morphologically very similar but have a larger corolla (over 10 mm long).

5. **Phaulopsis angolana** S.Moore in J.B. 18: 229 (1880), as *Phaylopsis*; C.B. Clarke in F.T.A. 5: 84 (1899), as *Phaylopsis*; M. Manktelow in Symb. Bot. Ups. 31(2): 99 (1996); Lebrun & Stork, Enum. Pl. Afr. Trop. 4: 498 (1997); Ensermu in F.E.E. 5: 385 (2006). Type: Angola, Golungo Alto, Queta, Catomba, *Welwitsch* 5175 (BM!, holo.; K!, iso.)

Perennial herb with prostrate or decumbent stems with erect flowering branches; stems to 50 cm long, rooting at lower nodes, when young sparsely to densely puberulous with downwardly directed hairs. Leaves ovate to elliptic, largest 3–7.5 × 1.8–4.2 cm, apex subacuminate to subacute, base cuneate to attenuate, decurrent to the up to 2.5 cm long petiole, subglabrous to sparsely sericeous-puberulous, densest along veins. Inflorescences terminal and lateral, ovoid-ellipsoid, 1–3 cm long; axis puberulous; axis bracts to 1.5 cm long, smaller upwards, indumentum as leaves; cyme bracts pale green, broadly ovate to orbicular, to 10 × 8 mm, subacute to rounded, ciliate with broad glossy hairs to 2 mm. Calyx 7–9 mm long of which the basal tube ± 1 mm, ciliate on edges and veins with broad glossy hairs to 2 mm long, often also puberulous; dorsal lobe ovate to elliptic, 2–3.5 mm wide, subacute to rounded; ventral lobes lanceolate, same length or slightly shorter than dorsal; lateral lobes linear, 2–3 mm shorter than ventral. Corolla white, 7–8.5 mm long; tube straight, 5–6.5 mm long of which the cylindric part about half; throat only slightly widened, 1–2 mm in diameter at apex; lips 2–3.5 mm long, lobes oblong, 1–1.5 mm long, rounded to retuse; lower lip with or without retrorse hairs. Filaments glabrous; anthers included in throat or slightly exserted, ± 0.5 mm long, without or with minute appendages. Capsule ± 6 mm long. Seed ± 1.5 mm in diameter.

UGANDA. Bunyoro District: Sonso River, Nov. 1935, *Eggeling* 2308!; Toro District: Burahya County, Kibale National Park, Kanyawara, 15 July 1994, *Poulsen* 675!; Masaka District: Sese, Bugala, 3 June 1932, *Thomas* 5!

KENYA. North Kavirondo District: Kakamega Forest, south of Yala River, 21 Jan. 1970, *Faden* 70/11!

TANZANIA. Bukoba District: Minziro Forest Reserve, 2 July 2000, *Bidgood et al.* 4791! & 18 July 2001, *Festo et al.* 1641!

DISTR. U 2, 4; **K** 5; **T** 1; Guinea, Sierra Leone, Liberia, Ghana, Nigeria, Cameroon, Equatorial Guinea, Gabon, Central African Republic, Congo (Brazzaville), Congo-Kinshasa, Ethiopia, Angola

HAB. Wet evergreen forest, often along streams, in clearings and along paths, swamp forest, also in secondary forest; 1000–1550 m

SYN. *Micranthus angolanus* (S.Moore) Kuntze, Rev. Gen. Pl. 2: 493 (1891); Hiern, Cat. Afr. Pl. Welwitsch 4: 811 (1900)

6. **Phaulopsis pulchella** *M.Manktelow* in Symb. Bot. Ups. 31(2): 150 (1996); Lebrun & Stork, Enum. Pl. Afr. Trop. 4: 499 (1997). Type: Tanzania, Zanzibar, Chwaka, *Faulkner* 2351 (K!, holo.; BR!, iso.)

Perennial herb with creeping and rooting branched stems with erect flowering branches, forming small patches; stems to 25 cm long and flowering branches to 10 cm tall, when young sericeous-puberulous or sparsely so with downwardly directed hairs, with band of longer hairs at nodes. Leaves ovate (rarely elliptic or suborbicular), largest 0.7–3.3 × 0.5–1.9 cm, apex acute to rounded, base cuneate to truncate, not or only slightly decurrent to the 2–8(–15) mm long petiole, beneath sparsely puberulous along veins, above with sparse broad glossy hairs. Inflorescences much reduced with flowers seemingly axillary and solitary or in 2-flowered cymules, 0.5–7 cm long; axis bracts similar to vegetative leaves; cyme bracts absent or much reduced, foliaceous, to 4 mm long. Calyx 6–7.5 mm long of which the basal tube 1.5–2 mm, puberulous on edges and veins with broad glossy hairs to 1 mm long; dorsal lobe narrowly ovate, 1.5–2.5 mm wide, acute; ventral and lobes similar, linear-lanceolate, ± 1 mm shorter than dorsal. Corolla white, 9–12 mm long; tube straight or slightly curved, 6–7 mm long of which the cylindric part about half; throat only slightly widened, 1–2 mm in diameter at apex; lips 3–5 mm long, lobes obovate, 2–3.5 mm long, rounded to truncate; lower lip without retrorse hairs. Filaments glabrous; anthers included in throat, ± 1 mm long, without appendages. Capsule 5–6 mm long. Seed ± 1.5 mm in diameter.

TANZANIA. Zanzibar, Chwaka, Aug. 1891, *Sacleux* 1619! & 5 June 1962, *Faulkner* 3047!; Pemba, Msala Island, 22 Dec. 1930, *Greenway* 2769!

DISTR. **Z**; **P**; not known elsewhere

HAB. Dry deciduous forest and thicket on raised coral reefs; near sea level to 25 m

NOTE. To my knowledge the only member of the *Acanthaceae* which is endemic on Zanzibar and Pemba.

26. SATANOCRATER

Schweinf. in Verh. Zool. Bot. Ges. Wien 18: 676 (1868); G.P. 2: 1085 (1876); Lindau in Nat. Pflanzenfam., Nachtr. zum II–IV: 305 (1897); C.B. Clarke in F.T.A. 5: 68 (1899); Thulin in Nord. J. Bot. 24: 385 (2007)

Haemacanthus S. Moore in J. Bot. 37: 63 (1899)

Shrubby herbs or shrubs; stems and leaves with scattered to dense subsessile peltate scale-like glands; cystoliths conspicuous. Leaves opposite, entire. Flowers solitary, axillary; bracts like vegetative leaves; bracteoles persistent. Calyx large, inflated, longitudinally 5-ribbed or almost winged, shortly 5-lobed, with peltate scales. Corolla hairy and with peltate scales, contorted in bud, subactinomorphic to distinctly 2-lipped; cylindric part of tube short or long, throat long; lobes 5, spreading or reflexed, subequal or middle in lower lip reduced to a tooth. Stamens 4, subequal or didynamous, inserted at base of throat, exserted or included in throat; free part of filaments fused at base and running like two ridges to base of tube; anthers 2-thecous, thecae linear-oblong, rounded at base and apex, no spurs. Ovary with 2 ovules per locule; style filiform, hairy throughout; dorsal stigma lobe reduced to a small tooth, ventral lobe large linear, flattened. Capsule 4-seeded, clavate, glabrous; retinaculae strong, straight. Seed discoid, densely covered with thick clavate to almost globose hairs and with peltate scales.

4 species in tropical Africa north of the Equator. One disjunct between Guinea and the border regions between Sudan and NW Ethiopia and three in the Horn of Africa.

Corolla mauve to purple, subactinomorphic with spreading lobes; stamens included in throat, didynamous; leaves glabrous beneath; calyx symmetric .... 1. *S. ruspolii*
Corolla pale orange to red, zygomorphic with 4 erect or reflexed lobes and middle lobe in lower lip tooth-like; stamens exserted, subequal; leaves hairy beneath, at least on veins; calyx asymmetric (curved dorsally) with ventral lobes longer than dorsal .... 2. *S. paradoxus*

1. **Satanocrater ruspolii** (*Lindau*) *Lindau* in Nat. Pflanzenfam., Nachtr. zum II–IV: 305 (1897); Ensermu in F.E.E. 5: 353 (2006); Thulin in Fl. Somalia 3: 396, fig. 269 (2006) & in Nord. J. Bot. 24: 386 (2007). Type: Ethiopia, Milmil, *Riva* 1068 (FT, lecto.; selected by Thulin (2007: 386))

Shrubby herb or shrub to 1(–2) m tall; young stems puberulous to sericeous on two sides, other two sides glabrous or with a few hairs, peltate scales scattered to dense, pale yellow. Leaves pale yellow, with petiole 2–5 mm long or lamina decurrent to stem; lamina ovate to elliptic or broadly so, largest 1–4 × 0.7–2.2 cm, apex subacute to broadly rounded, base truncate to attenuate, whitish tomentellous along midrib above and along edges of petiole, otherwise glabrous, peltate scales scattered to completely covering surface. Pedicels 0.2–2(–4) cm long below bracteoles and 2–3(–5) mm above, sparsely sericeous and with scattered to dense peltate scales; bracteoles linear-lanceolate, 1–6 mm long, rarely elliptic and up to 14 mm long. Calyx 18–28 mm long of which the triangular acuminate lobes 4–7 mm, glabrous to sparsely sericeous-puberulous and with scattered to dense peltate scales, densely whitish sericeous on edges of lobes. Corolla mauve to purple, pubescent and with scattered peltate scales; cylindric part of tube 15–25 mm long, strongly expanded throat 15–25 mm long and 15–20 mm in diameter apically, lobes reniform, 15–21 × 12–15 mm, broadly rounded to truncate with crenate edge. Stamens included in throat, didynamous, free part of filaments 3–5 and 8–10 mm long, glabrous; anthers linear-oblong, 3–4 (longer pair) and 4–5 mm long. Ventral stigma lobe 2–3 mm long. Capsule clavate, 12–15 mm long, with scattered peltate scales. Seed (immature) ellipsoid, ± 5 mm in diameter, with short almost globose hairs. Fig. 40, 8–10, p. 265.

KENYA. Northern Frontier District: 37 km SW of Malka Mari, Dida Wachuf, 20 Jan. 1972, *Bally & Radcliffe-Smith* 14913!
DISTR. **K** 1; Ethiopia, Somalia
HAB. *Acacia-Commiphora* bushland on crystalline limestone rock; ± 900 m

SYN. *Ruellia ruspolii* Lindau in Ann. Ist. Bot. Roma 6: 69 (1896)

NOTE. The anthers of the longer pair of stamens are sometimes sterile.

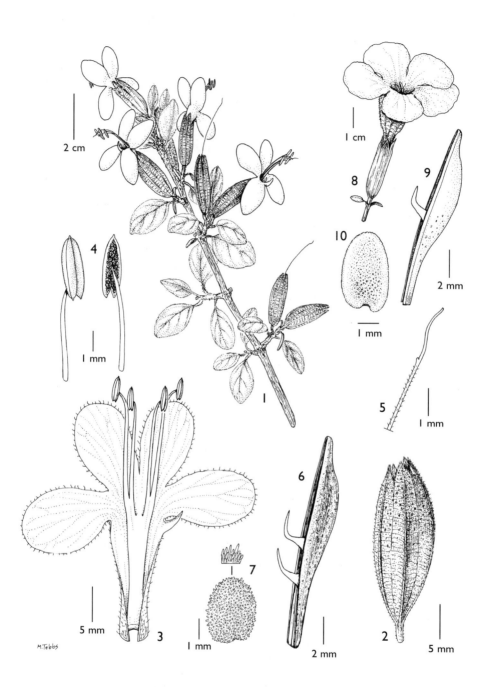

FIG. 40. *SATANOCRATER PARADOXUS* — **1**, habit; **2**, calyx; **3**, corolla opened up; **4**, stamens; **5**, style and stigma; **6**, capsule; **7**, seed with detail of "indumentum". *SATANOCRATER RUSPOLII* — **8**, calyx and corolla; **9**, capsule; **10**, seed. 1–7 from *Bally & Radcliffe-Smith* 14581, 8 from *Burger* 3330, 9–10 from *Bally & Radcliffe-Smith* 14913. Drawn by Margaret Tebbs.

2. **Satanocrater paradoxus** (*Lindau*) Lindau in Nat. Pflanzenfam., Nachtr. zum II–IV: 305 (1897); Turrill in Ic. Pl.: t. 2982 (1913); K.T.S.L.: 608 (1994); Lebrun & Stork, Enum. Pl. Afr. Trop. 4: 502 (1997); Ensermu in F.E.E. 5: 395 (2006); Thulin in Fl. Somalia 3: 397 (2006) & in Nord. J. Bot. 24: 387 (2007). Type: Ethiopia, Dschacorsa [Giacorsa], *Riva* 420 (FT, lecto., selected by Thulin (2007: 387))

Shrub to 1.5(–2.5) m tall; young stems whitish pubescent to tomentellous with thick "sausage-shaped" hairs, peltate scales scattered, pale yellow. Leaves with petiole to 1 cm long but lamina often decurrent to stem; lamina elliptic to subcircular, largest 1–5 × 0.8–3.7 cm, apex subacute to truncate (rarely retuse), base attenuate, decurrent, subglabrous to whitish tomentose, usually densest along veins and always hairy at least on veins beneath, peltate scales scattered to dense pale yellow. Pedicels 2–7 mm long below bracteoles and 2–7 mm above, puberulous to tomentellous and with scattered peltate scales; bracteoles linear, 5–10(–15) mm long. Calyx 18–28 mm long of which the triangular acute to acuminate lobes 3–5 mm, asymmetric (curved dorsally) with ventral lobes longer than dorsal, sparsely to densely puberulous and with scattered to dense peltate scales, densely whitish sericeous on edges of lobes. Corolla pale orange to orange red or bright red, sparsely pubescent and with stalked capitate glands, with scattered peltate scales; tube slightly to strongly curved, widened almost from base, 13–20 mm long ventrally (to base of tooth) and 25–35 mm dorsally, 6–9 mm in diameter apically; lobes: two dorsal erect, two lateral in lower lip reflexed, central in lower lip a dark red hardened tooth, dorsal and lateral subcircular, 12–15 × 13–16 mm, with crenate edge, tooth 2–4 mm long. Stamens exserted, subequal, free part of filaments 15–25 mm long, with scattered capitate glands; anthers linear-oblong, 3–5 mm long, thecae slightly diverging. Ventral stigma lobe ± 2 mm long. Capsule clavate, 20–22 mm long, with scattered peltate scales. Seed (immature) ellipsoid, ± 5 mm in diameter, with dense thick clavate hairs. Fig. 40, 1–7, p. 265.

KENYA. Northern Frontier District: 18 km SW of Mandera, 13 Dec. 1971, *Bally & Radcliffe-Smith* 14581! & 31 km SW of Malka Mari, 21 Jan. 1972, *Bally & Radcliffe-Smith* 14916! & 48 km on Ramu–El Wak road, 10 May 1978, *Gilbert & Thulin* 1630!
DISTR. **K** 1; Ethiopia, Somalia
HAB. *Acacia-Commiphora* woodland and bushland on crystalline limestone rock; 250–950 m

SYN. *Ruellia paradoxa* Lindau in Ann. Ist. Bot. Roma 6: 69 (1896); Schweinfurth & Volkens, Liste Pl. Somalis: 15 (1897)

27. WHITFIELDIA

Hook. in Bot. Mag. 71: t. 4155 (1845); Nees in DC., Prodr. 11: 220 (1847); G.P. 2: 1085 (1876); C.B. Clarke in F.T.A. 5: 65 (1899); Evrard & Demillecamps in Belg. J.Bot. 125: 89 (1992); Manktelow *et al.* in Syst. Bot. 26: 104 (2001)

Stylarthropus Baill. in Bull. Soc. Linn. Paris 2: 822 (1890) & Hist. Pl. 10: 437 (1890)

Erect or scandent shrubs; cystoliths conspicuous or not, elongated to ± dot-like; stems rounded or subquadrangular. Leaves opposite, with strongly raised midrib, with scattered sessile scale-like glands. Flowers single, aggregated into terminal racemiform or paniculate cymes; bracts large or small, not strobilate, often whitish and caducous; bracteoles large, often whitish. Calyx large, petaloid, often whitish, deeply divided into 5 subequal linear-oblong sepals. Corolla large, puberulous and/or glandular on the outside, basal tube curved just below throat, long and linear or shorter, throat conspicuous; lobes 5, contorted in bud. Stamens 4 (but see note under *W. stuhlmannii*), subequal or didynamous, inserted near base of throat, exserted or included; filaments linear; anthers 2-thecous, oblong, thecae subequal, rounded at both ends. Ovary 2-locular, 4-ovulate; style sparsely hairy, sometimes articulated just above ovary; stigma of 2 capitate lobes. Capsule 2–4-seeded, clavate, with solid contracted base, laterally flattened, glabrous. Seed discoid, glabrous, with concentric ridges and jagged edge.

12 species in tropical Africa. Most diverse in the rainforests of the Guineo-Congolian Region.

1. Corolla pure white, cylindric tube 3.5–5 cm long, much
     longer than throat, lobes spreading or reflexed, 1.8–2.8 cm
     long; stamens much exserted, anthers with dense capitate
     glands; calyx 2.5–4.2(–4.7) cm long . . . .   3. *W. elongata*
   Corolla white with dark purple lobes in lower lip, cylindric
     tube 0.9–1.8 cm long, lobes erect in upper lip, reflexed
     in lower, 0.6–1 cm long; stamens held under upper lip,
     anthers with long curly hairs; calyx 1.3–2.7 cm long 2
2. Young branches finely puberulous with simple hairs;
     cylindric corolla tube 1.5–1.8 cm long; stamens clearly
     didynamous . . . .   2. *W. stuhlmannii*
   Young branches with floccose stellate hairs; cylindric
     corolla tube 0.9–1.2 cm long; stamens subequal . . . .   1. *W. orientalis*

1. **Whitfieldia orientalis** *Vollesen* in K.B. 59: 123, fig. 1 (2004). Type: Tanzania, Lindi District: Rondo Plateau, *Bidgood et al.* 1458 (K!, holo.; C!, CAS!, DSM!, K!, NHT!, iso.)

Erect spindly brittle-stemmed shrub to 2(–3) m tall; young stems brown or dark brown, with sparse to dense floccose stellate hairs which rub off easily, glabrescent. Leaves ovate to elliptic, largest 5–21 × 2–7.5 cm, apex acute to acuminate (rarely obtuse), base attenuate (rarely cuneate to truncate), decurrent to the 0.5–4(–5.5) cm long petiole, margin subentire to distinctly crenate-dentate, glabrous or with sparse (rarely dense) stellate hairs on veins and edges. Cymes simple, racemiform or with two additional cymes from lowermost pair of bracts, 1–7 cm long; peduncle 0.2–4 cm long; axes with indumentum as stems; lowermost pairs of bracts foliaceous, upper caducous and white; pedicels (below bracteoles) (2–)3–5 mm long; bracteoles white or greenish white, ovate to elliptic, 4–11 × 3–7 mm, subglabrous to densely glandular-pubescent with curly glands to 2 mm long. Calyx white, 1.7–2.7 cm long of which the basal tube 1–3 mm, sparsely to densely glandular-pubescent with glands to 2 mm long; lobes strap-shaped, acute to rounded. Corolla tube and lobes in upper lip pure white, lobes in lower lip dark purple, glabrous or puberulous to pubescent with curly hairs to 1 mm long; cylindric tube 0.9–1.2 cm long and 3.5–5 mm in diameter, throat funnel-shaped or slightly gibbose, 1.5–2 cm long and 1–1.4 cm in diameter at apex; lobes erect in upper lip, reflexed in lower, triangular, 7–10 × 4–6 mm, subacute. Stamens held under upper lip, subequal; filaments 1.4–1.7 cm long, glabrous; anthers narrowly oblong, curved, 3–4 mm long, all functional, connective and sides of thecae with long curly hairs. Capsule 1.7–2 cm long, 2-seeded, seed-bearing part broadly ovoid, ± 1 cm long. Seed 6–8 mm long, broadly ellipsoid.

KENYA. Kilifi District: Chasimba, 26 Aug. 1971, *Faden* 71/767!; Kwale District: Shimba Hills, Mwele Mdogo Forest, 6 Jan. 1988, *Luke* 893B! & 12 Sep. 1997, *Luke* 4729!
TANZANIA. Tanga District: Pangwe, Kange, 15 Jan. 1956, *Faulkner* 1795!; Morogoro District: Nguru Mts, Liwale Valley, Manyangu Forest, 27 March 1953, *Drummond & Hemsley* 1842!; Lindi District: Rondo Plateau, Rondo Forest Reserve, 10 Feb. 1991, *Bidgood et al.* 1458!
DISTR. **K** 7; **T** 3, 6–8; not known elsewhere
HAB. Evergreen and semi-deciduous lowland forest; near sea level to 800(–1000) m

SYN. [*Whitfieldia stuhlmannii sensu* C.B.Clarke in F.T.A. 5: 68 (1899) quoad *Heinsen* 17, *Holst* 2673 & *Volkens* 2396; T.T.C.L.: 18 (1949); Iversen in Symb. Bot. Ups. 29(3): 162 (1991); Robertson & Luke, Kenya Coastal Forests 2: 84 (1993); K.T.S.L.: 610 (1994); Ruffo *et al.*, Cat. Lushoto Herb. Tanzania: 11 (1996), *non sensu stricto*]

NOTE. C.B.Clarke (l.c.) included the coastal East African material into *W. stuhlmannii* without any comment but stated that the species has stellately hairy young branches, a unique character in Whitfieldia. But Lindau – when describing *Stylarthropus stuhlmannii* – clearly states that the type (since destroyed) is subglabrous as is indeed all the Congo material (including the neotype) seen at BR and also the three Tanzanian collections cited below under that species. C. B. Clarke's approach has been universally followed ever since, but there is no doubt – as already suspected by Evrard & Demillecamps in Belg. J. Bot. 125: 92 (1992) – that we are dealing with two distinct species.

2. **Whitfieldia stuhlmannii** (*Lindau*) *C.B.Clarke* in F.T.A. 5: 68 (1899); Evrard & Demillecamps in Belg. J.Bot. 125: 92 (1992); Lebrun & Stork, Enum. Pl. Afr. Trop. 4: 508 (1997). Type: Congo-Kinshasa, Isange–Semliki, *Stuhlmann* 2963 (B†, holo.); Bobata–Kamengo, *Gille* 286 (BR!, neo, selected by Evrard & Demillecamps, l.c.)

Scrambling or scandent shrub to 2 m tall, forming large tangles; young stems green, finely puberulous in two bands, soon glabrous. Leaves ovate to elliptic, largest 15–18 × 5.5–8 cm, apex subacuminate, base attenuate, decurrent to the 1–4 cm long petiole, margin indistinctly crenate, glabrous. Cymes simple, racemiform or with two additional cymes from lowermost pair of bracts, 5–10 cm long; peduncle 1.5–4 cm long; axes with indumentum as stems; bracts caducous, green or with whitish basal part, lowermost pairs sometimes foliaceous; pedicels (below bracteoles) 1–2 mm long; bracteoles with green basal part and white towards apex, ovate to elliptic, 7–10 × 2–5 mm, glabrous. Calyx whitish at base, pale pink upwards, 1.3–1.8(–2.2 in fruit) cm long of which the basal tube 1–2 mm, densely glandular-pubescent with glands to 1 mm long; lobes linear, subacute. Corolla tube and lobes in upper lip pure white, lobes in lower lip dark purple, pubescent with curly hairs to 1 mm long; cylindric tube 1.5–1.8 cm long and 2–4 mm in diameter, throat gibbose, 1.7–1.9 cm long and 1–1.2 cm in diameter at apex; lobes erect in upper lip, reflexed in lower, narrowly triangular, 6–7 × 3–4 mm, subacute to rounded. Stamens held under upper lip, didynamous; filaments ± 1.7 and 1.9 cm long, glabrous; anthers narrowly oblong, curved, ± 3 (? non-functional) and 4.5 mm long, connective and sides of thecae with long curly hairs. Capsule ± 2 cm long, 2-seeded, seed-bearing part broadly ovoid-ellipsoid, ± 1 cm long. Seed ± 8 mm long, broadly ellipsoid.

TANZANIA. Kigoma District: Gombe Stream National Park, Upper Mkenke Valley, 29 Apr. 1992, *Mbago* 1032!; Mpanda District: Mt Livandabe [Lubalisi], 3 June 1997, *Bidgood et al.* 4285! & Ntakatta Forest, 14 June 2000, *Bidgood et al.* 4678!
DISTR. **T** 4; Congo-Kinshasa
HAB. Tall closed evergreen *Newtonia-Pterygota* forest, riverine forest; 800–1300 m

SYN. *Stylarthropus stuhlmannii* Lindau in E.J. 20: 11 (1894)
    Whitfieldia seretii De Wild. in Ann. Mus. Congo Bot., ser. 5, 3: 267 (1910). Type: Congo-Kinshasa, Duru–Rungu, *Seret* 739 (BR!, holo.; BR!, iso.)
    Whitfieldia seretii De Wild. var. *elliptica* De Wild. in Ann. Mus. Congo Bot., ser. 5, 3: 479 (1912). Type: Congo-Kinshasa, Yanga, *Jespersen* s.n. (BR!, holo.)

NOTE. In two of the Tanzanian collections the two short stamens seem to have non-functional anthers with one theca much shorter than the other.

3. **Whitfieldia elongata** (*P.Beauv.*) *De Wild. & T.Durand* in Bull. Soc. Roy. Bot. Belg. 38: 110 (1899); C.B. Clarke in F.T.A. 5: 66 (1899); Th. & H. Durand, Syll. Fl. Cong.: 421 (1909); De Wildeman in B.S.B.B. 41(3): t. 29 (1913); F.P.S. 3: 181 (1956); K.T.S.: 19 (1961); White, F.F.N.R.: 383 (1962); Heine in F.W.T.A., ed. 2, 2: 398 (1963); E.P.A.: 935 (1964); Heine in Fl. Gabon 13: 34, pl. 6 (1967); Richards & Morony, Checklist Mbala Distr.: 236 (1969); U.K.W.F.: 591 (1974); Champluvier in Fl. Rwanda 3: 492 (1985); Iversen in Symb. Bot. Ups. 29(3): 162 (1991); Evrard & Demillecamps in Belg. J.Bot. 125: 90 (1992); K.T.S.L.: 610 (1994); U.K.W.F., ed. 2: 270 (1994); Lebrun & Stork, Enum. Pl. Afr. Trop. 4: 507 (1997); Friis & Vollesen in Biol. Skr. 51(2): 456 (2005); Ensermu in F.E.E. 5: 395 (2006). Type: "Benin", *Palisot de Beauvois* s.n. (G, holo.).

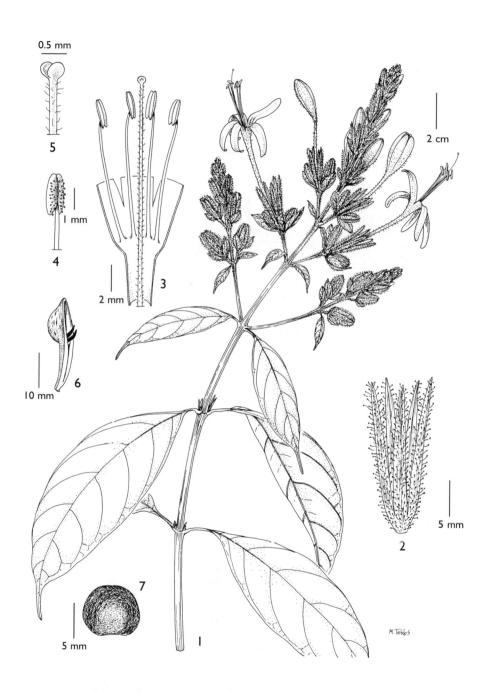

FIG. 41. *WHITFIELDIA ELONGATA* — **1**, habit; **2**, calyx; **3**, corolla tube opened up with stamens and style; **4**, anther; **5**, apical part of style and stigma; **6**, capsule; **7**, seed. 1 & 3 from *Drummond & Hemsley* 4718, 2 & 4–5 from *Bidgood* 4261, 6–7 from *Newbould* 4277. Drawn by Margaret Tebbs.

Erect or (usually) scrambling or scandent shrub to 3(–5 if scandent) m tall; young stems green, subglabrous to crisped-puberulous in two bands, soon glabrous. Leaves elliptic or narrowly so (rarely ovate), largest 9.5–28 × (2.5–)3.5–11 cm, apex acuminate to cuspidate, base attenuate, decurrent to the 1–4(–7) cm long petiole, margin entire, glabrous or sparsely puberulous along midrib. Cymes simple, racemiform or with two additional cymes from lower 1–2 pairs of bracts or with 2 racemes from upper pair of leaves to form a paniculate inflorescence 5–25 cm long; peduncle (1–)2–5 cm long; axes with indumentum as stems or upwards with sparse to dense sticky stalked capitate glands; bracts caducous, white, or lowermost pairs foliaceous; pedicels (below bracteoles) 1–4(–7) mm long; bracteoles white, elliptic, 10–17 × 5–10 mm, subglabrous to densely glandular-puberulous. Calyx white, 2.5–4.2(–4.7) cm long of which the basal tube 2–5 mm, glandular-pubescent with glands to 1 mm long; lobes linear to narrowly elliptic, acute. Corolla pure white or with tube greenish towards base, cylindric tube 3.5–5 cm long and ± 2 mm in diameter, throat funnel-shaped, 0.7–1.2 cm long and 7–8 mm in diameter at apex, pubescent with curly hairs to 1 mm long; lobes spreading or reflexed, elliptic or narrowly so, 1.8–2.8 × 0.5–0.8 cm, subacute to rounded. Stamens exserted, didynamous; filaments 1.8–2.8 and 2.3–3 cm long, glabrous; anthers purple, narrowly oblong, curved, 3–4 mm long, connective and sides of thecae with dense short-stalked capitate glands. Capsule 2.3–3.5 cm long, 2-seeded, seed-bearing part broadly ovoid-ellipsoid, 1.2–1.5 cm long. Seed 7–9 mm long, broadly ellipsoid to circular. Fig. 41, p. 269.

UGANDA. West Nile District: Amua, Otze Forest Reserve, 5 Dec. 1947, *Dawkins* 299!; Bunyoro District: Budongo Forest, Dec. 1934, *Eggeling* 1552!; Masaka District: NW side of Lake Nabugabo, 9 Oct. 1953, *Drummond & Hemsley* 4718!

KENYA. North Kavirondo District: Bukaro, 21 Oct. 1927, *McDonald* 1103!; Kwale District: Shimba Hills, Mwele Mdogo Forest, 6 Feb. 1953, *Drummond & Hemsley* 1146! & Mrima Hill, 11 Dec. 1969, *Bally* 13711!

TANZANIA. Mwanza District: Geita, Kigasi, Maisome Island, 10 Aug. 1962, *Carmichael* 890!; Mpanda District: Kungwe-Mahali Peninsula, Lubungwe River, 11 July 1958, *Juniper & Newbould* 103!; Uzaramo District: Pugu Hills, 14 Sep. 1971, *Wingfield* 1785!; Pemba, Ngezi Forest, Dec. 1983, *Rodgers et al.* 2651!

DISTR. U 1–4; K 3, 4 (see note), 5, 7; T 1, 3, 4, 6, 7; P; Nigeria, Cameroon, Bioko, Central African Republic, Congo-Kinshasa, Rwanda, Sudan, Ethiopia, Angola, Zambia

HAB. Evergreen lowland and lower montane forest (including secondary forest), usually on edges, in clearings and along streams, riverine forest; 10–1500 m, but around Nairobi – where originally introduced – to 1650 m

SYN. *Ruellia elongata* P.Beauv., Fl. Owar. 1: 46, t. 26 (1806)
    Dipteracanthus elongatus (P.Beauv.) Nees in DC., Prodr. 11: 140 (1847); Benth. in Hooker, Niger Fl.: 478 (1849)
    Whitfieldia longifolia T.Anderson in J.L.S. 7: 27 (1863), *nom. illeg.*; S. Moore in J.B. 18: 229 (1880), as "*longiflora*"; C.B. Clarke in F.T.A. 5: 66 (1899); Hiern, Cat. Afr. Pl. Welw. 4: 811 (1900); F.W.T.A. 2: 248 (1931). Type: Fernando Po, *Mann* 198 (K!, lecto.). See Note.
    W. elongata (P.Beauv.) De Wild. & T.Durand var. *dewevrei* De Wild. & T.Durand in Bull. Soc. Roy. Bot. Belg. 38: 111 (1899). Type: Congo-Kinshasa, Mobanga, *Dewevre* s.n. (BR, holo.)
    W. tanganyikensis C.B.Clarke in F.T.A. 5: 67 (1899). Types: Uganda, Buddu, *Scott Elliot* 7528 (BM!, K!, syn.) & Zambia, "Moero Plateau", *Carson* 22 (K!, syn.)
    W. elongata (P.Beauv.) De Wild. & T.Durand var. *abbreviata* Mildbr. in B.J.B.B. 17: 88 (1943). Type: Congo-Kinshasa, Kabasha, *de Witte* 2162 (BR!, holo.; BR!, iso.)

NOTE. T. Anderson (l.c.) cites "*Ruellia longifolia* P.Beauv. in Fl. Owar. 1: 45 " as basionym for his *Whitfieldia longifolia*. But there is no such name in Flora Owariensis. He cites *Ruellia elongata* as a synonym, but with *R. longifolia* P.Beauv. being a fictitious name this is the correct name.
    Cultivated and possibly naturalized in the grounds of the National Museum, Nairobi, e.g. *Mwangangi* 1836! It has also (pers. obs.) become naturalized in hedges at Makerere University, Uganda.

28. CHLAMYDACANTHUS

Lindau in E.J. 17: 109 (1893) & in E. & P. Pf. IV, 3b: 343, fig. 137 (1895);
Manktelow *et al.* in Syst. Bot. 26: 104 (2001)

Perennial or shrubby herbs; cystoliths conspicuous, ± dot-like; stem rounded or with weak longitudinal ridges. Leaves opposite, subentire to crenate. Flowers single or usually 2–3 (the third often only as an aborting bud), these cymules enveloped by two large partly fused foliaceous bracts and these again aggregated into a racemiform cyme; bracteoles present, filiform to linear. Calyx deeply divided into 5 equal linear lobes. Corolla puberulous outside; basal tube curved, widening gradually into the throat; lobes 5, contorted in bud, 3 in lower lip deflexed, 2 in upper lip erect. Stamens 4, subequal, inserted at base of throat, slightly exserted; anthers 2-thecous, narrowly oblong, slightly curved, thecae subequal, at same height, rounded at both ends. Ovary 2(?–4)-ovulate; style puberulous; stigma of 2 capitate lobes. Capsule 2(?–4)-seeded, clavate with contracted stalk-like base, laterally flattened, breaking irregularly from base at maturity; retinaculae thin. Seed discoid, glabrous with concentric ridges and jagged edge.

NOTE. Winkler in F.R. 9: 523 (1911) states that the ovary is 2–4-ovulate and the capsule 2- or 4-seeded. In the now plentiful East African material I have only seen 2 ovules and 2 seeds.

2 species, one in Madagascar and one on the East African coast.

Chlamydacanthus lindavianus *H. Winkl.* in F.R. 9: 523 (1911); T.T.C.L.: 6 (1949); Iversen in Symb. Bot. Ups. 29(3): 160 (1991); Robertson & Luke, Kenya Coastal Forests 2, Checklist Vasc. Pl.: 81 (1993); Ruffo *et al.*, Cat. Lushoto Herb. Tanzania: 3 (1996), as *Chlamydocardia*. Type: Tanzania, Tanga District: Siga Hills, *Winkler* 4288 (WRSL, holo.)

Shrubby herb or shrub to 1 m tall; young stems glabrous to puberulous. Leaves ovate, largest 5–22 × 2.5–9 cm, apex subacute to subacuminate, base gradually attenuate or abruptly narrowed and then attenuate, decurrent to the (0.2–)1–3 cm long petiole, puberulous or sparsely so along midrib beneath, otherwise glabrous. Cymes 3–23 cm long, unbranched or with lateral cymes from 1(–2) lower nodes; flowers 2, often with a third undeveloping bud; peduncle 0.5–5(–7) cm long; axes sparsely puberulous; main axis bracts subulate, 2–4(–8) mm long (lowermost pair rarely elliptic and up to 1 cm long); peduncles of cymules 1–7 mm long; cymule bracts 1.2–2 × 1–1.8 cm, fused ⅓ up on lower side, free on upper side, dark green, broadly elliptic to orbicular, acute to truncate and apiculate; pedicels 2–3 mm long, puberulous and with stalked capitate glands; bracteoles filiform, (2–)4–8 mm long, indumentum as pedicels. Calyx 6–8 mm long, with sparse to dense stalked capitate glands and with sparse hairs near apex of lobes, divided to 0.5–1 mm from base, lobes linear, acute. Corolla white to cream (rarely with maroon lines); tube 7–10 mm long, bent 1–3 mm above base; lobes 2–3.5 mm long, triangular, obtuse. Stamens held under upper lip, filaments 4–6 mm long, glabrous, anthers 1–1.5 mm long, glabrous. Capsule 9–11 mm long, glabrous, densely pustulate-rugose, seed-containing part broadly ovoid. Seed broadly cordiform in outline, 5–6 × 5–6 mm. Fig. 42, p. 272.

KENYA. Kilifi District: Mwarakaya, 13 Dec. 1990, *Luke & Robertson* 2625!; Kwale District: Shimba Hills, Mwadabara, 17 Oct. 1991, *Luke* 2941! & Marenji Forest Reserve South, 2 June 2000, *Luke et al.* 6273B!
TANZANIA. Tanga District: Amboni Caves, Mkulumuzi River, 1 Aug. 1953, *Drummond & Hemsley* 3592!; Morogoro District: Kimboza Forest Reserve, 9–17 July 1983, *Rodgers et al.* 2600!; Lindi District: Lake Lutamba, Litipo Forest Reserve, 4 Nov. 1984, *Mwasumbi & Mponda* 12664!
DISTR. **K** 7; **T** 3, 6, 8; not known elsewhere
HAB. Undergrowth in semi-evergreen or deciduous coastal forest; 10–250(–600) m

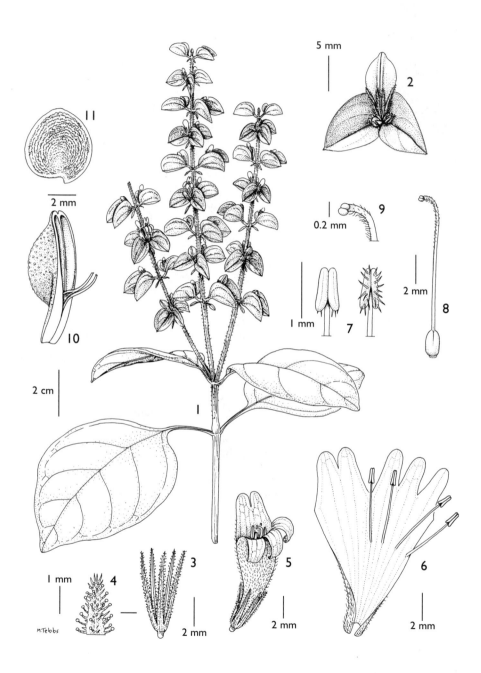

FIG. 42. *CHLAMYDACANTHUS LINDAVIANUS* — **1**, habit; **2**, cymule with young capsule and two buds; **3**, calyx; **4**, calyx indumentum; **5**, corolla; **6**, corolla opened up; **7**, anther, dorsal and ventral view; **8**, ovary, style and stigma; **9**, stigma; **10**, capsule; **11**, seed. 1–2 & 7 from *Luke* 6273B, 3–6 & 8–9 from *Geilinger* 1283, 10–11 from *Polhill* 4913. Drawn by Margaret Tebbs.

SYN. *Chlamydacanthus dichrostachyus* Mildbr. in N.B.G.B. 11: 823 (1933); T.T.C.L.: 6 (1949); Rodgers *et al.*, Conserv. Val. and Stat. Kimboza Forest Reserve Tanzania: 51 (1983). Type: Tanzania, Mahenge, Ruaha, *Schlieben* 2269 (B†, holo.; BM!, HBG!, iso.)

29. LANKESTERIA

Lindl., Bot. Reg. 31, Misc.: 86 (1845); G.P. 2: 1083 (1876); C.B. Clarke in F.T.A. 5: 69 (1899); Manktelow *et al.* in Syst. Bot. 26: 104 (2001)

Pernnial or shrubby herbs or shrubs; cystoliths conspicuous; stems subquadrangular. Leaves opposite, entire to crenate. Flowers single, aggregated into strobilate spiciform cymes; bracts conspicuous; bracteoles linear. Calyx deeply divided into 5 subequal linear sepals. Corolla contorted in bud, finely puberulous and/or glandular on the outside, with a long linear cylindric tube which is only slightly expanded into a short indistinct throat; limb spreading, subequally 5-lobed. Stamens 2, sometimes with 2 minute staminodes, inserted dorsally at base of throat, exserted, filaments much flattened basally, narrowly so or terete towards apex; anthers 2-thecous, narrowly oblong, thecae equal or subequal, rounded at both ends. Ovary 2-locular, 2–4-ovulate; style finely hairy; stigma of 2 capitate lobes. Capsule 2-seeded, clavate, with solid contracted basal part, laterally flattened, glabrous; retinaculae strong, hook-like. Seed discoid, covered with long hygroscopic hairs.

7 species in Tropical Africa and Madagascar.

Corolla bright orange, the cylindric tube 3–4 cm long and the lobes
  8–11 mm long; capsule 10–12 mm long; inflorescence axis,
  bracts and calyx with stalked capitate glands . . . .   1. *L. elegans*
Corolla white, the cylindric tube 1.1–1.9 cm long and the lobes
  4–6 mm long; capsule 6 8 mm long; inflorescence axis, bracts
  and calyx without glands . . . .   2. *L. alba*

1. **Lankesteria elegans** (*P.Beauv.*) *T.Anderson* in J.L.S. 7: 33 (1863); Lindau in P.O.A. C: 368 (1895); C.B. Clarke in F.T.A. 5: 70 (1899); Mangenot, Icon. Pl. Afr. I.F.A.N. 4: 87 (1957); Heine in F.W.T.A., ed. 2, 2: 407 (1963) & in Fl. Gabon 13: 96 (1966); Lebrun & Stork, Enum. Pl. Afr. Trop. 4: 492 (1997); Friis & Vollesen in Biol. Skr. 51(2): 448 (2005); Ensermu in F.E.E. 5: 437 (2006). Type: "Benin", *Palisot de Beauvois* s.n. (G, holo.; BM!, iso.)

Erect (rarely scrambling) shrubby herb or shrub to 2 (?–4) m tall; young stems green, sparsely puberulous on two sides, older stems greyish to pale brown. Leaves elliptic, largest 8–15 × 2.5–6 cm, apex cuspidate, base attenuate, decurrent to the 1–3 cm long petiole, glabrous (rarely with a few hairs on midrib). Inflorescences terminal or also two axillary from uppermost pair of leaves, 3–15 cm long; peduncle to 1(–4) cm long; axis finely puberulous and with dense stalked capitate glands; bracts green, minutely puberulous and with scattered capitate glands (densest towards base), ovate-elliptic, 1.8–2.5 × 0.8–1.3 cm, acute and mucronate; bracteoles linear to narrowly triangular, 3–4 mm long, with similar indumentum. Calyx 6–9 mm long of which the basal tube 1.5–2.5 mm, finely puberulous or sparsely so and with dense stalked capitate glands; lobes linear-lanceolate, cuspidate, 4–6.5 mm long. Corolla bright orange, finely puberulous and with dense stalked capitate glands; cylindric tube 3–4 cm long; throat 1–2 mm long; lobes obovate, 8–11 × 6–8 mm, truncate and irregularly toothed apically. Filaments 4–5 mm long, puberulous; anthers 1.5–2 mm long, connective with hairs and subsessile glands. Capsule 10–12 mm long; stipe ± 5 mm long. Seed 4–5 mm in diameter.

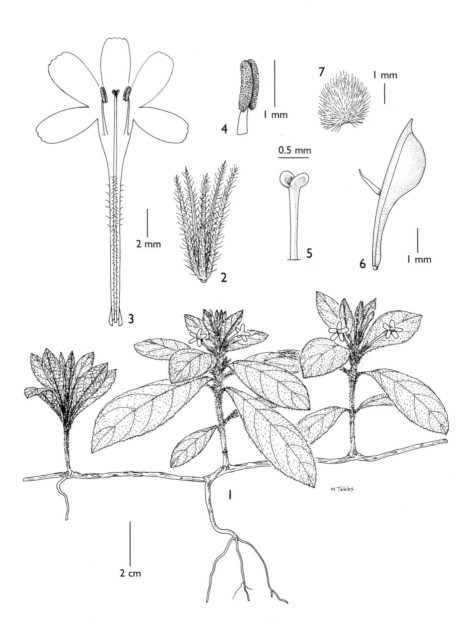

FIG. 43. *LANKESTERIA ALBA* — **1**, habit; **2**, calyx; **3**, corolla opened up with stamens and style; **4**, anther; **5**, apical part of style and stigma; **6**, capsule; **7**, seed. 1 & 6–7 from *Drummond & Hemsley* 1126, 2–5 from *Polhill* 4875. Drawn by Margaret Tebbs.

UGANDA. West Nile District: Zoka Forest, 3 March 1952, *Leggat* 70!; Bunyoro District: Budongo Forest, Nyakafunjo, 2 Jan. 1951, *Dawkins* 687!; Toro District: Kabango, 20 Jan. 1932, *Hazel* 134!

DISTR. **U** 1, 2; Sierra Leone, Liberia, Ivory Coast, Ghana, Benin, Nigeria, Cameroon, Gabon, Central African Republic, Congo (Kinshasa), Sudan, Ethiopia

HAB. Dense evergreen lowland and lower montane forest, often along tracks and in clearings, secondary forest; 850–1150 m

SYN. *Justicia elegans* P.Beauv., Fl. Owar. 1: 84, t. 50 (1807)
    Eranthemum elegans (P.Beauv.) Roem. & Schult., Syst. Veg., ed. 15, 1: 174 (1817); Nees in DC., Prodr. 11: 447 (1847)

2. **Lankesteria alba** *Lindau* in E.J. 38: 68 (1905); Iversen in Symb. Bot. Ups. 29(3): 161 (1991); Robertson & Luke, Kenya Coastal Forests 2: 83 (1993); Ruffo *et al.*, Cat. Lushoto Herb. Tanzania: 8 (1996). Type: Tanzania, Lindi District: Rondo Plateau, *Busse* 2576 (B†, holo.; EA!, iso.)

Perennial or shrubby herb, erect or more often with prostrate sometimes branched stems which root at the nodes and produce further erect stems; erect stems to 25 cm tall, prostrate stems to 50 cm long, when young sparsely to densely puberulous to pubescent, older stems with papery peeling bark. Leaves elliptic to obovate or narrowly so, largest 3–14 × 1–4.5 cm, apex subacuminate to broadly rounded, base attenuate, decurrent to the stem or with an up to 5 mm long petiole, sparsely pilose-pubescent, densest along veins with lamina sometimes glabrous. Inflorescences terminal, 1–5(–7) cm long, subsessile; axis pilose-pubescent; bracts caducous in fruit, dark green or suffused with brown or purple, glossy, pilosc-pubescent or sparsely so and pilose-ciliate, elliptic, 1–2 × 0.5–1.3 cm, acute to subacute and mucronate; bracteoles linear to narrowly triangular, 3–5 mm long, with similar indumentum. Calyx 6–8(–9) mm long of which the basal tube 0.5–1 mm, puberulous and pilose-ciliate; lobes linear, cuspidate, 5–7(–8) mm long, dorsal up to 1 mm longer than the rest. Corolla pure white, finely puberulous and with scattered stalked capitate glands; cylindric tube 1.1–1.9 cm long; throat 1–2 mm long; lobes obovate, 4–6 × 2–3.5 mm, truncate to retuse and irregularly toothed apically. Filaments 1–2.5 mm long, glabrous; anthers ± 1 mm long, connective and back of thecae with dense subsessile glands. Capsule 6–8 mm long; stipe ± 3 mm long. Seed ± 2.5 mm in diameter. Fig. 43, p. 274.

KENYA. Lamu District: Utwani Forest Reserve, Mambasasa, 16 Oct. 1957, *Greenway & Rawlins* 9351!; Kilifi District: Arabuko–Sokoke Forest, 13 Dec. 1990, *Luke & Robertson* 2623!; Kwale District: Shimba Hills, Mwele Mdogo Forest, 5 Feb. 1953, *Drummond & Hemsley* 1126!

TANZANIA. Tanga District: Amboni, Ukereni Hill, 12 Oct. 1918, *Peter* 60717!; Uzaramo District: Pande Forest Reserve, 13 July 1982, *Hawthorne* 1257!; Lindi District: Lake Lutamba, 8 Apr. 1935, *Schlieben* 6251!

DISTR. **K** 7; **T** 3, 6, 8; Mozambique

HAB. Dry deciduous and semi-deciduous lowland forest and thicket with *Cynometra*, *Scorodophloeus* and *Manilkara*, riverine forest; 25–500 m

INDEX TO ACANTHACEAE (Part 1)

New names validated in this part

Acanthopale macrocarpa *Vollesen* **sp. nov.**
Brillantaisia richardsiae *Vollesen* **sp. nov.**
Dyschoriste kitongaensis *Vollesen* **sp. nov.**
Dyschoriste sallyae *Vollesen* **sp. nov.**
Dyschoriste tanzaniensis *Vollesen* **sp. nov.**
Dyschoriste trichocalyx (*Oliv.*) *Lindau* subsp. **verticillaris** (*C.B.Clarke*) *Vollesen* **comb. nov.**
Epiclastopelma marroninus *Vollesen* **sp. nov.**
Hygrophila albobracteata *Vollesen* **sp. nov.**
Hygrophila richardsiae *Vollesen* **sp. nov.**
Mellera congdonii *Vollesen* **sp. nov.**
Mellera insignis *Vollesen* **sp. nov.**
Ruellia burttii *Vollesen* **sp. nov.**
Ruellia richardsiae *Vollesen* **sp. nov.**
Thunbergia subgen. **Stellatae** *Vollesen* **subgen. nov.**
Thunbergia austromontana *Vollesen* **sp. nov.**
Thunbergia barbata *Vollesen* **sp. nov.**
Thunbergia citrina *Vollesen* **sp. nov.**

Thunbergia heterochondros (*Mildbr.*) *Vollesen* comb. nov.
Thunbergia minziroensis *Vollesen* sp. nov.
Thunbergia mufindiensis *Vollesen* sp. nov.
Thunbergia napperae *Mwachala, Malombe & Vollesen* sp. nov.
Thunbergia racemosa *Vollesen* sp. nov.
Thunbergia reniformis *Vollesen* sp. nov.
Thunbergia richardsiae *Vollesen* sp. nov.
Thunbergia schliebenii *Vollesen* sp. nov.
Thunbergia tsavoensis *Vollesen* sp. nov.
Thunbergia verdcourtii *Vollesen* sp. nov.

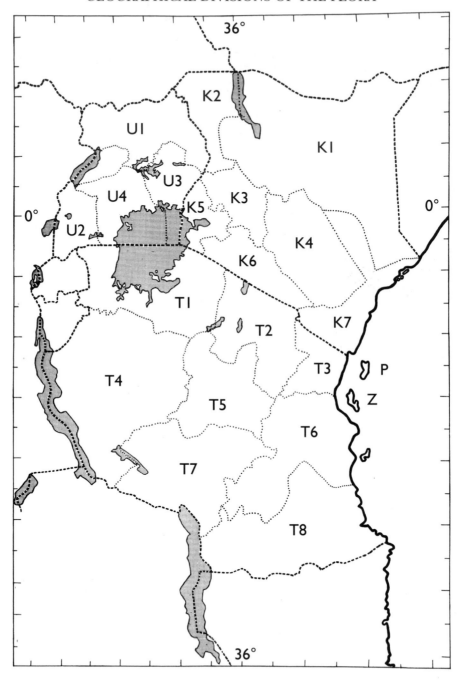

PLANTS PEOPLE
POSSIBILITIES

First published in 2008 by
Royal Botanic Gardens, Kew
Richmond, Surrey, TW9 3AB, UK
www.kew.org

ISBN 978 1 84246 217 1

British Library Cataloguing in Publication Data
A catalogue record for this book is available from the British Library

Design and typesetting by Margaret Newman,
Kew Publishing, Royal Botanic Gardens, Kew.

For information or to purchase all Kew titles please visit
www.kewbooks.com or email publishing@kew.org

All proceeds go to support Kew's work in saving the world's plants for life

LIST OF ABBREVIATIONS

A.V.P. = O. Hedberg, Afroalpine Vascular Plants; **B.J.B.B.** = Bulletin du Jardin Botanique de l'Etat, Bruxelles; Bulletin du Jardin Botanique Nationale de Belgique; **B.S.B.B.** = Bulletin de la Société Royale de Botanique de Belgique; **C.F.A.** = Conspectus Florae Angolensis; **E.J.** = A. Engler, Botanische Jahrbücher für Systematik, Pflanzengeschichte und Pflanzengeographie; **E.M.** = A. Engler, Monographieen Afrikanischer Pflanzen-Familien und Gattungen; **E.P.** = A. Engler, Das Pflanzenreich; **E.P.A.** = G. Cufodontis, Enumeratio Plantarum Aethiopiae Spermatophyta; in B.J.B.B. 23, Suppl. (1953) et seq.; **E. & P. Pf.** = A. Engler & K. Prantl, Die Natürlichen Pflanzenfamilien; **F.A.C.** = Flore d'Afrique Centrale (*formerly* F.C.B.); **F.C.B.** = Flore du Congo Belge et du Ruanda-Urundi; Flore du Congo, du Rwanda et du Burundi; **F.E.E.** = Flora of Ethiopia & Eritrea; **F.D.-O.A.** = A. Peter, Flora von Deutsch-Ostafrika; **F.F.N.R.** = F. White, Forest Flora of Northern Rhodesia; **F.P.N.A.** = W. Robyns, Flore des Spermatophytes du Parc National Albert; **F.P.S.** = F.W. Andrews, Flowering Plants of the Anglo-Egyptian Sudan *or* Flowering Plants of the Sudan; **F.P.U.** = E. Lind & A. Tallantire, Some Common Flowering Plants of Uganda; **F.R.** = F. Fedde, Repertorium Speciorum Novarum Regni Vegetabilis; **F.S.A.** = Flora of Southern Africa; **F.T.A.** = Flora of Tropical Africa; **F.W.T.A.** = Flora of West Tropical Africa; **F.Z.** = Flora Zambesiaca; **G.F.P.** = J. Hutchinson, The Genera of Flowering Plants; **G.P.** = G. Bentham & J.D. Hooker, Genera Plantarum; **G.T.** = D.M. Napper, Grasses of Tanganyika; **I.G.U.** = K.W. Harker & D.M. Napper, An Illustrated Guide to the Grasses of Uganda; **I.T.U.** = W.J. Eggeling, Indigenous Trees of the Uganda Protectorate; **J.B.** = Journal of Botany; **J.L.S.** = Journal of the Linnean Society of London, Botany; **K.B.** = Kew Bulletin, *or* Bulletin of Miscellaneous Information, Kew; **K.T.S.** = I. Dale & P.J. Greenway, Kenya Trees and Shrubs; **K.T.S.L.** = H.J. Beentje, Kenya Trees, Shrubs and Lianas; **L.T.A.** = E.G. Baker, Leguminosae of Tropical Africa; **N.B.G.B.** = Notizblatt des Botanischen Gartens und Museums zu Berlin-Dahlem; **P.O.A.** = A. Engler, Die Pflanzenwelt Ost-Afrikas und der Nachbargebiete; **R.K.G.** = A.V. Bogdan, A Revised List of Kenya Grasses; **T.S.K.** = E. Battiscombe, Trees and Shrubs of Kenya Colony; **T.T.C.L.** = J.P.M. Brenan, Check-lists of the Forest Trees and Shrubs of the British Empire no. 5, part II, Tanganyika Territory; **U.K.W.F.** = A.D.Q. Agnew (or for ed. 2, A.D.Q. Agnew & S. Agnew), Upland Kenya Wild Flowers; **U.O.P.Z.** = R.O. Williams, Useful and Ornamental Plants in Zanzibar and Pemba; **V.E.** = A. Engler & O. Drude, Die Vegetation der Erde, IX, Pflanzenwelt Afrikas; **W.F.K.** = A.J. Jex-Blake, Some Wild Flowers of Kenya; **Z.A.E.** = Wissenschaftliche Ergebnisse der Deutschen Zentral-Afrika-Expedition 1907–1908, 2 (Botanik).

FAMILIES OF VASCULAR PLANTS REPRESENTED IN
THE FLORA OF TROPICAL EAST AFRICA

The family system used in the Flora has diverged in some respects from that now in use at Kew and the herbaria in East Africa. The accepted family name of a synonym or alternative is indicated by the word "see". Included family names are referred to the one used in the Flora by "in" if in accordance with the current system, and "as" if not. Where two families are included in one fascicle the subsidiary family is referred to the main family by "with".

PUBLISHED PARTS

Foreword and preface
*Glossary
Index of Collecting Localities

Acanthaceae
   Part 1
*Actiniopteridaceae
*Adiantaceae
Aizoaceae
Alangiaceae
Alismataceae
*Alliaceae
*Aloaceae
*Amaranthaceae
*Amaryllidaceae
*Anacardiaceae
*Ancistrocladaceae
Anisophyllaceae — as Rhizophoraceae
Annonaceae
*Anthericaceae
Apiaceae — see Umbelliferae
Apocynaceae
   *Part 1
*Aponogetonaceae
Aquifoliaceae
*Araceae
Araliaceae
Arecaceae — see Palmae
*Aristolochiaceae
Asparagaceae
*Asphodelaceae
Aspleniaceae
Asteraceae — see Compositae
Avicenniaceae — as Verbenaceae
*Azollaceae

*Balanitaceae
*Balanophoraceae

*Balsaminaceae
Basellaceae
Begoniaceae
Berberidaceae
Bignoniaceae
Bischofiaceae — in Euphorbiaceae
Bixaceae
Blechnaceae
*Bombacaceae
*Boraginaceae
Brassicaceae — see Cruciferae
Brexiaceae
Buddlejaceae — as Loganiaceae
*Burmanniaceae
*Burseraceae
Butomaceae
Buxaceae

Cabombaceae
Cactaceae
Caesalpiniaceae — in Leguminosae
*Callitrichaceae
Campanulaceae
Canellaceae
Cannabaceae
Cannaceae — with Musaceae
Capparaceae
Caprifoliaceae
Caricaceae
Caryophyllaceae
*Casuarinaceae
Cecropiaceae — with Moraceae
*Celastraceae
*Ceratophyllaceae
Chenopodiaceae
Chrysobalanaceae — as Rosaceae
Clusiaceae — see Guttiferae
Cobaeaceae — with Bignoniaceae
Cochlospermaceae

Papaveraceae
Papilionaceae — in Leguminosae
*Parkeriaceae
Passifloraceae
Pedaliaceae
Periplocaceae — see Apocynaceae (Part 2)
Phytolaccaceae
*Piperaceae
Pittosporaceae
Plantaginaceae
Plumbaginaceae
Poaceae — see Gramineae
Podocarpaceae
Podostemaceae
Polemoniaceae — see Cobaeaceae
Polygalaceae
Polygonaceae
*Polypodiaceae
Pontederiaceae
*Portulacaceae
Potamogetonaceae
Primulaceae
*Proteaceae
*Psilotaceae
*Ptaeroxylaceae
*Pteridaceae

*Rafflesiaceae
Ranunculaceae
Resedaceae
Restionaceae
Rhamnaceae
Rhizophoraceae
Rosaceae
Rubiaceae
  Part 1
  *Part 2
  *Part 3
*Ruppiaceae
*Rutaceae

*Salicaceae
Salvadoraceae
*Salviniaceae
Santalaceae
*Sapindaceae
Sapotaceae
*Schizaeaceae
Scrophulariaceae

Scytopetalaceae
Selaginellaceae
Selaginaceae — in Scrophulariaceae
*Simaroubaceae
*Smilacaceae
Sonneratiaceae
Sphenocleaceae
Strychnaceae — in Loganiaceae
*Surianaceae
Sterculiaceae

Taccaceae
Tamaricaceae
Tecophilaeaceae
Ternstroemiaceae — in Theaceae
Tetragoniaceae — in Aizoaceae
Theaceae
Thelypteridaceae
Thismiaceae — in Burmanniaceae
Thymelaeaceae
*Tiliaceae
Trapaceae
Tribulaceae — in Zygophyllaceae
*Triuridaceae
Turneraceae
Typhaceae

Uapacaceae — in Euphorbiaceae
Ulmaceae
*Umbelliferae
*Urticaceae

Vacciniaceae — in Ericaceae
Valerianaceae
Velloziaceae
*Verbenaceae
*Violaceae
*Viscaceae
*Vitaceae
*Vittariaceae

*Woodsiaceae

*Xyridaceae

*Zannichelliaceae
*Zingiberaceae
*Zosteraceae
*Zygophyllaceae

FORTHCOMING PARTS

Acanthaceae
  Part 2
Apocynaceae
  Part 2

Asclepiadaceae — see Apocynaceae
Commelinaceae
Cyperaceae
Solanaceae

Editorial adviser, National Museums of Kenya: Quentin Luke
Adviser on Linnaean types: C. Jarvis

Parts of this Flora, unless otherwise indicated, are obtainable from:
Royal Botanic Gardens, Kew, Richmond, Surrey TW9 3AB, England. www.kew.org or www.kewbooks.com

* only available through CRC Press at:
UK and Rest of World (except North and South America):
CRS Press/ITPS,
Cheriton House, North Way, Andover, Hants SP10 5BE.
e: uk.tandf@thomsonpublishingservices. co.uk

North and South America:
CRC Press,
2000NW Corporate Blvd, Boco Raton, FL 33431-9868,
USA.
e: orders@crcpress.com

Information on current prices can be found at www.kewbooks.com or www.tandf.co.uk/books/